A new introduction to physics

Model of the sculpture *Nuclear Energy* by Henry Moore,
commissioned by the University of Chicago to commemorate
the birth of the Atomic Age and to be placed on the site
of the first atomic pile

W Ashhurst B Sc M Sc(Tech) F Inst P

Formerly Head of the Physics Department, Epsom College

A new introduction to
PHYSICS

John Murray Albemarle Street London

Printed in Great Britain by Jarrold and Sons Ltd, Norwich

0 7195 2054 1 boards edition
0 7195 2055 X limp edition

Preface

This book—a revised version of *Introduction to Physics*—uses SI units exclusively. In carrying out the revision I have introduced much new material which has, in some places, necessitated a reorientation of the subject matter.

All the current O-level physics syllabuses are adequately covered, including the Nuffield Physics syllabus (with the exception of the astronomy section), and numerous references are made to the Nuffield texts—the *Teachers' Guide* (T.G.) and the *Guide to Experiments* (G. to E.).

I am much indebted to many friends for their help and advice: Mr A. W. Blackaby has looked through the manuscript; Mr L. R. Middleton has again given considerable help at all stages of production; Mr J. P. Stephenson has read through the whole of the proof; and the staff of John Murray have once more been helpful and cooperative. To all these I extend my sincere thanks.

Sources of special and original information in the text have been acknowledged, but I should be grateful to hear of any that may inadvertently have been omitted.

After due consideration, and also because of lack of space, I have not included any 'new-type' questions at the ends of the chapters, as such questions are undoubtedly more useful and practical when available in a specially prepared publication (for example, *Objective and Completion Tests in O-level Science*—General Editor, Ernest Clarke, published by John Murray).

Worthing, April 1971   W.A.

Acknowledgments

Grateful acknowledgment is made to the examining boards listed below for their permission to reproduce questions from past papers and, furthermore, for their special kindness in allowing the units in the questions to be changed to SI. Any consequent adjustments have been kept to a minimum:

Joint Matriculation Board (JMB)
Oxford and Cambridge Schools Examination Board (O & C)
Oxford Delegacy of Local Examinations (O)
Senate of the University of London (L)
Southern Universities' Joint Board for School Examinations (S)
University of Cambridge Local Examinations Syndicate (C)

The illustration on the cover is *Supernovae* by Victor Vasarely; it is reproduced by courtesy of the artist and the Trustees of the Tate Gallery.

Thanks are due to the following who have kindly permitted the reproduction of copyright photographs:

Thanks are due to the following who have kindly permitted the reproduction of copyright photographs:

frontispiece, figure 2.4, page 47, 10.5, page 191, 28.10, 29.4, 29.5, 29.7, 36.13, 39.13, 39.14, 39.22, 42.12, 43.16, US Information Service; page 1, 2.1, 13.13, 37.10, National Physical Laboratory; 1.2, Crown copyright, Controller HMSO (NEL); 1.3, Moore & Wright Ltd; page 7, Prof. M. H. F. Wilkins, Biophysics Laboratories, Drury Lane, London; 2.2, McMaster Christie Studios Ltd; 2.3a, Dr Erwin Müller; 2.3b, 14.2, 17.1, 24.13, 40.4, 41.3, 41.10, 42.5, Leybold-Heraeus Ltd; 2.5, ICI Ltd (Dyestuffs Division); page 17, Shell Chemical Co. Ltd; 3.1, page 69, Paul Popper Ltd; 3.2, Fox Photos Ltd; 7.12, 12.4, 30.5, Morris Laboratory Instruments Ltd; 7.15, Davy Ashmore Ltd; 7.25, Radio Times Hulton Picture Library; 7.33 Negretti & Zambra Ltd; 8.1, 9.4, 20.5, 20.6, 20.9, 20.24, 34.2b, Griffin & George Ltd; 8.2, 9.2, 16.23, 28.5, 42.6, Philip Harris Ltd; 8.8, 9.8, PSSC *Physics* (1965) D. C. Heath & Co., Lexington, Mass.; 8.14, A. M. Lock & Co. Ltd; 9.6, *Exploring Physics, Book 5* by T. Duncan (1970), John Murray; 9.7, Drylund; 11.7, 41.16, 43.10, Crown copyright, Science Museum, London; 13.8, Editor, Unilever Educational Booklets; page 121, NASA; 16.5, 16.6, Dorman Long (Steel) Ltd; 16.7, Baird & Tatlock Ltd; 16.8, Central Press Photos; 16.9, British Oxygen Co. Ltd; 16.10, US Steel Corporation; 18.6, J. Allan Cash; 19.3, 41.32b, Crown copyright, HMSO; 19.9, Sunday Express; 19.10, Kodak Ltd; 19.13, Association for Applied Solar Energy; 19.14, American Army; 19.19, Alfol Insulation Ltd; 20.7, 20.25, Unilab Science Teaching Equipment; 20.12, from film *Reflection and Refraction,* Educational Services Inc., Watertown, Mass.; 20.14, 20.15, 20.21, *Ripple Tank Studies of Wave Motion* by Llowarch, Clarendon Press, Oxford; 20.16, Crown copyright, HMSO, courtesy Hydraulics Research Station, Wallingford; 20.20, Education Services Inc.; 21.1, 28.9, Westinghouse Ltd; 24.18, Stone-Chance Ltd; 24.23, Harper & Row Inc.; 24.26, Wide World Photos and Associated Press Ltd; 26.11, Degenhardt & Co. Ltd, for Carl Zeiss, West Germany; 27.1, 27.4, *Geometrical & Physical Optics* by Dr R. S. Longhurst, Chelsea College of Science; 27.2a C. J. R. Lillicrap; 27.2b, *Light* by Daish, EUP; 27.5, 27.6, John Fowler & Partners Ltd; page 255, Electricity Council; page 275, GEC International Nickel Ltd; 31.17, British Electrical Development Association; 32.15, Dubillier Condensor Co. Ltd; 32.16, Wingrove & Rogers Ltd; 32.19, Voltage Engineering Corporation, Boston, Mass.; 33.19, J. A. Crabtree & Co. Ltd; 33.20a, Ed. Holme & Co. (1931) Ltd; 33.23, 35.4, 37.14, White Electrical Instrument Co. Ltd; 36.12, Allis-Chalmers Manufacturing Co., Wisconsin; 37.17, Crompton Parkinson Ltd; 39.12, GEC-AEI Telecommunications Ltd; 39.16, EI du Pont de Nemours; 39.20, Director, Central Electricity Research Laboratories; 39.21, The C. A. Olsen Manufacturing Co., Ohio; page 379, Science Journal; 41.5, 41.6, EFVA; 41.9, 41.11, 41.24, 41.25, Teltron Ltd; 41.18, Kodak Research Laboratories; 41.19, 41.33, GEC of America; 41.32, Mullard Ltd; 42.2, 42.20, 42.21, 43.17, 43.18, 43.19, 43.20, 43.21, 43.23, UK Atomic Energy Authority; 42.9, Cavendish Museum; 42.10, Dr Kapitza; 42.11, AERE, Harwell; 43.13, Nuclear Physics Research Laboratory, Liverpool; 43.14, Lawrence Radiation Laboratory, University of California, Berkeley; 43.22, Derby & Midland Warp Knitting Co. Ltd, & UKAEA.

Thanks are also due to the following for permission to base diagrams on material from their publications: figure 7.2, *Exploring Physics* Book 1 by T. Duncan (1971), John Murray; figures 42.7 and 42.16, *Practical Modern Physics* by T. Duncan (1967), John Murray; appendix 8, May & Baker Ltd.

Contents

Part One INTRODUCTION

The caesium clock. The atomic beam chamber (right) incorporates a double beam system to give an improved accuracy of 4×10^{-12} (approximately 0.4 microseconds per day). The racks contain control and measuring equipment

1 Fundamental units and measurements

Scientific investigation and study springs from a desire to discover and to understand the working of the world and of the heavenly bodies comprising the universe.

Complete understanding is the ultimate goal, but this seems to be farther away with each forward step we take. Man always regards the unknown as a challenge: an unquenchable thirst for knowledge seems to be an essential part of his character.

The scientist takes part in this quest for knowledge, which has been going on from early times, and which is now being pursued at an ever-increasing tempo with its consequent greater and greater effect upon the entire populace of the world. No matter whether or not science is to be our vocation, it will, nevertheless, exert a dominating influence upon our lives, and for this reason alone we should at least know its language and its limitations: we should be able to appreciate the modern trends and should try to understand, as far as possible, just where science is leading us.

The immense fields of scientific knowledge have to be subdivided to admit of their being treated as single studies. The boundaries between these subdivisions are not well defined, nor should we try to make them so.

Physics is the study of the fundamental properties of matter and of energy. Physicists seek to discover the laws governing the behaviour of matter under varying conditions. These fundamental discoveries are handed on to others, such as engineers or technologists, who apply them, so that they become of service to the community. In the last chapter of this book, the story is briefly told of how physicists have discovered a means of obtaining energy from the atom.

Fundamental units

The physicist is responsible for the standardization of all measurements used in our daily lives. Control is necessary so that there will be uniformity. When we buy a kilogramme of apples at any shop in the world, we expect to get the same mass of apples. In the interests of the community, the Government of the United Kingdom controls 'weights' and measures. The weights (or, more correctly, masses) used by shopkeepers are periodically examined by representatives of the Government (Weights and Measures officials). Their accurate sets of masses must, in turn, be compared with some other standard masses, and so on, until finally comparison is made with one standard mass.

Establishing a system of measurement

When we measure an amount of any physical

quantity, we do not do so in absolute terms, but we compare its size with some already specified amount of the same quantity. An object weighing two kilogrammes is twice as heavy as a standard mass called a kilogramme. In other words, all *measurements* are *comparisons* or *ratios*.

Some physical quantities are dependent upon other physical quantities: velocity depends upon distance and time, hence we can measure velocity in terms of a distance standard and a time standard.

We find that all physical quantities can be expressed in terms of a few basic quantities. We can establish, by mutual agreement, any number of these basic quantities but, for simplicity, we keep the number as small as is practicable. Such basic quantities must be

    (i) capable of measurement to a high degree of accuracy,
    (ii) readily reproducible,
    (iii) precisely specified,
    (iv) universally acceptable.

The units employed in all measurements are now internationally agreed. They are controlled by the Conférence Générale des Poids et Mesures* and some important changes have come into effect in recent years.

Metric units†

There are two systems of units based on the metric standards. These are the Centimetre–Gramme–Second (cgs) and the Metre–Kilogramme–Second (mks) systems. The former is now obsolescent: the latter was adopted internationally in 1959. It is known as the 'Système International d'Unités' (SI). To the three units of mass, length and time, four other units are added to complete the system.

In 1901, Giorgi pointed out that the 'practical' electrical units, the ampere, the volt, the watt, etc. could be linked with the metric system very conveniently by using mks units.

---

* An international organization, with a membership of over sixty countries, created to control scientific standards. The headquarters and laboratories are at Sèvres near Paris.

† The Foot-Pound-Second (fps) system of units is based on British Standards and is now obsolescent. Parliament has decreed that the metric system be gradually adopted in the U.K. (In May 1965, the President of the Board of Trade expressed the hope that in ten years' time most industrial concerns would have effected the change.)

SI units

The following are the SI basic units.* All but the *mole* have been officially adopted. It is likely that this also will be adopted soon.

Metric standards

Length—the metre (m)

The metre is 1 650 763.73 wavelengths† of a specified orange line in the spectrum of the krypton atom.

This unit was adopted in January 1962, and replaced the International Prototype Metre, which was the distance between two marks on a bar of platinum-iridium. This bar is preserved at the Bureau International and a copy is in the custody of the *National Physical Laboratory*. It is likely that the obsolete standard will continue to be used as a working standard.

Mass—the kilogramme (kg)

This standard, the International Prototype Kilogramme 1901, is in the form of a cylinder of platinum-iridium kept at the Bureau International at Sèvres. Copies of this standard are kept by countries who were signatories to the Metric Convention of 1875. The British copy is kept at the N.P.L.

Time—the second (s)

The second is the time interval for a fixed number of periods of a particular radiation from the caesium atom.

Thus unit was adopted by the Conférence Générale at its October 1968 meeting in place of the unit based upon astronomical time. It has been used for some years as the working standard at the N.P.L.

Electric Current—the ampere (A)

This couples the above three units with electricity. (*See* page 347 for the definition of the ampere.)

Temperature—the kelvin (K)

This fixes the size of the degree and the absolute zero of temperature (page 128). The kelvin is the same as the degree Celsius.

---

* The precise definitions are given in Appendix 1.

† For wavelength, *see* page 96.

*Luminous Intensity—the candela (*cd*)*

This fixes with precision a standard of luminous intensity equivalent to,the old *standard candle*. (*See* page 195.)

*Amount of Substance—the mole (*mol*)*

This is a new basic unit which is related to and replaces the 'gramme-equivalent'.

The mole is the amount of substance containing the same number of elementary units as there are carbon atoms in 0.012 kg of carbon-12.

Supplementary units

The two following supplementary units have been long in use but have only recently been adopted officially.

*Angle (Circular Measure)—the radian (*rad*)*

The radian is the angle subtended by an arc of length r at the centre of a circle of radius r. It is approximately equal to 57°.

*Solid Angle—the steradian (*sr*)*

The steradian is the solid angle at the centre of a sphere subtended by an area of r^2 on the surface of the sphere of radius r. The total solid angle about a point is therefore 4π steradians.

These two supplementary units are independent of the other basic units: they are dimensionless.

Other Physical Quantities

All physical quantities other than the above can be measured in units dependent upon these nine basic quantities and are called *derived* units.

Fractions and multiples of SI units

The advantage of a metric system is obvious to people with an elementary knowledge of the decimal system. The strong recommendation is that SI units are used along with the prefixes given in Appendix 4.

It is further suggested that only multiples of 10^3 are used, i.e. 10^3, 10^6, 10^{-3}, etc. This means that eventually the *centimetre* will be discarded and we shall use the multiples *millimetre* and *kilometre*.

Additional units still in common use

There are a few units, not strictly SI units, which are likely to remain in use for some considerable time, e.g. *electron-volt* (*see* Chapter 43), *kilowatt-hour* (*see* Chapter 35), *litre* (now fixed at $1.000\ 028 \times 10^{-3}$ m^3: originally defined as the volume of 1 kg of pure water at 4° C), *hectare* (10^4 m^2), *tonne* (metric 'ton' = 10^3 kg).

Measurements

We have mentioned that a major part of the work of the physicist is concerned with measurement—the science of metrology. The basic units have to be adapted for measuring both small and large fractions of them.

The measurement of mass

A mass is pulled to the earth by the force of gravity. Masses are compared by the different gravitational pulls upon them at the same place (*see* Chapters 3 and 10). There are a few different types of instrument (balance) used for this purpose.

The *spring* balance uses the principle that the extension of a loaded spring is proportional to the load (*see* page 20). The scale on this type of balance has to be calibrated by attaching known masses or applying known forces.

The *beam* balance is the common type and an unknown mass on one scale-pan is balanced against known masses placed on the other pan. The pans hang from the ends of a rigid beam supported on a knife-edge at its centre.

The *lever* balance is a convenient direct-reading modification of the beam balance. A lever system adjusts itself into an equilibrium position when a load is placed on the pan. The position on a scale of a pointer attached to the lever system gives the mass directly.

Improved forms of balance are now available which are direct reading and enable masses to be measured to a high accuracy. A change in mass of one part in fifty million can be detected.

The measurement of length

Large distances are measured by triangulation. This method is employed by the surveyor and the astronomer. The unknown distance PA (Fig. 1.1), of a position P from a base-line AB is equal to $AB/\tan\theta$. The angle θ is measured by sighting P from the two ends of a base-line of measured

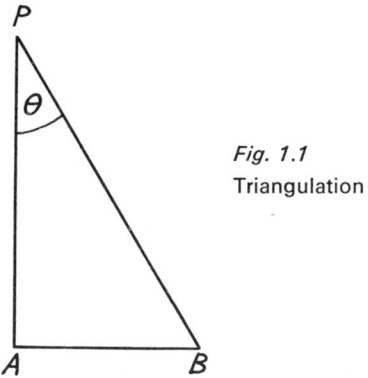

Fig. 1.1
Triangulation

length. The accuracy of the method depends upon the measurement of θ. This angle is made large by increasing AB. Thus base-lines of different lengths are used for measuring distances over a wide range. For planetary distances the base-line can be increased up to a length equal to the diameter of the earth and for stellar distances the diameter of the earth's orbit can be used. This is the largest possible base-line and it limits the farthest distance which can be measured by this method to about 10^{18} m. Other fundamentally different methods are employed for estimating the distances (10^{25} m) of remote galaxies.

The measurement of small distances involves the subdivision of the metre. This is done using an accurate screw. The screw is the all-important instrument in metrology. Fig. 1.2 shows a lathe incorporating a standard screw used by the engineering industry for calibrating other screws for use in precision instruments and machines.

The thread of a screw can be manufactured with remarkable accuracy. When a screw is turned through a complete revolution, its end moves a distance equal to the pitch of the screw. By means of a scale marked round the head of the screw, it

Fig. 1.2 Master correcting lathe

This screw-cutting lathe is installed in a closely temperature-controlled room at the National Engineering Laboratory, East Kilbride. It provides a service to industry in manufacturing lead screws of the highest standard. The accuracy is about one part in a million but the lead screw after manufacture is checked on a laser interferometer with a resolution of one in forty million

can be given a fraction of a turn. Thus, if the head shows 100 divisions round it, a rotation of one division moves the screw a distance of 1/100 of the pitch of the screw. This is the principle of the

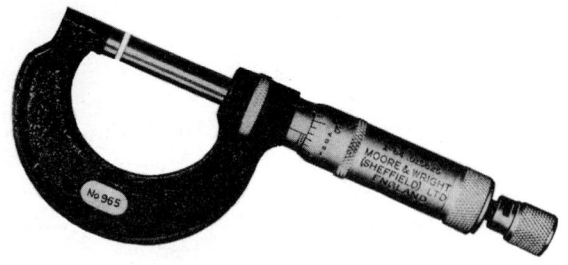

Fig. 1.3 Micrometer screw gauge

micrometer screw gauge (Fig. 1.3). The figure shows the scale along the screw and the divisions round the head of the screw.

For measurements of extremely small distances we resort to optical methods which, in effect, employ the wavelength of a particular coloured light as the unit of length. Such measuring instruments, which also incorporate an accurate screw, are called interferometers. An accuracy of the order of one part in 200 million is possible.

The measurement of time

The duration of an interval of time is usually measured with a pendulum clock or with a chronometer. Many electronic timing devices are now very accurate and relatively cheap.

Vibrating crystals, stimulated by electronic circuits, have been used at the N.P.L. for accurate time measurement for a number of years. The caesium clock, as mentioned above, has now been adopted as the basic unit: it is possible with it to measure time correct to 1 second in 3000 years. This is a much higher accuracy than is attainable by astronomical measurement. It is likely that greater accuracy still will be possible using a hydrogen maser.

QUESTIONS

1. Write a short account of the methods used in the accurate measurement of time.

2. Name and define the SI unit of length. What was the old standard? Enumerate some of the advantages of the new over the old standard.

3. Describe the importance of an accurate screw in the measurement of length.

4. Explain how the surveyor measures distances which are too large to be measured with his tape.

5. Suggest a possible standard for the future to replace the prototype kilogramme mass. Give reasons for your choice.

Part Two THE CONSTITUTION OF MATTER

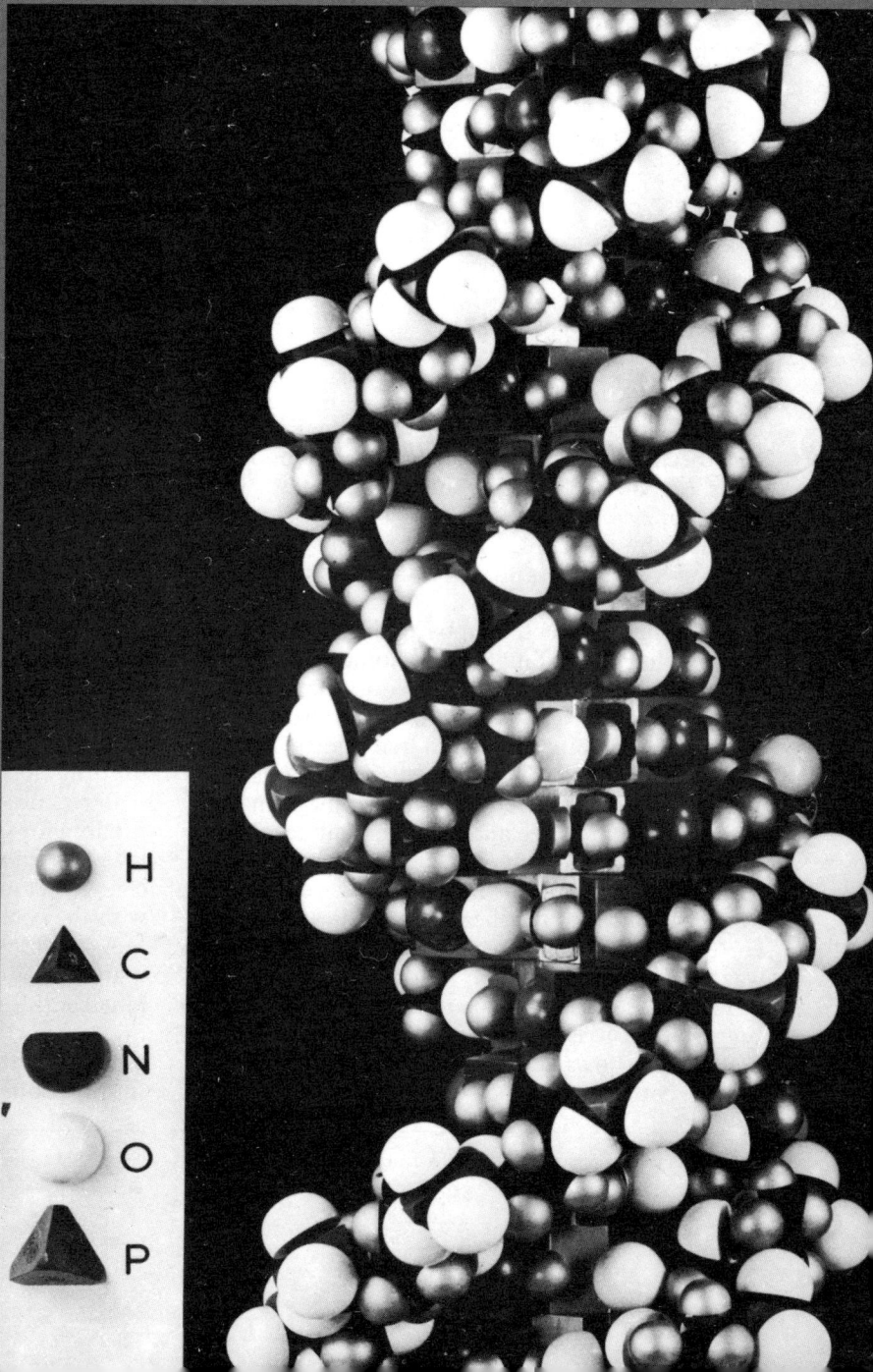

H

C

N

O

P

Photograph of the model of deoxyribonucleic acid (DNA) showing the positions of the constituent atoms. The structure of this substance was investigated by X-ray diffraction. The magnification in this reproduction is about 36 million times

2 Atoms and molecules

The architecture of matter

The particulate* theory

The constitution of matter has, throughout the ages, been a subject for speculation. The philosophers of Greece had their ideas, but these were often pure conjecture, not based upon fact nor produced by reasoning. There were, on the other hand, some attempts to arrive at conclusions by deduction.

The opinion accepted, apparently without need for thought, was that matter was homogeneous, i.e. the same throughout in every direction. Democritus, however, who lived in the fourth century B.C., developed a completely different idea of the universe. The bodies within the universe consisted of particles which were indivisible units of matter differing in shape, size and arrangement according to the substance. These particles were believed to be in perpetual motion; heavy atoms were driven together to form the heavenly bodies by huge vortex motions. The space between the particles was taken to be completely void.

There are certain properties of matter which suggest that it is particulate. The regular shape of crystals shows the building up of particles in regular patterns, like the stacking of spheres (Fig.

2.1). Information on the form of these structures is gained by observing the reflection of X-rays by the various layers of particles. This is more fully explained in Chapter 41.

This idea of the particulate nature of matter was not acceptable in the Middle Ages, particularly on religious grounds, but the seventeenth century saw a more favourable reception of it. Newton liked it and applied it to an explanation of Boyle's law. There was some idea of the existence of fundamental bricks of matter which could be arranged to produce different substances. Further, the formation of more complex units by the combination of units of different substances was suggested. Thus there grew up the first idea of *atoms** uniting to form *molecules*.

Bernoulli, in 1738, endowed on gaseous atoms random movement and went on to explain Boyle's law, identifying the pressure of a gas with the change in momentum of the corpuscles on collision with the enclosing walls.

A much clearer picture of the atomic nature of matter was created and developed during the nineteenth century; Dalton, in 1805, expounded his atomic theory accounting for chemical compounds as the combinations of atoms in definite simple proportions.

---

* Particulate—consisting of small particles.

* Atoms—the name given to the fundamental particles.

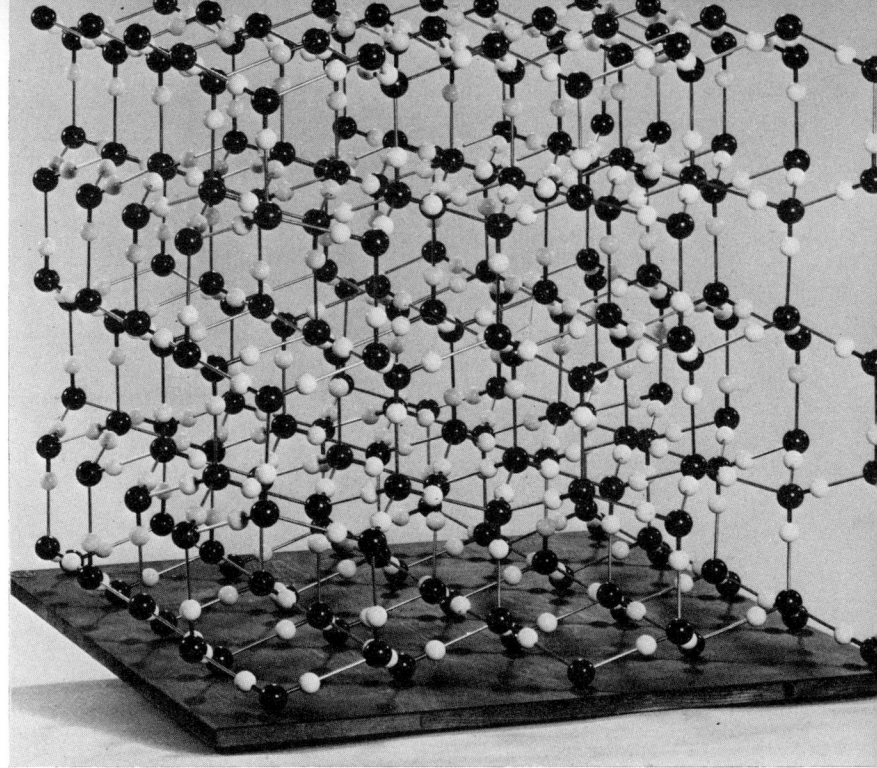

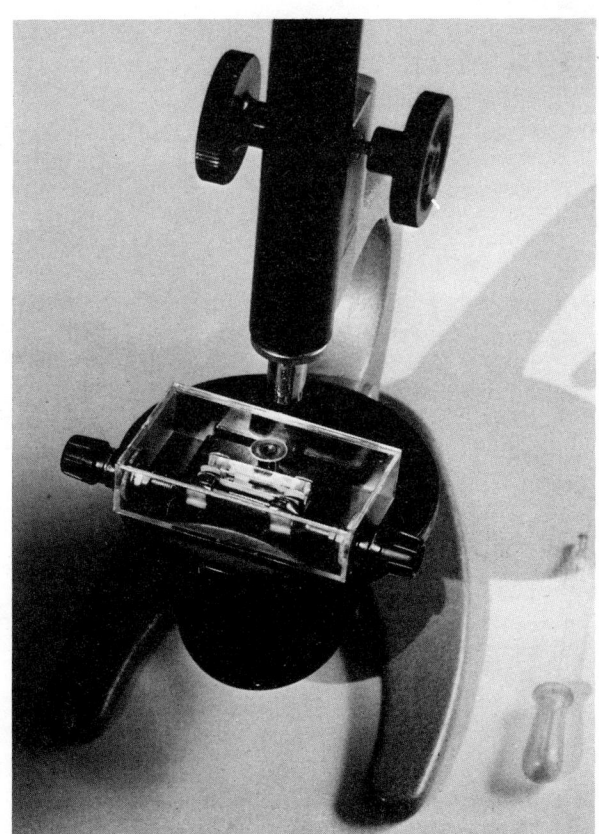

Fig. 2.2 Apparatus for observing Brownian movement

Brownian movement

Direct evidence of molecular motion in a liquid was first shown by Brown in 1861. He observed the movement, due to molecular bombardment, of particles* in a suspension, under a high-power microscope. The same effect in gases can be shown using smoke. The experiment is simple to perform using the standard apparatus now readily available (Fig. 2.2). The principle used in the apparatus is illustrated by the visibility of dust in the air which is revealed when a ray of sunlight is viewed in a direction at right angles to the ray. The small particles are seen as tiny specks of light. Their presence, but not their shape, is revealed. The Brownian movement shows the random paths of small particles caused by their bombardment by the surrounding molecules—water molecules in the case of an aqueous suspension, and oxygen and nitrogen molecules in the case of smoke. The observed particles are much more massive than the molecules, but the latter are very fast moving and this enables them to move the heavy solid particles.

Direct evidence of individual atoms is contained

---

* Brown examined pollen grains, but any very small particles, e.g. 'Aquadag' (colloidal graphite), may be used.

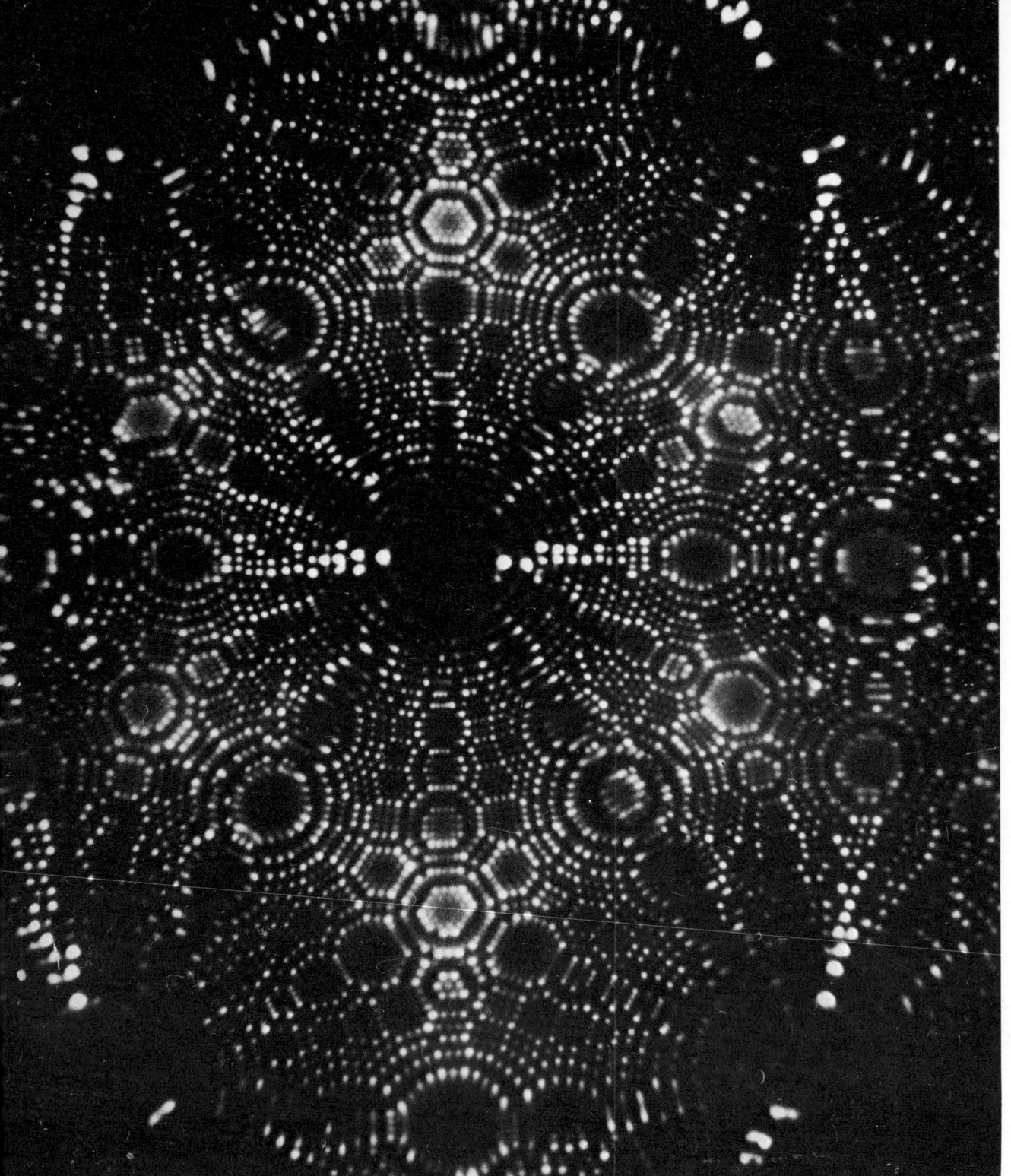

Fig. 2.3a Tungsten crystal photographed by Müller field-ion microscope
Each spot shows one atom: magnification 1 000 000 times

Fig. 2.3b Laboratory arrangement for field-ion microscope

The high voltage required is supplied by the van de Graaff generator seen on the right. The fluorescent screen is facing the camera

in the photograph Fig. 2.3*a* taken with the field emission microscope devised by Dr Müller. A laboratory demonstration of this can be given using the arrangement shown in Fig. 2.3*b*. The photograph shows 'shadow' images produced on a fluorescent screen by an electron beam in a highly evacuated tube. Dr Müller has more recently modified his original microscope so that individual atoms can be selected, separated and studied. He calls this an 'atom-probe field-ion microscope'.

The latest advance to be recorded is that of a microscope capable of photographing individual atoms of uranium and thorium and actually estimating their size. This is the work of a team headed by Dr A. Crewe. It is a combination of the field-ion microscope, the electron microscope and the mass spectrometer (Fig. 2.4).

Further abundant evidence resulting from recent discoveries has been obtained during this century. Advances in the study of radioactivity have led to the development of several techniques which reveal the presence of single atoms or molecules. A single *a*-particle, which is a helium atom, is detected using a spinthariscope* or an ionization

* Spinthariscope—*see* page 402.

Fig. 2.4 Individual atoms of uranium

One of the first photographs of individual atoms in a molecular chain, taken at Chicago University using a new-type electron microscope (magnification up to 5 000 000 times). The arrow shows two individual atoms of uranium

chamber (Geiger counter). The tracks of atoms can be made visible in the cloud chamber or in a photographic layer. The study of crystal structure by X-ray analysis has enabled us to discover details of the packing of atoms in a crystal pattern and the dimensions of the arrangement or lattice.

Size of molecules

There are several methods of determining the size of an individual molecule, but most of these are too difficult to understand at this stage. There is, however, a simple method of estimating the size of certain molecules. The molecules of oleic acid are long and they spread over the surface of water so thinly that the layer consists of single molecules all standing on end. Because a single drop of the acid will cover a large area, a very small quantity must be used if the area covered is to be small and measurable. The acid* is diluted with alcohol (about 1/200) and a drop of the solution dropped on the still surface of water covered with a thin sprinkling of lycopodium powder. The extent to which the oleic acid has spread is revealed by the clearing of the powder from the region covered. (This is due to the acid and not to the alcohol.) By finding the volume of acid in a single drop, obtained by measuring the volume of a number of similar drops, the thickness of the layer can be calculated. This will be an estimate of the molecular size, assuming the layer is one molecule thick. Of course, the experiment does not give a very accurate result, but it does give an *estimate* of the size of a molecule. This method can be developed to give the size of an atom, which is found to be of the order of 2×10^{-10} m or 2Å :† roughly speaking fifty million atoms arranged in one line would have a length of only 1 cm.

It is difficult to conceive the enormous number of molecules in 1 cm³ of matter, but physicists and chemists have devised methods of measuring these quantities. In Chapter 41 you will read of methods of determining the masses of atoms. The mass of a hydrogen atom is 1.67×10^{-24} g; under standard conditions of temperature and pressure, 1 cm³ of any gas contains 2.7×10^{19} molecules. This number of molecules cannot be

appreciated. Wembley Stadium in London holds 100 000 people, so that ten such crowds make one million. But twenty-seven million million million is a fantastic number. The earth's population is about two thousand million, so that if there were seven times as many earths as there are people on the earth, the total population would just about equal the number of molecules in 1 cm³ of gas.

The three states of matter— the kinetic theory*

A substance may exist in three forms, solid, liquid and gas. Solids retain their shape, but liquids and gases take the form of the containing vessels.

In all three states the atoms or molecules are in motion: they possess energy of motion or *kinetic* energy. Heat is this kinetic energy. Between atoms there are forces of attraction and the effectiveness of these forces controls the freedom of the atoms. We have much evidence, however, some of which we shall deal with later, to show that this force of attraction does not become effective until the distance between the atoms is small. When this distance becomes extremely small there is a force of repulsion. These forces arise from the electrical charges within the atoms.

Solids

In solids the atoms are constrained in their positions and are not free to move about as are those in liquids and gases. In a crystal (most solids have a crystalline structure), the atoms or groups of atoms are arranged in a definite pattern which gives the crystal its characteristic form. The atoms are not still, but in a state of agitation. The interatomic forces keep them in their definite positions, but they are vibrating fiercely. Supplying heat to a solid increases the rate and extent of vibration of the atoms—a point which will be dealt with more fully later.

The volume of a solid changes very little when it becomes liquid and, hence, the packing of the molecules must be similar in the two states. Evidence shows that the molecules in both solids and liquids are densely packed and give little room for free movement—they are practically incom-

---

* Other liquids, e.g. engine oil, stearic acid, can be used. For details, see Physical Science Study Committee book of experiments from which this method is taken.

† Angstrom unit.

---

* We return to this topic in Chapter 12.

fusus fr. fundere - to pour.

pressible, a very large increase in pressure being required to produce a minute decrease in volume. Each atom in a solid is dominated and kept in position by the surrounding atoms. The molecules in a liquid are able to push their way among the densely packed surrounding molecules and so move throughout the liquid. The energy required by the molecules in a solid to gain this freedom is called 'latent heat of fusion'. Likewise the conversion from liquid to gas requires energy to separate the molecules and so to give them more freedom. This is provided by the 'latent heat of vaporization'.

Liquids

The molecules in a liquid are free to move about but cannot, except under special circumstances, escape through the surface. Any molecule at the surface is attracted by molecules which are on one side of it only, and so exert a resultant inward force. To overcome this inward force, a molecule must have a high speed in the upward direction. Some of the fastest moving molecules may have just the necessary speed at the correct moment and so they will leave the liquid to become vapour molecules. This is evaporation.

These molecular forces and movements give rise to surface effects which we shall study later (page 115).

Gases

The very great increase in volume when a liquid changes into a gas means that the molecules become widely separated and, at ordinary pressures, the intermolecular forces are negligible. Thus the molecules in a gas have complete freedom of movement and are confined only by the walls of the containing vessel. Their impacts with the walls constitute the pressure which a gas exerts.

Structure of atoms

In this chapter the ideas of the eighteenth and nineteenth centuries have been presented which have led to the acceptance of a particulate theory of matter. Atoms were considered to be indivisible units of *pure* substances. Less than one hundred of these substances (elements) were identified. A new era in scientific history commenced at the end of the nineteenth century when atoms were shown to be built up of other more fundamental units. The study of the structure of the atom is the theme of twentieth-century physics and an introduction to this study is given in the last few chapters of this book. Suffice it for the present to give here the general outlines which may be of immediate help in understanding some fundamental principles and concepts.

An idea of the structure of an atom can be gathered by imagining it to be like a miniature solar system. The atom is not solid but has a very open structure and is dynamic—its parts are moving. Practically the whole mass of an atom is concentrated at its centre within a tiny core called the *nucleus* which carries a positive electric charge. Orbiting round the nucleus are small light particles, called *electrons,* each carrying an equal negative charge. The total negative charge on the electrons is equal to the positive charge on the nucleus so that the atom as a whole is neutral. The 'size' of the atom is the diameter of the outer orbiting electrons.

Density

At some time or other we have probably been caught out with the question, 'Which is heavier, a kilogramme of lead or a kilogramme of cork?' If, however, we are asked which is heavier, lead or cork, a justifiable answer would be lead. It

Fig. 2.5 Not a block of stone but a large block of urethane foam, density about 22 kg/m³

DENSITIES (kg/m^3)					
Aluminium	2.70×10^3	Zinc	7.1×10^3	Methylated spirit	0.83×10^3
Copper	8.93×10^3	Steel	$7.7–7.9 \times 10^3$	Olive oil	0.92×10^3
Gold	19.32×10^3	Brass	$8.4–8.9 \times 10^3$	Turpentine	0.87×10^3
Iron	7.87×10^3	Marble	$2.5–2.8 \times 10^3$	Cork	$0.22–0.26 \times 10^3$
Lead	11.37×10^3	Sand (silver)	2.63×10^3	Paraffin wax	0.88×10^3
Mercury	13.56×10^3	Glycerine	1.26×10^3	Air	1.29
Platinum	21.50×10^3	Glass	2.5×10^3	Hydrogen	0.09
Silver	10.50×10^3	Ice	0.92×10^3	Oxygen	1.43

The density of water at 4 C is 1×10^3 kg/m^3.

would be assumed that the question referred to equal volumes of lead and cork. Thus, we should always state the mass of a certain volume and, naturally, we quote the mass of unit volume. This we call the 'density'.

Density is mass of unit volume

It is measured in kilogrammes per cubic metre.

$$\text{Density} = \frac{\text{mass}}{\text{volume}}$$

Accurate knowledge of the densities of substances is very important, not only in pure science, but also in engineering and in everyday life.

Density of a solid

To determine the density of a substance we have to find the mass and the volume of a specimen of it. If the substance is a solid, and supplied in a regular form such as a rectangular block, cone, or sphere, we can weigh it and calculate its volume from its measurements. If the solid is of irregular shape, however, its volume has to be found by displacement. If it is immersed in water contained in a measuring cylinder, the increase in the reading will give its volume. If the solid is large, a large measuring cylinder will have to be used, and a small change in the reading with such a cylinder will not be very accurate. In such cases a displacement can should be used. This is a cylindrical vessel of appreciable cross-section provided with an overflow spout. On filling with water, the excess runs out until the surface is just level with the spout (Fig. 2.6). The immersion of the solid causes the displaced water to overflow, and this is collected in a small measuring cylinder by which its volume is more accurately found.

The displacement can is sometimes termed a

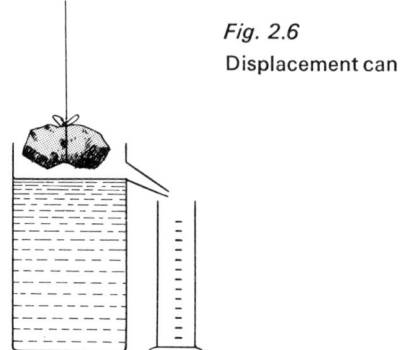

Fig. 2.6
Displacement can

Eureka can, from the famous legend associated with Archimedes. Archimedes, 287–212 B.C., was one of the famous Greek philosophers. He was supposed to have been commissioned by Hiero, the King, to find out whether a gold crown contained any base metal. The King had supplied the gold for its making, but he had an idea that the goldsmith had mixed in some base metal and removed an equal mass of gold, thereby retaining the same weight for the finished crown. The problem presented to Archimedes was to discover the presence of the base metal without destroying the crown. He realized that he could do this by comparing the densities of the crown and of pure gold. The legend relates that Archimedes noticed the rise in the level of the water in his bath when his body was submerged, and so conceived the idea of measuring the volume of the crown by displacement. So thrilled was he with his idea that he ran from his bath crying, 'Eureka!', meaning, 'I have found it!'

Density of a liquid

The density of a liquid can be easily and quickly measured by first weighing an empty beaker and then, after adding a known volume of liquid measured in a measuring cylinder or pipette, weighing it again.

The weighings, in density experiments, can conveniently be done with a spring balance of the compression type (household spring balance). By working with large quantities, accuracy equal to that obtained using the chemical balance can be achieved.

Density of a gas

The density of a gas is obtained by first weighing a flask from which all the air has been evacuated. For this purpose the flask must be provided with a rubber stopper with glass tube and tap. The flask is then filled with the gas and re-weighed. From the two weighings the mass of the gas is obtained. Its volume is that of the flask, which is found by measuring the volume of water which just fills it.

An alternative method is to pump air by means of a foot pump into a large plastic container having a volume of about 0.02 m^3. The mass of the air it will contain is large enough to enable it to be determined by weighing on a lever balance. The volume of the air at atmospheric pressure is obtained by releasing it in stages and collecting it over water in a measuring vessel. By this simple method the mass of the excess air is released and so the density of air at atmospheric pressure can be determined. (*See* Nuffield *G. to E.*, I, No. 18.)

It should be noted that the density of a gas depends upon its pressure (*see* Boyle's law, page 141).

Relative density

One m^3 of water weighs 10^3 kg.* If the density of

---

* This is not a coincidence but arises from the fact that the kilogramme was originally equated to the mass of a litre of water.

any substance is greater than 10^3 kg/m^3 it will sink in water. Thus, it is often convenient to know the ratio of the density of a substance to that of water. This quantity is called the relative density.*

$$\text{Relative density} = \frac{\text{density of substance}}{\text{density of water}}$$

$$= \frac{\text{mass of substance}}{\text{volume of substance}} \times \frac{\text{volume of water}}{\text{mass of water}}$$

If the substance and the water have the same volume, this equation reduces to:

$$\text{relative density} = \frac{\text{mass of substance}}{\text{mass of an equal volume of water}}$$

It will be seen that relative density is a number —the same number as the density when measured in grammes per cubic centimetre. For example, the density of aluminium is $2.7 \times 10^3 \text{ kg/m}^3$, or 2.7 g/cm^3. Its relative density is 2.7.

The density bottle

A convenient and much-used way of finding the relative density of a liquid is by using a density bottle. This is an ordinary small bottle with a ground-glass stopper. The stopper has a fine hole through its centre (a substitute can be made by filing a groove down the side of the stopper of a scent-bottle). If the bottle is filled with liquid and the stopper inserted, excess liquid can escape by way of the hole in the stopper. In this way the bottle can be completely filled. By weighing a bottle empty, and then filled with liquid, the mass of the liquid can be found. The mass of an equal volume of water can similarly be found. The ratio of these masses gives the relative density.

---

* This is equivalent to the term formerly used, *specific gravity*.

QUESTIONS

1. Describe Brownian motion and explain precisely what it shows.

2. Discuss the evidence there is in support of a molecular theory of matter.

3. The molecular theory of matter seems to indicate a mixture of perfect order and perfect disorder. Discuss this statement.

4. Describe a simple experimental method of estimating the size of a large molecule.

5. Describe the mechanism causing a change of state (*a*) from solid to liquid, (*b*) from liquid to gas.

6. What comparisons would you use in attempting to explain to a friend: (i) the size of a molecule, (ii) the number of molecules in 1 cm^3 of gas?

7. Account for the facts that (*a*) a solid changes little in volume when it changes into liquid, (*b*) the volume of a liquid increases greatly when it changes into a gas.

8. Explain how the kinetic theory accounts for:
(*a*) Brownian motion.
(*b*) The heat produced when a bullet is stopped suddenly by a target.
(*c*) The fact that liquids expand more than solids when heated.   (O & C)

9. Consider the following experiments:
(*a*) A beaker is filled completely with water. A spoonful of common salt is added slowly. The salt dissolves and the water does not overflow.
(*b*) One drop of a solution of stearic acid placed in the centre of a sheet of water in a dish rapidly spread out into a circular film of area about 50 cm^2 on the surface of the water. The drop contained approximately 10^{-5} cm^3 of stearic acid.
State carefully what can be deduced from each of these experiments concerning the molecular theory of matter.
Describe an experiment to provide evidence of the movement of molecules in *either* a liquid *or* a gas.
Discuss in general terms the way in which the molecular theory of matter accounts for the characteristic differences between solids, liquids and gases.   (C)

10. A rectangular block of metal has a mass 2.7 kg and a volume of 900 cm^3. Find the density of the metal.
A second block is made of the same metal with the same external dimensions but containing an internal cavity of volume 600 cm^3. What happens when this block is put into water? Give your reason.   (C)

11. The density bottle can be used to find the density of a liquid with great accuracy. What precautions must be taken to attain such accuracy?
How would you use a density bottle to find the density of steel in the form of small ball-bearings?

12. At standard temperature and pressure the densities of air, nitrogen and oxygen, relative to that of hydrogen, are respectively 14.4, 14.0 and 16.0. Assuming that air is a mixture of oxygen and nitrogen only, calculate the percentage by volume of each gas.   (O)

13. A cube of wood of side 5.0 cm has embedded within it a copper sphere. If the loaded cube floats on water so that it is just totally submerged, find the volume of the sphere. (Density of wood $=0.85 \times 10^3$ kg/m^3; density of copper$=8.9 \times 10^3$ kg/m^3.)

14. A platinum ring with a single diamond has a total volume of 0.436 cm^3 and a mass of 5.37 g. If the densities of platinum and diamond are 21.5×10^3 and 3.4×10^3 kg/m^3 respectively, find the size of the diamond in carats, given 1 carat$=0.200$ g.

Part Three MECHANICAL ENERGY, MECHANICS OF BODIES AT REST

The old and the new Forth bridges. Their construction reveals many fundamental principles of mechanics employed to obtain strength using the minimum of material

3 Force, mass, weight

Notion of a force: pushes and pulls

We are familiar with some of the effects of forces. We can exert *pushes* (Fig. 3.1) and *pulls* (Fig. 3.2): we must apply a push through something rigid, although a pull can act through a non-rigid connector such as a rope or a wire. But in physics we also meet with forces which do not make contact and act through space.

Invisible forces—fields of force

Gravity is an invisible force by which a body is pulled to the earth, but there is nothing attached to the body by which the force is applied. We have something similar in the case of magnetic attraction. One magnet attracts (or repels) another magnet, but there is nothing visible pulling them together or separating them (Fig. 3.3). This

Fig. 3.1 Snowplough clearing roads in Scotland—force is acting from behind, i.e. push

Fig. 3.2 Derrick extracting aluminium core from world's largest solid rocket motor near Miami—force is acting from above, i.e. pull

Fig. 3.3 Invisible forces

ability of a force to act through space we say is due to the body creating round itself a 'sphere of influence' which we call a 'field'. Thus round a magnet is a *magnetic field*, while round a material body or mass is a *gravitational field*. The strength of a field depends upon the size of what is producing it. A large mass produces a powerful field. Fields of different masses react with each other. When one mass is near enough to another, their fields react and they attract each other. A mass on the earth's surface is in the strong gravitational field of the earth and the mass experiences a force of attraction to the earth's centre. The mass pulls on the earth with a similar pull which, however, is quite ineffective upon such a massive body as the earth. The result of the interaction between the gravitational fields of two masses is a force of attraction between them. The strength of a gravitational field is measured by the force which would act upon a mass of 1 kg placed in the field. This is similar to the definition of an electric field as equal to the force acting upon a unit charge placed in the field.

Inertia

We all know that if a book is removed from the bottom of a pile of books by a quick jerk, the pile remains undisturbed. Similarly, a block of wood *A* (Fig. 3.4) can be given a sharp blow to displace block *B*. The blocks on *B* finish up undisturbed on block *A* which is now in the position originally occupied by block *B*.

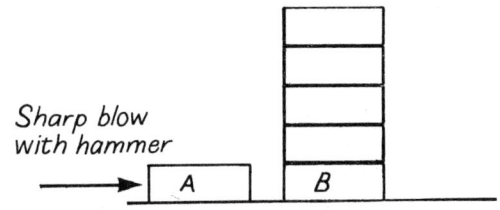

Fig. 3.4 Experiment to illustrate inertia

A piece of paper under a glass of water can be snatched away without spilling the water. In the theatre this 'trick' is often performed by pulling a tablecloth from under crockery arranged on it.

We say that the above effects are possible because of the *inertia* of the objects.

A force is necessary to produce motion, but the force must be large enough and sustained. The piece of paper, in being snatched quickly, acts on the glass with a force arising from friction for a very short interval of time before movement starts and so the effect is insufficient to move the glass. In all these cases the sliding surfaces must be smooth.

A well-known demonstration of inertia simply requires a very heavy ball or other object A suspended by a thread X (*see* Fig. 3.5). A similar

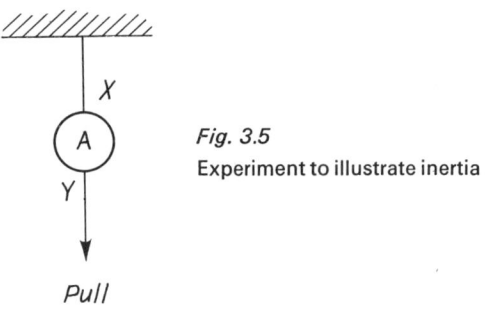

Fig. 3.5

Experiment to illustrate inertia

piece of thread Y (which could be even stronger than X) is attached to the underside of A. A sharp pull on the free end of the string causes a break in the section Y, but a steady pull, gradually increasing in magnitude, breaks the section X. This is due to the large inertia of the ball A.

The inertia of a body is represented by the opposition it offers to starting it in motion or to stopping its existing motion. This has nothing to do with friction and concerns the effort required to move a body on a frictionless horizontal surface.

To produce exactly the same motion of two bodies requires different forces: i.e. if forces F_1 and F_2 acting on two bodies produce in them the same velocity in the same time, then F_1/F_2 is the ratio of their *inertias*. This is how Newton saw the problem.

The inertia of a body is measured by its mass.

Rotation of a body

We have just mentioned inertia as it concerns the linear motion or translation of a body. We shall see later (Chapter 10) that the inertia of a body which is rotating presents a different problem and the mass of the body does not in itself express the *rotational* inertia. This depends not only on the mass but on the distribution of the mass about the axis of rotation.

Measurement of mass

A mass is measured by comparing it with the standard mass—the kilogramme. A familiar method of comparing masses is with a beam balance, as has been mentioned on page 4.

The pull of gravity is proportional to the mass.

There are other ways of comparing masses which do not depend upon gravity. It will be seen in Chapter 9 that Newton's second law of motion provides a simple link between force, mass and acceleration. Hence, masses can be compared by the ratio of the accelerations produced by the same force. This can be done by comparing the accelerations produced on two masses as they are pulled over a horizontal smooth surface by an elastic band stretched by a constant amount.

Also masses can be compared by oscillation. If a mass M (Fig. 3.6) is attached to springs S_1 and

Fig. 3.6 Comparison of masses by oscillation

S_2 which have their other ends fixed at A and B, the periodic time of M when set in oscillation on a smooth horizontal surface depends on the mass of M.

Measurement of a force

Force can be conveniently measured with a spring balance or even with a length of elastic. Within limits, the extension is proportional to the pull. Compression-extension spring balances are available which measure both pulls and pushes.

The weight of a body is the pull of gravity on the body.

The pull of gravity on a kilogramme mass is called a *kilogramme force* (kgf). This is sometimes used as a unit of force (a 'gravitational' unit), but it is not a good unit for it depends upon the pull of gravity, i.e. on the strength of the gravitational field. This varies slightly over the earth's surface. Obviously, we must adopt a unit of force which is independent, not only of the position on the earth's surface, but of the position anywhere in the Universe. The *newton* (N) is one such unit and it is the SI unit. It is defined in Chapter 9.

1 kgf is about 9.8 N.

We shall measure all forces in newtons and use spring balances graduated in newtons. A 100 g mass hanging from a string produces a tension in the string of about 1 N.

Equilibrium

When a body is still, it is said to be in equilibrium. As every body on the earth's surface is acted on by gravity, there must be at least one other force acting on all bodies which are at rest, e.g. a book on a table is in equilibrium under the action of gravity pulling downwards and the table pushing upwards. This upward force of the table on the book is the gravitational pull of the book on the earth. The two pulls, one, the pull of the earth on the book and the other, the pull of the book on the earth, are equal and opposite. The effects of the equal forces are very different because of the great difference in the mass of the earth and the mass of the book.

If a person pulls on a rope which has its end securely tied to a hook in·a wall, equilibrium is produced by the hook pulling on the rope with a force equal and opposite to that exerted by the person. This idea is pursued further in Chapter 9 when we deal with Newton's laws of motion.

A force has an effect in every direction except at right angles to the direction in which it acts; e.g. although the weight of a body acts vertically downwards, a body will slide down an inclined plane, hence the weight must have an effect along the plane. The weight of a body will not cause it to move along a horizontal plane.

Parallelogram of forces

If two forces act on a body in different directions, they produce the same effect as would be produced by one force, called the 'resultant'. The resultant can be found by construction or by calculation. The two forces are represented by lines *OA* and *OB* (Fig. 3.7). These lines are drawn

parallel to the forces in the same sense or direction and their lengths are drawn to scale to represent the sizes of the forces. If the parallelogram is constructed having *OA* and *OB* adjacent sides, then

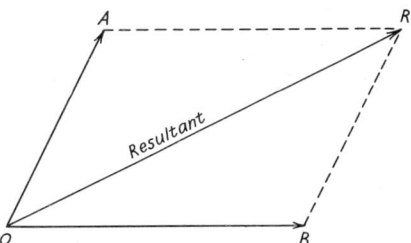

Fig. 3.7 Parallelogram of forces
OR is the resultant of *OA* and *OB*

the resultant is represented in direction and size (magnitude) by the diagonal *OR*.

This can be demonstrated experimentally using the apparatus in Fig. 3.8 which is known as a statics board. It consists of a vertical blackboard

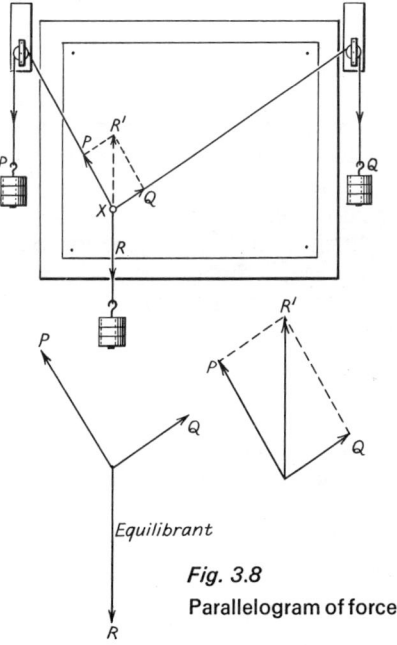

Equilibrant

Fig. 3.8
Parallelogram of forces

to which adjustable pulleys are affixed. One end of each of three strings is attached to a ring *X*. On the other ends of the strings are attached masses. By passing two of the strings over the pulleys, as shown, the ring is allowed to take up its equilibrium position. If the pulleys are assumed to be frictionless, the magnitude of the forces P, Q and R keeping the ring in equilibrium will be given by the attached loads. The directions of the strings are marked on the board, and lines of lengths corresponding to the magnitudes of forces P and Q are constructed as shown. By completing the parallelogram, with P and Q as adjacent sides, the diagonal R' is found to be equal and opposite to R. The force which produces equilibrium with forces P and Q is said to be the *equilibrant* of forces P and Q. The equilibrant of two forces is equal and opposite to their resultant.

When forces act along the same line, their resultant is the sum of the forces if they act in the same direction, or their difference if they act in opposite directions.

Vector and scalar quantities

Any quantity which acts along a specific direction, i.e. any quantity which has direction as well as magnitude, is called a *vector* quantity. Such a quantity may be represented by a straight line or vector, the length of which indicates its magnitude, and the direction of which gives its line of action. An arrow on the line indicates what is sometimes called the *sense* of the vector. Examples of vector quantities are force, velocity, acceleration, magnetic field, etc. They must always be combined by the parallelogram rule.

Quantities which do not have direction but only magnitude, e.g. time, temperature and speed, are called *scalar* quantities, and are combined by simple algebraic addition.

Resolution of forces

A force OR may be replaced by two forces which form adjacent sides of the parallelogram having OR as diagonal, e.g. OA and OB, or OP and OQ. These forces are called the components of OR (Fig. 3.9). This is the converse of the parallelogram of forces.

It is sometimes convenient to think of the effect of a force such as OR, as being due to two forces OP and OQ acting at right angles. As OP will have no effect along OQ, OQ must represent the total effect of OR along this direction. OQ is called the resolved part of OR along OQ and is equal to $OR \cos \theta$.

$$\cos \theta = \frac{OQ}{OR} \quad \left(\frac{base}{hypot}\right)$$

$$\therefore OQ = OR \cos \theta$$

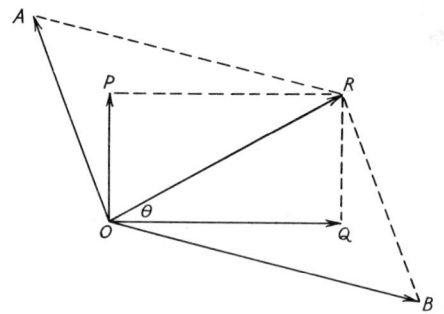

Fig. 3.9 Resolution of forces
Force OR can be replaced by forces OA and OB or OP and OQ

A garden roller provides a good practical illustration of the resolution of forces. It is much less effective to pull a roller than to push it. A push F (Fig. 3.10a) applied to the handle can be resolved into a horizontal force F_1, propelling the roller, and a downward force F_2, causing the roller to sink into the ground. If, however, the roller is pulled (Fig. 3.10b), the vertical component F_2 is now upward and the roller tends to be lifted off the ground. The weight of the roller is not shown in the diagrams, but it will be seen that the vertical component works with or against the weight, and the magnitude of this component depends upon the inclination of the handle. The theory is strik-

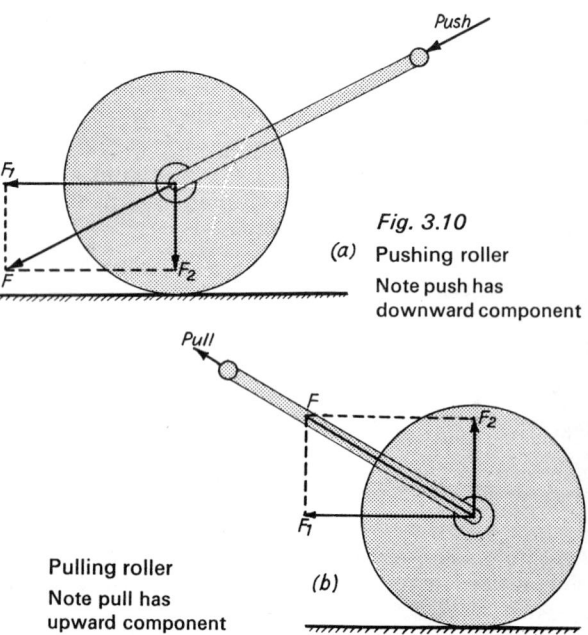

Fig. 3.10
(a) Pushing roller
Note push has downward component

Pulling roller
Note pull has upward component
(b)

ingly illustrated by pushing and pulling a perambulator on a sandy beach.

Forces producing equilibrium

We have seen that when three forces P, Q and R are in equilibrium, R is equal and opposite to the resultant of P and Q. R must act through the intersection of P and Q; hence, all three forces in equilibrium must act through one point (Fig. 3.11).

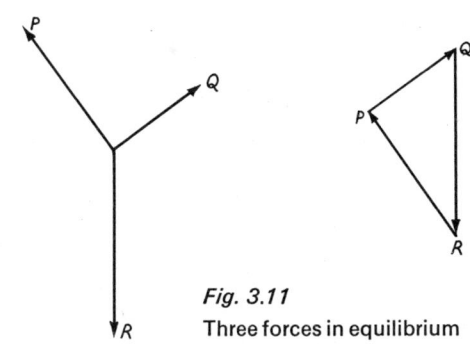

Fig. 3.11
Three forces in equilibrium

The lengths of the vectors **P**, **Q** and **R** are such that they can form the sides of a triangle.

When a body (Fig. 3.12) is in equilibrium under the action of three forces, we know that:

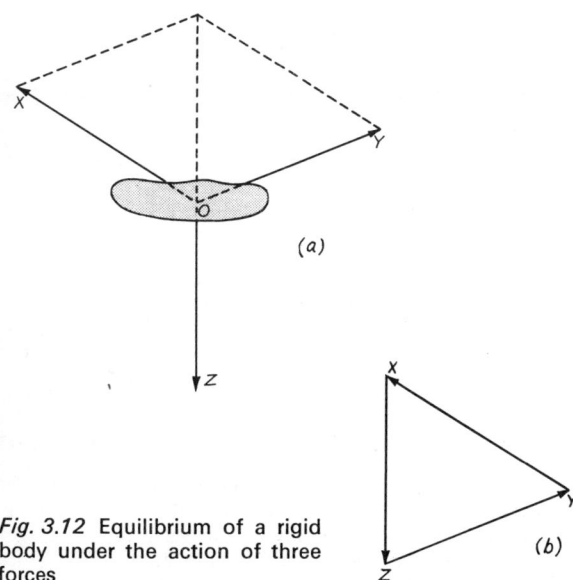

Fig. 3.12 Equilibrium of a rigid body under the action of three forces

1. the forces are concurrent, i.e. their lines of action meet at a point,
2. the vectors representing **X**, **Y** and **Z** form the sides of a triangle.

In the diagram **X**, **Y** and **Z** meet at O. A triangle can be constructed, as shown, with sides parallel to the forces **X**, **Y** and **Z**, and of lengths proportional to their magnitudes.

Triangle of forces

If a body is acted upon by three concurrent forces which can be represented in magnitude and direction by the three sides of a triangle, taken in order, then the body is in equilibrium.

The converse of this is also true, and is stated:

When a body is in equilibrium under the action of three co-planar forces, they can be represented in magnitude and direction by the three sides of a triangle, taken in order.

Suppose that a heavy horizontal bar AB (Fig. 3.13a) has one end A supported at a wall, and the other by a wire fastened to the wall at C. The reaction R of the wall upon the end A must pass

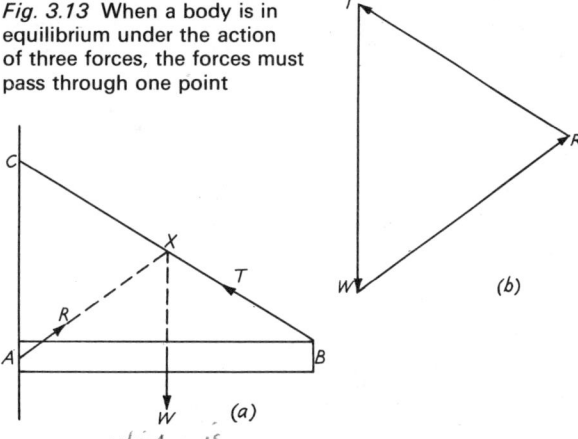

Fig. 3.13 When a body is in equilibrium under the action of three forces, the forces must pass through one point

through X, which is the intersection of the lines of action of the other two forces acting on AB, viz. its weight W and the tension T in the wire BC. The three forces W, R and T must form a closed triangle (Fig. 3.13b). The magnitudes of R and T can be found when W is known. In this example $R = T$ approximately.

Examples 1 and 2 serve to show the application of the above principles.

Example 1

An unknown mass M (too heavy to be weighed on the spring balance S) hangs from a string which has its upper end fixed at X, Fig. 3.14. The spring balance S,

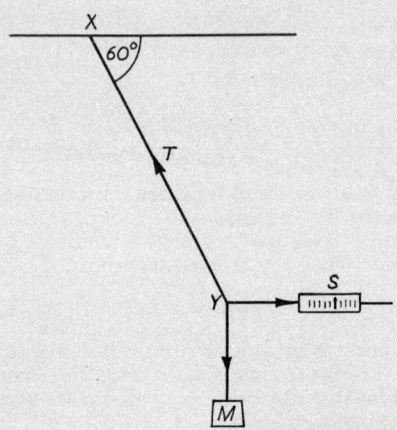

attached to the string at Y, pulls the string XY aside until it is inclined at $60°$ to the horizontal. If the spring balance reads 5 N, find the value of M.

In this example, the directions of the three forces and the magnitude of one, S, are known. To draw the triangle (Fig. 3.15), we start with S, i.e. the force which we know in magnitude and direction, and make AB 5 units long. From B we draw W ($=9.8M$), but the length of this

vector is unknown. We now work from A again and draw T making an angle of $60°$ with S. The intersection of W and T at C gives the lengths of W and T, and so their magnitudes on the scale we used to draw S. Alternatively,

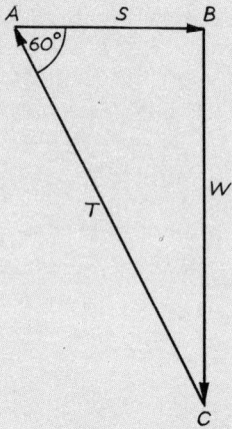

the magnitudes can be found by using trigonometry. We have,

$$W = S \tan 60° = 5\sqrt{3} = 8.7 \text{ N}$$

Hence mass ≃ 0.87 kg.

The two diagrams must not be confused. The first is the force diagram which shows the forces acting on the body. The second is a graphical solution of the problem and does not show how and where the forces act.

QUESTIONS

[Where not given assume g = 10 m/s² or N/kg.]

1. Find by construction and by calculation the resultant of two forces 5 and 10 N acting at an angle of $60°$.

2. The *difference* between two vectors of the same kind (e.g. velocities) of magnitude 4 and 3 units is a vector of magnitude 2 units. Draw a scale diagram to illustrate this and use it to find the magnitude of the sum of the original vectors.

(O & C)

3. A trolley is pulled along by exerting a force of 20 N on a handle inclined at $30°$ to the horizontal. What horizontal force would push the trolley?

4. A balloon cable makes an angle of $70°$ with the ground. If the upthrust on the balloon is 10 000 N what is the horizontal force of the wind on the balloon?

5. A uniform ladder weighing 40 kg and 3.0 m long rests against a smooth wall with its base 1.0 m from it. Find graphically the magnitude and direction of the reaction at the base.

6. Explain, with the help of diagrams, why it is easier to pull a roller than to push it.

7. A boy pushes a motor-cycle weighing 150 kg up a hill of gradient 1 in 20. What force is needed?

8. State a set of conditions for the equilibrium of three non-parallel forces acting in one plane.
   Explain the rule for resolving a force into two components in given directions at right angles to one another.
   A body of mass 10 kg is placed on a smooth plane inclined at $30°$ to the horizontal. Find the horizontal force which must be applied in order just to keep it at rest on the plane.

(O)

9. State conditions for the equilibrium of a body acted on by three non-parallel coplanar forces.
   A non-uniform beam AB, 4 m long and of mass 20 kg, is smoothly hinged at B and held horizontally by a light cord AC which is joined to a point C 4 m vertically above B. The direction of the reaction of the hinge at B is $30°$ to the horizontal. Find the tension in the cord AC, and also the

distance of the centre of gravity of the beam from the point A. (O)

10. Explain what is meant by (a) a parallelogram of forces, (b) the law of moments for a body in equilibrium. Describe a simple experiment in each instance to illustrate your answer.

$ABCD$ is a square of side 40 cm. Forces of 5 N and 8 N act at A towards B and D respectively. Determine, by a scale drawing or otherwise, the magnitude of the resultant force and its moment about the point C. (L)

11. State the parallelogram law for forces. Describe an experiment to illustrate it and show how the law may be applied to resolve a force into two perpendicular components.

A nail projects horizontally from a vertical wall and a cord attached to its head is pulled at an angle of 30° to the wall, with a force of 120 N. By a scale drawing, or otherwise, find (a) the force tending to bend the nail, (b) the force tending to pull it out of the wall. . (L)

12. State the triangle law for forces and describe an experiment to illustrate it.

A ventilating window has a mass of 2.0 kg and has a smooth hinge along its lower horizontal edge. A horizontal cord, attached to its upper edge, holds the window at an angle of 30° with the vertical. Make a scale diagram, indicating a triangle of forces. Hence, or otherwise, determine (a) the direction and magnitude of the reaction at the hinge, (b) the tension in the cord. (Assume the centre of gravity of the window to be at its middle point.) (L)

13. What is meant by (a) the resultant, (b) the equilibrant, of two forces acting at a point? Illustrate your answers by reference to the parallelogram of forces.

Show how a force of 120 N acting at a point may be replaced by two, acting through the same point and inclined at 30° and 45° on either side of it. Find by a scale drawing the value of the two components.

A load of 150 kg is hauled up a smooth plane inclined at 30° to the horizontal, by a rope parallel to the plane. Draw the triangle of forces and hence or otherwise find the pull on the rope in newtons. (L)

14. Define the resultant and the equilibrant of two forces acting at a point.

Find, by calculation, or by a scale diagram, the magnitude and direction of the resultant of forces of 5.4 N and 7.2 N acting at right angles. Describe how you would check your answer experimentally. (C)

15. A wooden block of mass 15.0 kg rests on an inclined plane 5 m long, the upper end of the plane being 1 m higher than the lower end. The block is attached to a string lying parallel to the plane, and passing over a pulley wheel at the top of the incline. If a vertical downwards force is applied to the other end of the string, find the least force that would be needed to move the block if there were no friction.

It is found in practice that a downwards force of 58 N is needed to move the block up the plane. Calculate (a) the coefficient of friction between the block and the plane, (b) whether the block will slide down the plane if there is no tension in the string. (C)

16. State the theorem of the parallelogram of forces. Explain how you would find the combined effect of two forces acting at a point.

A sack of coal of mass 50 kg hangs at the end of a rope. The sack is pulled aside by a horizontal force acting at the lower end of the rope. If the rope makes an angle of 20° with the vertical, find the tension in the rope. (C)

17. Show how a force can be resolved into two components at right angles to each other and state the law on which your answer is based. How would you verify the law experimentally?

A horse tows a barge along a canal by means of a rope. Assuming that the rope is horizontal, that the tension in it is 2000 N and that it makes an angle of 30° with the direction of travel, find (a) the force causing motion, (b) the force tending to draw the barge towards the bank.

Explain how the force (b) is resisted. (O & C)

18. Explain how you would find the resultant of two forces acting at a point.

A steerable barge is drawn at a steady speed along the middle of a straight canal by a horizontal rope attached to a horse on the towpath. If the rope makes an angle of 20° with the direction of motion, and the tension is T, what is (a) the force causing motion, (b) the force with which the barge is pulled towards the bank?

How is the barge prevented from moving towards the bank? Illustrate your answer by a sketch. (O & C)

19. Explain how a force may be resolved into two components in given directions and state the law on which your construction is based. Why is it often convenient to take the components at right angles?

When pulling a garden roller of mass 100 kg along a level lawn a man exerts a force of 500 N at 30° to the ground. Find graphically or by calculation (a) the force causing the roller to move forwards, (b) the vertical force of the roller on the ground while being pulled. (O & C)

20. Explain the terms *moment of a force about a point*, *centre of gravity*, *resultant force*.

A uniform rod of mass 10 kg rests with one end A on a rough horizontal plane and the other B against a smooth vertical wall so that the rod makes an angle of 30° with the vertical.

Draw a diagram showing the forces acting at the ends and through the centre C of the rod and find graphically, or otherwise, their magnitude and direction. (O & C)

21. State and explain the conditions of equilibrium of three non-parallel coplanar forces.

Describe how you would test the conditions experimentally.

A mass of 10 kg is supported from a point C on a string 3 m from the end A and 4 m from the end B. A and B are 5 m apart and AB is horizontal. Find by drawing or calculation the tension in AC and BC. (O & C)

22. What are the conditions for the equilibrium of three coplanar forces acting at a point?

A lead ball of mass 10 kg hangs from the lower end of a long wire. What steady horizontal force must be exerted on the ball to deflect the wire 30° from the vertical? What is then the tension in the wire?

Describe how you would test this experimentally. (O & C)

23. State the theorem of the parallelogram of forces and explain how you would make an experimental test of its validity.

A uniform ladder 24.0 m long, mass 40 kg, rests in equilibrium with one end A against a smooth vertical wall. The lower end B rests on the ground and is at a perpendicular distance of 12.0 m from the wall. Draw a diagram showing the forces acting on the ladder and find graphically or otherwise the forces acting at A and B.   (O & C)

24. State what is meant by the triangle of forces and describe, with full explanation, an experiment to illustrate it.

A uniform girder AB, of 100 kg mass, makes an angle of 30° with the ground which is horizontal. The end A rests on the ground and the end B is attached by a rope to point C vertically above A so that the angle ABC is 60°. Find, graphically or otherwise, the tension in the rope and the force exerted by the ground on the girder.   (L)

25. Explain what is meant by the components of a force and show that a single force can have any number of pairs of components.

A body of mass 100 kg is held at rest on a smooth plane inclined at 30° to the horizontal by a force applied horizontally. Give a diagram showing the forces acting on the body and find the magnitude of the applied force. What is the magnitude and direction of the least force necessary to keep the body at rest on the plane?

What would be the work done in moving the body 5.0 m up the plane if there were a frictional force of 100 N?   (L)

26. State the conditions for the equilibrium of a body under the action of three non-parallel coplanar forces. A uniform ladder 5.0 m long and of mass 40 kg rests in equilibrium with one end A against a smooth vertical wall and the other end B on the ground at a horizontal distance of 3.0 m from the wall.

Draw a diagram showing the forces acting on the ladder and find graphically or by calculation the forces acting at A and B.   (O & C)

4 Moments

Moment of a force

In addition to, or alternative to, moving a body, a force may cause or tend to cause a body to rotate. This turning effect of a force is called its *moment*. It is commonly known as *leverage*.

Many tools and implements are designed to make use of leverage. Early experience of the turning effect is gained from the working of a see-saw. Handles on doors are put as far away as convenient from the hinges; pliers have long handles; a hammer has a long shaft.

These examples all lead to one conclusion, that a rotation about a pivot requires the least effort when this is applied as far away from the pivot as possible. We note also, that it is better to apply the effort in a direction at right angles to the line joining the pivot and the point where the effort is applied, e.g. we open a door by a pull applied at right angles to the door.

We see then, that in order to produce turning, we must consider not only the size of the applied force or the effort, but also its distance from the pivot. The quantity taking into account both these factors is the *moment* of the force.

> The moment of a force about a fixed point or axis is the turning effect of the force about that point and is measured by the product of the force and the perpendicular distance from the point to the line of action of the force.

Moment of F about $P = F \times d$ (Fig. 4.1)

A moment is measured in N m.

> The moment of a force about a point in its line of action is zero.

(The length of the perpendicular is then zero, i.e. $d = 0$.)

When a body is counterpoised on scales the

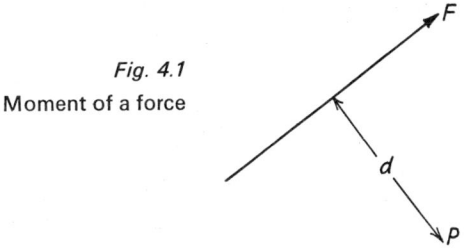

Fig. 4.1
Moment of a force

moment of the weight of the body about the central pivot must be equal and opposite to that of the 'masses' on the scale-pan.

The principle of moments

When a body is in equilibrium the sum of the clockwise moments is equal to the sum of the anti-clockwise moments about any point.

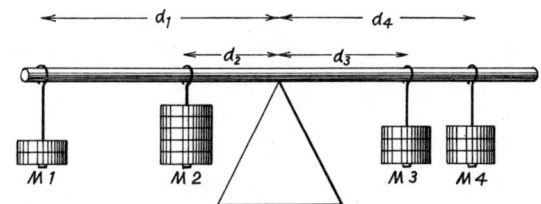

Fig. 4.2 Balancing of a beam
Sum of clockwise moments = sum of anti-clockwise moments

The principle of moments can be verified experimentally.

A metre scale is first balanced on a knife-edge (fulcrum), Fig. 4.2. Loads are suspended from the scale and their positions adjusted until the scale is in equilibrium in a horizontal position. A table of results is drawn up as shown below, the last two columns of which are found to be the same.

The principle of moments will explain why two boys can balance each other on a see-saw, provided the larger boy sits nearer the pivot than the smaller one. It is the principle involved in the use of levers of all types. The moment of the applied force about the pivot is always equal to that of the load about the pivot. Several types of lever are illustrated in Fig. 4.3.

The chemical balance and domestic scales use arms of equal length, but the steelyard (Fig. 4.4) uses unequal arms. By this means a heavy load can be weighed by altering the position of a small mass on a long arm.

$9.8 \times M_1$ /N	d_1 /m	$9.8 \times M_2$ /N	d_2 /m	$9.8 \times M_4$ /N	d_3 /m	$9.8 \times M_4$ /N	d_4 /m	Sum of clockwise moments $9.8(M_3d_3 + M_4d_4)$ /N m	Sum of anti-clockwise moments $9.8(M_1d_1 + M_2d_2)$ /N m

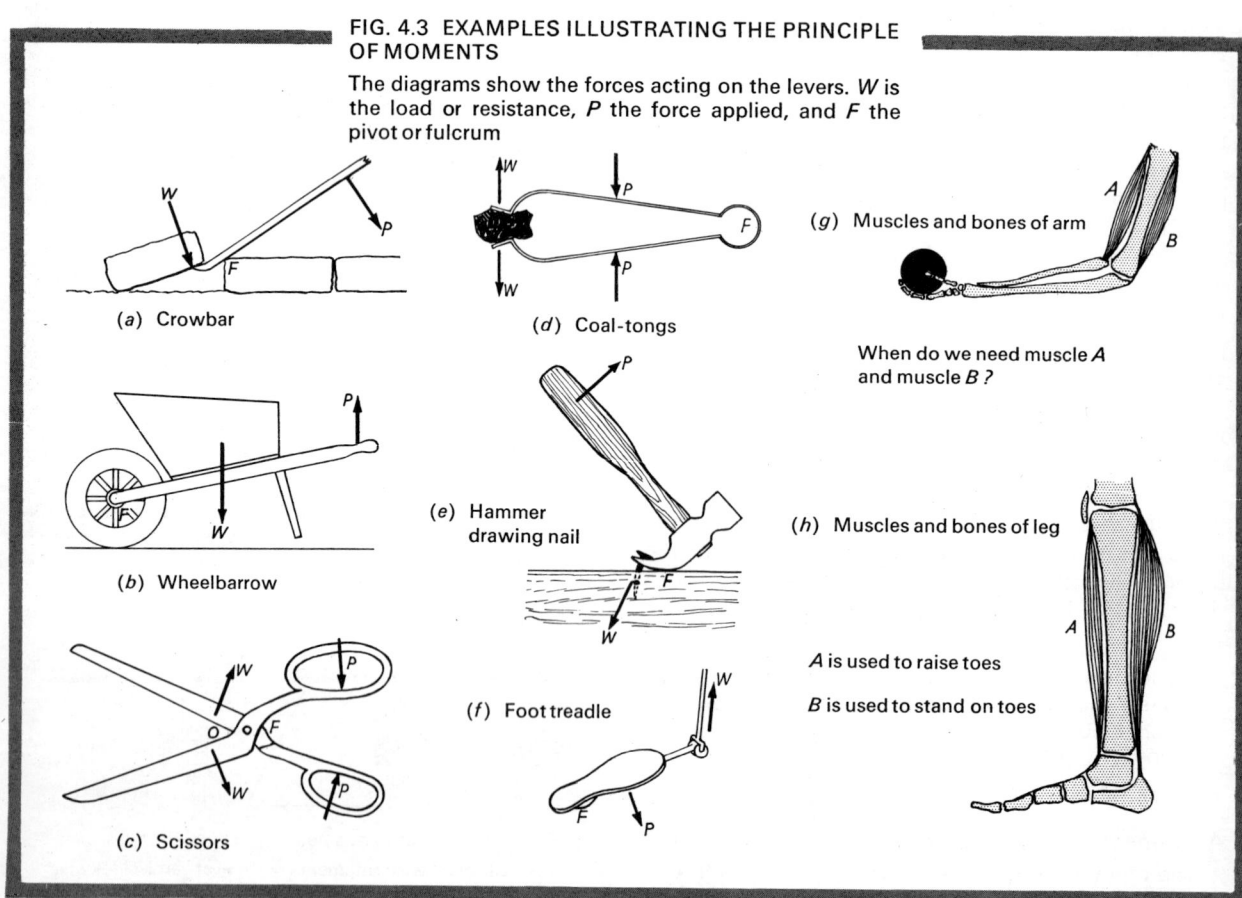

FIG. 4.3 EXAMPLES ILLUSTRATING THE PRINCIPLE OF MOMENTS

The diagrams show the forces acting on the levers. W is the load or resistance, P the force applied, and F the pivot or fulcrum

(a) Crowbar

(b) Wheelbarrow

(c) Scissors

(d) Coal-tongs

(e) Hammer drawing nail

(f) Foot treadle

(g) Muscles and bones of arm

When do we need muscle A and muscle B ?

(h) Muscles and bones of leg

A is used to raise toes

B is used to stand on toes

If we remember that the whole weight of a body acts at its centre of gravity, an interesting experiment can be done to find the mass of the metre scale. First balance the scale horizontally and carefully locate its centre of gravity. Then suspend a 50-g mass from one point near to one

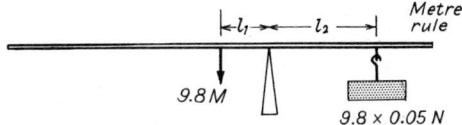

Fig. 4.6 'Weighing' a metre scale

end and alter the position of the fulcrum until a balance is restored (Fig. 4.6). The moments of the forces about the fulcrum must be equal, hence,

anti-clockwise moment=clockwise moment

$$9.8M \times l_1 = 0.05 \times 9.8 \times l_2$$

$$M \text{ (in grammes)} = \frac{l_2}{l_1} \times 50$$

where l_1=distance of c.g. of scale from fulcrum,
  l_2=distance of 50-g mass from fulcrum.

Parallel forces

If a loaded metre scale (Fig. 4.7) is suspended from a spring balance, then we note:

reading of balance=$(M_1 + M_2 + M)$ 9.8 (newtons)

where M_1 and M_2 are the loads, and M is the mass of the scale.

Fig. 4.4 The steelyard

Centre of gravity

If a metre scale is suspended from a spring balance by a string passing through a hole at its midpoint, the scale hangs in a horizontal position. Gravity is acting on each little section of the scale with a downward force 9.8 m (Fig. 4.5). The resultant

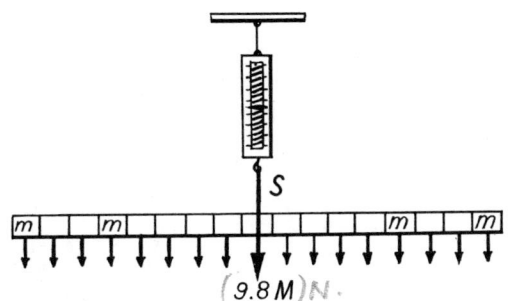

Fig. 4.5 The weight of a body acts through its centre of gravity

of all these small forces must be the force 9.8 M equal and opposite to S, the upward pull of the string. We conclude then that the total effect of gravity on the scale is the force 9.8 M acting at the centre. M is the mass of the scale and the point at which it acts is called the 'centre of gravity'.

The centre of gravity of a body is the point at which the whole weight of the body acts.

If a metre scale is not pivoted at its centre of gravity, its weight will have a moment about the fulcrum and the scale will rotate.

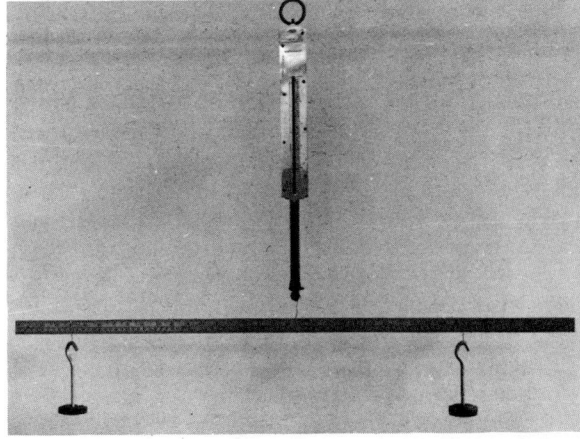

Fig. 4.7 Balancing a beam

Upward force = sum of downward forces.

This experiment can be carried a stage further by suspending the loaded scale from two spring balances (Fig. 4.8). Again the result is the same.

Fig. 4.8 Parallel forces acting on a beam
Sum of upward forces = sum of downward forces

Conclusion for any body in equilibrium:

Sum of upward forces = sum of downward forces.

If the beam is not uniformly loaded, the two spring balances will not read the same. The readings can be calculated by taking moments about any two points.

Couples

A couple consists of two equal forces acting in opposite directions along parallel lines.

Rotation of any body is really due to a couple. A single force acting on any body cannot by itself produce rotation. The reaction at the pivot along with the force constitute a couple and this produces rotation. If we consider a simple case like the balanced beam in Fig. 4.7, the scale is in equilibrium under the action of two opposite couples. The spring balance is pulling upwards with a total force of $9.8(M_1 + M_2 + M)$. Thus there is an anticlockwise couple of $9.8M_1 \times l_1$ and a clockwise couple of $9.8M_2 \times l_2$.

When we open a door, the rotation of the door is produced by a couple consisting of the force we exert and a reaction at the hinge.

The moment of a couple is the same about any axis.

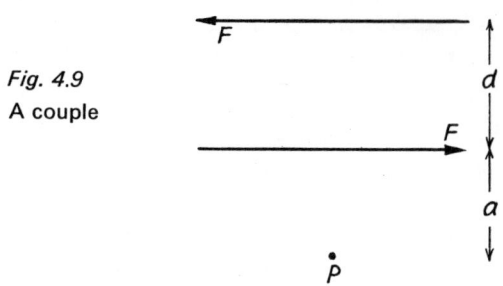

Fig. 4.9
A couple

Consider the resultant moment of the two forces F about an axis through P, Fig. 4.9.

$$\text{Moment about } P = F \times (a+d) - F \times a$$
$$= F \times d.$$

d is sometimes called the arm of the couple.

Equilibrium

The conditions under which a body is in equilibrium depend on the position of the centre of gravity. If a block of wood is tilted, it will topple over if the vertical line through the centre of gravity passes beyond the edge about which it is tilted.

The stability of a body depends on how far it may be tilted without toppling. Greater stability is obtained by making the base large and keeping the centre of gravity as low as possible (Fig. 4.10).

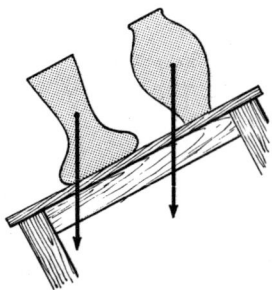

Fig. 4.10
Stability
Which vase will topple?

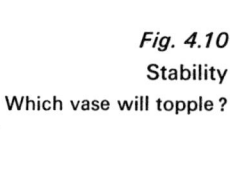

Types of equilibrium

1. *Stable*. A body is in stable equilibrium if, after a slight displacement from a state of rest, it falls back to its undisplaced position, e.g. a block resting on one of its faces. Tilting in this case raises the centre of gravity.

2. *Unstable*. A body in unstable equilibrium is one that if displaced moves farther away from its

initial position, e.g. a pencil standing on its point —a slight movement here, lowers the centre of gravity.

3. *Neutral.* A body in neutral equilibrium, after displacement by rolling, remains stationary, e.g. a ball on a billiard-table—the centre remains always at the same horizontal level.

If a body is pivoted about its centre of gravity, it is in neutral equilibrium. The flywheel and air-screw are examples of this.

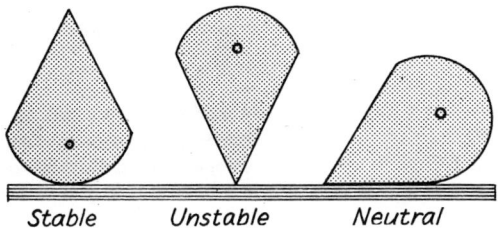

Stable   Unstable   Neutral

Fig. 4.11 Equilibrium

A cone with hemispherical end, e.g. a top, illustrates the three types of equilibrium

Position of the centre of gravity

We have defined centre of gravity (*see* page 29) as the point at which the whole weight of the body acts. The position of the centre of gravity is important for the stability and balance of a body, and it can be found by construction in a number of cases.

The centre of gravity of a triangular sheet of cardboard or metal can easily be found geometrically. Imagine the triangle *ABC* divided into strips cut parallel to *BC* as shown in Fig. 4.12.

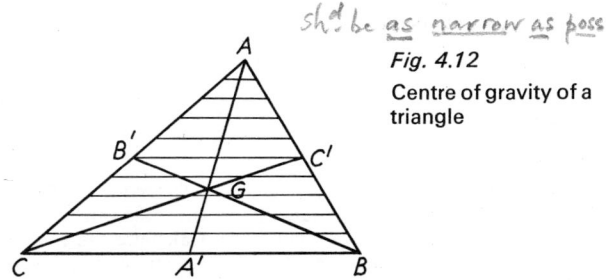

sh⁴ be as narrow as poss.

Fig. 4.12

Centre of gravity of a triangle

The centre of gravity of each strip is at its centre, hence the centre of gravity of the whole triangle must be on the line *AA'* joining *A* to the mid-point of *BC*, i.e. the median. Similarly, the centre must be on *BB'* and *CC'*, the other medians.

Hence, the centre of gravity of the triangle must be at *G*, the point where the three medians meet.

It can be shown geometrically that the centre of gravity is one-third of the way along a median, i.e. $AG = 2\,A'G$, etc.

If a body is suspended or pivoted, it always hangs with its centre of gravity vertically below the point of support. This must be so as there are only two forces acting on it (the weight and the tension in the string or reaction at the support) and these must be equal and opposite to produce equilibrium. The centre of gravity of a body can be found experimentally using this fact. For example, if a sheet (lamina) of card or metal be suspended by a single string, and in front of it, from the same support, is a plumb-line, then the centre of gravity must be along this line (Fig. 4.13).

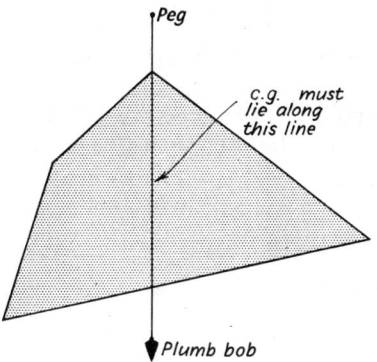

Peg

c.g. must lie along this line

Plumb bob

Fig. 4.13 Centre of gravity of a lamina

By suspending from any other point, the exact position of the centre of gravity is located at the intersection of the two plumb-line directions.

A modification of this method is to balance the sheet, when in a horizontal plane, on a knife-edge, e.g. a bevelled ruler.

We have seen that the position of the centre of gravity determines the stability of a body. The tight-rope walker has to be able quickly to alter the position of his centre of gravity. His equilibrium is unstable, and his centre of gravity must be kept vertically above the rope. If, by slight movement, it moves to one side, he has, by certain movements, to bring it back over the rope. He makes this process easier by using, for example, an umbrella, which he can quickly swing from one side to the other. The walker is steadier when he

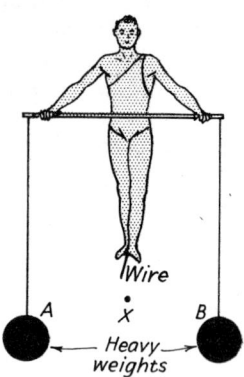

Fig. 4.14 Stable equilibrium
on the tight-rope

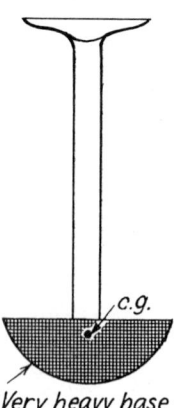

Fig. 4.15 Ashtray
Cannot be overturned

Fig. 4.16 Toy parrot
Can be perched on a rod and is
stabilized by lead in the end of its
tail

bends, and so lowers his centre of gravity—a tall person is at a disadvantage in performing on a rope. Another aid very often adopted is to carry a long heavy pole. This makes his sideways movements slower. If a tight-rope walker carried a bar as shown in Fig. 4.14, with two heavy masses *A* and *B* suspended from the ends, his equilibrium would be stable and he would be capable of doing anything on the rope; in fact, with such an arrangement anybody could walk the rope. The equilibrium would be stable because the heavy masses would bring his centre of gravity *X* below the rope.

There are numerous examples (*see* Figs. 4.15 and 4.16) in which stability is obtained by masses so placed as to bring the centre of gravity into the best position.

QUESTIONS

[g = 10 m/s².]

1. A uniform bar 5.0 m long is pivoted at its centre. Loads of 10 kg and 20 kg are attached at points 1.0 m and 2.0 m respectively from one end. At what distance from the other end must a load of 20 kg be attached to restore equilibrium?

2. A crowbar 2.0 m long is levered about a point 0.20 m from its tip. What is the smallest force which must be applied to displace a load of 100 kg at the tip?

3. A uniform beam 5.0 m long and mass 50 kg is supported at its end on two trestles. A man weighing 100 kg stands on the beam 1.0 m from one end. Find the reactions at the trestles.

4. State the principle of moments as applied to a body at rest under the action of a number of forces acting on the same plane.

From the opposite ends of a uniform plank, 4.0 m long and mass 24 kg, masses of *x* kg and 30 kg respectively are suspended. The plank rests in a horizontal position when supported at a point 0.50 m away from the end carrying the *x* kg mass. Find the mass of *x*.

5. Define the *moment of a force about an axis*.
State the principle of moments and describe an experiment which verifies it.

A uniform metre ruler balances horizontally on a knife-edge at the 20 cm graduation when a 200 g mass is hung from the 5 cm graduation. What is the mass of the ruler? What additional mass hung from the 60 cm mark will make the ruler balance at the 50 cm graduation?   (O)

6. Define the *moment of a force about an axis*, and describe an experiment which tests the principle of moments.
A uniform horizontal bar *AB*, 3.0 m long and of mass 6.0 kg, is supported at both ends. It supports a load of 3.0 kg at a point on the bar 1.0 m from the end *A*. Find the thrust exerted on each support.   (O)

7. A pole 5 m long and of mass 120 kg lies on the ground. The centre of gravity is 2 m from the thicker end. Draw a diagram to show the forces acting on the pole when the thinner end is lifted by a vertical force. Calculate the force required to begin to lift this end. Give the reason why a larger force will be required to begin to lift the thicker end from the ground.   (C)

8. A man rolls a cylindrical drum of beer, mass 48 kg, up a ramp of slope 1 (vertically) in 8 (along the ramp). Find, by taking moments about the line of contact between the drum and the ramp, the least force the man needs to exert to maintain the drum in equilibrium. Draw a diagram showing this

force and the others acting on the drum, in equilibrium.
(O & C)

9. Define *moment of a force about an axis* and describe an experiment to illustrate the principle of moments.

A uniform beam *AB*, 4.0 m long and mass 50 kg, rests horizontally on two supports placed 0.50 m from *A* and *B* respectively. The beam carries a load of 75 kg at a distance of 1.50 m from *A*. Find the reactions at the supports. How far must the support near *B* be moved in order to make the reactions at the supports equal? (L)

10. Draw a labelled diagram to illustrate the structure of a spring balance. State why a spring balance gives a different value for the weight of a body in different parts of the world, whereas a common balance will always give the same value.

A balance with equal arms, each 20 cm long, has a 10 g mass hanging from one end and an unknown mass *M* from the other. The balance is in equilibrium when a small rider weighing 1 g is 15.8 cm from the pivot on the same side as *M*. Calculate the mass of *M*. (C)

11. What is meant by the *moment of a force about a point*?

State the principle of moments and briefly explain two everyday applications.

In a faulty chemical balance the scale pans are of unequal mass. The instrument has no other defect. An unknown mass *M* in one pan is balanced by a 50 g mass in the other. When the masses are interchanged it is necessary to add 10 g to the pan containing the 50 g mass in order to restore the balance. Calculate *M*. (O & C)

12. Explain the expressions *principle of moments, principle of conservation of energy*. Show how the first principle follows from the second for the case of a simple lever in equilibrium about a frictionless fulcrum.

A uniform stick can be balanced on a knife-edge 10 cm from one end when a mass of 200 g is hung from that end. When the knife-edge is moved 5 cm further from that end the mass has to be moved to a point 8.75 cm from the knife-edge to obtain a balance. Find the length of the stick and its mass.
(O & C)

13. Define *moment of a force about a point* and explain the meaning of centre of gravity, stable and unstable equilibrium

A rod *AB* 50 cm long has been bored out for a distance of 20 cm from one end *A*. If the bored out part weighs half as much per unit length as the solid part and the total mass is 0.80 kg, find the position of the centre of gravity of the whole rod and draw a diagram showing the values of the effective

forces and their lines of action when the rod is balanced on a fulcrum. (O & C)

14. A garden roller of 100 kg mass and 25 cm radius is being pulled up a 10 cm step. Make a drawing showing the forces acting on the roller as it leaves the lower level and indicate the direction of pull which requires the smallest force. Calculate this force. (O & C)

15. What do you understand by (*a*) the *moment of a force about a point*, (*b*) the *principle of moments*?

Find the smallest value of the tangential force which can cause a uniform cylinder 1.0 m in diameter and weighing 50 kg to roll up a 20 cm step.

Draw a sectional diagram to show the line of action of the force. (O & C)

16. Define *centre of gravity* of a body.

Show that a body freely suspended from a fixed point rests with its centre of gravity vertically below the point of suspension.

A uniform rod *AB* of length 120 cm and mass 2 kg is suspended horizontally by two vertical strings, one attached to *A* and the other to a point on the rod 20 cm from *B*. Calculate the least mass which, suspended from the point *B*, will make one of the strings become slack. (O & C)

17. Define *centre of gravity* and explain how the c.g. of a body made up of two parts can be calculated, given the c.g. and weight of each of the parts.

A uniform square metal plate of 12 cm sides has a circular hole of area 4π cm^2 cut out of it, the centre of the hole being 3 cm from the centre of the plate. Calculate the position of the c.g. of the remainder of the plate. (O & C)

18. Define *centre of gravity*. Describe how you would determine the position of the centre of gravity of a thin cardboard sheet of irregular outline.

A uniform bar of mass 40 kg is pivoted freely at one end. Find the horizontal force which must be applied to its other end to hold the bar at an angle of 30° to the vertical. (O)

19. Define *centre of gravity*.

Describe how you would find experimentally the position of the centre of gravity of a thin sheet of cardboard.

The four legs of a laboratory stool touch the ground at the corners of a square of side 40 cm. When tilted on two legs through an angle of 30°, the stool overturns. Assuming that the centre of gravity lies on the vertical axis of symmetry of the stool, find its position. (O)

5 Energy, work, power

Energy

Energy is a term used freely in everyday life, but its meaning is obscure. It is used in many slogans, e.g. 'Bread for Energy', 'For extra energy eat . . .'.

Although many of these popular phrases use the term incorrectly, they all associate energy with vitality, action, and the ability to do things. Energy is sometimes defined as the *capacity to do work*.

Simple examples of such action are lifting objects from a lower to a higher level, making a car or a train move, winding up a spring, hammering a piece of lead, shaking up a quantity of small objects in a box.

In all these cases there is a transference of energy. Energy can exist in different forms and when there is a transfer from one form to another, some useful work may be done. An electric motor can be made to lift a load by winding up a rope to which the load is attached. The mass is lifted by the transfer of electrical energy. A similar change can occur with a steam-engine: the chemical energy in the fuel is transferred, first into heat, and then into mechanical energy. Hammering a piece of lead causes it to get hot. The mechanical energy of the hammering is changed into heat.

Work

Work to the scientist has a more precise meaning than is attached to the word used in ordinary conversation. Work is associated with action such as the lifting of a load. The hoisting of bricks from the ground to a platform involves the transfer of energy and we say work is done. Work involves the action of a force through a distance.

Work is *not* done unless there is movement in the direction in which the force is acting. A person holding a brick stationary may get fatigued, but he is not doing work upon the brick. The fatigue arises from muscular action, etc. The brick could be supported on a table in the same position without any transfer of energy to it, hence no work would be done. Likewise no work is done when a mass is moved over a smooth horizontal table. The weight is acting at right angles to the direction of motion.

Work is done when energy is transferred. Work is done when the point of application of a force is moved.

If we consider bricks being lifted from the ground to a platform, we know that twice as much work will be done in lifting twice the number of bricks. Likewise twice as much work will be done in lifting the same bricks through twice the height. Hence, the work will depend upon the load or force and the distance.

Work=force×distance moved in the direction of the force.
SI unit of energy (work)=the joule.
One joule of work is done when the point of application of a force of one newton is moved through one metre in the direction of the force.

Forms of energy

Potential energy

Potential energy is that possessed by a body by virtue of its position, e.g. still water at a high level, a pile-driver ready to be released.

Consider a mass on the floor and an equal and similar mass resting on a table. In allowing the upper one to fall and come to rest alongside the lower one, useful work could be obtained from it. Thus the two masses are identical only when they are side by side. The upper mass is said to have possessed extra energy called 'potential energy'.

Another form of potential energy is that possessed by a strained body, e.g. a stretched piece of elastic, a wound-up clock spring.

The potential energy of a body is the work done in raising the body to its present position or state.

Kinetic energy

Kinetic energy is energy of motion.

A moving body possesses kinetic energy and can do work in being brought to rest. The motion may be *translatory*, e.g. a train, a bullet, or *rotational*, e.g. a flywheel.

Heat

Heat is a form of energy and may result by transfer from chemical, kinetic or other forms of energy. Heat is really the kinetic energy of the molecules, a point which is dealt with fully in Chapter 14.

Chemical energy

When fuels are burned their chemical energy is transferred to heat and work may be done. The energy of the human body is obtained from the chemical energy of the food which is digested and transferred to heat and kinetic energy.

These transfers provide the work involved in our daily activities. The body of the average adult requires at least 12 MJ of energy per day. A workman in a heavy industry requires three times this amount.

M = Weight of person?

Radiation

Energy from the sun reaches us on earth by means of radiation. The radiant energy is absorbed and transferred to other forms of energy.

Electricity

Electrical energy is easily converted to other forms of energy, such as kinetic energy, heat, and radiation.

Conservation of energy*

Energy cannot be created or destroyed, but only transferred from one form to another.

Two examples of energy transfer†

1. *Electricity from Coal*

COAL	→ STEAM →	TURBINE →	GENERATOR →	ELECTRICAL APPLIANCE
Chemical energy	Heat	Kinetic energy	Electrical energy	Heat or Kinetic energy etc.

In these changes energy is lost at every stage—not destroyed—but lost from the useful chain. Only a part of the heat developed by the coal produces steam, much being lost to the surroundings. Friction in machines produces heat which is lost energy.

2. *Heat from Falling Water*. The water at the bottom of a waterfall is at a higher temperature than that at the top. In falling, the water loses potential energy which is transformed into kinetic energy. When it reaches the bottom and comes to rest, the water has lost all the potential energy it derived from its high position, and the kinetic energy which has replaced this potential energy is transformed into heat (*see* Chapter 14).

By considering such energy changes, we see that all the energy on earth (perhaps with the exception of that possessed by radioactive substances and cosmic radiation) has originally come from the sun. The energy of the sun's rays is absorbed by plants which, in the passage of time, decay and form coal.

The water in rivers from which we obtain energy to drive turbines has been 'lifted' by the process of evaporation (from heat to potential energy) and deposited as rain on the high mountains to drain down as rivers (potential energy to kinetic energy).

---

* This requires some modification for nuclear energy when the constitution of matter is changed. *See* Chapter 45.

† For numerous examples of energy transfer *see* Nuffield *T.G.*, II, Chapter 8.

Power

Power is rate of working.
The SI unit of power is the watt which is one joule per second.

Watt, the inventor of the steam-engine, thought that it would be a good idea to quote the power of an engine in terms of that of a large horse. So we got the *horsepower* which became the British unit of power. This is now replaced by the watt for all purposes.

1 British horsepower = 746 watts.

The metric horsepower is 735 watts. This unit is defined as the power required to lift a mass of 75 kg through a vertical distance of one metre in one second. (It is not an SI unit and so it is obsolescent.)

The power of a machine is not a constant: a machine can be driven at different rates. The power of a car varies with the number of revolutions of the engine per minute, which depends on the number of explosions of petrol-air mixture per minute. The power increases with the revolutions per minute to a maximum and then falls off.

QUESTIONS

[g = 10 m/s².]

1. A man pushes a box of mass 100 kg up a plank 2.0 m long which is inclined at 30° to the ground. Calculate the work done.

2. How much useful work does a horse do in pulling a canal barge 200 m if the tow-rope makes an angle of 20° with the canal and the tension in it is 2500 N?

3. The water for a workshop supply has to be pumped up to a tank 30 m high. If 5.0 m³ are used every hour, find the hp of the engine necessary. Assume that it has an efficiency of 100%.

4. How long will it take a lift driven by a 0.10 MW motor to raise 5000 kg of coal from a mine 1000 m deep?

5. Define the terms *kinetic energy*, *potential energy*, *power*.
   A lift, together with its load, is of mass 3000 kg. It is raised through a vertical distance of 20 m in 5 s, the motion being uniform except for a very short period of acceleration at the start and of retardation at the end. Calculate (*a*) the work done in setting the lift in motion, and (*b*) the work done in raising it to the top of its journey.   (O)

6. Define *work*, *power*, *efficiency of a machine*.
   Explain why a machine can never have an efficiency of 100%.
   A machine having an efficiency of 30% is used to raise a load of 120 kg through a height of 50 m in 90 s. Calculate (*a*) the useful work done per minute on the load, (*b*) the energy supplied per minute to the machine, (*c*) the wasted power.   (L)

7. Define *force*, *work*, *power*, and give a unit in which each is measured.
   A truck of mass 1000 kg is hauled 100 m in 2 minutes up a slope of 1 in 4 (measured along the slope) by a rope attached to an engine at the top. What is the work done by the engine, and what minimum power must it develop? [Neglect frictional forces.]   (S)

8. Define *work* and *power*. Name and define one unit in which each is measured.

State the principle of work as applied to a machine. What is meant by the *efficiency* of a machine?
A force of 25 000 N applied parallel to the slope of a moving staircase causes it to move steadily upwards when fully loaded with 55 passengers of average mass 70 kg. The staircase has a vertical rise of 12 m and a length measured along the slope of 30 m. Find (*a*) the efficiency of the arrangement, (*b*) the power required when the speed is 0.80 m/s.   (L)

9. Define *work*, *energy*, *power*. In what units are they measured on the f.p.s. system?
   A train of mass 300 000 kg runs at a constant speed of 15 m/s down an incline of 1 in 100 without brakes and with the steam shut off.
   What is the frictional force acting along the plane so as to oppose motion?
   Calculate the power the engine must develop in order to draw the train at the same speed on the level.   (O & C)

10. What is meant by the *efficiency* of a machine? Explain how you would attempt to measure its value for any machine you choose.
    Explain with a diagram the action of the hydraulic press.
    If the diameters of two pistons are 1 cm and 24 cm and the former is worked by a lever of mechanical advantage 8, what is the mechanical advantage of the complete machine assuming 100% efficiency?
    Why, in practice, is the efficiency not 100%?   (O & C)

11. Explain by reference to *three* examples what is meant by the transformation and conservation of energy.
    10^5 kg of water per minute flow over a waterfall of average height 50 m. How much energy is available per minute? If this energy is used to drive a turbine of efficiency 60%, what power is developed?   (C)

12. A car of mass 1000 kg travels at 40 m/s and starts to climb a slope of 1 in 20. What extra power must it develop in order to keep moving at 40 m/s?   (C)

13. A mass of 4 kg is attached to the end of a string wrapped round the rim of a flywheel on the axle of a dynamo which

is connected to an electric lamp. A short time after release the mass falls at a constant rate of 1 m/s. If the dynamo is 60% efficient calculate the power rating of the lamp. (JMB)

14. Distinguish between the terms *work* and *power*.

A force of 500 N is used to operate a machine and the force moves through 5.0 m in 2.0 s; in this time the machine raises a load of 500 kg through 0.40 m. Calculate (*a*) the applied power, (*b*) the efficiency of the machine.   (C)

15. Define *work*, *power* and *efficiency*.

A crane which is 60% efficient raises a mass of 800 kg through a vertical height of 6.0 m in 10 s. Calculate the power at which the crane is working.   (L)

16. Explain the terms *kinetic energy*, *work*, *power*. Name and define the units in which they are measured in the SI system.

A car of mass 1000 kg ascends a hill of slope 1 in 20 at a steady speed of 25 m/s. What power must be developed by the engine in addition to that required to maintain this speed on a level road?   (O & C)

17. A person walks 300 metres on the level at an average speed of 2.0 m/s, each step in the walk being 1.0 m long. Assuming that the energy of walking consists solely in giving the moving leg a kinetic energy of 14.0 joules at each step calculate the work done and the power used.

Give an account of the actions of friction and the relative amounts of energy expended when a person (*a*) walks along a level road, (*b*) skates the same distance at the same average speed over a frozen pond.   (C)

18. Give the meaning of the terms *mass, acceleration, force, work, power*.

A trailer is attached to a car through a spring balance. When the car and trailer are travelling at a steady speed of 45 km/h on a level road, the balance reading is 120 N. What power is being used in towing the trailer?   (O & C)

6 Machines

From the point of view of mechanics, a machine is an arrangement by which work can be done conveniently on a load or against a resistance, e.g. lifting a load by the exertion of an effort less than the weight of the load. The lever is a very simple machine. With a crowbar a heavy mass can be lifted by exerting a small force on the end of the crowbar.

It is apparent, therefore, that the ratio of the load to the effort will be an important quantity in studying machines. We give this a special name.

$$\text{Mechanical advantage} = \frac{\text{load*}}{\text{effort}}$$

Of course, in the crowbar, the effort is moved through a much greater distance than the load. We have,

$$\text{velocity ratio} = \frac{\text{distance moved by effort}}{\text{distance moved by load}}$$

Considering the work done by the machine, we have:

work done by the machine =
$$\text{load} \times \text{distance moved by load}$$

and

work done on the machine =
$$\text{effort} \times \text{distance moved by effort.}$$

From what we have said, in the previous chapter,

---

* A *load* is strictly a *mass*, but we use the term to mean '*the resistance offered by the load*'. In the expression for the mechanical advantage, *load* means '*opposing force exerted by load*' and so it has the same units as the *effort*, and the mechanical advantage is a number.

about the impossibility of creating energy, the following *Principle of Work* seems clear:

The work done by a machine (output) can never exceed that which is done on the machine (input).

In fact, the output is always less than the input because some energy is always lost by friction. This lost energy is transformed into heat:

Work done on machine =
work done by machine + work done in moving the parts of the machine + work done in overcoming friction.

It is a disadvantage to lose much energy by friction, for the machine will then not do as much work. We measure the efficiency of a machine by the ratio of the useful work got out to the total work put in.

$$\begin{aligned}
\text{Efficiency} &= \frac{\text{output}}{\text{input}} \\
&= \frac{\text{work done on load}}{\text{work done by effort}} \\
&= \frac{\text{load} \times \text{distance moved by load}}{\text{effort} \times \text{distance moved by effort}} \\
&= \frac{\text{mechanical advantage}}{\text{velocity ratio}}
\end{aligned}$$

The efficiency is very often expressed as a percentage.

The inclined plane

In the process of loading a lorry, one often sees packing cases being pushed up an inclined board fixed from the lorry floor to the ground. This method is easier than lifting the goods on to the lorry and is, we think, the method used by the ancients in building the pyramids. The sloping board is an inclined plane, a very simple machine.

By resolving the weight of the mass M (Fig. 6.1)

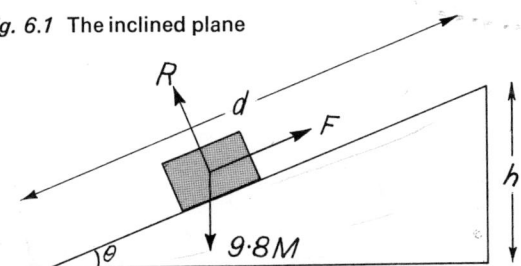

Fig. 6.1 The inclined plane

along the plane, we find the force F needed to push the load upwards is

$$9.8M \sin \theta = 9.8M \cdot \frac{h}{d}$$

Note that the force R with which the plane acts on the load is neglected. This acts normal to the plane so its effect along the plane is zero.

$$\text{Mechanical advantage} = \frac{\text{load}}{\text{effort}}$$

$$= \frac{9.8M}{F} = \frac{9.8M}{9.8M \cdot \frac{h}{d}} = \frac{d}{h}$$

But also note,

work done on load
$$= \text{weight} \times \text{height raised} = 9.8M \cdot h$$

work done by effort
$$= \text{force} \times \text{distance} = F \cdot d$$

Hence, by using the inclined plane, a smaller force F is required, but this has to be exerted through a bigger distance. In practice, the mechanical advantage is less than d/h because of friction which has been neglected in this calculation. Friction, which always acts against the motion, would act down the plane.

The screw

The screw is, in effect, a spiral inclined plane. One revolution raises or lowers the screw by a distance

called the pitch of the screw. In the screw-jack and screw-press (Fig. 6.2) the effort is exerted by a long bar. For the former, the work done in one

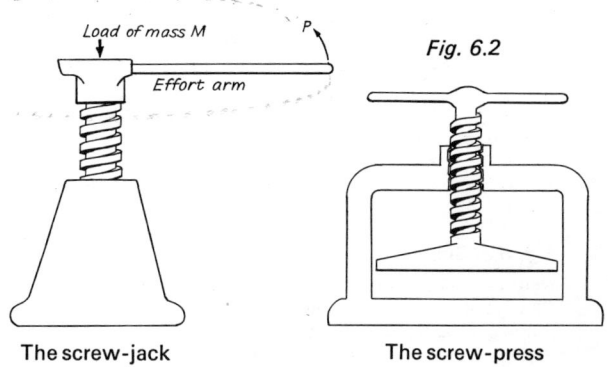

Fig. 6.2

The screw-jack   The screw-press

revolution is $F.2\pi l$, where l is the length of the bar. The load is lifted through a height h, the pitch of the screw. Thus:

$$\text{work done by effort} = \text{work done by load}$$

$$F.2\pi l = 9.8M \cdot h$$

This assumes 100 per cent efficiency, but in practice the efficiency of a screw is very small.

It can be shown that if the efficiency of a machine is less than 50 per cent it cannot run backwards. We see, therefore, that if the efficiency of a screw-jack were greater than 50 per cent, the load would turn the screw and lower itself as soon as the effort ceased. In the same way, the screw-press would not be able to maintain its pressure unless held by the effort.

Gear wheels

Gear wheels consist of cog wheels mounted on shafts which can be engaged to transmit the rotation. By engaging a wheel with many cogs on one shaft, and one with few cogs on another shaft, the rate of rotation may be changed. In the gear-box of a car, Fig. 6.3, the rotation of the engine-shaft is transmitted to the driving-shaft by way of cog wheels and the countershaft. Moving the gear lever of the car changes the position of C and D, thereby causing different cog wheels to engage so that the rate of rotation of the driving-shaft is changed.

Gear wheels alter the mechanical advantage and velocity ratio of a machine. The change in these

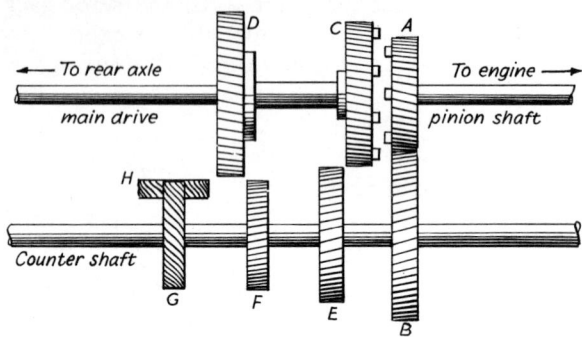

Fig. 6.3 Simple form of gear-box

Neutral position is shown. *A* and *B* are always engaged
Change of gear alters distance between *C* and *D* :
  low gear = *D* and *F* engaged
  second gear = *C* and *E* engaged
  top gear = *A* and *C* engaged
  reverse gear = *D, H,* and *G* engaged

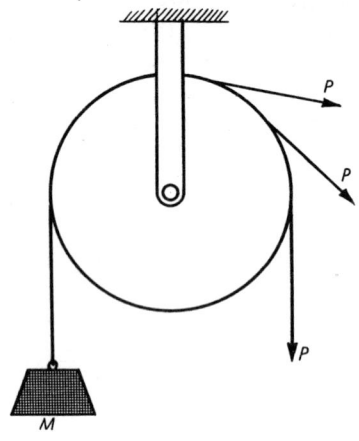

Fig. 6.4 Single pulley to change direction of effort
P = 9.8 *M* for any position of *P* if friction is negligible

quantities is easily noted when riding a bicycle. The relative speeds of the back wheel and pedals depend on the number of cogs on the wheels round which the chain passes.

The cycle is an interesting machine, for it has a mechanical advantage and velocity ratio both less than one.

The distance moved by the pedals is much less than that moved forward by the cycle. The effort exerted on the pedal is much greater than the force required to push the cycle.

The velocity ratio can be altered by use of gears. When going uphill the velocity ratio is raised by using a lower gear.

Pulleys

A single fixed pulley (Fig. 6.4) is a means of changing the direction of the effort so that it can be applied more conveniently. The velocity ratio is 1 and, if the pulley is frictionless, the mechanical advantage is also 1. In practice, therefore, the effort always exceeds the load by an amount equal to the friction of the pulley.

A spring balance introduced at *P* will show a constant reading while mass *M* is raised steadily irrespective of the direction of the pull *P*.

A single movable pulley, however, has a velocity ratio of 2 for when the effort *F*, in Fig. 6.5, lifts the end of the string 2 m, the load is raised 1 m. Neglecting friction and the mass of the block,

the tension in the string is constant, hence the load of mass *M* is supported by upward forces *F* + *F*

$$\therefore F = \frac{9.8M}{2}$$

You're only supporting half the weight.

Pulley systems

There are two pulley systems in common use: the block and tackle and the Weston differential pulley.

The block and tackle. In this pulley system each block contains one or more pulleys mounted on the same axle. Though it is customary to have the same number of pulleys in each block, there may be one less in the lower than in the upper block.

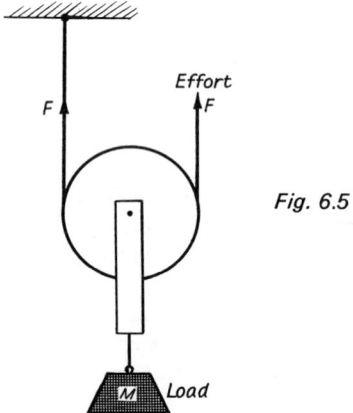

Effort
F
F

Fig. 6.5

Load

a 3-pulley block.

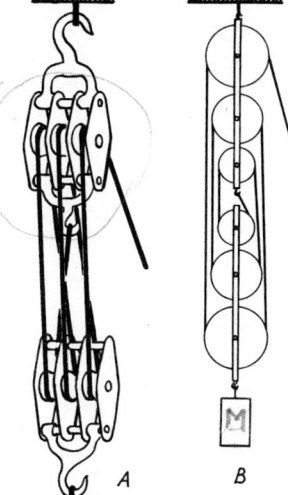

Fig. 6.6
Block and tackle

Fig. 6.6 shows a system with two 3-pulley blocks. *B* is a diagrammatical method of drawing it.

Assuming the pulleys to have no friction, and neglecting the weight of the lower block, the tension in the rope is the same throughout its length. Hence, the load *M* (in kg) is supported by six ropes each having a tension *F*, therefore

$$9.8M = 6F$$

If the load is raised 1 m, there will be 1 m of slack in each of the six ropes, hence the effort will have to be moved 6 m to raise the load 1 m.

$$\text{Mechanical advantage} = \frac{9.8M}{F} = \frac{6F}{F} = 6$$

$$\text{Velocity ratio} = \frac{\text{distance effort moves}}{\text{distance load moves}} = 6$$

The efficiency of a pulley system can be made quite high if the pulleys are well lubricated. Reduction of the efficiency is introduced because the lower block must always be raised with the load. This causes the useless expenditure of energy.

A point to be noticed about the block and tackle is that the individual pulleys on the same block must be capable of independent rotation. Examine the system and you will notice that 6 m of rope must pass round the last pulley while only 1 m is passing round the first.

Weston's differential pulley system. In this pulley system (Fig. 6.7) the lower block is a simple single pulley, while the upper block is a double pulley, in which the two pulleys have slightly different radii and are fixed rigidly together. An endless chain, not a rope, is used, and the upper pulleys are provided with notches which engage the links of the chain and so prevent slipping. The lower pulley must be smooth to allow the chain to slip. With this simple and neat arrangement, the mechanical advantage can be made large.

Consider one complete revolution of the upper block. The effort will move a distance $2\pi b$, where *b* is the radius of the larger pulley, and the loop passing round the lower pulley will be shortened by this amount. But, at the same time, a length of chain $2\pi a$, where *a* is the radius of the smaller pulley, will be added to the loop from the smaller member of the upper block. Thus, the net reduction in length of the loop will be $2\pi b - 2\pi a$. This will cause a lift of the load equal to *half* $(2\pi b - 2\pi a)$ as *there are 2 ropes*

$$\frac{2\pi b - 2\pi a}{2} = \pi(b-a)$$

Hence, the velocity ratio is

$$\frac{2\pi b}{\pi(b-a)} = \frac{2b}{b-a}$$

This system has an efficiency less than 50 per cent, and will not run backwards. This is of great advantage in an engineering workshop for

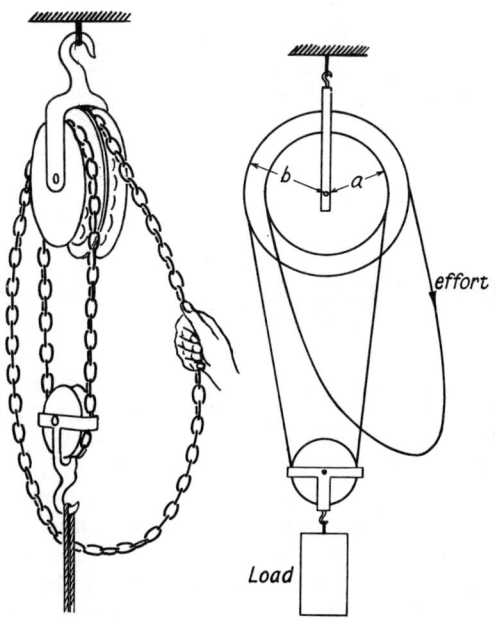

effort

Load

Fig. 6.7 Weston's differential pulley system

such jobs as the mounting of a heavy component of a machine.

To lower the load, the system has to be wound down.

Friction

Consider a book lying on the bench. If it is pushed by a small force parallel to the bench, it will not move. It must therefore be acted upon by some equal and opposite force to keep it in equilibrium. But this equalizing force must act only when the disturbing force acts. Further, no matter how the force is varied, provided it is still small, the book does not move. Thus, the opposing force must be self-adjusting and always equal to that tending to move the book. The force which opposes motion is friction. By this simple experiment we discover that the force of friction adjusts itself to be equal to the push. It cannot, however, exceed a certain value, for the book will eventually move if a large enough push is given.

The laws of friction can be investigated by using the arrangement shown in Fig. 6.8. *B* is a

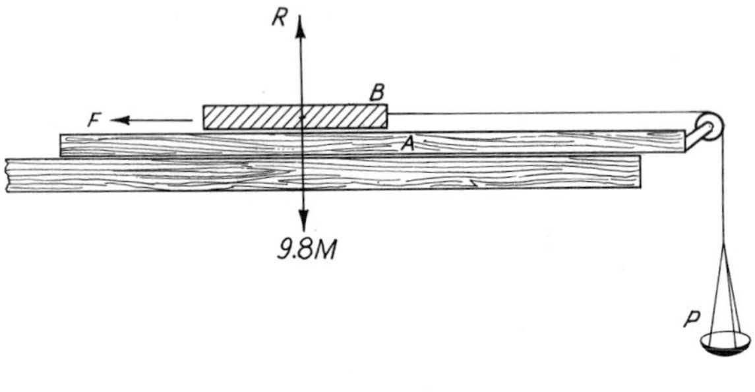

Fig. 6.8 Apparatus for experiments on friction

M = mass of block in kg
R = normal reaction in N
F = limiting friction in N

rectangular block of wood which can be pulled along a horizontal board *A* by a piece of string attached to it. The string passes over a pulley and has a scale-pan *P* attached at its other end. Both the block and board should have plane uniform surfaces. By loading the scale-pan, the force necessary just to produce motion of the block is found. (The mean of several experiments.) This force is called the 'limiting friction'. The block is

then turned on to another face of different area and the limiting friction again determined.

A better arrangement which gives more consistent results is shown in Fig. 6.9.* The board

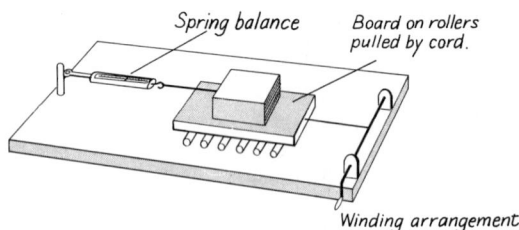

Fig. 6.9 Apparatus for experiments on friction

is moved and the block maintained stationary. Steady movement of the board is produced by a simple winding gear and the block is held stationary by a string attached to a spring balance. The reading of this balance is the frictional force.

These experiments are repeated with (*a*) the block loaded, and (*b*) the sliding surfaces changed, e.g. by covering them with sandpaper.

The following conclusions can be made. Limiting friction is:

(i) opposed to the force tending to produce motion;
(ii) independent of the area in contact;
(iii) proportional to the weight of the block + load;
(iv) dependent on the character of the surfaces.

When the block is on the point of moving, the normal reaction *R* of the board on the block and the limiting friction *F* acting along the surface are proportional, i.e.

$$\frac{F}{R} = \text{a constant, } \mu.$$

This constant is called the 'coefficient of friction'. When the board is horizontal the normal reaction is equal to the weight of the block.

Consider now the board inclined to the horizontal at an angle θ (Fig. 6.10). The normal reaction

$$R = 9.8M \cos \theta,$$

hence,

$$F = \mu R = \mu 9.8M \cos \theta.$$

---

* Nuffield *G. to E.*, II, No. 44.

The friction acts up the plane when the body is slipping downwards, and down the plane when it is being pulled upwards.

When the block rests on a plane inclined at an angle θ, the component of the weight down the plane is $9.8M \sin \theta$. If θ is slowly increased, the block will be on the point of sliding down the plane when

$$9.8M \sin \theta = \mu 9.8M \cos \theta$$

hence,

$$\tan \theta = \mu$$

θ is called the 'angle of friction'.

The determination of the coefficient of friction makes an interesting experiment, but it is not capable of giving very consistent results under normal laboratory conditions. The apparatus in Fig. 6.10 is convenient, but the surfaces of the block and of the board must be carefully prepared. The coefficient of friction varies considerably for different surfaces.

A surface is said to be *smooth* when the coefficient of friction is zero. The resultant reaction between two surfaces, one of which is smooth, is always normal to the surface at the point of contact.

The study of the laws of friction leads to the important topic of lubrication. When lubricated, two solid surfaces slide more easily over each other. The surfaces are kept apart by a thin layer of lubricant. The physical chemist has much improved lubrication by providing us with lubricants having long hydrocarbon molecules which will maintain a surface layer under great pressure.

In a machine, friction results in a loss of energy which, if transformed into heat, may cause the *seizing up* of the moving part. It is reduced by lubrication. The oiling system of a motor-car engine is very important for smooth running. A constant stream of oil is forced round the moving parts.

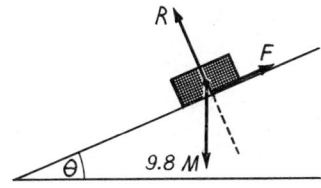

Fig. 6.10 Body in equilibrium on inclined plane
mass M in kg
forces R and F in N

We must not overlook the use we make of the force of friction. A car uses a friction cone or plate in the clutch which connects the engine to the main shaft rotating the wheels. In braking also, the car is stopped by the force of friction between the brake-lining and the drum on the wheel. Moreover, a car would not be able to move on the road were it not for the friction between the tyres and the road surface (Fig. 6.11). Neither

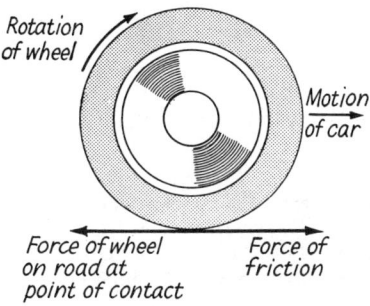

Fig. 6.11 Force of friction is responsible for forward motion of car

would we be able to walk. To move forward, we exert a backward force with the foot on the ground, and it is the opposing force of friction which is responsible for the forward force producing motion. If we step off a low platform on

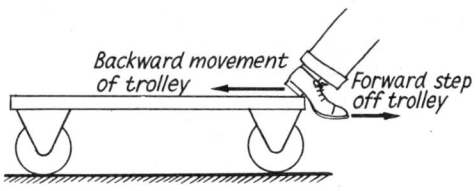

Fig. 6.12 Action and reaction

wheels (Fig. 6.12), the platform is pushed backwards.

The force of friction is the reaction to the exerted force. This is really only one aspect of a more universal law given to us by Newton (*see* Chapter 9).

Rolling friction: ball-bearings

When one body rolls over another, the friction is much less than when sliding occurs. The effort required to pull a railway coach is only a small fraction of that which would be necessary if the wheels were locked. Sliding friction is caused by small irregularities on the surfaces becoming interlocked. The motion causes these projections to be torn away and, thus, to produce wearing. Rolling friction eliminates this. There is, however, some sliding friction at the bearings of a wheel. For this reason they are sometimes made of soft metal, which lowers the friction. Lubrication also reduces sliding friction. The advantage of ball-and roller-bearings is that they replace sliding by rolling friction at the axle.

QUESTIONS

[$g=10$ m/s².]

1. How would you determine experimentally the mechanical advantage of a screw jack?

A screw jack, with efficiency 40%, is used to lift a load of 200 kg. If the pitch of the screw is 0.75 cm and the effort arm 0.50 m long, find the effort.

2. The handle, length 30 cm, of a windlass turns a gear wheel with 11 teeth. This wheel meshes with another gear wheel which has 66 teeth and is mounted on the axle of the winding drum, radius 7.5 cm. Sketch the apparatus and calculate its velocity ratio.

What is the minimum effort required to raise 1000 kg weight if the efficiency of the windlass is 70%? If the operator was not strong enough to raise the load how might he be able to do so by using a single pulley in conjunction with the windlass? (JMB)

3. A single string pulley system with two pulleys in each block is fitted up in the laboratory. The effort needed to lift various loads is shown below.

Load	0.100	0.200	0.300	0.400	0.500	0.600 kg
Effort	0.43	0.66	0.89	1.11	1.34	1.57 N

Draw a graph to show the relation between these quantities and from it find the effort to lift the lower block alone. Sketch the system and deduce its velocity ratio. What is the efficiency of the machine when the load is 350 g? (JMB)

4. Deduce a formula for the velocity ratio of a wheel and axle.

A windlass of efficiency 60% is used to lift a bucket of water of mass 20 kg from a well 15 m deep. Find the effort, the distance it moves and the work it does, given: length of handle=30 cm; diameter of drum=15 cm.

What effect, if any, would be produced upon the mechanical advantage, the velocity ratio and the efficiency of the machine by oiling the bearings of the drum? Give reasons for your answers. (JMB)

5. Draw a diagram of a simple form of screw jack or press, and explain how the velocity ratio of this machine may be calculated.

How would you measure its efficiency experimentally? In a certain jack, the screw has a pitch of 0.5 cm and the effort is applied at right angles to an arm 50 cm long. If the efficiency is 30%, find the effort needed to raise a load of 400 kg. (O)

6. Define work, energy, power.

Describe a simple sheaved pulley system with a velocity ratio of 4.

If the efficiency of this machine is 60%, find:
(a) the effort required to raise a load of 150 kg;
(b) the work done by a man using this machine to raise this load through a vertical distance of 3.0 m;
(c) the time it takes him to do this, if his average rate of working is 150 W. (O)

7. Define the terms mechanical advantage, velocity ratio, and efficiency as applied to a machine.

Give a diagram of some simple form of screw jack or screw press, and explain how the velocity ratio of such a machine can be calculated.

If, in a simple screw machine, the velocity ratio is 210 and the efficiency 13%, find the load that can be moved by an effort of 30 N. (O)

8. Give the meaning of the terms mechanical advantage, velocity ratio and efficiency applied to a simple machine (e.g. a lever) and find the relation between them.

The brake of a bicycle is operated by two levers connected by rods and hinges which do not stretch. The first lever has a velocity ratio of 4.0 and a mechanical advantage of 3.2. The second lever has a velocity ratio of 5.0 and a mechanical advantage of 3.5.

What will be the velocity ratio and efficiency of the combined system of levers if the connecting rods and hinges are frictionless? Deduce the mechanical advantage. (O & C)

9. A test is carried out on a simple machine of velocity ratio 3 and it is found that an effort of 240 N moves a load W which rises steadily through a vertical distance of 2 m. If the efficiency of the machine at this load is 75%, determine (a) the mechanical advantage, (b) the value of the load W, and (c) the energy lost (wasted work) in the performance of this test. (L)

10. Give the meaning of the terms mechanical advantage, velocity ratio and efficiency of a simple machine, and find the relation between them.

In a bicycle the chain wheel has 44 teeth, the driven wheel 16 teeth, the radius of the crank is 17.5 cm and that of the rear wheel 35 cm. What is the velocity ratio? Assuming 88% efficiency what is the mechanical advantage? (O & C)

11. Draw a diagram of a pulley hoist (block and tackle) whose velocity ratio is 4. Describe how you would determine the mechanical advantage of the hoist when it is used for lifting a bucket of sand.

A pulley hoist with a velocity ratio of 4 is used to lift a load of 100 kg through a vertical height of 8.0 m. The effort required is 350 N. Calculate (a) the work done by the effort, (b) the efficiency of the hoist. (C)

12. Define *coefficient of friction*.

A body of mass 16 kg moving with an initial speed of 4.0 m/s across a horizontal surface experiences a steady force of friction of 40 N. Calculate (*a*) the coefficient of friction, (*b*) the distance travelled by the body before it comes to rest. (C)

13. Describe a wheel and axle and deduce a formula for its velocity ratio.

A certain wheel and axle has a velocity ratio of 6. Its efficiency is 75% when the load is 450 kg. Calculate (*a*) the least force necessary to lift this load, (*b*) the work done by the effort when the load of 450 kg is raised through a vertical distance of 3 m. (JMB)

14. Describe an experiment to determine the limiting frictional force between a wooden block and a horizontal table top. How would you expect the value of this force to vary as the load on the block is gradually increased? (L)

15. Define the *velocity ratio*, *mechanical advantage*, *efficiency* of a machine and derive the relation between them.

Briefly describe a machine having a mechanical advantage (*a*) greater than unity, (*b*) less than unity, (*c*) unity.

In a wheel and axle used for lifting heavy loads the ratio of the radii is 8 : 1 and the efficiency is 75%. Calculate the effort required to raise 200 kg. (O & C)

16. Illustrate, by referring to a simple pulley system, the meaning of the terms *machine, velocity ratio, mechanical advantage*.

Two men are rolling a 100 kg oil drum up a plane inclined at 30° to the horizontal, by the following method. The men stand at the top of the plane and each pulls a rope fixed to the plane at the top, passing down the plane, under and half-round the drum, then, parallel to the plane, up to the man. What is (*a*) the least force that each man must exert, and (*b*) the least work each does in rolling the drum a distance of 4 m up the plane? (O & C)

17. Draw a diagram of a block and tackle pulley system for which the velocity ratio is 5 and mark on it the load and the effort.

Describe the experiment you would carry out to obtain values for plotting a graph to show how the efficiency of such a machine changes with load.

An effort of 500 N just raises a certain load attached to a block and tackle pulley system with two pulleys in each block. Find the work done by the effort in raising the load through a vertical distance of 5 m. (JMB)

Part Four   FUNDAMENTAL PROPERTIES OF FLUIDS

The American Bathyscaphe *Trieste II* built for the exploration of the depths of the oceans. It is designed to withstand the tremendous pressure at depths well in excess of 10 000 metres in the Pacific Ocean

7 Pressure in fluids

Pressure

If we are carrying a heavy parcel by the string tied round it, we find this rather painful because the string exerts a large pressure locally on our fingers. We overcome this discomfort by some such means as wearing gloves. This 'spreads out' the load over a bigger area of skin and does not hurt so much. The load is the same, but not the area over which it acts.

Large caterpillar tracks are used on heavy farm machines and army tanks to prevent their sinking into soft earth. The heavy weight of the machine is distributed over a large area, so that the pressure is not very great.

Foundations of buildings cover a bigger area than the bottom row of bricks. This spreads out the load, reduces the pressure, and so prevents sinking.

The large feet of an elephant enable it to walk on soft ground.

The idea of pressure can be appreciated by considering the depression caused in a large piece of plasticine by a heavy brick. If the brick is placed on a layer of plasticine of uniform thickness, the depression produced depends upon which face of the brick is the base. This shows that, for the same load, the pressure is greater the smaller the base.

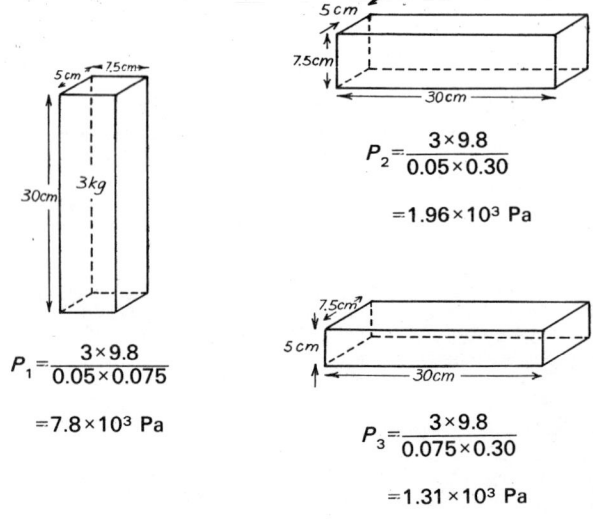

$$P_1 = \frac{3 \times 9.8}{0.05 \times 0.075}$$

$$= 7.8 \times 10^3 \text{ Pa}$$

$$P_2 = \frac{3 \times 9.8}{0.05 \times 0.30}$$

$$= 1.96 \times 10^3 \text{ Pa}$$

$$P_3 = \frac{3 \times 9.8}{0.075 \times 0.30}$$

$$= 1.31 \times 10^3 \text{ Pa}$$

Fig. 7.1 Pressure is force per unit area

48

A person who walks into a bog is advised to lie flat on his back so as to spread the weight of his body over a large area. Likewise rescuers working on thin ice put wide planks of wood or ladders on the surface thereby spreading the load of their bodies over a large surface area of ice.

Pressure is the force acting over unit area.

The SI unit of pressure is called the pascal (Pa) and is equal to one newton per square metre.

$$1 \text{ pascal} = 1 \text{ N/m}^2 \quad viz. \ | \ Nm^{-2}$$

For convenience, pressure is sometimes measured in terms of the length of a column of mercury (*see* p. 61). Thus we use mHg, cmHg, and mmHg. It then follows

$$1 \text{ atmosphere pressure} = 101\,325 \text{ Pa}$$

$$\text{pressure} = \frac{\text{force (thrust*)}}{\text{area}}$$

The difference between force and pressure is clearly demonstrated with the simple apparatus† (Fig. 7.2).

A flat box contains a plastic bag provided with two tubes as shown. The top surface of the box has two cut-out platforms A and B, one four times

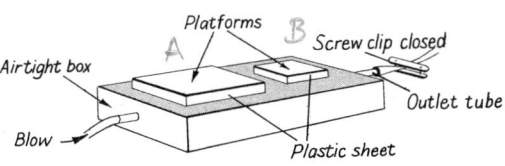

Fig. 7.2 Apparatus to show relation between pressure and force

the area of the other. If loads of 0.5 kg are put on each of the platforms and the pressure increased in the bag, the larger platform A will rise up before the smaller one B. If a 2 kg load is placed on A and a 0.5 kg load on B increasing the pressure causes both platforms to rise simultaneously. The pressure in the bag is the same all over and so we see that the upward forces are in the ratio of the areas of A

---

* Thrust is another name given to a push force.
† Nuffield *G. to E.*, I, page 103.

and B. Further simple experiments with different loads can be carried out which illustrate always that

force = pressure × area.

Pressure in liquids

A deep-sea diver has to be equipped with a very strong steel suit or jointed case. Every part of this enclosure must be strong because the pressure is large at great depths. If this pressure acted downwards only, the diver would be crushed on the sea bottom.

The enclosure is strong all over because

the pressure at any point in a liquid is exerted equally in all directions.

This is illustrated by a ball retaining its spherical shape when submerged in a liquid. In the laboratory it may be demonstrated, using a can, filled with water, which has equal-sized holes in the bottom and at the side near the bottom. Water issues from each hole at the same rate (Fig. 7.3).

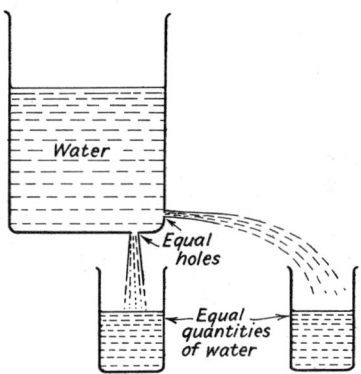

Fig. 7.3 Experiment to illustrate pressure is equal in all directions

If several equal holes, on the same horizontal level are drilled in a can, it is found that the water issues from each hole at the same rate. This shows that

the pressure is the same at all points on the same horizontal plane in a liquid.

This can be demonstrated by other methods, one of which is shown in Fig. 7.4. As the funnel covered with a piece of rubber is lowered into the tank of liquid, the pressure on it increases, as is shown by the liquid in the U-tube gauge falling

Force = pressure × area

= blow (constant) × area

∴ force is directly proportional to area.

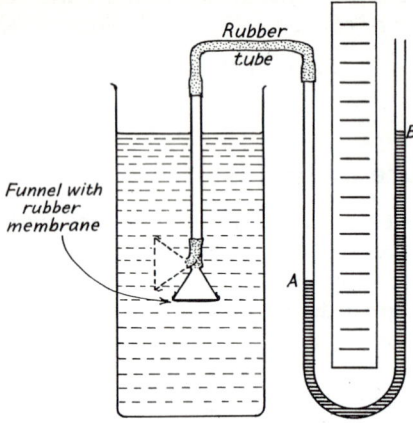

Fig. 7.4 Pressure equal in all directions

in limb *A* and rising in limb *B*. If the funnel end is twisted at right angles or moved about in the same horizontal plane, the gauge remains steady. This apparatus shows roughly the increase in pressure with depth.

As the pressure on the surface of a liquid must be everywhere atmospheric, the surface of a liquid at rest is horizontal.

The variation of pressure with depth is more precisely investigated in the following manner.

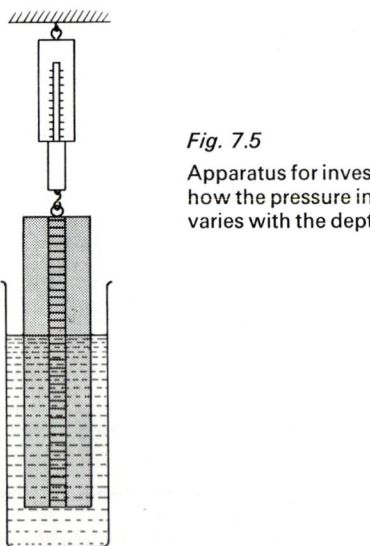

Fig. 7.5

Apparatus for investigating how the pressure in a liquid varies with the depth

A rectangular block of wood, loaded with lead so that it will sink in most liquids, is suspended from a spring balance (Fig. 7.5). When the base is submerged in a liquid, the reading of the balance is

reduced. This reduction is the upward force (pressure × area of base) acting on the base. By taking readings at different depths, it is found that the upthrust and hence the pressure, varies with the depth of the base below the surface. Using different liquids it is found that in all cases pressure/depth is a constant for the particular liquid, which shows that

the pressure at a point below the surface of a liquid is proportional to the depth.

The actual value of the pressure within a liquid is the force exerted on the base of a column of the liquid of unit area cross-section extending to the liquid surface. The pressure at *P* (Fig. 7.6) is the

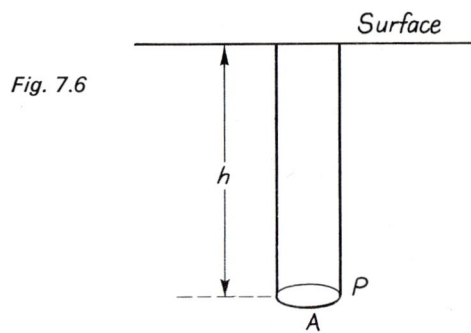

Fig. 7.6

weight of the liquid column of height *h* divided by the cross-sectional area *A*.

$$\text{Hence, pressure at } P = \frac{hA\rho \times 9.8}{A}$$

$$= 9.8h\rho \text{ or } h\rho g$$

The pressure will be measured in newtons per square metre when *h* is in metres and ρ, the density of the liquid, is in kg/m³. The value of the gravitational field *g* is 9.8N/kg and it is the necessary factor to bring the force into newtons (*see* Chapter 3).

In the above experiment the value of the constant, viz. pressure/depth is, therefore, 9.8 × density of liquid (in kg/m³).

The depth *h* must be measured vertically.

The pressure is very often expressed as a length of a liquid column, e.g. in centimetres of mercury. This is only for convenience, and a conversion into the more orthodox units can easily be made (Example 1).

In calculating the total pressure below the surface of a lake, the atmospheric pressure must not be forgotten (Example 2).

Example 1

The height of a mercury barometer is 76.0 cm when atmospheric pressure is normal. What is this pressure in newtons per square metre? Density of mercury $= 13.6 \times 10^3$ kg/m³.

Atmospheric pressure $= 76.0$ cm of mercury

$$= h \rho g$$
$$= 0.76 \times 13.6 \times 10^3 \times 9.8 \text{ N/m}^2$$
$$= 1.013 \times 10^5 \text{ N/m}^2$$

Atmospheric pressure is 1.013×10^5 N/m² (or Pa)

Example 2

Calculate the total pressure at a depth of 100 m below the surface of the sea on a day when the barometer registers 76 cm of mercury. (Density of (i) sea-water is 1.03×10^3 kg/m³, (ii) mercury is 13.6×10^3 kg/m³.)

Pressure due to sea-water $= \dfrac{100 \times 1.03 \times 10^3}{13.6 \times 10^3}$ m of mercury

$$= 7.57 \text{ m of mercury}$$

Total pressure
$$= 0.76 + 7.57$$
$$= 8.33 \text{ m of mercury}$$
$$= 8.33 \times 13.6 \times 10^3 \times 8.9 \text{ N/m}^2$$
$$= 1.11 \times 10^6 \text{ N/m}^2$$

The total pressure is 8.33 mHg or 1.11×10^6 N/m² (or Pa)

The U-tube

We are now in a position to investigate the well-known statement—'A liquid finds its own level.'

Consider the liquid levels in the U-tube *ABCD* (Fig. 7.7). The pressure at *B* must equal that at *C* on the same horizontal plane and in the same liquid. Hence the heights of the columns *AB* and *CD* must be equal. Further, since the pressures are equal, and pressure depends only on depth and density, the positions of the liquid surfaces at *A* and *D* must be independent of the cross-sections and inclinations of the tubes. The apparatus in Fig. 7.8 illustrates this.

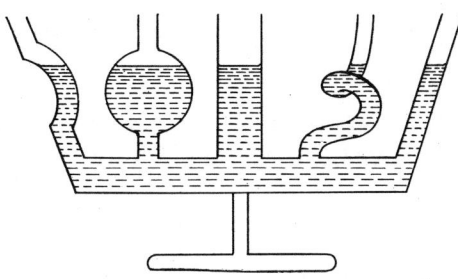

Fig. 7.8 Communicating vessel
Illustrating that level of liquid is independent of cross-section and inclination of tube

Thus,

pressure depends on depth and is independent of the shape of the vessel.

As the pressures are balanced at *B* and *C* in Fig. 7.7, the levels *A* and *D* will be unequal if the limbs of the U-tube contain liquids of different densities.

This principle can be employed to measure the relative density of a liquid. With a U-tube, some arrangement must be used to prevent the two liquids mixing if they tend to do so. The most satisfactory method is to have a short index of mercury *BC* separating the two liquid columns *AB* and *CD* (Fig. 7.9). These columns are adjusted

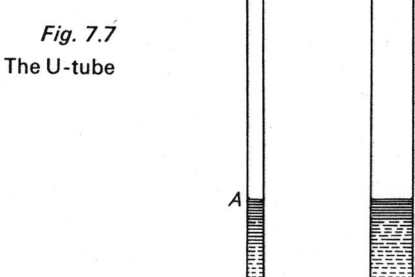

Fig. 7.7
The U-tube

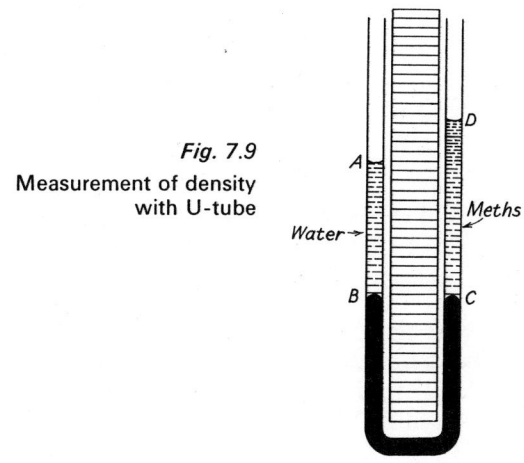

Fig. 7.9
Measurement of density with U-tube

in length, by adding liquid to one of them, until the mercury levels at B and C are in the same horizontal plane. Then we have:

$$\left\{\begin{array}{l}\text{pressure at } B \text{ due to}\\ \text{water column } AB\end{array}\right\} = \left\{\begin{array}{l}\text{pressure at } C \text{ due to}\\ \text{meths column } CD\end{array}\right.$$

$$9.8h_1\rho_1 = 9.8h_2\rho_2$$

where h_1 and h_2 are the heights in metres of the columns above plane BC, and ρ_1 and ρ_2 are the densities.

Hare's apparatus, Fig. 7.10, provides a more satisfactory method. In this case the U-tube is

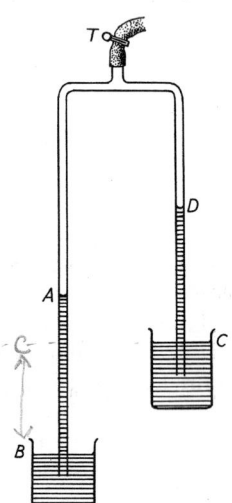

Fig. 7.10
Hare's apparatus

inverted, the ends dipping into beakers containing, for example, water and methylated spirit. On opening the clip T on the side tube, and 'drawing out' some air, the liquids are forced up the tubes by atmospheric pressure.

Pressure at A = pressure at D
= pressure of air left in space ATD

Pressure at B = pressure at C
= atmospheric pressure

Pressure due to column AB
= pressure due to column CD

$$h_1\rho_1 = h_2\rho_2$$

Note that the levels of the liquid surfaces at B and C need not be in the same horizontal plane. This follows since the pressure of the air column BC is so small that atmospheric pressure at B = atmospheric pressure at C.

Measurement of pressure

The U-tube can be employed to measure pressure. For example, if the limb A of a U-tube containing water be connected to the gas main, the level in A will be lower than that in B, Fig. 7.11.

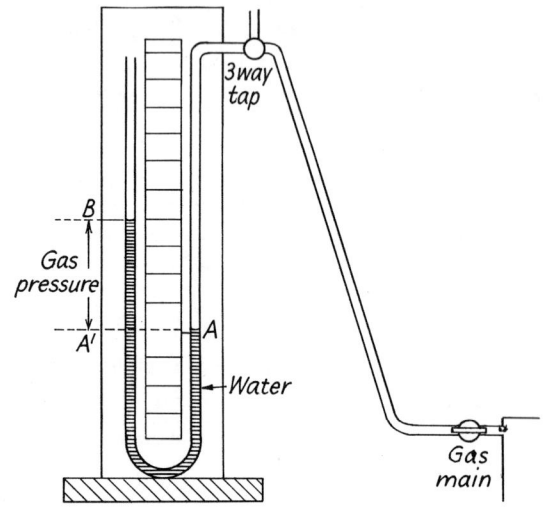

Fig. 7.11 Pressure in gas main
Water U-tube manometer is attached directly to gas main. (Using mercury instead of water in the manometer, pressures up to about 0.1 N/m² can be measured)

Pressure of gas = pressure at A
= pressure at A'
= atmospheric pressure on B
  + pressure due to column of water $A'B$

In this way it is found that the pressure in the gas main is a few centimetres (usually about 15) of water above that of the atmosphere.

Employed in such a manner, the U-tube is called a 'manometer'. For larger pressure variations mercury with its larger density may be used in the manometer.

To measure high pressures, a U-tube containing mercury, and with one limb closed, is employed. This is a closed manometer. Here, the theoretical basis is quite different, for the enclosed air in the tube will change its volume according to Boyle's law (*see* page 141). Pressure gauges which read directly are usually of the Bourdon type (Fig. 7.12).

Fig. 7.12 Bourdon-type pressure gauge
An increase in pressure causes a flexible tube to uncoil and thus move the pointer round the scale.

Transmission of pressure

A liquid is practically incompressible: a tremendous pressure is required to diminish its volume by a very small amount. If an enclosed liquid is subjected to a pressure, this is transmitted equally throughout the liquid. If a syringe is filled with water and the aperture then closed, pressing down the piston causes an increased pressure throughout the water and the syringe is liable to burst at any point.

The equal transmission of pressure in all directions is demonstrated by an ordinary watering-can. When the can is tilted, a pressure is developed in the rose by the head of water and the water issues from each hole with the same force.

In the laboratory, the experiment is usually demonstrated with a syringe having a number of holes in a spherical end. Forcing down the piston produces a jet of equal strength from each hole. This is the underlying principle of hydraulic machines of all types.

Consider two cylinders (Fig. 7.13) with pistons of cross-section 1 cm² or 1×10^{-4} m² and 100 cm² or 1×10^{-2} m² respectively, joined by a tube and filled with a liquid. A force of 10N on the piston in A produces a pressure of $1 \times 10^5 \mathrm{N/m^2}$ in the

$10^{-4} = 0001$ or $\frac{1}{10,000}$

$\therefore 10^{-4} m^2 = 1 cm^2$

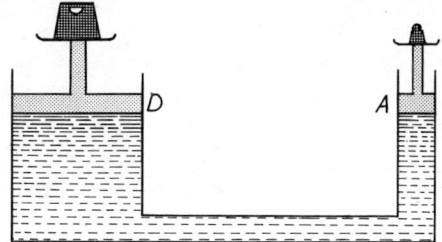

Fig. 7.13 Equal transmission of pressure in a liquid in all directions

liquid. This is transmitted to D where, acting over the piston of area 1×10^{-2} m², it exerts an upward force of 1,000N. Note, however, that if the piston in A is lowered 1 m, that in D will be raised only 10^{-2} m. Thus:

$$\text{work done on } A = \text{force} \times \text{distance}$$
$$= 10 \times 1 \text{ J}$$
$$\text{work done by } D = 10^3 \times 10^{-2}$$
$$= 10 \text{ J}$$

This is assuming there is no loss of energy.

The hydraulic press

The hydraulic press (Figs. 7.14 and 7.15) consists of a heavy ram attached to the piston of a large cylinder. A high pressure is transmitted to the liquid in this cylinder by a piston worked by a small force in a small cylinder or, alternatively, by connecting the large cylinder to the water main in which the pressure is fairly high. This pressure is transmitted through the liquid to the ram of larger area. The force exerted upwards by the ram is consequently very large. In order to raise it

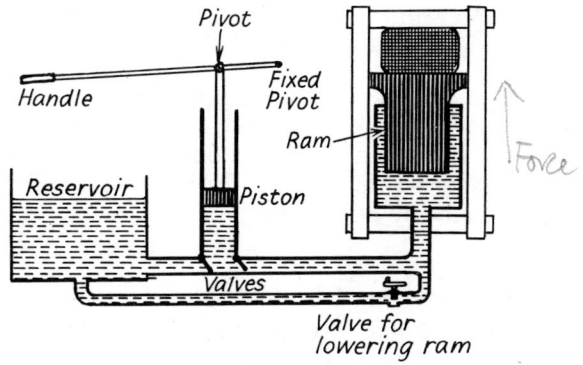

Fig. 7.14 The hydraulic ram

Fig. 7.15 This integrated forging facility comprises a 2000-ton drawdown-type hydraulic press and two manipulators; all motions of the press and manipulators are remotely controlled by one operator

through an appreciable distance, the small cylinder is fitted with valves to allow the intake of extra liquid.

A similar arrangement is employed in the hydraulic jack, car lift and hoist.

The Cartesian diver

The Cartesian diver is an interesting toy which uses the principle of transmission of pressure. It consists of a small hollow glass figure, or bulb, with a small opening, which is partly filled with water so that it just floats. It is floated in a tall jar filled with water and covered with a sheet of rubber. If this is pressed in, the excess pressure is transmitted through the water to the diver. More water enters it, compressing the air it contains. This increases its mass without changing its volume, and it sinks. Releasing the pressure causes water to leave and the diver to rise. The experiment can be carried out in a corked bottle. Pressing on the sides of the bottle is sufficient to cause the diver to sink.

The principle of Archimedes

Fluid pressure is responsible for the upthrust or buoyancy of a body immersed in a fluid. For a body to float on a liquid, e.g. a piece of wood or cork on water, it must be buoyed up by an upward force. The existence of such an upthrust is very noticeable to us when having a bath. We can lift our legs quite easily when seated in water.

This problem was first investigated by Archimedes, who explained how to calculate the size of the upthrust. His result is expressed in his principle.

When a body is totally or partially immersed in a fluid, it experiences an upthrust equal to the weight of the mass of fluid displaced.

(The term 'fluid' means gas or liquid.)

To demonstrate this experimentally, suspend a large stone or other solid from a spring balance. Note the reading, and then lower the stone into water. The reading of the balance will decrease until all the stone is immersed, when it will remain steady. The difference in the initial and final readings will give the upthrust which is the weight of that water having the same volume as the stone. This can be shown in another way by lowering the stone into a displacement can. The water which the stone displaces overflows and, when weighed, is found to have a weight equal to the upthrust.

Fig. 7.16 Downward reaction = upthrust

Decrease in reading of spring balance = increase in reading of household balance

If a body, hung from a spring balance, is lowered into a beaker of liquid standing on a household balance (Fig. 7.16) the decrease in reading of the spring balance is equal to the increase in reading shown by the household balance. This illustrates Newton's third law— action and reaction are equal and opposite (*see* Chapters 3, 6 and 9).

A convincing demonstration of the principle is carried out with the 'cylinder and bucket'. A solid cylinder just fits and fills a bucket. Hooks are provided so that the cylinder may be suspended below the bucket, as shown in Fig. 7.17. The

Fig. 7.17 Cylinder and bucket apparatus

apparatus, with cylinder below bucket, is weighed. If it is small enough, a chemical balance is used; if large, a spring balance is more convenient. The cylinder is then immersed in water. (With a chemical balance the beaker containing the water is supported on a bridge which fits over the scale-pan, the feet resting on the base of the balance.) Balance is again restored by just filling the bucket with water. The volume of the water added is equal to that of the cylinder.

Volume of bucket = volume of water displaced by cylinder.

Mass of water filling bucket = mass of water displaced by cylinder.

But,

    weight of water filling bucket = upthrust.

Therefore,

    upthrust = weight of water displaced by cylinder.

The experiment can be repeated, using other liquids.

Explanation of the principle

[handwritten: Proof of Archimedes]

We have seen that the pressure at a point in a fluid varies with the depth of fluid above it. If we consider a vertical cylinder (Fig. 7.18) of cross-

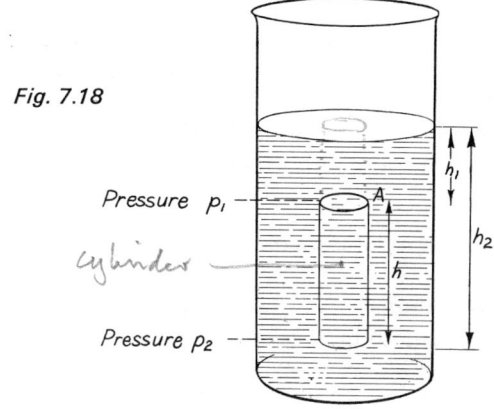

Fig. 7.18

[labels in figure: Pressure p_1; cylinder; Pressure p_2; A; h_1; h_2; h]

section A in a fluid of density ρ, the downward force over the top surface of the cylinder will be Ap_1, while the upward force on the base will be Ap_2. Hence, the resultant upward force (upthrust) on the cylinder will be $Ap_2 - Ap_1$

$$= A(p_2 - p_1)$$

But $p_1 = h_1 \rho \times 9.8$ (*see* earlier in this chapter)

and $p_2 = h_2 \rho \times 9.8$.

$\therefore$ upthrust $= A\rho(h_2 - h_1) \times 9.8$

$$= A\rho h \times 9.8.$$

But $Ah =$ volume of cylinder

and $Ah\rho =$ mass of fluid displaced.

$\therefore$ upthrust = mass of fluid displaced $\times 9.8$

$$= \text{weight of fluid displaced.}$$

[handwritten margin note: regrettable use of 'p' and 'ρ' — pressure — specific gravity]

There is another way of arriving at the principle theoretically.

Consider the shaded volume of the liquid contained in the vessel shown in Fig. 7.19. This

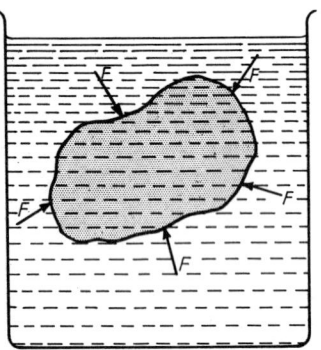

Fig. 7.19

portion of liquid is in equilibrium with the remainder of the liquid in the vessel. Hence, the resultant of all the forces exerted by the rest of the liquid on the shaded portion must be equal and opposite to the weight of the shaded liquid. If this liquid is replaced by a solid of the same volume, then the remainder of the liquid must act on the solid with an upward force equal to the weight of the liquid it has displaced.

Measurement of density

The principle of Archimedes provides a simple method of finding the value of the density of either a solid or a liquid.

Density of a Solid

A solid is first hung from a spring balance reading in newtons. The reading of the balance is then taken with the solid immersed in water. From these two readings the density can be calculated.

Reading of spring balance
  (in N) for solid in air $=a$

Mass of solid (in kg) $=\dfrac{a}{9.8}$

Reading of spring balance
  for solid immersed in water $=b$

Upthrust (in N) $=a-b$

Weight of equal volume of water
  (in N) $=a-b$

Mass of equal volume of water
  (in kg) $=\dfrac{a-b}{9.8}$

Volume of solid (in cubic metres)
  (mass of 1 m³ of water $=10^3$ kg) $=\dfrac{a-b}{9.8}\times 10^{-3}$

$$\text{Density (in kg/m}^3) = \frac{\text{mass of solid}}{\text{volume of solid}}$$

$$=\frac{\dfrac{a}{9.8}}{\dfrac{a-b}{9.8}\times 10^{-3}}$$

$$=\frac{a}{a-b}\times 10^3$$

Density of a Liquid

The density of a liquid is found from the readings of the spring balance (i) with the stone in air, a: (ii) with the stone immersed in water, b: (iii) with the stone immersed in the liquid, c.

By subtracting the second and third weighings, in turn, from the first, the upthrusts in water and liquid are found. These are the weights of water and liquid each having a volume equal to that of the solid. ✳

Using a similar method of calculation as above for a solid:

Upthrust = weight of liquid
  displaced (in N) $=a-c$

Mass of liquid displaced
  (in kg) $=\dfrac{a-c}{9.8}$

Volume of water displaced i.e.
  volume of solid (in m³) $=\dfrac{\text{mass of water}}{\text{density of water}}$

$$=\frac{a-b}{9.8}\times\frac{1}{10^3}$$

$$\text{Density of liquid (in kg/m}^3) = \frac{\text{mass}}{\text{volume}}$$

$$=\frac{\dfrac{a-c}{9.8}}{\dfrac{a-b}{9.8}\times\dfrac{1}{10^3}}$$

$$=\frac{a-c}{a-b}\times 10^3$$

Flotation

When a body floats in a liquid, it is in equilibrium, and is acted upon by two forces only, its weight and the upthrust. These must be equal and opposite. Hence, for flotation,

weight of body = upthrust
= weight of liquid displaced

✳ change of weight = upthrust
upthrust = weight of fluid displaced
relative upthrusts = relative densities

Thus, a body sinks in a liquid until it displaces a mass of liquid equal to its own mass. Cork, which has a very small density, floats with only a little of its volume submerged, whereas ice, with a density of approximately 9/10 that of water, floats with 9/10 of its volume beneath the water. This has accounted, in the past, for many disasters at sea. Boats sailing near to icebergs have crushed their hulls on the submerged parts.

Consider a body floating in water with 0.8 of its volume immersed. If

mass of body (in kg) $=M$
volume of body (in m³) $=V$
∴ mass of water displaced (in kg) $=M$
volume of water displaced (in m³)$=0.8V$
    But the density of water $=10^3$ kg/m³

∴ $$\frac{M}{0.8V}=10^3$$

Density of solid (in kg/m³)$=\dfrac{M}{V}$

$$=0.8\times10^3$$

Hence, if a body floats in water with a fraction x of its volume immersed, its density is $x\times10^3$ kg/m³.

Hydrometers

Attach a small piece of lead to the bottom of a wooden rod of uniform cross-section. This will make it float upright. Put it in different liquids and note the depth of immersion in each case. This is conveniently done if a scale is attached to the rod.

In every case,

weight of rod=weight of liquid displaced
        =density of liquid×cross-section of rod×length of rod immersed×9.8
        =density×volume immersed×9.8
        =$\rho V\times9.8$.

For several liquids we get,

$$\rho_1V_1=\rho_2V_2=\rho_3V_3$$

Therefore

$$\text{density}\propto\frac{1}{\text{volume immersed}}$$

Such a rod can be calibrated to read the density of a liquid directly, and is called a 'hydrometer'.

A small change in the density of the liquid causes a small change in the volume of the rod immersed. If the cross-section of the rod is large, this will cause only a small change in depth. On the other hand, if a thin rod is used, a large change in depth will result, and so a more accurate reading of the density can be taken. But, as most of the rod will always be immersed, this unused length can, so to speak, be coiled up to form a bulb. Fig. 7.20 shows the usual pattern of hydrometer.

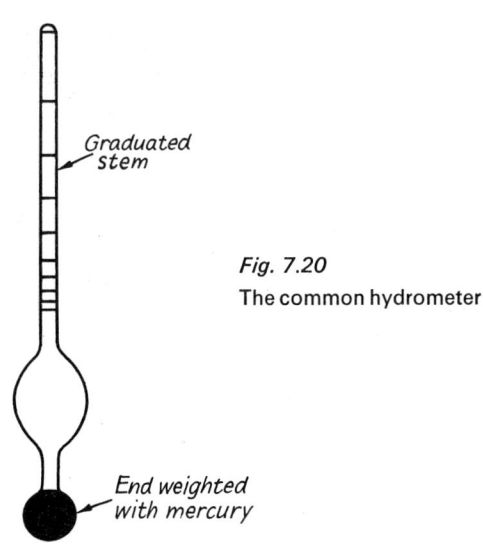

Fig. 7.20
The common hydrometer

Hydrometers are usually made of glass or of brass. Their sensitivity depends upon the cross-section of the stem. A hydrometer with a thick stem is used to cover a wide range of densities. Very often, sensitive hydrometers with thin stems are sold in sets, each one made to cover a small part of the range covered by the complete set.

Commercially, hydrometers are used extensively. The state of charge of a lead plate accumulator is indicated by the relative density of the acid. For this purpose a special hydrometer is used. A hydrometer of special range for the density of milk is usually termed a 'lactometer'. The density of milk reveals the relative amount of water present. In testing spirits, a hydrometer (Sikes) with special scale is used. From tables the reading is converted into the proof number of the spirit. The range of a Sikes hydrometer is changed by attaching masses to the bulb. Special test sets are usually equipped with a thermometer, for temperature changes cause alteration in the density.

How to make a simple hydrometer from a drinking-straw*

Close one end of a drinking-straw by dipping it into molten sealing-wax. Cut a thin ring from a piece of narrow-bore rubber tubing and slip it on the straw. Then load the straw, by dropping in seven or eight lead shots, so that it floats upright in water. Adjust the rubber ring so that, when the straw is floating, the ring lies on the surface of the water. Remove the straw and measure the length immersed. Now float the straw in the liquid whose relative density is to be measured (e.g. paraffin or brine) and repeat the above adjustment and measurement.

Let h_1 = length of straw immersed in water
and h_2 = length of straw immersed in liquid

Then, since the hydrometer displaces equal masses of water and liquid,

$$M = \rho_w V_w = \rho_l V_l$$

$$\text{relative density} = \frac{\rho_l}{\rho_w} = \frac{V_w}{V_l}$$

relative density of liquid

$$= \frac{\text{volume of water displaced}}{\text{volume of liquid displaced}}$$

$$= \frac{h_1 \times \text{area of cross-section of straw}}{h_2 \times \text{area of cross-section of straw}}$$

$$= \underline{\frac{h_1}{h_2}}$$

Ships

A ship lies deeper in the water the more cargo it takes on board. Compulsory marking of the safe loading level on the side of a boat became law by Act of Parliament in 1876. This was introduced to check unscrupulous owners who were overloading boats deliberately to render them unseaworthy, and so obtain insurance compensation in case of loss. The line is marked by an agent of Lloyd's, and is commonly referred to as the Plimsoll line, after the man who introduced the Bill into Parliament. Other marks (Fig. 7.21) show the levels in fresh water and in sea-water at average summer and winter temperatures.

---

* Due to L. F. Ennever.

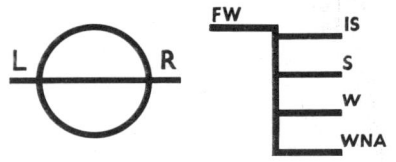

Fig. 7.21 Plimsoll line

LR = Lloyd's register	*S* = summer in temperate
FW = fresh water	latitudes
IS = Indian summer	*WNA* = winter in north
W = winter	atlantic

Airships: balloons

Although we used the term 'fluid' in stating Archimedes' principle, we have not mentioned upthrust or buoyancy in gases. As all objects on the earth are displacing air, they must be acted upon by upthrusts. This is so, but, as most objects are much denser than air, the upthrusts are relatively small and so do not make much difference.

An airship or balloon rises when the upthrust exceeds it weight. The weight of the balloon is the sum of the weights of the envelope and accessories and the gas enclosed. This is made less than the weight of the air displaced by filling the envelope with a very light gas. Hydrogen has been used for this purpose but with disastrous results. (Large British and German airships filled with hydrogen have been destroyed by fire with heavy casualties). Helium is suitable as it does not burn, but it is not quite as effective as hydrogen.

At ground-level there is usually an appreciable upthrust, but, as the balloon gets higher and the density of the air gets less, equilibrium is attained when the upthrust is equal to the total weight of the balloon. It does not go any higher unless ballast is thrown overboard. To descend, gas is released through valves.

The problem presented by balloons is not so simple as that outlined above. The pressure of the air decreases with height and consequently, as the balloon ascends, the gas inside tends to expand. The difference in pressure between the gas inside and the air may be so great as to burst the balloon. Usually, when a balloon is taking off, the envelope is only partially filled with gas at ground-level. At the lower pressure of high altitudes this gas is sufficient to extend the envelope fully.

Ballooning today is a popular sport for a small

number of people. The type of balloon used resembles a very early type whereby the lift is produced not by using a gas of lower density than the air, but by heating the air within the envelope. This is achieved by means of a burner mounted below a large aperture at the base of the envelope and thus sufficient upthrust or lift is obtained.

Pressure in gases

Atmospheric pressure
Surrounding the earth is an envelope of air which extends for some kilometres above its surface. We live, therefore, at the bottom of this thick layer of air. In many ways, the effects produced resemble

Fig. 7.22 Can before and after collapsing by atmospheric pressure

the same at all depths, because of the incompressibility of water.

The pressure of the air at sea-level is the weight of a column of air of unit cross-section extending upwards indefinitely. This pressure amounts to about 10^5 Pa (N/m^2). We do not feel burdened by this high pressure upon our bodies because the pressure is exerted equally in all directions, and the organs within our bodies are also at the same pressure.

The effect of atmospheric pressure is easily demonstrated by reducing the pressure inside a limited space. Perhaps the most striking of such experiments is to arrange for atmospheric pressure to crush a large can (Fig. 7.22). The air inside a large oil drum is driven out by steam from a small quantity of water boiling vigorously in it. After some minutes of boiling, the heat source is removed and the bung inserted. On cooling, the steam filling the can condenses, leaving the space inside at a low pressure. Under an external pressure of 10^5 Pa the can collapses in a spectacular manner. Other experiments are illustrated in Fig. 7.23.

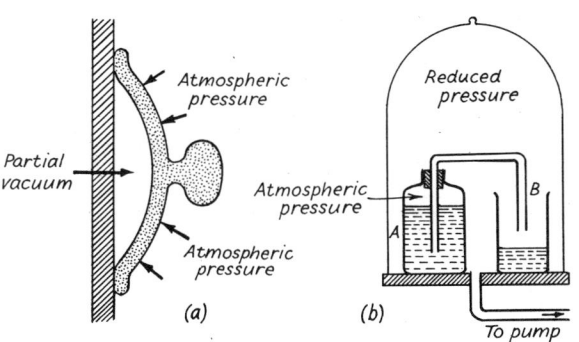

Fig. 7.23 Experiments to demonstrate the pressure of the atmosphere
(*a*) the 'sucker'; (*b*) water is forced from *A* into *B*

those beneath a liquid. The molecules of the gases comprising the air have mass and, therefore, weight. A gas, unlike a liquid, is easily compressible, hence the layers of air nearest the earth's surface are denser than those higher up: the density of the air decreases with altitude. This reduction in density at great heights demands special design of aircraft for flying at high altitudes. The pilots have also to use special breathing apparatus. Although the pressure at great depths in the sea is very high, the density of the sea is practically

The vacuum
Experiments demonstrating air pressure usually involve the creation of a partial vacuum. It was not until the seventeenth century that a reasonably good vacuum was produced. A common statement of those times was, 'Nature abhors a vacuum', and it was concluded that a vacuum was impossible to produce. Several simple observations led to this conclusion. For example, it was noted that when a jar was filled with water and inverted in a bowl of water, the jar remained filled. We know

Fig. 7.24 Statue to von Guericke in Magdeburg

Fig. 7.25 Guericke's original pump and hemispheres

now that no water will run out of an inverted tube, closed at the top, unless it is over 10 m long.

Although the first vacuum was produced by Torricelli (*see* next page), most of the pioneer work was carried out by von Guericke (Figs. 7.24 and 7.25). He employed a simple pump with two valves, and with it performed many ingenious experiments, the best known being that reputed to have been carried out before the Emperor Ferdinand III and a great assembly at Regensburg in 1654.

A large, hollow copper sphere was made in two well-fitting hemispheres. These were greased and fitted together, the sphere so formed being then evacuated, using the simple air pump. It required two teams of horses pulling against each other to separate them.

The cycle pump

In the cycle pump, air is compressed in the cylinder A by the piston P, Fig. 7.26a. This is provided with a leather washer W, of the shape shown. When there is compression in A, pressure on the sides of the washer produces an air-tight fit. On withdrawal of the piston, the pressure in A

is reduced and atmospheric pressure forces air round the edge of the washer. The air which is compressed in A passes through the valve, Fig. 7.26b, into the tyre when it exceeds the pressure of the air in the tyre. This tyre valve consists of a short length of rubber tube round a metal tube. In the metal tube is a small hole. The air under

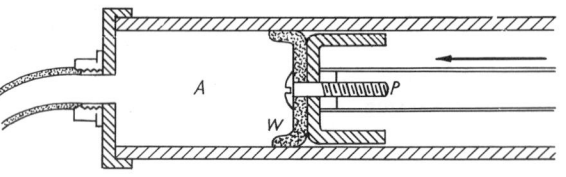

(*a*) The cycle pump

Fig. 7.26   ATMOSPHERIC   HIGH
            PRESSURE   PRESSURE

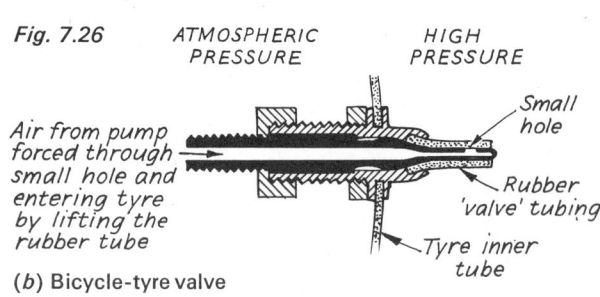

Air from pump forced through small hole and entering tyre by lifting the rubber tube

Small hole

Rubber 'valve' tubing

Tyre inner tube

(*b*) Bicycle-tyre valve

pressure from the pump enters the metal tube from the inside, lifts the rubber and passes into the tyre. When no pressure is applied from the inside of the metal tube, the pressure within the tyre keeps the rubber tube pressed hard against the small hole.

The rotary air pump
The common form of vacuum pump today is the oil-sealed rotary pump. Fig. 7.27 shows one form of such a pump in section. The rotary vane DD_1 revolves about an axis asymmetrically placed with respect to the cylindrical chamber of the pump. As D_1 rotates, the volume of the cavity V increases, and so gas is extracted from R. The vane D_1 is

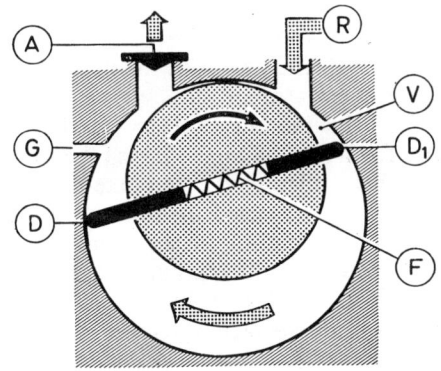

Fig. 7.27 The rotary air pump

pressed against the wall of the chamber by the spring F. When the opposite vane D has reached the opening R, the gas extracted by D_1 is enclosed. From this point, its volume is decreased until it is finally ejected through the exhaust valve at A.

Modern forms of this type of pump have a refinement to prevent the re-entry of condensible vapours such as water vapour.

Measurement of atmospheric pressure
We have already seen that pressure is conveniently measured using a U-tube as in Fig. 7.10. If an air pump is attached to one limb of the manometer, it is found that the differences in levels of the liquid surfaces in the two limbs increases, as the pump is worked, up to a maximum which is about 0.76 m for mercury and about 10 m (0.76×13.6) for water. The experiment can more easily be carried out

using the arrangement in Fig. 7.28. This is really the equivalent of a U-tube manometer, the reservoir being, in effect, the second limb.

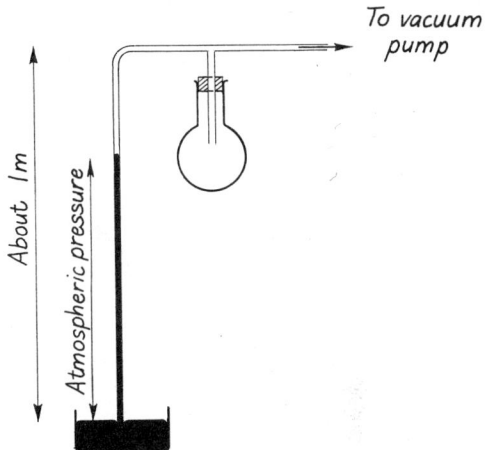

Fig. 7.28 Measurement of atmospheric pressure
The flask is inserted to give the apparatus additional volume. This makes the reduction in pressure slower and prevents mercury being drawn up the tube too quickly and so overflowing into the pump

The barometer
This is the name given to a device for measuring atmospheric pressure. The first type was made by Torricelli. He filled a glass tube, about a metre long, sealed at one end, with mercury, and inverted the open end, temporarily closed by the thumb, below the surface of mercury in a bowl. The mercury column fell until its length was about 76 cm, leaving a vacuum at the top of the tube.

Assembling a simple barometer
A reservoir of clean, dry mercury is prepared, along with a clean thick-walled glass tube closed at one end. The tube is filled with mercury to within a centimetre of the top. The thumb is placed firmly over the open end, and the tube inverted and re-inverted once or twice. This causes the air bubble to pass along the tube, so collecting the small air bubbles which stick to the glass walls. The tube is then filled completely with mercury and inverted, the thumb being pressed on the end while it is held below the surface of the mercury in the reservoir. The thumb is removed, and the

tube clamped in a vertical position. The length of the mercury column above the surface of the mercury in the reservoir gives barometric pressure.

This height above the mercury in the reservoir is maintained when the tube is tilted, and is independent of the length and shape of the tube used (Fig. 7.29). With a tube shorter than about

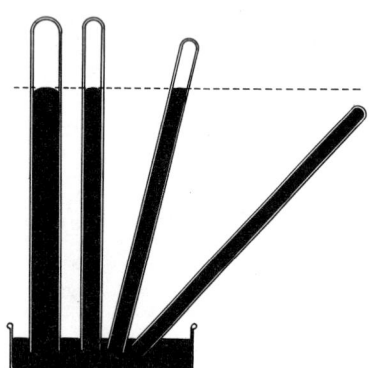

Fig. 7.29 The barometer

When the tube is tilted the mercury remains at the same horizontal level

75 cm, the mercury fills the tube. Torricelli was a student of Pascal, who concluded that this column of mercury was supported by atmospheric pressure acting on the surface of the mercury in the reservoir, and forecast that the length of the column would change from day to day and decrease at higher levels. This he tried by taking his barometer to the top of a church spire, but without positive result. Later, one of Pascal's relations, at his request, observed an appreciable fall in the mercury level when the barometer was taken up a mountain.

Von Guericke constructed a water barometer, probably without knowledge of Torricelli's experiment. This water barometer was about 11 m long and was arranged through the floors in his house. The height was observed by a figure floating on the water surface which showed above the roof on fine days.

To prove conclusively that the mercury column of a barometer is supported by atmospheric pressure on the surface of the mercury in the reservoir, the pressure on the mercury surface can be reduced. If the apparatus shown in Fig. 7.30 is used, reducing the pressure with an air pump causes the mercury column to fall until,

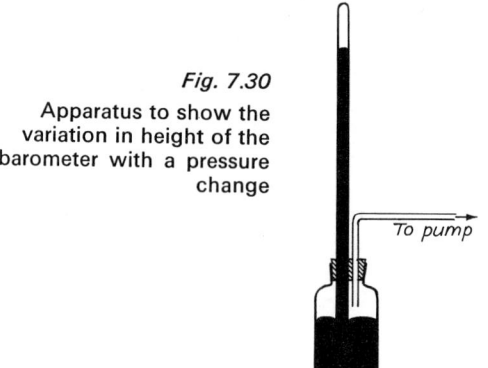

Fig. 7.30

Apparatus to show the variation in height of the barometer with a pressure change

To pump

when the pressure is zero, the mercury levels in the tube and in the reservoir are the same.

The Fortin barometer

Measurement of the height of the mercury column of a barometer is made difficult by a change in the level of the mercury surface in the reservoir resulting from the change in the height of the column. This difficulty is overcome in the Fortin barometer, which is used in most laboratories to give an accurate reading of the atmospheric pressure. The bottom of the mercury reservoir is made of leather (Fig. 7.31) and, by means of the screw touching its undersurface, the level of the mercury

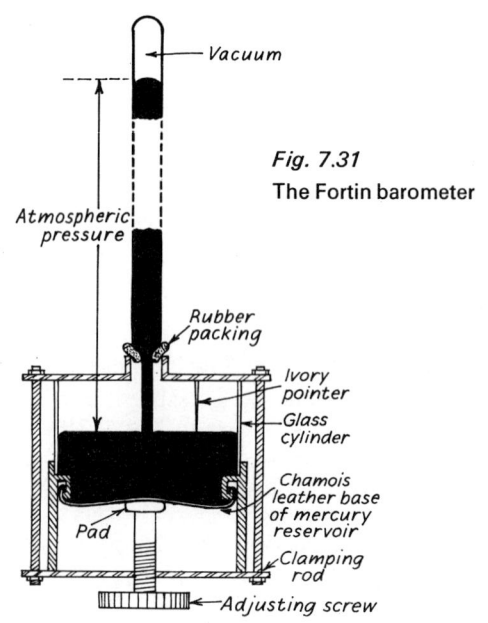

Vacuum

Atmospheric pressure

Fig. 7.31

The Fortin barometer

Rubber packing

Ivory pointer

Glass cylinder

Chamois leather base of mercury reservoir

Pad

Clamping rod

Adjusting screw

surface can be altered so that it is always in the same horizontal plane. This is arranged by adjusting the surface so that it just touches the tip of an ivory pointer. The scale is fixed with its zero at the same level as the tip of the pointer.

The aneroid barometer

This direct-reading barometer is very common. It consists of a small corrugated metal drum, partially evacuated (Fig. 7.32). As it is made of thin metal, a spring is required to prevent its collapsing under atmospheric pressure. Small changes in pressure cause fluctuations in its volume, and these are magnified by a system of levers. In some instruments, the pressure variations are automatically recorded on paper fastened round a drum rotated by clockwork (*see* Fig. 7.33).

A type of altimeter still used in some aircraft is an aneroid barometer. It is calibrated to read height instead of pressure. Before a plane takes off,

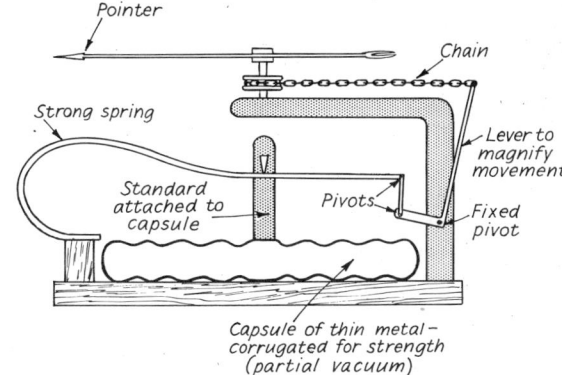

Fig. 7.32 The aneroid barometer

Fig. 7.33 The barograph

The photograph shows the automatic recording of pressure. The fluctuations are produced by nine evacuated capsules.

the altimeter has to be set to read correctly, because the reading depends upon the atmospheric pressure which, in turn, depends upon the height of the airfield above sea-level as well as the climatic conditions.

Meteorology

The meteorologist* collects the barometric readings from observation posts all over the country and, where possible, from points out at sea. A chart is constructed, and points having the same atmospheric pressure at a particular time are joined by lines—*isobars*. In this way, a weather map or synoptic chart is prepared. A depression consists of a region of low pressure, and this shows on a weather map as a series of closed isobars, the pressure decreasing towards the centre. From maps prepared at frequent intervals, the passage of the depression can be traced across the country. By studying the forms and motions of depressions, the meteorologists can predict weather changes.

The siphon

The siphon provides a convenient means of emptying liquid from a high tank which is not fitted with a draining bung. It consists simply of an inverted U-shaped tube such as *CB* in Fig. 7.34.

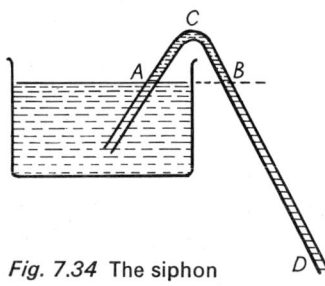

Fig. 7.34 The siphon

The conditions for the working of the siphon are important:

(i) The tube must initially be full of liquid.
(ii) The end *D* must be below the level *A*.
(iii) The height of *C* above *A* must usually be less than the height of a barometer containing the same liquid (but *see* below).

If the tube is full of liquid when the end *D* is opened, the pressure inside the tube will be greater than that outside and the liquid column *BD* will flow out. But there is cohesion* between the molecules of a liquid which gives it *tenacity* and, instead of the column breaking at *B* or *C*, it remains intact and a continuous ejection of the liquid takes place. The tenacity of some liquids, e.g. water, is much reduced by the presence of air in the liquid. In such cases atmospheric pressure keeps the liquid under pressure and so prevents the breaking of the column. A siphon using air-free liquid will work in a vacuum and a mercury siphon has been made to work with *C* about 100 cm above *B*.

Hence, we conclude that any liquid must possess tenacity to be siphoned. Air pressure acting on a liquid of low tenacity due to the presence of small air bubbles, will prevent the liquid from breaking and so assist a siphon to function.

The lift pump

The old village pump uses atmospheric pressure to raise the water from the well. A simple lift pump is shown diagrammatically in Fig. 7.35. It

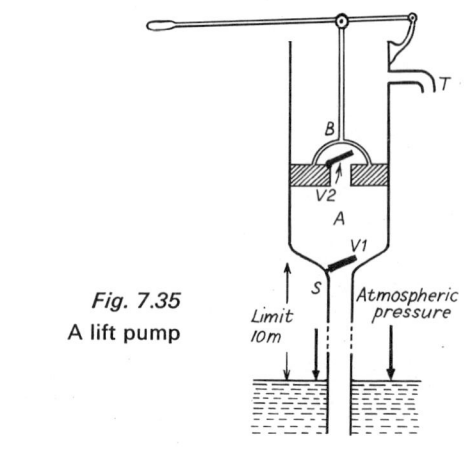

Fig. 7.35
A lift pump

---

   * A unit of pressure still used by meteorologists is the *bar*. This is not an SI unit. It is defined as 10^5 N/m² and is approximately equal to normal atmospheric pressure. (At.P. $=1.013 \times 10^5$ N/m² or Pa. *See* Chapter 7.)
   The millibar is the unit in which the isobars are graduated on the 'weather' maps (synoptic charts). The hectobar (100 bars) is used in industry.

* *See* Chapter 12. Mutual attractions between molecules.

has two valves, both opening upwards, one on the supply pipe dipping into the well and the other in the piston.

Upstroke. Pressure in A reduced: valve V_1 open; V_2 closed. Atmospheric pressure on water in well forces water into section A of the cylinder.

Downstroke. Pressure in A increased: valve V_1 closed; V_2 open. Water in A passes through valve V_2 into B above piston.

On the next upstroke the water in section B is lifted by the piston and delivered at T.

Note. 1. Water is delivered only on the upstroke, and merely runs out at T, i.e. it is not forced out.

2. Atmospheric pressure has to force water up the supply pipe S, hence the length of S cannot exceed 10 m. In practice, this length is much less than 10 m owing to imperfect valves, etc. preventing the pressure in A from being made very small.

Very often, to make a sufficiently good air seal at the piston, water has to be poured into B before the necessary reduction in pressure in A can be obtained. This is commonly known as 'priming' the pump.

The force pump

In the force pump (Fig. 7.36) the piston has no valve, the second valve V_2 being on a side-tube B from the cylinder A.

Upstroke. Pressure in A decreased: valve V_1 open; V_2 closed. Atmospheric pressure on water in tank forces water into A.

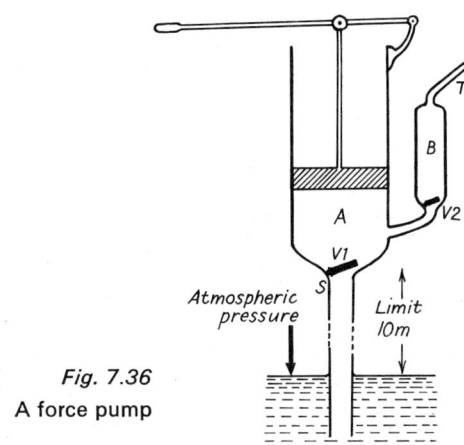

Fig. 7.36
A force pump

Downstroke. Pressure in A increased: valve V_1 closed; V_2 open. Water is forced through valve V_2.

The force at which the water is sent along tube T depends on the pressure applied at the piston. Thus, although the valve V_1 must be less than about 10 m above the water-level in the tank, the water ejected can be forced to any height. The pump delivers only on the downstroke.

QUESTIONS

[$g = 10$ m/s^2 or 10 N/kg.]

1. Define *pressure* and obtain a value for the pressure due to a liquid of density d kg/m^3 at a depth h m below its surface. How would you demonstrate experimentally the variation of pressure with depth?

The upper end of a vertical glass tube is connected by a rubber tube to a pump and its lower end is placed below the surface of water contained in a beaker. After the pump has been in action the level of the water inside the tube is 2.72 m above the level in the beaker. Assuming the height of the mercury barometer to be 76 cm, what is the pressure, in cm of mercury, in the space above the water in the tube?
[The density of mercury is to be taken as 13.6×10^3 kg/m^3.]
(L)

2. Distinguish between force and pressure.

How would you show experimentally that the pressure due to a liquid at a point below its surface is proportional to the depth?

A U-tube of cross-sectional area 1.0 cm^2 contains mercury which rises a short distance up each limb. 12.4 cm^3 of water are poured in one limb and 7.0 cm^2 of liquid of relative density 0.80 in the other. What is the difference in level between the surfaces of the water and the liquid? (The relative density of mercury is 13.6.)
(L)

3. Draw a labelled diagram to show any one arrangement, using balanced columns of liquid, for comparing the densities of oil and water. What disadvantages, if any, arise from the use of very narrow columns rather than wider ones, and from non-uniform columns rather than uniform ones?

Prove a formula for the calculation of the result from the measurements you would make.

Show on your diagram two points marked P, which have the same fluid pressure and which are at the same level; and two points, marked Q, which have the same fluid pressure but which are at different levels. Explain how you chose the positions of these points.
(C)

4. A water manometer reads 9.0 cm when connected to the gas main at the bottom of a hill. What is its reading when connected to the main at a vertical distance of 100 m up the hill?
[Density of air, 1.29 kg/m^3; density of coal-gas 0.75 kg/m^3.]
(O & C)

5. Define *pressure at a point* in a liquid.

Derive an expression for the pressure in N/m^2 at a depth h m below the surface of a liquid of density ρ kg/m^3 when the pressure of the atmosphere is 0.76 m of mercury.

An open-ended water-manometer connected to the gas main reads 10 cm at the bottom of a hill. What does it read when connected to the same gas supply at the top of the hill 150 m higher in a vertical direction? Explain how and why the second reading differs from the first.

[Density of air, 1.3 kg/m^3; density of coal-gas, 0.2 kg/m^3; density of mercury 13.6×10^3 kg/m^3.]　　　　(O & C)

6. Define *pressure at a point* and *thrust on a horizontal surface* in a fluid.

Calculate the thrusts on the top and bottom ends of a solid vertical cylinder of length 10 cm and radius 2 cm, when the top end is 5 cm below the surface of a liquid of relative density 1.5. Deduce the net upthrust on the cylinder.

Describe how you would check the value for the net upthrust by experiment, if the relative density of the cylinder exceeds 1.5.　　　　(O & C)

7. Explain how to calculate the thrust on a horizontal surface at a depth h cm below the surface of a fluid of relative density 7.

A solid right circular cylinder of length 10 cm and radius 2 cm has its axis vertical and its top end is 15 cm below the surface of a fluid of relative density 1.3. Calculate the thrust on (a) the upper, (b) the lower end of the cylinder due to the fluid.

From your results deduce the loss in weight of the cylinder on immersion in the fluid.　　　　(O & C)

8. The areas of two pistons in a hydraulic press are 4.0 cm^2 and 500 cm^2 respectively. If the piston in the former is pressed downwards with a force of 100 N, what will be the upward thrust of the larger piston?

9. Describe with a sketch Hare's inverted U-tube apparatus for finding the relative density of a liquid. Indicate clearly the measurements taken and show how the result is obtained from them.

In such an experiment to determine the relative density of methylated spirit, the length of the column of water was 11.6 cm and that of the methylated spirit 14.5 cm. Calculate the relative density of the liquid. If the length of the water column was then altered to 15 cm, what would be the height of the column of methylated spirit?　　　　(L)

10. Draw a diagram of a hydraulic press and explain its action.

The area of the small piston is 2 cm^2 and the area of the large piston is 30 cm^2. If the small piston is worked by a lever which has a velocity ratio of 4, calculate the velocity ratio of the complete press.　　　　(C)

11. Explain what is meant by a machine and define its efficiency.

Describe the hydraulic press and explain its action. The diameters of the ram and plunger of a hydraulic press are 14 cm and 2 cm respectively, and the plunger is actuated by a lever of velocity ratio 10. Find the total velocity ratio of the arrangement. Find also its efficiency if a force of 55 N applied to the lever causes the transmission of a pressure of 150 N/cm^2 to the ram.　　　　(L)

12. The height of the mercury barometer is 76 cm. What would be the height of (a) a water barometer and (b) a glycerine barometer (relative density 1.26)?

13. How would you demonstrate the principle of Archimedes?

An alloy is made of 3 parts by volume of gold to 1 of brass. What will be its density if the densities of gold and brass are 19 and 8×10^3 kg/m^3 respectively?

14. A piece of stone is suspended from a spring balance which reads 4.0 N. The stone (density 1.8×10^3 kg/m^3) is then submerged in water. What is the new reading of the balance?

15. State the *law of flotation*.

A weighted wooden rod floats vertically in water with 20 cm of its length submerged. What length will be submerged when it is placed in (a) brine (relative density 1.08) and (b) paraffin oil (relative density 0.85)?

16. An airship has a volume of 100 000 m^3. The envelope and structure weigh 80 000 kg. What extra load can it raise when it is filled with hydrogen?

[Density of air = 1.29 kg/m^3; density of hydrogen = 0.09 kg/m^3.]

17. Describe an experiment to find the relative density of a liquid by using Archimedes' principle. State *two* possible sources of error.

The envelope of a small balloon of mass 20 g contains 40 dm^3 of hydrogen. Given that the density of the hydrogen is 0.090 kg/m^3, and of the air 1.29 kg/m^3, find (a) the upthrust of the air upon the balloon, (b) the force which must be applied to the balloon to keep it stationary in still air.　　　　(JMB)

18. State the principle of Archimedes, and explain the bearing it has on the ascent of a balloon.

A uniform aluminium cylinder 8 cm long and with cross-sectional area 2.5 cm^2 hangs by a thread from a spring balance. The cylinder is lowered gradually, with its long axis vertical, into a vessel of water. What is the reading of the balance (a) when the cylinder is half immersed, (b) when the cylinder is totally immersed?

[Take the relative density of aluminium to be 2.6.]　　　　(O)

19. Describe a method, which does not entail the measuring of volumes or the use of a hydrometer, of finding the relative density of a liquid.

A piece of iron of mass 156 g and density 7.8×10^3 kg/m^3 floats on mercury of density 13.6×10^3 kg/m^3. What is the minimum force required to submerge it?　　　　(JMB)

20. State Archimedes' principle.

A piece of cork is held below the surface of some water in a vessel. State and explain carefully what is observed when the cork is released.

Find the least mass of copper wire which, when wrapped round a cork of mass 7.5 g, will cause it to submerge completely in water.

[The relative density of cork is 0.25 and that of copper 8.8.]　　　　(O)

21. State the principle of Archimedes and describe an experiment to illustrate it.

A common hydrometer is constructed by putting 10 g of mercury into a glass tube of external diameter 1 cm closed at one end. When placed in water, the hydrometer sinks to a depth of 20 cm.

Calculate (a) the mass of the hydrometer, (b) the additional depth the hydrometer sinks when placed in a liquid of relative density 0.8. (O & C)

22. State the principle of Archimedes and explain how you would deduce it theoretically for a body of simple shape.

Blocks of lead and aluminium each weighing 200 g are suspended one from each end of a straight rod 25 cm long, of negligible mass. The rod is balanced on a knife-edge. Explain why the equilibirum is destroyed when the blocks are allowed to hang, totally immersed, in water and calculate where the knife-edge must be placed so as to restore equilibrium. What is then the force acting on the knife-edge?
[Relative density of lead 11.3; relative density of aluminium 2.7.] (O & C)

23. How would you show experimentally that when a solid is completely immersed in a liquid (a) the upthrust exerted by the liquid on the solid equals the weight of liquid displaced by the solid, (b) the solid exerts an equal downthrust on the liquid?

A solid has a mass of 100 g and density 0.8×10^3 kg/m^3. What will it appear to weigh when totally immersed in a liquid of density 0.6×10^3 kg/m^3? What fraction of its volume will be immersed when it floats in water? (S)

24. You are provided with a piece of solid rubber of mass about 20 g and having a relative density of approximately 1.1×10^3 kg/m^3. Describe any *one* method you would use to obtain an accurate value of the density of the rubber. Show how you would calculate the result.

A hollow sealed copper globe weighing 90 g contains 800 cm^3 of air. It is completely submerged in a tank of water by means of a weightless thread which is attached to the bottom of the tank. Calculate (a) the volume of the copper of which the globe is made, (b) the tension in the thread.
[Density of the copper $= 8.9 \times 10^3$ kg/m^3; density of the air $= 1.3$ kg/m^3.] (C)

25. State the principle of Archimedes and describe how you would verify it experimentally.

A balloon of volume 10 m^3 is filled with hydrogen of density 0.09 kg/m^3. Neglecting the envelope, calculate the tension of the string which tethers the balloon to the ground.
[Density of air $= 1.29$ kg/m^3.] (O & C)

26. Describe, with the aid of a diagram, the common hydrometer and explain its use. What would be the effect of (a) increasing the cross-section of the stem, and (b) increasing the mass?

A test-tube loaded with lead shot weighs 600 g. It floats vertically with 12 cm immersed in water. Find (a) its area of cross-section, (b) what length will be immersed when placed in a liquid of relative density 1.2. (L)

27. State the principle of Archimedes and describe an experimental proof of its validity.

The envelope of a balloon is of 500 m^3 capacity and it is filled with hydrogen of density 0.09 kg/m^3. What gross mass can the balloon support in air of density 1.29 kg/m^3?

If the envelope were filled with helium of density double that of hydrogen by what amount would the buoyancy be reduced? (O & C)

28. State Archimedes' principle and describe an experiment to verify it.

A uniform closed glass tube, weighted internally at one end, has a mass of 50 g. When it floats vertically in water 10 cm of its stem is exposed. When it floats vertically in a liquid of relative density 1.25, 20 cm of its stem is exposed.

Calculate the area of cross-section of the tube. (JMB)

29. Draw and describe in detail the following devices and explain how they are used: (a) the density bottle, (b) the common hydrometer.

A certain hydrometer consists of a loaded cylindrical bulb of length 5 cm and outside diameter 2 cm with a stem of diameter 1 cm. The mass is 22 g. To what point on the stem does it sink in (a) water, (b) sulphuric acid of relative density 1.16? (O & C)

30. Draw a diagram of a common hydrometer. What is the effect on the hydrometer of (a) increasing the cross-section of the stem, (b) increasing the volume of the bulb?

A common hydrometer weighs 20 g. The volume of the bulb is 10 cm^3 and the cross-section of the stem is 0.5 cm^2. How far will the liquid level be above the bulb when the hydrometer floats in a liquid of density 1.12×10^3 kg/m^3? (O & C)

31. Draw a common hydrometer suitable for measuring the density of liquids denser than water. Mark the scale graduations approximately. (The distances need not be calculated exactly.) Explain why it floats at different depths in different liquids.

A hydrometer with a uniform stem and weighted bulb weighs 36 g. It floats vertically in water with 5 cm of stem above the surface. When floating in a liquid of density 1.20×10^3 kg/m^3, 10 cm of stem are above the surface. Find the area of cross-section of the stem. (C)

32. Describe how you would find the relative density of salt solution by floating a weighted test-tube in it. How would you make the test-tube into a simple hydrometer?

A metal can of cross-section 25 cm^2 is loaded with sand until it floats upright in copper sulphate solution of relative density 1.20 with 10 cm of its length below the surface of the liquid. What alteration in its mass would be necessary to make it float at the same mark in methylated spirit of relative density 0.84? (JMB)

33. Describe a simple hydrometer, and explain the physical principles on which its use depends.

The volume of an angler's float is 20 cm^3, its mass is 8.5 g, and the line hanging from the float is loaded with 5.5 g of lead shot. Find the volume of the float remaining above water. Find also the vertical pull exerted by a fish which submerges the float completely.
[Take the relative density of lead to be 11.0.] (O)

34. Describe how you would set up a simple mercury barometer.

The density of mercury is 13.6×10^3 kg/m^3. What is standard pressure (760 mmHg) expressed in N/m^2?

35. How would you make a mercury barometer in the laboratory?

Describe and explain how the barometer reading would change if (a) the barometer were slightly tilted, (b) it were carried down a mine, (c) the temperature of the laboratory were to rise.

If the height of the column is 75 cm, what would be its height if sulphuric acid (relative density=1.70) were used instead of mercury (relative density=13.6)?   (JMB)

36. Describe how you would demonstrate experimentally:
    (a) that the atmosphere exerts a considerable pressure;
    (b) that this pressure is exerted uniformly in all directions.
The screen face of a highly evacuated television tube measures 40 cm by 30 cm. Find the thrust on this face when the height of the barometer is 75 cm.

[Take the relative density of mercury to be 13.6.]   (O)

37. Define *pressure*. Describe an experiment which shows that the atmosphere exerts a pressure.

A man of mass 65 kg rides a bicycle of mass 10 kg. The pressure of the air in the tyres is 1.9×10^5 N/m^2, and the atmospheric pressure is 1.0×10^5 N/m^2. Two-thirds of the total weight is supported by the back tyre. Find the area of each tyre in contact with the road.   (O)

38. Describe fully a barometer with which atmospheric pressure can be measured to 0.1 mmHg. Draw a diagram to illustrate your answer.

Calculate by how much the barometer reading, in centimetres of mercury, diminishes with each 100 metres ascent above sea-level if the average density of the atmosphere is 1.29 kg/m^3.

[Density of mercury, 13.6×10^3 kg/m^3.]   (O & C)

39. Describe how you would set up a simple mercury barometer.

How would you demonstrate that the mercury column is supported by atmospheric pressure?

How would the readings of the barometer be affected if there were a little air in the space above the mercury?

How would you detect the presence of a small quantity of air without consulting another barometer?   (O)

40. Water is pumped from a well about 5 m deep into a tank about 15 m above the ground. Draw a diagram of a suitable pump and explain its action.

If the mass of the water raised 20 m per minute is 25 kg, how much work is done every minute if the pump is $33\frac{1}{3}\%$ efficient and what is the power of the operator?

Explain how the tank could be emptied by a siphon arrangement.   (JMB)

41. Define *work*, *energy*, and *power*.

Describe a lift pump and explain its action.

Calculate the power of a lift pump which raises 2400 kg of water per minute through a vertical height of 5.0 m.   (O)

42. Describe a cycle-tyre pump, and explain how it works. What is the total mass of air in a vessel of volume 2500 cm^3 when the pressure inside it is 200 cm of mercury? What fraction of this mass will escape if the vessel is opened to the atmosphere?

[Take the atmospheric pressure to be 75 cm of mercury, and the density of air at this pressure to be 1.2 kg/m^3.]   (O)

43. Describe how you would make a simple barometer and explain how you would calculate from its reading the pressure of the atmosphere in N/m^2. Sketch the essential parts of an aneroid barometer and compare its merits and defects with those of a mercury barometer.

A mercury barometer reads a pressure of 750 mmHg in a stationary lift. What reading will it show when the lift is moving downwards with an acceleration of 1.0 m/s^2, and why? How would you expect an aneroid barometer to behave in the same circumstances?   (O & C)

Part Five MECHANICAL ENERGY, MECHANICS OF BODIES IN MOTION

Vapour trails produced by condensation from the jets of
a team of RAF Hunters

8 Motion

Velocity

When we travel along a road in a car we often note the time taken to go a certain distance. We then divide the distance by the time to get the average speed. On the journey the direction of motion changes due to bends in the road. Although the distance travelled may be 25 kilometres, the destination may be only 20 kilometres from the starting position measured 'as the crow flies'. If we combine *direction* with *speed* then we get *velocity*—a vector quantity.

Velocity is the rate of change of distance measured in the direction of motion.

A change in velocity is produced either by a change in speed or by a change in direction.

Speed is rate of change of distance.

it is a scalar quantity.

$$\text{Speed} = \frac{\text{distance}}{\text{time}}$$

If the speed is constant but the direction of motion is constantly changing, the velocity is not constant. We can find the 'instantaneous' velocity by dividing the distance travelled in a *very* short interval of time by the duration of the interval.

A change in velocity produces an acceleration.

Experiments on motion

Much experimental work can be done on this topic using trolleys and ticker-tape vibrators.* A long flat board is required as a track (Fig. 8.1).

---

* The trolleys and ticker-tape vibrators were developed by the P.S.S.C. and have been adopted by the Nuffield project. They are now being made in England. Frictionless pucks can also be employed to study the laws of motion. They are not so convenient, but give better results. (*See* Nuffield literature, e.g. *T.G.*, IV, page 39.)

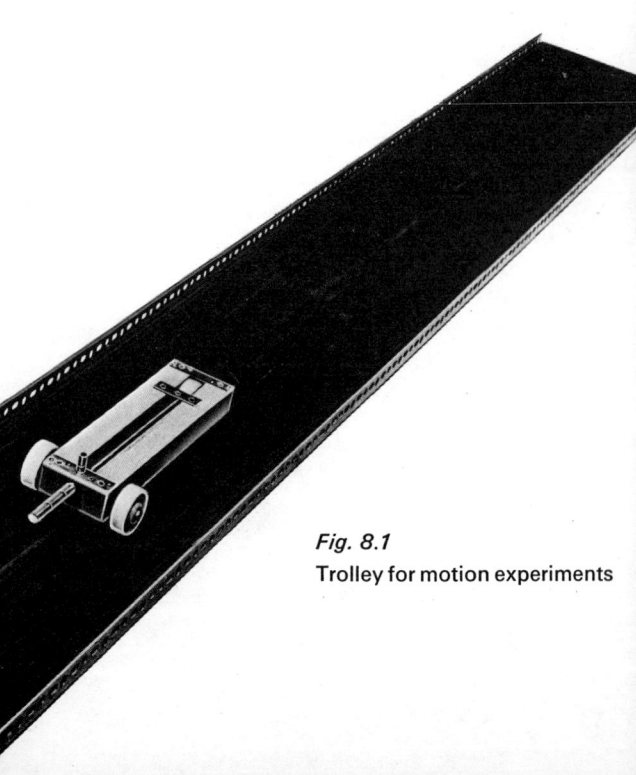

Fig. 8.1
Trolley for motion experiments

is the distance travelled in 0.02 s, thus 50 × distance between dots is the velocity in centimetres per second averaged over 0.02 s. If the tape for such a run is cut across at each dot and the lengths mounted as shown in Fig. 8.3, the tops of the tapes lie on a straight line. This shows a regular increase in speed, i.e. a constant acceleration.

The trolley may also be accelerated by pulling it along a level track. This can be done by attaching an elastic band or a spring to the trolley. A constant pull is obtained by keeping the extension of the elastic constant. (This method is developed in the next chapter. The technique is fully described in the P.S.S.C. and Nuffield literature.)

Acceleration is the rate of change of velocity.
(Units m/s^2; cm/s^2.)

Acceleration is a vector quantity, but we often assume it to be rate of change of speed.

Graphical representation

The motion of a body can be conveniently represented graphically. For a body travelling with constant velocity, the graph of *distance travelled*

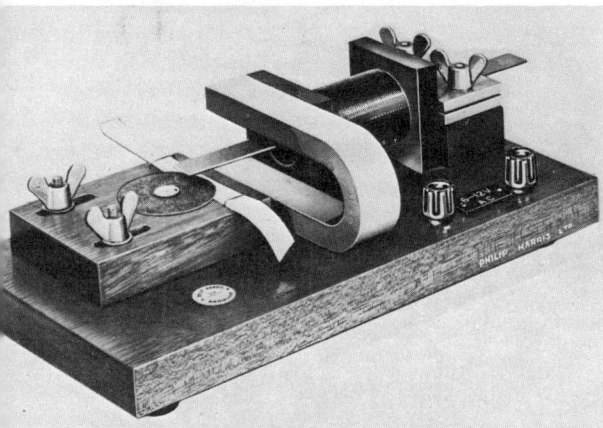

Fig. 8.2 Vibrator for timing

The vibrator (Fig. 8.2) is a similar device to that described in Chapter 40. Ticker-tape is drawn along under the end of a vibrating iron strip. The point of a screw on the end strikes the tape every fiftieth of a second and makes a mark by pressing on a disc of carbon paper inserted between the contact screw and the tape. The disc of carbon paper revolves as the tape passes it. The distance between successive dots measures the distance the tape has moved in 0.02 seconds.

Timing the motion of a trolley is carried out by attaching the end of the tape to the trolley so that the tape is pulled through the vibrator. If a trolley moves with a constant speed, the recorded dots are equally spaced. To obtain a constant speed, the track must be sloping slightly downhill to compensate for friction and air resistance.

Acceleration

If the track is given an additional slope, the speed of the trolley will increase. The distance between the recorded dots will increase progressively. This

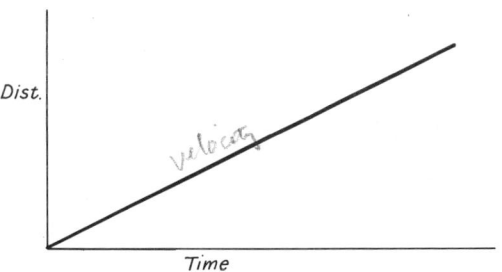

(a) Constant velocity

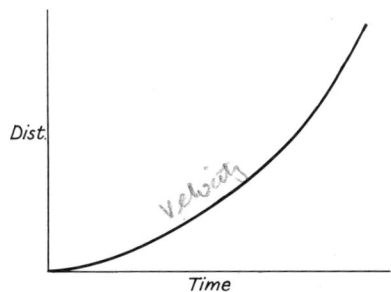

(b) Constant acceleration

Fig. 8.4 Distance/time graphs

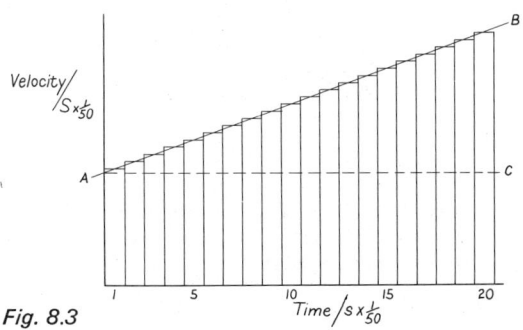

Fig. 8.3

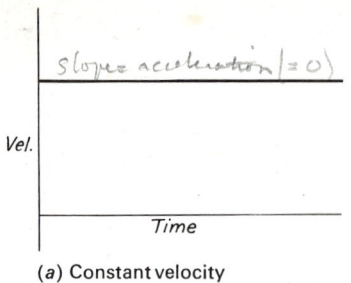

(a) Constant velocity

(b) Constant acceleration

Fig. 8.5 Velocity/time graphs

against *time* is a straight line through the origin (Fig. 8.4a). The slope of the graph is the rate at which the distance is increasing, i.e. the velocity.

The *distance/time* graph for a body moving with constantly increasing velocity, i.e. uniform acceleration, is a curve (parabola) as shown in Fig. 8.4b.

The *velocity/time* graph is a straight line when the acceleration is constant. For zero acceleration the line is parallel to the time axis (Fig. 8.5a). Acceleration is rate of change of velocity, and so it is given by the slope of the line (Fig. 8.5b).

The graph also gives the distance travelled. The mean velocity, $\frac{u+v}{2}$, is the average height of the line above the time axis. Hence, the value of the distance travelled is represented by the area under the graph (Fig. 8.6), viz. $\frac{u+v}{2} \times t$.

We conclude then:

Velocity is represented by the slope of the *distance/time* graph.

Acceleration is represented by the slope of the *velocity/time* graph.

Distance travelled is represented by the area under the *velocity/time* graph.

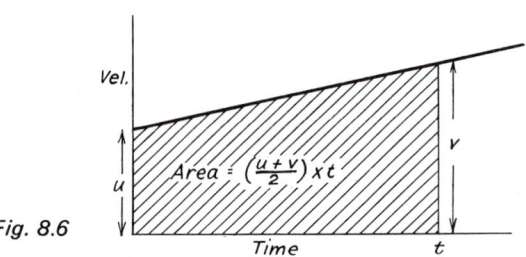

Fig. 8.6

If the acceleration is not constant, i.e. for a body moving with irregular motion, the slopes of the *distance/time* and the *velocity/time* graphs at any particular time give the instantaneous values of the velocity and acceleration respectively. The distance is still represented by the area under the *velocity/time* graph (Example 1).

Example 1

An electric train starts from rest with an acceleration of 0.5 m/s², which it maintains for 40 s. Its speed is kept con-

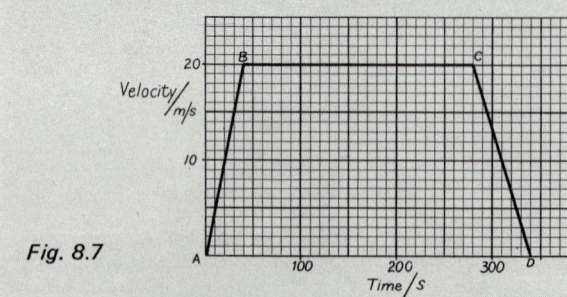

Fig. 8.7

stant for 240 s, and then it is brought to rest with constant retardation in 60 s. Find the distance between the stops.

In 40 s the velocity attained will be 20 m/s. ✳

The *velocity/time* graph is shown in Fig. 8.7.

AB represents the first part of the journey where there is an acceleration, *BC* represents the second part of the journey where the velocity is constant, *CD* represents the third part of the journey where there is a retardation. The distance travelled is the area under this graph measured in units of area given by the product of the units of the two scales.

Area under $AB = \frac{1}{2} \times 40 \times 20 = 400$ m
Area under $BC = 20 \times 240 = 4800$ m
Area under $CD = \frac{1}{2} \times 20 \times 60 = 600$ m
Total distance $= 5.8$ km.

The total distance travelled by the train is 5.8 km.

✳ $\text{acceleration} = \dfrac{\text{change in velocity}}{\text{time}} = \dfrac{v-u}{t}$

∴ ch. in velocity $=$ acceleration $\times$ time

$v - 0 = 0.5 \times 40$ m/s.

$v = 20$ m/s

Uniformly accelerated motion

The commonest example of uniformly accelerated motion is that of a body falling under the effect of gravity, air resistance being neglected; the acceleration is then 9.8 m/s². (This is the mean value of the acceleration g due to gravity on the earth's surface.)

The relationship between force and acceleration is contained in Newton's second law and is discussed in the next chapter.

The teachings of the Greek philosophers, which were accepted in the seventeenth century, stated that heavy bodies fell to earth quicker than light ones. These statements were proved incorrect by Galileo. He concluded that any difference in in the rate of fall was due to the resistance of the air. The experiments of Galileo and his students on the motion of bodies falling under the effect of gravity, were possibly carried out from the Leaning Tower of Pisa.

In a vacuum, all bodies fall at the same rate. Newton deduced this from his theory of gravitation and verified it experimentally by allowing a guinea and a feather to fall together in an evacuated tube.* The effect of the air on a falling object is to limit its velocity (*see* Chapter 13).

Equations for uniformly accelerated motion

Consider a body starting at zero time with an initial velocity u and moving with a constant acceleration a. If the velocity after time t is v, final velocity = initial velocity + acceleration × time

$$v = u + at \qquad (1)$$

The average velocity over time t will be $\dfrac{u+v}{2}$ hence

distance travelled = average velocity × time

$$s = \frac{u+v}{2} \times t \dagger \qquad (2)$$

The application of the above equations to the solution of numerical problems requires attention

---

* For apparatus see Nuffield Item 110.

† From equations (1) and (2) we can derive other useful equations

$$s = ut + \tfrac{1}{2}at^2 \qquad (3)$$

and $$v^2 - u^2 = 2as \qquad (4)$$

These are obtained by eliminating v and t from equations (1) and (2).

---

to be paid to the direction of the quantities, e.g. a change in sign of the velocity signifies a reversal of direction. Thus, a first step to take in solving such problems is to fix the positive direction. This can be better appreciated from Example 2.

Example 2

A stone is thrown vertically upwards with a velocity of 4.0 m/s from the edge of a cliff 96 m high. How long will it take for the stone to reach the ground at the foot of the cliff?
Acceleration due to gravity is 10 m/s².

In this example we know the distance s, the initial velocity u, the acceleration a, and we have fo find the time t. From the four equations, we choose the one which incorporates u, t, a and s. This is eqn. (3).

$$s = ut + \tfrac{1}{2}at^2$$

If we fix the cliff-top as origin, and the downward direction as positive, then, $u = -4.0$ m/s, $a = 10$ m/s², $s = 96$ m. Substituting:

$$96 = -4t + \tfrac{1}{2} \times 10 \times t^2$$
$$5t^2 - 4t - 96 = 0$$
$$(5t - 24)(t + 4) = 0$$
$$t = 4.8 \text{ or } -4$$

The time taken is 4.8 seconds.

If the stone had been thrown downwards, u would have been $+4.0$ m/s, and the result for t would have been smaller.

Determination of g

(a) *By free-fall*

The motion of a free-falling mass is determined by attaching the mass to the end of a length of ticker-tape which is pulled through a vibrator clamped in a vertical position. A record of the distance fallen in successive equal intervals of time is thus obtained. When the ticker-tape is cut and pasted up, as previously, a straight line graph is obtained, as Fig. 8.3. (The first few dots starting from rest are not included.)

The value of the acceleration due to gravity is the slope of the line AB, i.e. $g = \dfrac{BC}{AC}$

The distance BC and AC must be measured using the scales on the graph:

$$g \text{ (in m/s}^2) = \frac{BC \text{ (in m)}}{\text{no. of strips}} \times 50 \times 50$$

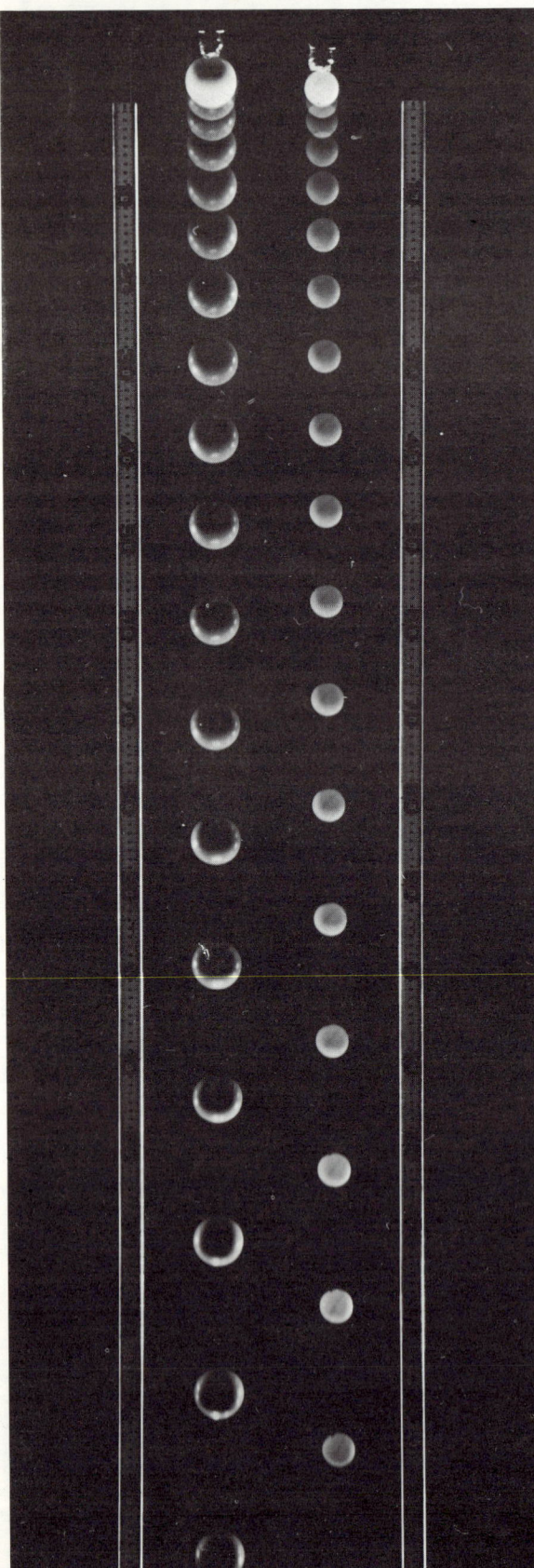

(b) *By other methods*

Other methods of measuring *g* are both instructive and interesting. They all depend upon the accurate measurement of a short interval of time.

A xenon discharge tube can be made to flash at regular intervals. The frequency of the flashes can be controlled electronically and registered on a dial. Using such a source of illumination, a falling object, e.g. a steel ball-bearing or a golf-ball can be photographed in consecutive positions at successive equal intervals of time (Fig. 8.8). If a metric rule is supported vertically and photographed alongside the falling ball, all the data for plotting a velocity-time graph as Fig. 8.3 are available. A similar result can be obtained by rotating a slotted disc or stroboscope in front of the camera lens while photographing a falling ball. In this case the time interval is obtained from the speed of rotation of the slotted disc and the number of slots.

An alternative method of calculating the value of *g* from the experimental results is to use the equation (3) at the foot of page 73 which gives the vertical distance *h* fallen from rest in time *t*, by putting the initial velocity $u=0$, $s=h$ and $a=g$.

$$h=\tfrac{1}{2}gt^2.$$

The time *t* in this equation can be determined using a scaler.* This is an instrument which counts electrical pulses electronically. It has been developed for counting the pulses caused by charged particles in the study of radioactivity, but it can count pulses fed into it at any rate. The cycles of the alternating current mains are suitable pulses. Switching devices are used to start the pulses as the ball is released and to stop them as the ball completes a measured fall.

Electric clocks are now available which will measure a short interval of time between switching on and off.

It is necessary to use quick-acting mechanisms with these electric timing devices otherwise large errors can be introduced.

Distance travelled from velocity-time graph

We have seen that the area under the *v/t* graph represents the distance travelled. It can be shown to hold for an irregular movement represented by

---

* For *g* with scaler see Nuffield *T.G.*, IV, page 58.

Fig. 8.8
Free fall

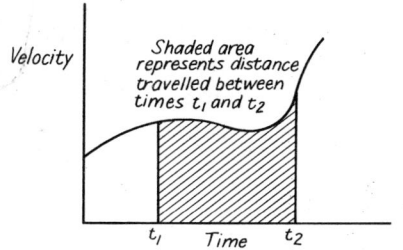

Fig. 8.9 Distance on velocity/time graph

a curve of any shape. The area has to be calculated in accordance with the scales used on the graph (Fig. 8.9).

Composition of velocities

A moving body may be subjected to two velocities simultaneously. If we walk forward at 2 m/s, along the corridor of a train travelling at 20 m/s, we approach our destination at 22 m/s. If the two velocities are not along the same direction, the resultant may be found using the parallelogram law. (Velocity is a *vector* quantity.)

Consider an aeroplane with a course set true north, flying at 60 m/s in a westerly wind blowing

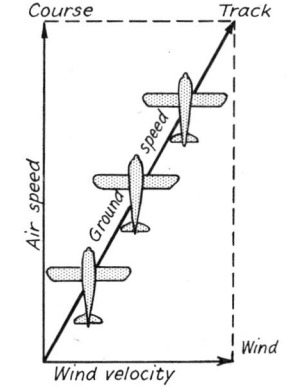

Fig. 8.10
Composition of velocities

at 30 m/s. The track of the plane will be east of north, the direction and magnitude of the resultant velocity being given by constructing the parallelogram (Fig. 8.10). The plane drifts to the east while it steers north.

Such practical problems as this are solved by the navigators of aeroplanes during flight (*see* Example 3).

Example 3

It is desired to fly from airfield A to airfield B, which lies 1000 km away in a direction 12° east of north. If the wind is blowing from due east with a speed of 30 m/s, find the necessary course if the airspeed of the plane is 150 m/s. Also find the estimated time of arrival at airfield B if the flight starts from A at noon.

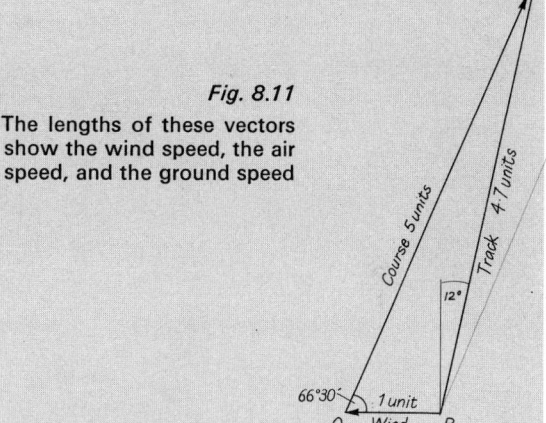

Fig. 8.11

The lengths of these vectors show the wind speed, the air speed, and the ground speed

This problem is solved by applying the triangle of vectors to the velocities (Fig. 8.11). We know the wind magnitude and wind direction, the airspeed and the track. The two unknowns are the course and the ground speed.

Draw **PQ**, the wind vector, 1 unit long and along the east–west direction.

From P draw the track, which is 12° east of north.

From Q draw a vector (using a compass) 5 units long ($150 \div 30$) to meet the track at R.

Then, **QR** gives the course and airspeed, and **PR** gives the track and ground speed.

By measurement:

PR=ground speed=4.7×30=141 m/s

Time for journey=$\dfrac{1000^2}{141 \times 60}$=118 minutes $\qquad (1000 \, Km \times 1000)m$

Estimated time of arrival=13.58 hours.

The course is 23° 30′ approx. east of north. The estimated time of arrival is 13.58 hours.

Projectiles

A stone thrown in a horizontal direction at 10 m/s from the top of a cliff retains its horizontal velocity as it moves downwards under the effect of gravity.

It falls vertically, precisely as it would do if released from the hand without a horizontal velocity. Thus, the resultant path is a combination of the two motions: at the end of the first second it has moved horizontally 10 m, and downward 4.9 m: after two seconds, these distances are 19.6 m and 20 m respectively. If the distances are plotted, the resultant path of the stone is found to be a parabola* (Fig. 8.12).

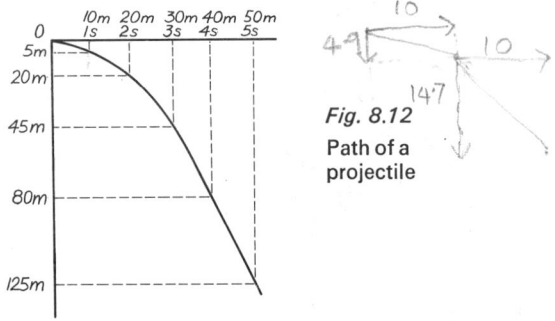

Fig. 8.12

Path of a projectile

Rifle bullets, golf-balls and other objects propelled into the air, all move in parabolic paths. A bullet drops 4.9 m below the line of fire in the first second, and the sights have to be adjusted to

---

* An interesting demonstration of this is the 'monkey and hunter' experiment which is well described in Nuffield *G. to E.*, III, No. 69A.

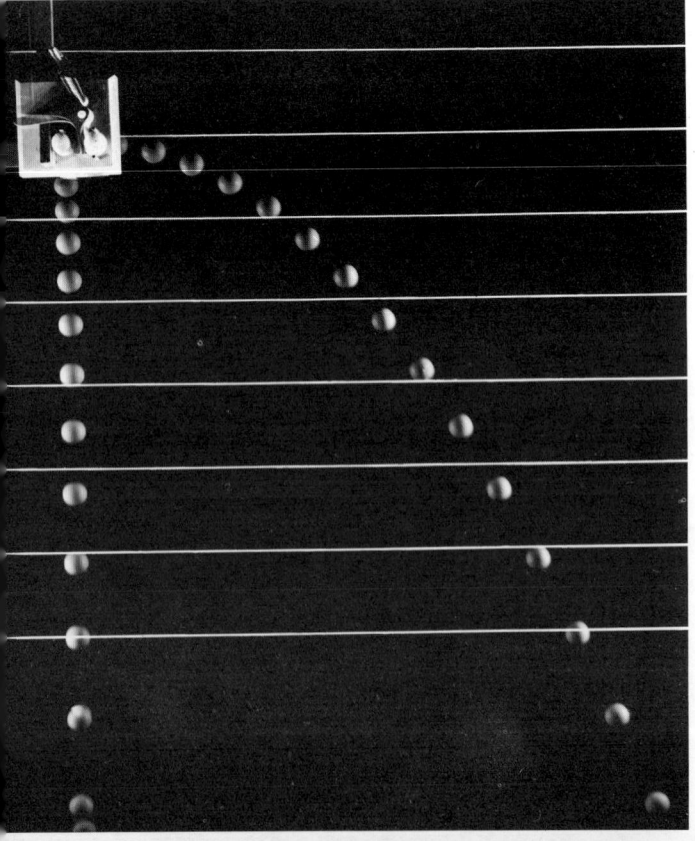

allow for this. Actually, air resistance causes a slight deviation from the parabolic path.

Fig. 8.13 shows a flash photograph of the motion of two balls which were released simultaneously, one from rest and the other with a horizontal velocity. The apparatus to produce this release is shown in Fig. 8.14.

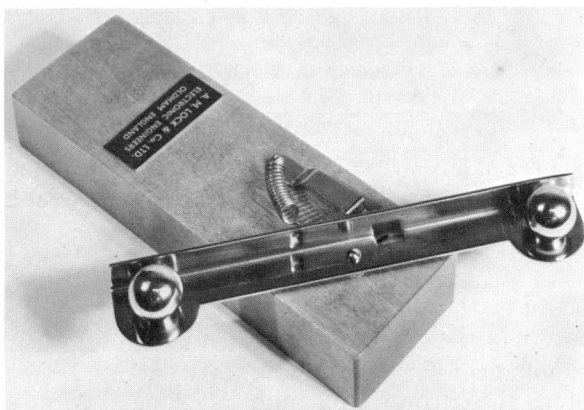

Fig. 8.14 Apparatus for simultaneous release of balls in vertical and horizontal directions—used to obtain Fig. 8.13

Example 4

A stone is thrown horizontally from the top of a cliff 40 m high, with a velocity of 10 m/s. Find its velocity on reaching the ground.

The velocity of the stone can be resolved into two components, horizontal and vertical. The horizontal velocity remains constant and is 10 m/s.

The vertical component begins at zero and increases continually. To find this velocity on reaching the ground, we use

$$v^2 = u^2 + 2as$$

Here $u = 0$, $a = 9.8$ m/s^2, $s = 40$ m

$$v^2 = 0 + 2 \times 9.8 \times 40$$

$$v = 28 \text{ m/s}$$

The final velocity is the resultant of the two velocities: 10 m/s horizontally and 28 m/s vertically. These must be combined by using the parallelogram law.

$$\text{Resultant} = \sqrt{(10^2 + 28^2)}$$
$$= 29.7 \text{ m/s}.$$

The direction is given by $\tan \theta = \dfrac{28}{10}$, i.e. $\theta = 70° \ 21'$.

The velocity at the ground is 29.7 m/s along a direction making an angle of 70° 21' to the horizontal.

Fig. 8.13 Vertical fall with and without horizontal motion

QUESTIONS

[$g = 10$ m/s^2.]

1. A body moving with a speed of 40 km/h commences to accelerate with an acceleration of 1.5 m/s^2. What will be its speed after (a) 6 s; (b) travelling a further 100 m?

2. A stone is dropped down a well 80 m deep. How long will it take to reach the bottom?

3. A stone is thrown (a) vertically upwards, (b) vertically downwards with a speed of 5 m/s from the top of a cliff 100 m high. How long will it take to reach the ground?

4. Explain what is meant by *uniform velocity* and *uniform acceleration*.
   Derive, from first principles, an expression for the distance travelled by a uniformly accelerated body moving from rest, in terms of the acceleration and the time taken.
   A trolley running down an inclined plane travels distances S in times T, starting from rest, as follows:

S/cm	0	2.4	9.6	21.6	38.4	60.0	86.4
T/s	0	0.2	0.4	0.6	0.8	1.0	1.2

   Show, graphically or otherwise, that the acceleration of the trolley is uniform and calculate its value. (L)

5. What is meant by the terms *uniform velocity* and *uniform acceleration*?
   A body moves with a velocity v after a time t given in the table below:

v/cm/s	0	2.3	4.6	6.9	9.2
t/s	0	1	2	3	4

   Plot a graph of velocity against time and use the graph to determine (a) the acceleration of the body, (b) the distance it has moved 3.5 s after starting from rest.
   What is (a) the kinetic energy, (b) the momentum of the body 4.2 s after starting from rest if its mass is 250 g and the acceleration remains constant? (JMB)

6. Define (a) acceleration, (b) uniform acceleration and explain under what conditions a body is uniformly accelerated.
   A freely falling body passes a certain point with a velocity of 30 m/s. Calculate its velocity 3 s later and find how far it will travel in this time. (O & C)

7. Define *speed*, *velocity* and *acceleration*.
   A particle slides under gravity down a smooth inclined plane of slope 30° to the horizontal. What is (a) its acceleration along the plane, (b) its speed along the plane when it has moved from rest a vertical distance of 4.0 m?
   How does the speed for the same vertical distance depend on the angle of inclination of the plane? Give a reason for your answer. (O & C)

8. Explain how the acceleration of a body at any instant, and the distance covered during any interval of time, can be found from a velocity–time graph.
   Draw the velocity–time graph for the motion of a body which is thrown vertically upwards with a velocity of 30 m/s. From the graph, or otherwise, find the greatest height reached and the time taken to reach its original level again. (O)

9. Define *uniform acceleration*.
   Describe an experiment which shows that a body acted on by a constant force moves with uniform acceleration.
   A vertical smoked glass plate is allowed to fall freely from rest past a horizontally vibrating rod which traces a wavy line on it as it falls. If the first 25 complete 'waves' occupy a distance of 20 cm, what is the frequency of vibration of the rod? (O)

10. Define *velocity* and explain how a velocity may be resolved into two components. State the law on which your construction is based.
   An aircraft flies horizontally at a constant air-speed of 360 km/h and in a straight line from a point A to a point B 120 km due east of A. There is a 80 km/h wind blowing from the north. Find by drawing or calculation (a) the time taken by the aircraft to fly from A to B, (b) the direction in which the aircraft points on the flight. Illustrate your answer by a diagram. (O & C)

11. What is meant by *uniform motion, uniformly accelerated motion in a straight line*?
   Briefly describe an experiment to show that for a body moving in a straight line with acceleration a, the distance travelled in a time t' from rest is $\frac{1}{2}at^2$.
   A stone is dropped from rest at the top of a mine shaft 180 m deep. After what time interval does it reach the bottom? With what vertical speed must a second stone be thrown after it 1.0 s later in order that it may reach the bottom at the same time? (O & C)

12. Define *displacement, velocity, acceleration*.
   A railway engine A travelling at a uniform velocity of 20 m/s passes another engine B travelling in the same direction on a parallel track at a speed of 6 m/s. If B has a uniform acceleration of 0.4 m/s^2, find graphically, or otherwise, the time that elapses before B passes A. What will be the speed of B when this occurs? (O & C)

13. Describe any experiment which will enable you to verify the formula connecting distance and time for a body starting from rest and moving with uniform acceleration in a straight line.
   Draw an accurate graph of velocity against time for a stone thrown vertically upwards with an initial velocity of 20 m/s. Calculate the greatest height reached. (O & C)

14. What is meant by *uniform acceleration, gravitational acceleration*?
   Describe a method of measuring gravitational acceleration.
   The speed of a train is reduced from 30 m/s to 15 m/s in a distance of 900 m on applying the brakes. How much further will the train travel before coming to rest assuming the retardation remains constant, and how long will it take to bring the train to rest after the application of the brakes? (O & C)

15. A body of mass 5 kg is allowed to fall from rest from a point 18 m above the surface of a planet where the acceleration due to gravity is 2.0 m/s^2. Determine and tabulate the distance S fallen in a time t using at least six values of t, none in excess of 4.5 s, and plot a graph of S as y-axis against t as

x-axis. From the graph determine, giving a full explanation:

    (*a*) the time of reaching the surface,

    (*b*) the velocity on reaching the surface.

From (*b*) calculate the kinetic energy just before impact and hence or otherwise determine the loss of potential energy of the body on reaching the surface.   (JMB)

16. Two vehicles *A* and *B* are approaching a crossroads. *A* is travelling due south at a speed of 6 m/s and *B* is travelling due east at a speed of 8 m/s. Find the apparent speed and direction of *B* as seen by an observer in *A*.

17. Examination of a ciné-film of the motion of a heavy ball, thrown upwards inclined to the vertical, showed that the ball travelled sideways a distance of 10 m in each successive second and that its height in successive seconds was 3, 25, 37, 39, 31, 13 m. Draw, on squared paper, a scale diagram to show the path of the ball in its flight. Estimate how long it was in flight, how high it rose and how far it went.

    Describe in general terms the vertical and horizontal motions of the ball and relate these to any forces acting on the ball. Assume air resistance to be negligible in this case, and then explain how the motion of the ball would have been affected by a strong wind blowing horizontally against the ball. Illustrate your answer by showing, on your scale diagram, a possible path for the ball.   (C)

9 Motion produced by forces; Newton's laws of motion

Motion and force

The study of the motion of bodies (kinematics) had its real beginnings with the experiments of Galileo, which were introduced in the preceding chapter. Newton, following on Galileo (he was born the year Galileo died), was able to build up a system of mechanics so comprehensive that its limitations* have been encountered only in the twentieth century. It was appreciated that force and motion were related, but it was this relationship which the Aristotelian philosophers had not understood and which was left to Galileo and Newton to explain.

A body opposes motion or change of motion. This property of a body is called its *inertia*.† Galileo and Newton had a clear understanding of the concept of inertia and realized that for linear motion the inertia of a body is measured by its mass.

A body, because of its inertia, requires a force to move it from rest or to change its motion. To move a body from rest or to bring it to rest or to change its speed, all require a force. This is the same as saying that a force acting on a body causes an acceleration. The Aristotelians associated *constant force* with *constant velocity*, whereas Newton correctly associated *constant force* with *constant acceleration*. It was thought that a body required a force acting on it to keep its velocity constant, but Newton explained that the force is necessary to overcome friction and air resistance. In order to maintain a constant velocity the engine of a car has to produce a force mainly to overcome air resistance and friction in the engine itself.

A change in velocity may be a change either in magnitude or in direction.

The first law of motion

Galileo considered a ball rolling down a slope *AB*, and up another slope *BC* (Fig. 9.1). Allowing for friction and air resistance, he appreciated that the ball should rise to *C* up slope *BC* or to *C'* up slope

---

* A different approach is necessary for bodies moving with speeds approaching that of light.
† Refer back to Chapter 3.

Fig. 9.1

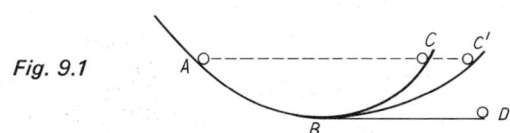

BC'. Thus he concluded that along a horizontal plane *BD*, the ball should continue with the same velocity indefinitely. Thus a body continues to move with a constant velocity if no force acts upon it. These ideas are summarized in Newton's first law of motion.

> Every body continues in a state of rest or of constant velocity unless it is acted upon by an external force. *Law of Inertia of Galileo*

Experimental demonstration

It is not easy to demonstrate this first law experimentally because of the difficulty of eliminating friction. A modern demonstration of frictionless motion is performed using 'pucks'. These consist of metal discs of radius about 8 cm and thickness about 1.5 cm. The underside has a circular cavity. A piece of solid carbon dioxide is put into this cavity and the vapour from it escapes by slightly lifting the puck. The puck is used on a large sheet of clean glass which is first carefully levelled (Fig. 9.2). The puck moves on the glass surface with nearly zero friction, as it is supported on a cushion of carbon dioxide vapour. When a puck is given a push, it continues to move across the glass surface maintaining a constant velocity.

The linear air track, described later in this chapter, can also be used for this demonstration.

We are today becoming better acquainted with this first law. We are witnessing the journeys of space vehicles which, when they move into regions in space where the earth's gravitational field is very small, travel along straight paths with virtually constant speed for there is no air to offer resistance. They are steered by forces produced by special rockets attached to their outside.

The second law of motion

The first law summarizes the effects of no resultant force acting on a body: the second law states what happens when the force is not zero.

To accelerate or to retard a mass it must be acted upon by a force. A constant force on a body produces a constant acceleration. But the same force will produce a larger acceleration on a smaller mass. Hence, force is related to mass and acceleration. Newton's second law relates them.

> The force acting on a body is proportional to its mass and to the acceleration produced.*

$$F \propto m \times a$$

Demonstration of the second law

This law can be approached experimentally by using a trolley and vibrator.† A trolley can be pulled along a track (corrected for friction) by an elastic band. The pull is maintained constant by maintaining the extension of the band constant. The acceleration can be obtained by the ticker-tape method. If the tape is cut and mounted as explained on page 71, a graph of *velocity* against *time* is obtained. This is a straight line and its slope gives the acceleration. If the pull is increased by a known fraction, e.g. by pulling with two similar bands each extended by the same amount as previously, the acceleration is found to be doubled, i.e. the acceleration is increased in the same proportion as the pull is increased.

A second experiment‡ can be performed in which the pull is kept constant while the mass of the trolley is increased, e.g. by mounting a second and similar trolley on the first.

The investigation can be made more precise by using the arrangement shown in Fig. 9.3. The trolley is pulled by a string to which is attached a scale-pan. Various loads can be placed in the pan

---

* Newton stated the law in terms of the momentum: *see* page 83.
† *See* Nuffield *G. to E.*, IV, No. 7.
‡ *See* Nuffield *G. to E.*, IV, Nos. 11 and 12.

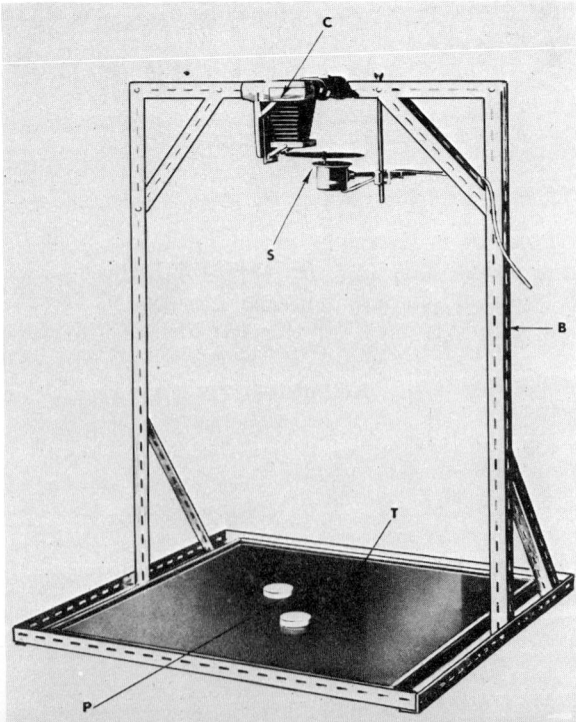

Fig. 9.2 Horizontal glass surface for motion of pucks

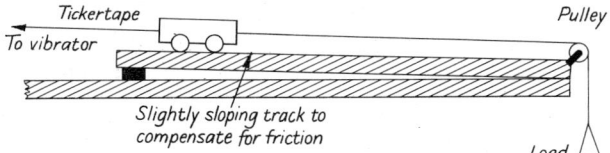

Fig. 9.3 Apparatus to show acceleration is proportional to force

to provide different pulls on the trolley. The trolley can also be loaded in order to change its mass.

Still more elaborate equipment, the linear air track (Fig. 9.4), is available in which friction is reduced by supporting a moving carriage on a cushion of gas. A long hollow bar of triangular cross-section is drilled with small holes along its length. Air is blown into the hollow bar and escapes through the holes. This escaping air supports a carriage of suitable design to rest on the track. The motion of the carriage is virtually frictionless. The apparatus also lends itself to collision experiments mentioned later in this chapter.

Further experiments can be performed with a trolley timing its motion with a scaler.*

The motion of a puck can be plotted by multiflash photography as it is drawn across the glass plate by an extended rubber band.†

Units of force

In the relation $F \propto ma$, the value of F is controlled

---

* *See* Chapter 42 for scaler. Nuffield *G. to E.*, IV, No. 9.
† Nuffield *G. to E.*, IV, No. 8.

by the values of m and a. Thus to define a force we need to fix a mass and an acceleration. In doing so we can make the constant of proportionality equal to unity. Hence,

$$\mathbf{F} = \mathbf{ma}.$$

The SI unit—*the newton* (N)
  The newton is that force which when acting upon a mass of 1 kg produces in it an acceleration of 1 m/s².

Gravitational unit—*the kilogramme force* (kgf)
  When a mass of 1 kg is in the earth's gravitational field at the earth's surface, it is acted upon by a force of 1 kgf which produces in it an acceleration of 9.8 m/s², hence

$$\mathbf{1 \ kgf = 9.8 \ N}$$

This is the constant we used in Chapter 3 in order to express forces in newtons. The kilogramme force and other gravitational units are now not used, but mention is made here in order to explain the presence of the constant 9.8.

The third law of motion

The second law states clearly that a body acted upon by a resultant force moves with an acceleration in the direction of that force. But this seems to be contradicted when we push against a wall and no motion results. In this case the wall pushes back with an equal force. Forces occur in pairs, one being equal and opposite to the other. Newton expressed this in what we call the third law.

  To every action there is an equal and opposite reaction.

This is the traditional way of stating Newton's third law. What it really should convey may perhaps be better understood in the following form.

  When a body is acted upon by a force, there is always another force, acting on another body, which is equal in magnitude but opposite in direction to the first force.

This means, simply, that if a body A acts upon a body B with a force in a certain direction, then B will act upon A with an equal and opposite force.

A body on the earth's surface falls under the effect of gravity. The earth acts upon the body with a force F which produces in the body an

---

* The cgs unit, now obsolete, was called a *dyne*.
  1 newton = 10⁵ dynes.

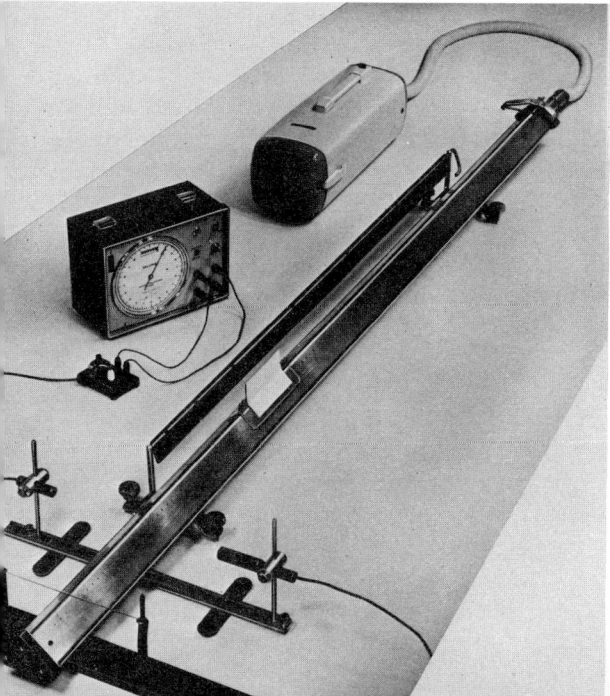

Fig. 9.4
Linear air track

acceleration g: the earth is also acted upon by a force F and suffers an acceleration *towards* the body. The acceleration will be inconceivably small compared with g because the mass of the body will be inconceivably small compared with the mass of the earth.

Example 1

A body of mass 4 kg is pulled along a smooth horizontal bench by a string which passes over a pulley and carries a 2-kg mass on its other end (Fig. 9.5). Find the acceleration of the system and the tension in the string.

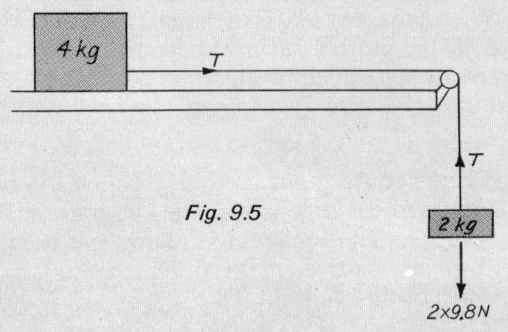

Fig. 9.5

We first find the acceleration of the system as a whole by applying to it Newton's second law. As there is no friction, no external force, other than its weight, acts on the 4-kg mass, and this has no effect in a horizontal direction.

The total force on the system is the weight of the 2-kg mass, i.e. 2×9.8 N.

We have,

$$F = ma$$

(m is total mass moving $= 6$ kg)

$$2 \times 9.8 = 6a$$

$$a = \frac{19.6}{6} = 3.27 \text{ m/s}^2$$

To find the tension we now consider the motion of either mass.

(*a*) Resultant force on 2-kg mass in the direction of its motion

$$= 2 \times 9.8 - T$$

Hence,

$$2 \times 9.8 - T = 2 \times \frac{19.6}{6}$$

$$T = 13.1 \text{ N}$$

or (*b*) Resultant force on 4-kg mass $= T$

Hence,

$$T = 4 \times \frac{19.6}{6}$$

$$= 13.1 \text{ N}$$

The acceleration of the system is 3.27 m/s^2 and the tension in the string is 13.1 N.

Example 2

A man of mass 100 kg stands on a weighing machine in a lift. The lift moves upward with an acceleration of 0.6 m/s^2 for a short time and, after moving with constant velocity for a brief period, is brought to rest with a retardation of 1.0 m/s^2. Find the reading of the weighing machine during the three phases of the motion.

The reading of the weighing machine is the force of the man on the machine, which is equal and opposite to the force of the machine on the man.

(*a*) For upward acceleration of 0.60 m/s^2,

Resultant force on man = force of machine on man (in newtons)   (upward)
    − weight of man (downward)

$$= R − 100 \times 9.8$$

Substituting in $F = ma$, we have,

$$R − 980 = 100 \times 0.60$$

$$R = 1040.$$

If the machine is calibrated in kg, it will read

$$\frac{1040}{9.8} = 106.1 \text{ kg}$$

(*b*) For constant velocity reading is

$$980 \text{ N or } 100 \text{ kg.}$$

(*c*) For retardation of 1 m/s^2, [method as for (a)]

Resultant upward force

$$= \text{force of machine on man} - \text{weight of man}$$
(upward)   (downward)

Substituting in $F = ma$

$$R' - 100 \times 9.8 = 100 \times -1$$

$$R' = 880$$

Expressed in kg, $R' = \frac{880}{9.8} = 89.9.$

The three readings of the weighing machine are 106.1, 100 and 89.9 kg or 1040, 980 and 880 N.

Momentum

We have just considered at some length Newton's second law which we expressed as $F=ma$. But Newton approached it in another way. He did not think of the mass and the acceleration as two separate quantities, but as one combined quantity.

Let us consider a force acting on a body for a time t. For example, we can imagine pushing a massive object, such as a car, along a level road. Neglecting any friction and air resistance, we find that the longer we maintain the push the greater will be the velocity of the car. On ceasing to push, the car continues to move with a constant speed. To stop the motion, an opposing force must be applied and maintained for a specific time. This force can be of any size, but the larger it is, the shorter will be the time required to bring the car to rest. We see, that the product *force × time* is the quantity which decides the *velocity* produced upon the *mass*. This can be investigated experimentally with trolleys pulled by elastic bands. Exerting the same force for the same time, first on one trolley and then on another of double mass,* produces a velocity in the second case only half that in the first. Hence we have

$$F \times t \propto m \times v$$

The product mass × velocity is the momentum of a body.

Newton stated the second law in the form:

The rate of change in momentum of a body is proportional to the resultant force acting on the body.

i.e.
$$F \propto \frac{mv - mu}{t}$$

mu is the initial and mv the final momentum after time t.

$$F \propto m\frac{(v-u)}{t}$$

$$F \propto ma$$

Impulse of a force

$$\mathbf{F \times t = mv - mu}$$

The product *force × time for which force acts* is

---

* A second trolley similar to the first can be inverted on the first to give one of double mass.

called the *impulse* of a force and it is this which determines the value of the velocity v imparted to an object of mass m. Impulse is appreciated by most sportsmen (although they may not be aware of it) for the 'follow-through' is important in most ball games. When driving in golf, the swing is carried through as is the drive in tennis; the footballer follows through with his foot to produce a long kick. The fundamental object is to keep the striking weapon in contact with the ball for as long a time as possible. By doing so, $F \times t$ is made large and a large velocity results.

Example 3

A pile-driver, of mass 150 kg, drops from a height of 2.0 m and is brought to rest in 0.20 s. What force is exerted on the pile ($g=9.8$ m/s²)?

We have
$$F \times t = mv$$

To find v, we use
$$v^2 = 2gs$$
$$= 2 \times 9.8 \times 2$$
$$v = \sqrt{39.2}$$
$$F \times \frac{2}{10} = 150\sqrt{39.2}$$

Therefore
$$F = 4300 \text{ N}$$

The force exerted is 4300 N.

Conservation of momentum

Newton's third law leads to an important relation concerning the momenta of bodies involved in impact.

When a body A strikes another body B, we know

force of B on $A = -$ force of A on B.

The negative sign shows the forces are opposite in direction,

i.e.
$$F_A = -F_B$$

Newtons second law gives
$$m_A a_A = -m_B a_B$$

or
$$m_A\frac{(v_A - u_A)}{t} = -m_B\frac{(v_B - u_B)}{t}$$

where u_A and u_B are the velocities of A and B before the forces began to act, and v_A and v_B are the velocities after the forces have been acting for time t.

$$m_A v_A - m_A u_A = -(m_B v_B - m_B u_B)$$

or $m_A u_A + m_B u_B = m_A v_A + m_B v_B$

which is:

**sum of initial momenta = sum of final momenta
in same direction.**

This is known as the *law of conservation of momentum.*

> The total momentum of any system in any direction remains unchanged after any interaction within it.

The simplest examples of this law are those in which the action takes place in a short interval of time. For example, in firing a rifle or a gun the recoil momentum is equal to the momentum of the projected missile. Likewise, if two vehicles collide and combine, as in the shunting of railway wagons, the sum of the momenta of the separate wagons is equal to that of the combination of the wagons.

Collision experiments

There are many experiments which can be performed to lead to the law of conservation of momentum. The velocities of two trolleys before and after collision on a friction-compensated track can be measured by using a ticker-tape timer. Tapes attached to the two trolleys are both run through the same vibrator using two carbon paper discs. One trolley is started off with a push and the second is started off a little later with a bigger push so that a collision takes place. The velocities of the trolleys before and after collision are obtained from the tapes.

This experiment can be carried out in a more elaborate way by using a linear air track (Fig. 9.4). In this case, where the track is horizontal, head-on collisions may be investigated. Two electric timers are required similar to the one shown in the foreground of the photograph.

Very pretty demonstrations of collisions can be performed with frictionless pucks, as described earlier in the chapter.

The technique of using the pucks is to push them gently across the glass surface so that they collide. The motion of the pucks is recorded by taking a succession of pictures by multiflash photography (Fig. 9.6).

Magnetic pucks can be used and these suffer perfectly elastic 'collisions' as they approach

Fig. 9.6 Collision of pucks by multiflash photography

closely and then mutually repel being of 'like' polarity.

The measurements for testing the law of conservation of momentum can be made from the flash photographs.

Direct and oblique collisions can also be studied using heavy spheres suspended on very long threads. The motions of such spheres can usually be measured more conveniently by observing their shadows cast by a light beam on to a horizontal screen.

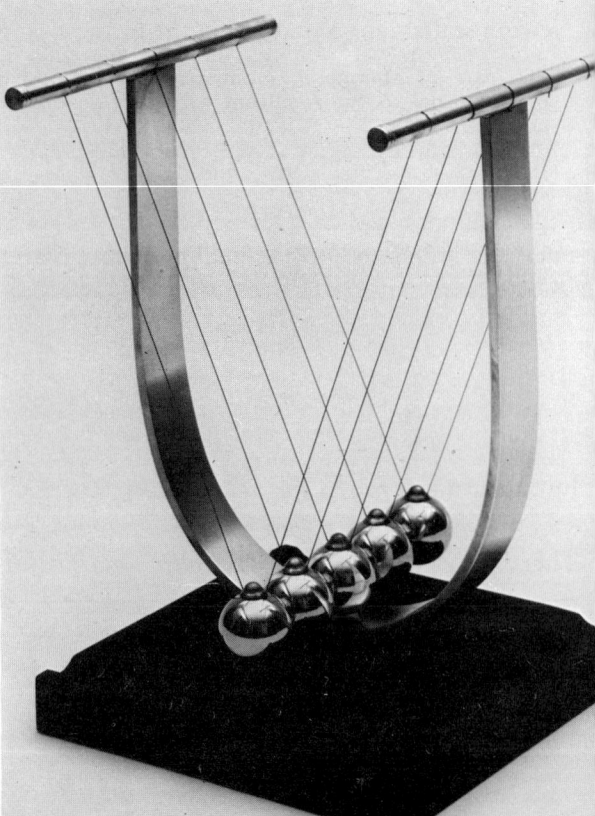

Fig. 9.7 Impact toy
An elaborate and sophisticated toy sold for the use of overworked executives to relieve their strain. The metal spheres are displaced and allowed to impact. Interesting results can be obtained

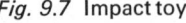

Application of conservation

An estimation of the speed of a rifle bullet can be made using the law of conservation of momentum. A bullet is fired into a suitable arresting block mounted on a carriage running on a friction-compensated horizontal track (linear air track). The velocity imparted to the carriage is measured by an electrical timing method. Knowing the mass of the carriage and that of the bullet, the velocity of the bullet can be calculated.*

Elastic and inelastic collisions

When two bodies, such as two balls, collide head-on, the surfaces at the points of contact are momentarily depressed. The rebound is caused by the correction of this distortion. Substances like glass and marble have high elasticities and their speeds of rebound are only slightly less than the initial speeds. In all collisions such as these there is a loss of kinetic energy. The lost K.E. is changed to heat. The higher the elasticity, the smaller is the heat generated. A collision in which K.E. is lost is said to be *inelastic*. A *perfectly elastic* collision is one in which no K.E. is lost and the relative speeds of the bodies are unchanged.

In both types of collision the law of conservation of momentum is obeyed.

Perfectly elastic collisions can be demonstrated using magnetic carbon dioxide pucks. When two of these pucks are pushed over a glass surface so that they engage in oblique collision, actual physical contact does not occur. This is because of magnetic repulsion. Oblique collision is beautifully recorded by multiflash photography (*see* Figs. 9.6 and 9.8).

* Nuffield T. G., IV, page 123.

This experiment is important as it illustrates molecular collisions in gases which are perfectly elastic collisions.

Collisions between charged particles may be recorded in cloud chamber photographs (*see* Chapter 42). Using the data obtained from these photographs, particulars of the changes resulting from the collisions can be deduced: the particles can be identified and their energies calculated.

Momentum is a vector quantity, and so oblique impact presents a more difficult case than direct impact. The momentum in one direction remains unchanged. In Fig. 9.9 the ball A strikes the

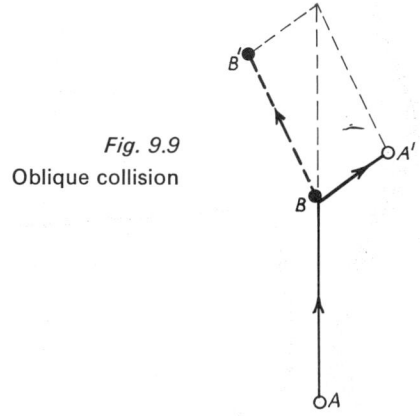

Fig. 9.9
Oblique collision

stationary ball B (of equal mass) and the balls occupy the position A' and B' after a short interval of time. It will be observed that the resultant momentum of the balls at A' and B' is along AB and can be shown to be equal to the initial momentum of A.

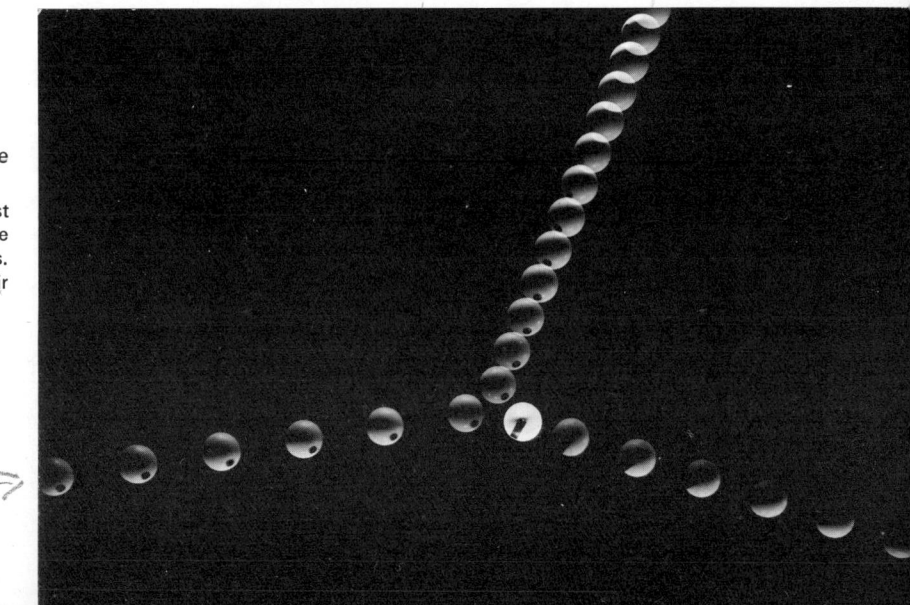

Fig. 9.8 Multiflash photograph of the collision between two billiard balls.

The ball with the stripe was initially at rest and was struck an off-centre blow by the spotted ball. The balls were of equal mass. What do you notice specially about their tracks after collision?

If the impact is a perfectly elastic one, the angle between the final directions of motion $A'BB'$ is a right angle.

These elementary principles apply to many examples, covering a wide range of magnitudes. Application to the collision of elementary particles is of special importance.

Example 4

A truck of mass 10^4 kg moving with a velocity of 2.0 m/s collides with a stationary truck of mass 6.0×10^3 kg. If both trucks move together after impact, what is their common velocity?

Applying the law of conservation of momentum,

$$m_1 u_1 + m_2 u_2 = (m_1 + m_2) \times v$$
$$10^4 \times 2 + 6.0 \times 10^3 \times 0 = (10 + 6.0)\, 10^3\, v$$
$$v = \frac{20}{16}$$
$$= 1.25 \text{ m/s}$$

The final velocity is 1.25 m/s.

Energy changes   Work = Mech. Energy = F × S

When a force F increases the momentum of a body from mu to mv, the kinetic energy of the body increases.

We have $F = ma$

and $2as = v^2 - u^2$ (page 73) $S = \dfrac{v^2 - u^2}{2a}$

But change in energy is Fs and so, from the above equations $Fs = ma\left(\dfrac{v^2 - u^2}{2a}\right)$

$$Fs = \tfrac{1}{2}mv^2 - \tfrac{1}{2}mu^2$$
final energy   *orig. energy*

When F is measured in newtons and s in metres, the kinetic energy is measured in joules.

Example 5

A gun of mass 60 000 kg fires a shell of mass 100 kg with a velocity of 500 m/s. Find the velocity of recoil of the gun.

Equating the momenta of the gun and shell we have,

$$60\,000 \times v = 100 \times 500$$

(Note the units of mass and velocity must be the same in all the terms.)

$$v = \frac{50\,000}{60\,000} = 0.83 \text{ m/s}$$

The velocity of recoil is 0.83 m/s.

Example 6

A bullet weighing 30 g and moving with a velocity of 400 m/s enters a suspended block of wood of mass 10 kg and becomes embedded in it. Find the heat developed within the wood.

Using the conservation of momentum to find the final velocity v of the system, we have $m_1 u_1 + m_2 u_2 = m_1 v_1 + m_2 v_2$

$$\frac{30}{1000} \times 400 = \left(10 + \frac{30}{1000}\right) \times v$$
$$v = 1.2 \text{ m/s}$$

Initial energy of bullet $= \tfrac{1}{2} \times \dfrac{30}{1000} \times 400^2$
$\tfrac{1}{2}mu^2$
$$= 2400 \text{ J}$$

Final energy of system $= \tfrac{1}{2}\left(10 + \dfrac{30}{1000}\right) \times 1.2^2$
$\tfrac{1}{2}mv^2$
$$= 7.2 \text{ J}$$

Loss of energy $= 2400 - 7.2$
$$= 2390 \text{ J}$$

The heat developed within the wood is 2390 J.

Example 7

A bullet weighing 16 g is forced normally through a block of wood 7.5 cm thick. If the initial velocity of the bullet is 300 m/s, and the final velocity with which it emerges from the block is 30 m/s, find the average force exerted by the wood on the bullet.

If the average force is F, then

$$Fs = \tfrac{1}{2}mv^2 - \tfrac{1}{2}mu^2$$
$$F \times \frac{7.5}{100} = \tfrac{1}{2} \times \frac{16}{1000} \times 30^2 - \tfrac{1}{2} \times \frac{16}{1000} \times 300^2$$
$$F \times 7.5 = 720 - 72\,000$$
$$F = -9.5 \times 10^3 \text{ N}$$

The negative sign shows that the force is acting in the opposite direction to the velocity.

The retarding force is 9.5×10^3 N.

Example 8

An engine weighing 2.0×10^5 kg is travelling at a speed of 40 km/h on a straight and level track. How far will it travel before being brought to rest if the frictional force of the brake is 200 N per 1000 kg.

Retarding force on engine
$$= \frac{2.0 \times 10^5 \times 200}{1000}$$

$$= 4 \times 10^4 \text{ N}$$

Substituting in
$$Fs = \tfrac{1}{2}mv^2 - \tfrac{1}{2}mu^2$$
$$= \tfrac{1}{2}m\left(v^2 - u^2\right)$$

$$-4 \times 10^4 \times s = \tfrac{1}{2} \times 2.0 \times 10^5 \left[0^2 - \left(\frac{4 \times 10^4}{60 \times 60}\right)^2\right]$$

$$s = 308 \text{ m}$$

The engine travels 308 m.

Newton's second law for a variable mass

In our investigation of Newton's second law, we have considered the motion of a body of fixed mass. In certain circumstances the mass of a body may change during motion. At the present time we are greatly concerned with rocketry and this provides an interesting case of variable mass. In launching a rocket, the upward thrust is provided by the change in momentum of the high-velocity molecules comprising the hot gases of the burning fuel.

The fuel is consumed in some of the rockets at the rate of over 3000 kg/s. Thus while the thrust is fairly steady for some minutes, the mass of the rocket falls appreciably. This produces an increase in the acceleration.

QUESTIONS

[Take $g = 10$ m/s^2 unless given differently in question.]

1. State Newton's laws of motion.
   A block of wood of mass 245 g is sent sliding along a rough horizontal table, the initial speed being 3 m/s. If the resistance to motion is 0.50 N, find the distance travelled, and the time taken, before the block comes to rest.
   [$g = 9.8$ m/s^2.] (O)

2. State Newton's laws of motion.
   Describe an experiment with a Fletcher's trolley, or other suitable apparatus, to demonstrate that the acceleration produced in a given body is proportional to the force acting on it.
   Calculate the average retarding force that must be exerted in order to bring a car of mass 600 kg, travelling at 20 m/s, to rest in 10 s. (O)

3. Define *newton* and *joule* and explain the relationship between the newton and the kilogramme force.
   A mass of 6 kg is pulled by a force of 0.30 N along a frictionless, horizontal plane. Calculate (*a*) the acceleration, (*b*) the velocity 3 s after starting from rest, (*c*) the distance gone and the work done in those 3 s. (L)

4. Explain the distinction between *mass* and *weight*. How may two masses be compared, using simple apparatus?
   A mass of 50 g hangs by a light string from the roof of a motor-car travelling along a level road. What is the inclination of the string to the vertical when the car is (*a*) accelerating, (*b*) decelerating, at a rate of 1.20 m/s^2?
   Illustrate your answer by diagrams. (O & C)

5. Explain the distinction between *mass* and *weight*.
   A mass of 10 g is held at rest at the centre of a smooth horizontal table 1.0 m in diameter. It is connected by a light string to a second mass of 10 g hanging at rest over the edge of the table. If the mass is released find its acceleration while it is still on the table. After how long does it reach the edge? (O & C)

6. Masses of 3 kg and 5 kg hang one at each end of a string which passes over a frictionless pulley. Calculate the velocity of the masses after they have moved from rest through 6 m. (C)

7. A body of mass 100 kg rests on a smooth horizontal surface. It is acted on by a horizontal force of 1.5×10^4 N for 0.100 s. Calculate (*a*) the acceleration of the body, (*b*) the speed attained by the body. What would have been the acceleration of the body if the coefficient of friction between it and the surface had been 0.05? (C)

8. An object of mass 5.0 kg is attached to a spring balance set up in a lift. What is the reading of the balance if the lift is (*a*) ascending at a steady velocity of 6.0 m/s, (*b*) accelerating downwards at 2.0 m/s^2? Explain clearly how you arrive at your answers. (C)

9. How would you find g in the acceleration due to gravity?
   What is the relation between force, mass and acceleration? If a force of 50 N acts on a mass of 8 kg for 10 s, how far will the mass be moved?
   Find the work done by the force. (JMB)

10. Define *velocity, uniform acceleration*.
    Describe an experimental method of determining the value of g.
    A string which passes over a light frictionless pulley, supports at its ends masses of 99 and 101 g. What is the acceleration of the system? Calculate the velocity when the string has moved 200 cm. (C)

11. What is meant by *force* and in what units is it measured in the SI system?
    If the greatest retarding force which can be applied to a car through the brake is numerically equal to its weight, find (*a*) the least distance, (*b*) the least time, in which a car travelling at 20 m/s on a level road can be brought to rest.
    Draw a diagram showing the direction in which a pendulum, hanging freely from the roof of the car, would point while the car is coming to rest. (O & C)

12. Describe how you would attempt to measure in the laboratory the displacement as a function of time for a body moving with uniform acceleration, e.g. a trolley running down an inclined plane.

Plot a graph of velocity against time for a stone thrown vertically upwards from the ground with velocity 30 m/s. Find from your graph, or by calculation, the greatest height reached by the stone.   (O & C)

13. Define *speed, velocity, acceleration.*

Describe and explain *two* different ways in which you could ascertain the acceleration of a railway compartment in which you were travelling along a straight horizontal line.

A car is moving along a level road at 40 km/h when the brakes are suddenly applied. A pendulum, suspended from the roof of the car, swings forward to make, after the oscillations have died away, an angle of 30° with the vertical; it stays in this position until the car comes to rest. Explain this and calculate the deceleration of the car and the distance in which it is brought to rest.   (O & C)

14. State Newton's laws of motion and define the terms *uniform velocity, momentum, force.*

A cricket ball of mass 200 g rises to a vertical height of 21 m and drops vertically into a fielder's hand when 1.0 m above the ground. The fielder brings it to rest in 0.10 s. Calculate (a) the speed of the ball as it reaches the fielder's hands, (b) the average force in newtons that must be applied in stopping it.   (O & C)

15. Explain the meaning of *momentum, force, kinetic energy, work.*

A motor-car has brakes which can exert a maximum retarding force equal in magnitude to the weight of the car.

Calculate (a) the shortest time, (b) the shortest distance, in which the car may be brought to rest on a level road from a speed of 40 km/h. What is its deceleration during the braking period?   (O & C)

16. State Newton's laws of motion.

Explain how the laws are illustrated in the operation of a rocket.

A rocket is held fixed so that it ejects material horizontally with a speed of 200 m/s. If it ejects a mass of 1.0 kg each second calculate the force in newtons needed to hold it fixed.

If the rocket were to be used to propel a sledge on a frozen lake, find, neglecting friction, the acceleration of the sledge at a time when the total mass was 160 kg.   (O & C)

17. When a force acts on a body, the body accelerates. To test this, a trolley acted on by a steady force was timed over its journey. The following table gives observations of s, the distance travelled in millimetres, and t, the time in seconds taken to travel this distance; it also shows the corresponding values of the square root of 2s.

s (mm)	15	78	208	400	630	820
t (s)	0.25	0.50	0.75	1.00	1.25	1.40
$\sqrt{2s}$	5.5	12.5	20.4	28.3	35.5	40.5

Plot the graph of $\sqrt{2s}$ against t, commencing both scales at the origin. Comment on the graph, and find the acceleration of the trolley. (*Note.* A straight line indicates a uniform acceleration, and a change in $\sqrt{2s}$ divided by the corresponding change in t gives the square root of the acceleration; if the line does not pass through the origin, it indicates a fault in the timing.)

How, in conducting such an experiment, would you arrange for the application of a steady force, and what method would you use for the timing?   (C)

18. A body of mass 0.5 kg is projected across a uniformly rough horizontal floor with an initial velocity of 4 m/s. It takes 2 s to come to rest. Determine (a) the retardation (negative acceleration), assuming it to be uniform, and (b) the change in momentum per second.   (L)

19. A block of wood placed on an inclined plane is observed to slide down the plane with uniform acceleration. The same block moving down a different inclined plane of the same slope is observed to move with uniform speed. Account for the different motions by referring to the forces acting in each case.   (C)

20. State the forces which act upon a submarine in the following circumstances: (a) when the submarine floats on the surface of the water; (b) when it is submerged at a mean depth of, for example, 100 m of water; (c) when it rests upon the sea-bed at a depth greater than 100 m. Explain how the submarine is able to alter the depth to which it is submerged and to rise to the surface of the water.

Compare the forces acting on the submarine with the forces which act upon a rigid gas-filled balloon when it is released from the ground and rises in the atmosphere. Discuss the vertical movement of the balloon as it rises to a great height.

A balloon of volume 300 m³ is filled with hot air of density half the density of the atmosphere. The mass of the balloon and its essential equipment is 100 kg. Calculate (a) the acceleration with which the balloon leaves the ground, when no extra load is carried, (b) the maximum extra load the balloon can just lift.

[The density of the atmosphere=1.29 kg/m³. Take $g=1000$ cm/s².]   (C)

21. The brakes of a car can exert a retarding force equal to 0.75 of the weight of the car. What is the shortest distance in which it can be stopped by the brakes when travelling at 15 m/s on a level road?   (O & C)

22. From the fourth century B.C. to the seventeenth century A.D. a force was conceived as that which maintained the velocity of a body constant, but Newton based his laws of motion on the idea that a force produces a *change* in the velocity of a body.

How would you attempt to explain the motion of a car moving with constant speed along a straight road, using pre-seventeenth-century ideas? Point out the fallacy in the argument and correct the explanation using Newtonian ideas.

Explain the motion of a ball thrown vertically upwards and caught on its return, by considering its velocity, acceleration and the force acting on it.   (JMB)

23. A vehicle is kept stationary in the air by the action of jet engines. What forces act on the vehicle and what are the causes of these forces? If the mass of the vehicle is 1.10×10^3 kg, calculate the additional engine thrust required to propel the vehicle vertically upwards with an acceleration of 0.80 m/s².   (C)

24. Explain the terms *vector* and *scalar*.

Which of the following are vectors: speed, velocity, momentum, force, energy?

A ball of mass 100 g falls from a height of 3.0 m onto a horizontal surface and rebounds to a height of 2.0 m. Calculate the change (*a*) in momentum, (*b*) in kinetic energy, of the ball when it strikes the surface. (O & C)

25. Define *momentum* and *kinetic energy*.

A trolley *A* of mass 2 kg travelling at 5 m/s collides with a stationary trolley *B* of mass 3 kg. After the collision the two travel on together at 2 m/s.

(*a*) What is the momentum of *A* before the collision?

(*b*) What is the momentum of *A* after the collision?

(*c*) Account for the change in momentum of *A*.

(*d*) What is the kinetic energy of *A* before the collision?

(*e*) What is the kinetic energy of *each* trolley after the collision?

(*f*) During the collision, the kinetic energy gained by *B* is less than the kinetic energy lost by *A*. How much kinetic energy is unaccounted for, and what has become of it? (O)

26. State the principle of conservation of momentum and describe an experiment to test its validity.

Good brakes on a car can exert a retarding force equal in magnitude to the car's weight. If the car is running freely along a level road at 45 km per hour when the brakes are suddenly applied, what time elapses before the car comes to rest and how far has it travelled in this time? (O & C)

27. Explain what is meant by momentum. What do you understand by the principle of the conservation of momentum? Describe an experiment to demonstrate this principle.

A trolley of mass 500 g travelling horizontally at 20 cm/s has a piece of Plasticine of mass 250 g dropped on to it from just above. The Plasticine sticks to the trolley. Calculate the final velocity of the trolley, if friction can be neglected. (JMB)

28. State the principles of conservation of momentum and of energy.

Two toy trucks each of mass 0.5 kg approach each other along a frictionless track at the same speed of 2.0 m/s relative to the track. They can be fitted with alternative sets of buffers which are *either* perfectly elastic, i.e. they return to the trucks all the kinetic energy they absorb, *or* completely inelastic, i.e. they return none of the energy they absorb.

Describe and explain what happens when the collision is (*a*) perfectly elastic, (*b*) completely inelastic.

What is the combined momentum of the two trucks in each case? What is the kinetic energy of each truck before the collision and what changes take place in the nature of the energy during the elastic collision? (O & C)

29. Give a reason why the engines of a rocket designed to launch a space-craft are made to give a thrust many times their own weight. Such a rocket, on vertical launch and giving a constant thrust, has an increasing acceleration as it goes up; suggest any *two* possible causes of this. (C)

30. You find yourself, fully dressed, in the middle of a perfectly smooth, level floor in a large empty room in an empty house. How would you get to the door? (O & C)

10 Circular motion and gravitation

Circular motion

Centripetal force

An object moving in a circular path has a tendency to leave its path and move along a tangent to it. This is noticed when a stone on the end of a piece of string is whirled round. If the string is released, the stone moves along a tangent to its circular path at the point of release. This principle is displayed in many sports, e.g. throwing the discus, throwing the hammer.

Newton's first law states that a body will continue to move along a straight line with a constant speed unless it is acted upon by a force. If

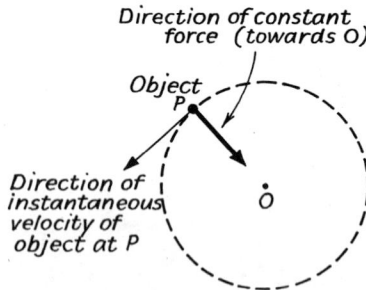

Fig. 10.1 Body moving in a circular path

the force is applied at right angles to the path of a moving body, it will change the direction of motion, but not the speed of the body. Further, if

this force is constant the path of the body will be circular (Fig. 10.1). The force is related to the mass, speed and radius of path. It is called the *centripetal* force.

$$\text{Centripetal force} \propto \text{mass}$$
$$\propto (\text{velocity})^2$$
$$\propto \frac{1}{\text{radius of path}}$$

There are many everyday examples of the utilization of this force. A centrifuge separates lighter from heavier particles in a suspension, e.g. cream from milk. The spin-drier extracts water from wet clothes by spinning them in a rapidly rotating perforated drum.

The motion of the planets round the sun and that of the satellites round their planets are approximately circular. In these cases the centripetal force is the gravitational attraction.

A cyclist turning

When a cyclist wishes to turn to the left, he must be acted upon by a force from right to left. He can provide this force himself by leaning over to the left. The friction between the tyres and the ground, together with the normal reaction at the ground, provide the resultant reaction R (Fig. 10.2). This force, together with the weight W of the cyclist (which is the only other force acting

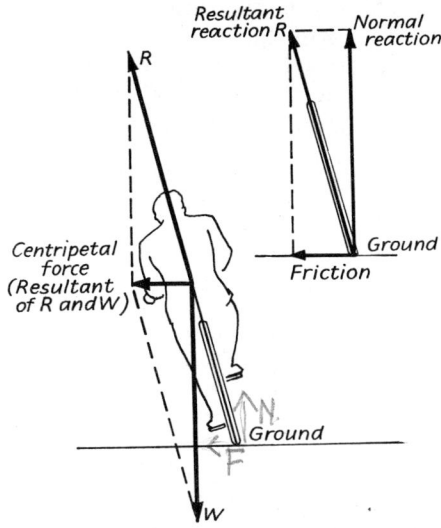

Fig. 10.2 Forces acting on a cyclist as he turns left

Due to this spinning there is a tendency for all bodies to leave the earth's surface. The gravitational attraction, however, far exceeds the centripetal force necessary to keep a body on the surface. At the poles no centripetal force is necessary: at the equator the centripetal force is a maximum. Thus the force acting on a body at the equator, i.e. its weight, is less than it is at the poles for two reasons, (i) the gravitational force is less due to the greater distance from the centre and (ii) the centripetal force is greater. If the rotational speed of the earth were increased by about 17 times it would disintegrate. For the same reason, a flywheel must not be driven too fast or it will burst. This has sometimes happened with fatal results.

excepting air resistance), will produce a resultant force which is the centripetal force.*

A car turning

For a motor-car to change direction, the turning of the wheels creates the necessary force at right angles to the direction of motion. The centripetal force of a train is provided by the reaction at the flanges of the wheels.

The banking of roads

In both the above cases the centripetal force causes wear, in the first case on the tyres and in the second on the wheel flanges. By banking the track at a bend, this wear is reduced. The centripetal force is provided by the horizontal components of the normal reactions R_1 and R_2 at the wheels (Fig. 10.3). In this case there is no sideways force to cause wear. The banking angle a varies according to the radius of the track and to the recommended speed of negotiating it.

The spinning of the earth about its N–S axis has produced a bulging at the equator: the earth is a slightly flattened sphere (an oblate spheroid).

* Because the frictional force, which provides the centripetal force acts at the ground and not through the centre of gravity of the cyclist, a couple is produced tending to roll the cyclist to the right.

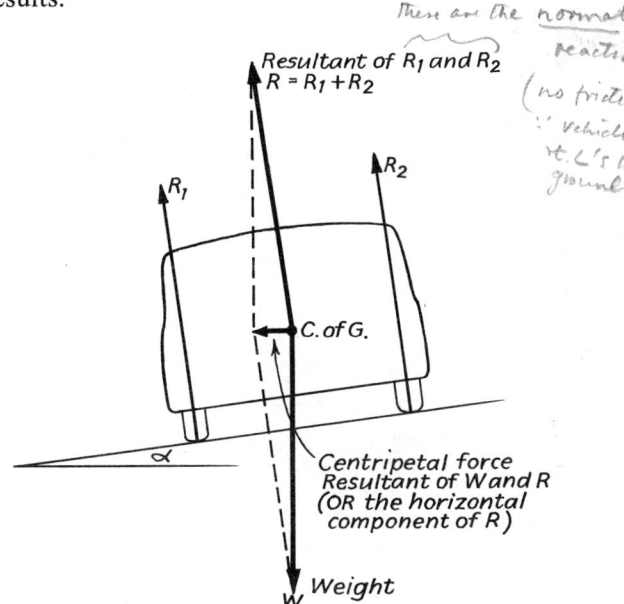

Fig. 10.3 Forces acting on a vehicle on a banked road

Newton's law of gravitation

In Chapter 3 we briefly mentioned Newton's law of gravitation when talking of the weight of a body. This we defined as the pull of gravity on a body and we said gravity was an example of a universal attraction between masses. In the seventeenth century great progress was made in astronomy and much was discovered about the solar system, but an explanation of planetary motion remained unknown although it was constantly sought. A fable is told of Newton, who solved this problem, that the idea came to him when he observed the falling of an apple from a tree in the garden of his home. The great idea was the coupling of planetary motion with the falling of the apple. The law states:

> **The force of attraction between any two masses is directly proportional to the product of the masses and inversely proportional to the square of the distance they are apart.**

This is expressed as

$$F \propto \frac{m_1 \times m_2}{d^2}$$

where m_1 and m_2 are the masses and d is the distance between them,

or
$$F = G \frac{m_1 \times m_2}{d^2}$$

where G is the gravitational constant.

The pull of gravity on a body of mass m on the earth's surface is

$$G \frac{m \times M_E}{R^2}$$

where M_E is the mass of the earth and R the radius of the earth.

But this force is the weight of the mass m and is equal to mg. Hence,

$$mg = G \frac{m \times M_E}{R^2}$$

$$g = G \frac{M_E}{R^2}$$

We can determine both g and G experimentally and, knowing R, we can calculate the mass of the earth.

The gravitational constant G can be determined in the laboratory by measuring the attraction between two large lead balls and two small balls mounted on the ends of a light rod suspended by a fine wire. The arrangement is a torsion balance, which is often adapted for measuring a very small force.

Planets and satellites
The planets move round the sun and the satellites round their particular planets in orbits which are approximately circular. In each of these cases there

is a force pulling on the orbiting body constraining it to remain in its orbit.

Galileo approached the problem in this simple way. Consider a body released at P (Fig. 10.4), a point above the earth's surface. It will move along

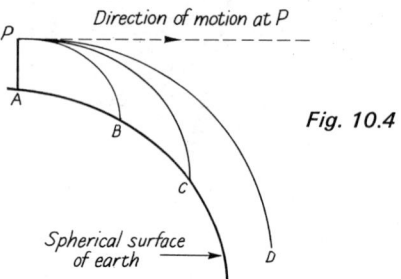

Fig. 10.4

the vertical line PA meeting the earth's surface at A. If it is given an initial velocity at right angles to PA, it will move along a path PB: when given a greater velocity it will take the path PC. When given a still greater and particular velocity it will move along the path PD and never reach the earth's surface. It will remain in a circular orbit about the earth.

Artificial satellites—space vehicles—rocketry
The above ideas of Galileo illustrate the procedure for putting an artificial satellite in orbit. The satellite is launched on an upward path until it reaches a predetermined height. It is then turned through 90° and given a particular speed appropriate to its height.

The propulsion of a space vehicle is effected by means of rockets. When a mass m is projected backwards from the vehicle with a velocity v, a forward momentum of mv is produced. The hot gases of the burning fuel consist of rapidly moving molecules which move backwards from the vehicle. The consequent change in momentum in the forward direction gives the vehicle a forward thrust. This thrust is required to move the vehicle on launching against the gravitational field. When it is turned into orbit, no further thrust is required after its requisite speed has been attained.

The gravitational attraction between a satellite and the earth is the centripetal force always acting at right angles to its motion. Earth satellites are put into orbit above the earth's atmosphere, otherwise air resistance would reduce their speeds and would also raise them to a high temperature by friction.

Fig. 10.5 Men in cabin under influence of 'zero *g*'

Artificial satellites circumnavigating the earth can be useful. They can 'observe' the changes in the earth's atmosphere and so assist in weather prediction. They can make measurements of radiations in space and signal the readings back to earth by short-wave radio. In this way telecommunications have been expedited and global television made possible.

Astronauts and pilots flying very fast aircraft must be physically fit, for their bodies have to undergo strain due to variation from the normal gravitational force. A space ship is given a large acceleration in order to increase its velocity to the required value. The resultant force may be increased to several times its normal value which causes a distortion of the body of the astronaut. While in orbit, however, his condition is such as would be experienced in a field of no gravitational pull, commonly known as 'zero *g*'. This produces weightlessness as is experienced by a body in free-fall. In this case all the gravitational pull is used as centripetal force* and so the vehicle does not move nearer the earth: it virtually floats in the gravitational field. Objects not fastened down in the cockpit float about. The American and Russian astronauts who have experienced 'zero *g*' for appreciable periods of time claim that they have not suffered ill-effects (Fig. 10.5).

When on a journey to the moon a space vehicle sets off from the vicinity of the earth with a high speed. Because of the earth's field it is slowed down as it ascends. This retarding force decreases as the vehicle gets farther away. Eventually, the effect of the moon's gravitational field is felt. This increases as the earth's field continues to decrease and the vehicle accelerates towards the moon. For a moon landing, a similar technique as is used for an earth landing is necessary, which involves the use of retro-rockets to reduce the speed to a few kilometres per hour.

* centripetal force (g) ≡ centrifugal force

net gravity = 0 ∴ you float around in the spaceship.

QUESTIONS

1. How can the velocity of a body be changed without changing its speed? Explain your answer.

2. A heavy object tied to one end of a piece of string is whirled round in a horizontal circular path above the head. If the speed of rotation is too big, the string will break. Explain this.

   Would the string have broken had it been shorter? Give your reason.

3. What effect has the rotation of the earth upon bodies on its surface?

4. Explain how it is possible to whirl a bucketful of water round in a vertical circle without spilling any of the water.

5. Explain the action of a spin-drier consisting of a perforated drum into which the wet clothes are first placed and which is then rapidly rotated.

6. When an artificial satellite is put into a circular orbit, what determines the radius of its orbit?

7. A small object is placed on a gramophone turntable. When near to the centre spindle the object rides steadily on the turntable, but when near to the edge it is thrown off. Explain this. How does the mass of the object affect its ability to resist being thrown off?

8. A small heavy object on the end of a piece of string is whirled round in a vertical circular path with a constant speed. Explain any variations in the tension in the string.

9. A piece of aluminium and a piece of gold are adjusted to balance one another on a very sensitive beam balance on a day when the barometer reads 760 mm. A fall in barometric height causes the balance to be upset but the balance can be restored by placing a large block of lead under the pan containing the gold. Explain these observations.   (O & C)

11 Simple harmonic motion

Characteristics

Another type of motion, in addition to the progressive motion along a straight line or a curve, is that which we call a *vibration* or *oscillation* in which a body moves backwards and forwards rhythmically. Such motion may be along a line as in the cases of a pendulum bob and a mass on the end of a spring, or it may be angular as in the case of a torsion pendulum.* We shall see later that sounds are produced by vibrating bodies which, in turn, cause the air molecules to vibrate so as to create a sound wave. In Chapter 20 we shall deal fully with vibration and the production of waves.

By studying the motion of an oscillating body such as a pendulum bob, we arrive at one or two conclusions:

1. The periodic time is a constant and is independent of the amplitude,†
2. The force acting is always directed towards the point about which the body is vibrating,
3. The velocity is a maximum, and the acceleration zero at the centre of the oscillation: the velocity is zero, and the acceleration a maximum at the extreme positions.

This type of motion is known as *simple harmonic motion* and is represented mathematically by the equation:

acceleration = − constant × displacement

The negative sign shows the acceleration is always opposite to the direction of the displacement increasing.

A body executes SHM along a straight line when its acceleration is always directed towards, and is proportional to its distance from, a fixed point on the straight line.*

The graph of the displacement against the time is a sinusoidal curve. Fig. 11.1 shows this and also the variations of the velocity and acceleration with time.

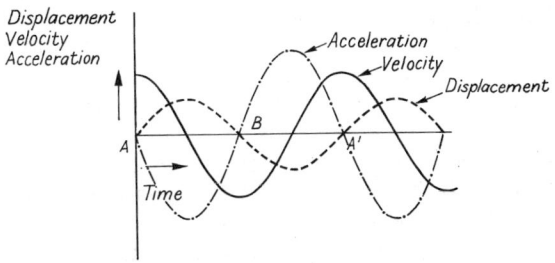

Fig. 11.1 Displacement/time graph for SHM

The variation of displacement with time can be demonstrated by many methods. One simple method is to use an inked pen or brush attached

---

* A torsional pendulum consists of a heavy object, such as a metal cylinder, suspended by a wire from the centre point of one of its ends, so that it hangs symmetrically. The object is rotated while the wire is kept vertical. On release, the torsion introduced in the wire maintains rotational oscillations in the body. Such a pendulum is used in some carriage clocks.

† The amplitude is the maximum displacement from the centre (undisplaced position).

---

* The angular oscillations of a body are harmonic when its angular acceleration is always in the direction of, and proportional to the angle it makes with, a fixed direction.

to a heavy pendulum bob to make a trace on paper placed below it and to draw the paper steadily along in a direction at right angles to the oscillations.

The voltage across the a.c. mains varies sinusoidally. This can be shown using a cathode ray oscilloscope (page 394) in which an electron beam, displaced vertically by the alternating voltage, is made to transverse the screen at a constant speed in a horizontal direction.

Diagrammatic representation

In Fig. 11.2, P is a point moving round a circle with a constant speed. If N is the foot of a per-

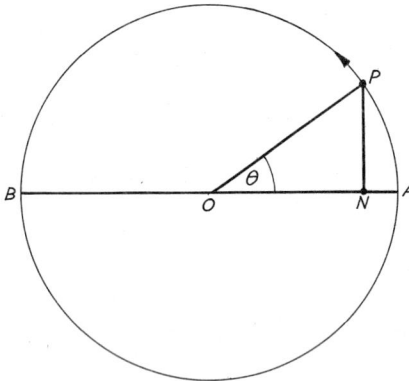

Fig. 11.2 Circle of reference—SHM

pendicular drawn from P on to a diameter AB, then N executes simple harmonic motion.

In order to describe the state of a cyclic variation, we speak of its *phase* (e.g. the phases of the moon). The angle θ measures the phase of a particle at N, taking zero phase to be at A. Thus the cycle repeats itself when the phase change is 360°.

The phase difference between A and A' (Fig. 11.1) is 360°. Displacement with a phase difference of 360° (or a multiple of 360°) are said to be *in phase*. When the phase difference is 180°, the displacements are said to be *out of phase*.

The *periodic time* is the time taken for a phase change of 360°. The *frequency* is the reciprocal of the periodic time and is the number of vibrations executed in unit time. The unit is the *hertz* (Hz) which is the number of vibrations (cycles) in one second.

When a harmonic vibration produces a wave motion (*see* Chapters 20 and 29), the distance

between points on this wave differing in phase by 360° is called a *wavelength*.

Energy changes

In Chapter 5, we have seen that two important types of energy are (1) energy of motion, called *kinetic* energy and (2) energy of position, called *potential* energy. A body possessing either of these can do useful work: it can, for example, lift a load.

A pendulum bob comes to rest twice in a complete cycle of changes. At these instants its kinetic energy is zero, but it is at its highest point and hence possesses its maximum potential energy. At the centre of its swing, the opposite occurs. It

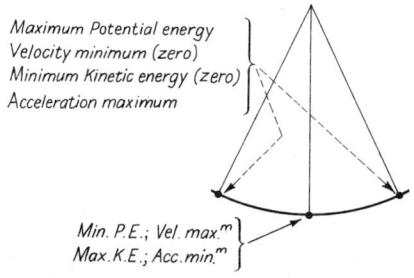

Fig. 11.3 Energy transfer by simple pendulum

has its maximum speed and hence its maximum kinetic energy, but it is at its lowest point and so has its minimum potential energy (Fig. 11.3).

Simple harmonic vibrations involve a continuous exchange of energy from potential to kinetic, the sum of the two forms always adding up to the same total (assuming no dying down of the vibrations).

In the case of electrical oscillations, such as the current variations in a tuned circuit, energy fluctuations again occur, but they are fluctuations of the energies of the magnetic and electric fields of the circuit.

Damping

The amplitude of a simple harmonic oscillation gradually decreases until the motion ceases. This reduction in the amplitude of any vibrating body is known as *damping*. The oscillations of the pendulum of a clock or the balance wheel of a watch are maintained by small impulses given to the pendulum or to the balance wheel just at the right time—a similar action to that which we use in pushing someone on a swing. The energy of the impulse given in the case of a watch is obtained from the main spring. Some watches now available are controlled by the oscillations of a very small 'tuning' fork which are maintained electrically using the energy from a small battery. In some cases damping of the harmonic vibrations is essential for a desired effect. When a musical instrument is played (*see* Chapter 30), a string or air column, or other system, e.g the taut skin of a drum, is made to execute simple harmonic vibrations and so to create a sound wave. If these vibrations were allowed to persist then the sounds from a succession of notes would all be jumbled together.

If the displacement of the vibrations is plotted against the time, a graph as shown in Fig. 11.4 is obtained. The dotted line joining the maximum

e.g. the surge of electric charge which occurs in a tuned circuit in electronics would quickly cease because of the damping caused by the electrical resistance if impulses of charge were not fed into the circuit.

Transfer of energy

If it is possible for a mass to vibrate in more than one manner; there may be a regular interchange of energy between the two modes of oscillation.

If a spring is loaded with a suitable mass and the mass displaced vertically to set it into S.H.M., after sometime the up-and-down motion of the mass will slowly turn into pendulum-like sideways motion. Any system which has two methods of oscillation will act in this way, the kinetic energy passing backwards and forwards from one type of oscillation to the other. Exchange is more rapid when the two oscillations have nearly equal periodic times.

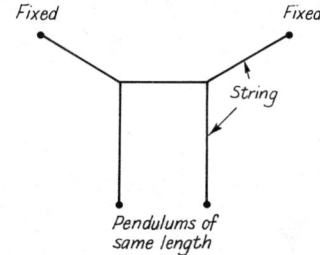

Fig. 11.5 Energy transfer between pendulums

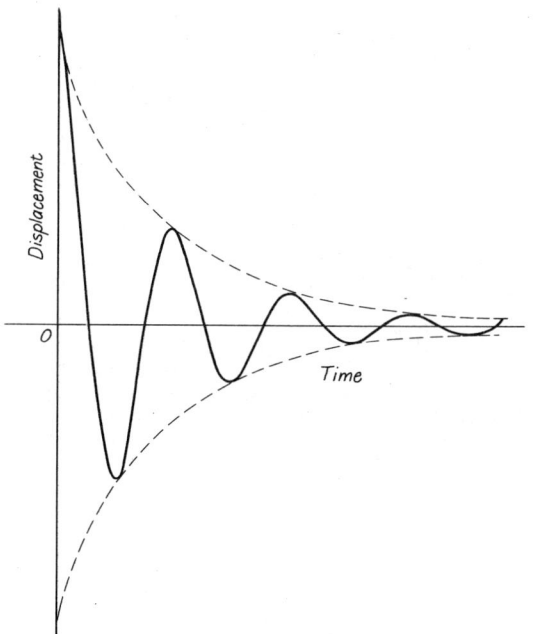

Fig. 11.4 Logarithmic damping

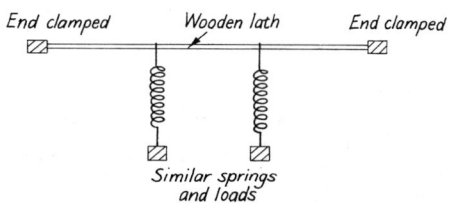

Fig. 11.6 Energy transfer between springs

displacements is of the same form as that on page 406 for radioactive decay. It is a logarithmic curve, i.e. one obtained by plotting the log of a number against the number.

Damping of a simple harmonic variation is an important factor which frequently occurs in physics,

This type of experiment can be demonstrated in many ways. Two similar pendulums or two similarly loaded springs attached to symmetrically placed points on a long horizontal wooden lath provide an interesting case which keeps watchers interested for quite a time (Figs. 11.5 and 11.6).

The simple pendulum

Galileo investigated the laws governing the oscillation of a pendulum.

Experiments with a simple pendulum reveal the following principles:

1. The periodic time or time period is independent of the material and mass of the bob. The periodic time is the time taken for a complete (to and fro) oscillation. In timing the oscillations, the stop-clock should be started as the bob passes its central position, for it is then moving with its greatest speed and can be timed most accurately.

2. The periodic time is independent of the amplitude, provided this is not too great. The amplitude is the distance the bob moves on each side of its undisplaced position measured in a horizontal direction.

3. The periodic time is proportional to the square root of the length of the cord (measured from the point of suspension to the centre of gravity of the bob).

If the periodic time of a simple pendulum is measured for various lengths, the graph of length against (periodic time)^2 gives a straight line through the origin.

The periodic time is given by the formula,

$$T = 2\pi \sqrt{\frac{l}{g}}$$

where l is the length, and g the acceleration due to gravity (9.8 m/s^2).

It should be noted that, because g varies over the earth's surface (*see* page 91), the period of a pendulum will differ also.

Pendulums

We are familiar with many forms of pendulum and, although they are used for a variety of purposes, there is one common reason for their use—the periodic time of oscillation is a constant.

A simple pendulum consists of a heavy sphere suspended by a light thread. The sphere is displaced slightly sideways and is then allowed to make small to and fro oscillations.

A conical pendulum uses similar apparatus but the sphere is made to move round in a circular path in a horizontal plane.

A compound pendulum consists of a bar, sometimes loaded, which is supported by a knife-edge about which it is allowed to make small oscillations.

No. of oscillations	Time/s	Time T for 1 oscillation/s	T^2/s^2	Length of pendulum/m
50	110.5	2.21	4.88	1.21
50	103	2.06	4.24	1.04
50	93	1.86	3.46	0.86
80	133	1.66	2.76	0.695
80	117	1.46	2.13	0.51
100	128	1.28	1.64	0.405
100	115	1.15	1.32	0.31

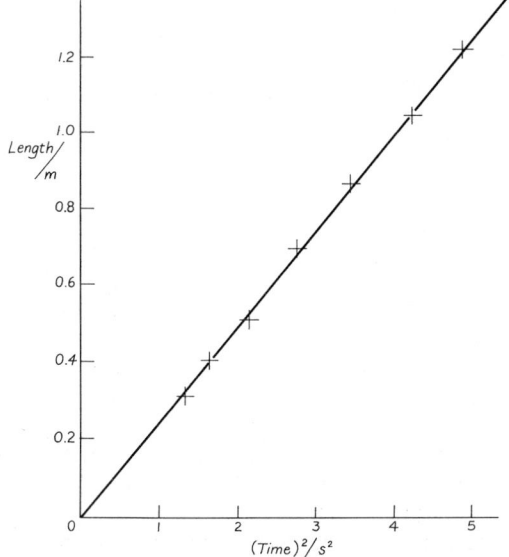

Fig. 11.8 Graph of length/period² for a simple pendulum

Experimental determination of g

The above readings were taken by observing the oscillations of a simple pendulum.

From the graph (Fig. 11.8) we have:

$$\frac{l}{T^2}=\frac{123.5}{5.0}$$

$$T=2\pi\sqrt{\frac{l}{g}}$$

$$g=4\pi^2\frac{l}{T^2}$$

$$=4\times3.14^2\times\frac{1.235}{5}$$

$$=9.80 \ \mathrm{m/s^2}$$

The acceleration due to gravity is 9.80 m/s².

QUESTIONS

1. Define *simple harmonic motion*.
   What is meant by *displacement, amplitude, periodic time, phase*?

2. What is the connection between periodic time and frequency of a vibrating body?

3. Compile a list of examples of bodies executing simple harmonic motion.

4. The equation for simple harmonic motion is:
   acceleration = −constant × displacement
Explain the significance of the negative sign.

5. Briefly describe the devices which execute simple harmonic motion which are used for the purpose of measuring time in clocks and watches.

6. What is a torsional pendulum? Give one practical use to which it is put. Describe the energy changes which it undergoes in executing one cycle.

7. A mass attached to a spiral spring is displaced and allowed to oscillate along a vertical line. Describe and explain the energy changes which it undergoes.

12 The kinetic theory of matter*

The motions of atoms and molecules

We now return to consider in greater detail the constitution of matter and molecular motion. In Chapter 2 we stated that molecules were always in motion and described how this motion is demonstrated for liquids and gases by the Brownian movement. Heat is the kinetic energy of the molecules; supplying heat to a body raises this kinetic energy. Temperature is a measure of the average energy of all the molecules. The velocities (and so the kinetic energies) of the molecules in a quantity of substance differ. There is transference of energy amongst themselves, but the the total energy remains constant, unless energy is received from outside.

The properties of molecules arise from the forces which act between them. These forces are electrical in nature and, although we cannot here delve deeply into the theory involved, we can form some working picture of what goes on.

Solids

Around an atom or molecule is a field of influence. The effect of this field is far greater than the effect due to a gravitational field obeying the inverse

square law of Newton. When two molecules approach closely (i.e. to within a distance of 10^{-10} m between centres) there is attraction. Molecule 2 is in the field of molecule 1, and is pulled inwards but, in addition to a force of attraction, there is one of repulsion which is less effective at 'large' distances, but more effective at 'small' distances. Thus as molecule 2 approaches molecule 1, the repulsion eventually dominates the attraction and the approach becomes a retreat. It will now be seen that the approach and retreat will be repeated and the molecule will vibrate. It can perhaps be made clearer by considering the sketch graphs (Fig. 12.1), which represent approximately the effect on molecule 2 due to molecule 1. It must be realized that molecule 2 has a similar but opposite effect on molecule 1. The curves show that the attraction falls off much more slowly with distance than does the repulsion. When the two effects are added to give the resultant curve, the force is zero for a separation of d. A resultant attraction occurs when the curve is above the x-axis and repulsion when below. The normal equilibrium position will be P, but the molecule will not be still, it will vibrate. The action can be compared to a floating object. When it is depressed below its equilibrium position, the upthrust is greater than its weight and it moves upward. When it is above its equilibrium position the

---

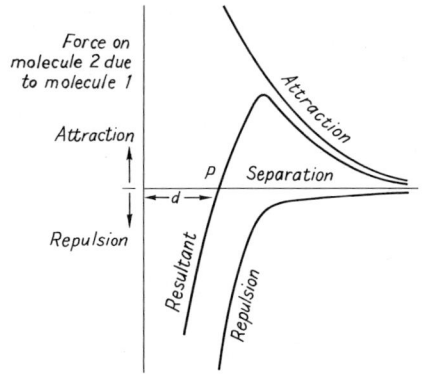

Fig.12.1 Forces between two molecules separated by distance x

upthrust is less than its weight and it falls. When depressed and released it rises beyond the equilibrium position and oscillates.

Let us next consider the energy or work required to bring one molecule near to another. At some distance there is a small force of attraction and so molecule 2 loses energy, which it continues to do at an increasing rate until it reaches position R. After that the rate of giving energy decreases until it reaches P when its energy has a minimum value. After that the resultant force is one of repulsion and so work has to be done on the molecule and this continues to increase. The sketch graph of the energy is shown in Fig. 12.2. Molecule

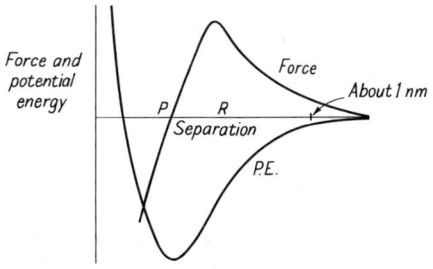

Fig. 12.2 Potential energy of molecule in vicinity of a .similar neighbouring molecule

2 oscillates about P just like a ball rolling up and down in the depression of the energy curve about P.

It must be appreciated that we have considered only the effect of one molecule on another. In reality each molecule is surrounded by millions of others and so each molecule affects its many neighbours and, in turn, is affected by them.

In a solid the molecules are so near together that they lie in each other's energy troughs and vibrate about these mean positions. They have insufficient energy to rise out of the trough, which means they cannot overcome the forces of attraction of their neighbours. Due to the interactions of the molecules, their energies are not all the same but vary over a wide range. The total energy possessed by a molecule is the sum of its kinetic and potential energies. There is fluctuation in the relative amounts as the molecule vibrates.

Adding heat to a solid increases the molecular energy and so the molecular activity. The vibrations are quicker and extend farther. This means that the mean position of the oscillation is farther out from the neighbouring molecule, due to the less steep slope of the energy curve on the outside than on the inside, i.e. about point P (Fig. 12.2). Eventually the addition of sufficient heat gives the molecules energy to become free. It can be visualized as giving sufficient energy to lift the molecules out of the energy trough. This is latent heat (page 151). The molecules now have freedom to move about—the solid has become liquid.

Liquids

The density of a liquid differs little from that of its solid, and hence the packing of the molecules must be approximately the same in both states.

In a liquid the molecules have freedom to move about, but cannot escape from the liquid except in certain circumstances: the Brownian movement (Chapter 2) confirms this molecular motion. The translatory motion of the molecules in a liquid, as distinct from the vibratory motion of molecules in a solid, leads to a bigger range of molecular speeds. Collisions cause changes in speeds but, as so many molecules are involved, for every molecule whose speed is increased, there is another whose speed is decreased. This condition of constant distribution of speeds amongst the molecules must exist because the total energy remains constant—the collisions are perfectly elastic (Chapter 9). A molecule within a liquid is attracted by all the molecules surrounding it, so that the effective resultant force acting on it is zero. A molecule near the surface, however, is only attracted by the molecules which are on its underside. These exert a resultant downward force. A molecule which happens to be near the surface of a liquid and possessing a high speed in the upward direction, may have the necessary energy to overcome the resultant downward pull of its neighbours. It then escapes and becomes a gaseous molecule with complete independence. This is evaporation. The escaping of the fastest moving molecules reduces the average energy of those remaining, which results in a fall in temperature. Thus *evaporation is accompanied by cooling.*

In Chapter 13 we shall study *surface tension,* an effect of molecular attraction or cohesion.

Gases

The mean separation of molecules in a gas is about ten times that of molecules in a liquid. There is no cohesion until one molecule approaches near to another, i.e. at S in Fig. 12.3. Its change in P.E. is then represented by $SPQQ'$. At Q' its velocity is zero and the repulsion causes a movement in the opposite direction. It loses some K.E. between P and S, but at S it still has sufficient K.E. left to escape. The collisions are of short duration and perfectly elastic so that the molecular motion is random.

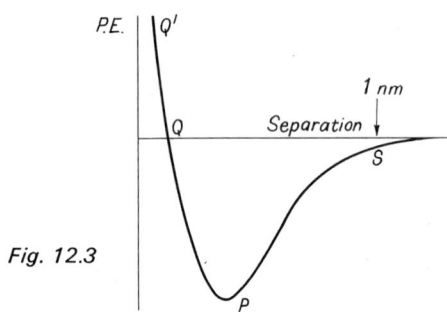

Fig. 12.3

The random motion can be illustrated with the mechanical model shown in Fig. 12.4. A glass cylinder about 25 cm long and 6 cm in diameter has a movable base in the form of a piston which is vibrated by a rod connected eccentrically (off-centre) to the shaft of an electric motor. In the tube are small balls (made of phosphor bronze) which, when the base is vibrating, are projected upwards. The balls suffer collisions and their random motion is clearly shown by the balls which are projected up the tube. If one or two balls are coloured differently from the others, the motion of these balls is easily followed. A few have upward speeds taking them very high. Increasing the motor speed gives the balls bigger speeds which is shown by a greater number rising high up the tube.

Fig. 12.4 Molecular action: kinetic theory model

The kinetic theory of gases

The kinetic theory enables us to derive theoretically many important relationships concerning gases. In applying these ideas it is necessary at the outset to appreciate that we make certain assumptions which are only permissible for a gas under ordinary conditions. These assumptions are:

1. The molecules in a gas have complete freedom and so their motion is random.
2. They are not influenced by their neighbours, i.e. there is no cohesion between the molecules.
3. They have negligible volume compared with the volume occupied by the gas.
4. They suffer perfectly elastic collisions.
5. They come within each other's influence for a very short interval of time compared with the time between collisions.

The above assumptions are very nearly correct for a gas under ordinary conditions, but they may not be so under extreme conditions. For example, under very high pressure the molecules are so near together that cohesion is appreciable.

Pressure exerted by a gas

The molecules in a gas strike the walls of the containing vessel. They suffer perfectly elastic collisions. It is interesting to note that, if these were inelastic collisions, then there would be a loss of kinetic energy and so the gas would quickly cool down. The impact of a single molecule causes an insignificant force, but the total effect of the many millions of collisions becomes appreciable. A molecule of mass m striking a surface with a velocity u perpendicular to the surface, rebounds with the same velocity u. The change in momentum is $2mu$. If n molecules strike unit area of the surface every second, the change in momentum is $2mnu$. But as this occurs in one second, the force (rate of change of momentum) is $2mnu$. As this is exerted over unit area, it is also the expression for the pressure.

Gas pressure can be demonstrated with the vibration apparatus described above. A light piston is added which rests on the balls. When these are set in motion, the piston is supported midway in the tube, indicating the force exerted by the combination of the impacts of the balls. Another way of illustrating this pressure is to drop a succession of balls upon a convex surface resting on a household scale (the upturned scale pan serves well). The effect of each individual ball is observed as a small momentary movement of the pointer. Dropping many balls in rapid succession produces a large force which remains steady so long as the shower of balls is maintained.

In the case of the molecules where millions of collisions occur each second, the combined effect of all the impacts is a steady force and so we have constant pressure exerted on the walls of the vessel containing a gas.

Example 1

Small balls of mass 0.002 kg are projected with a speed of 3.0 m/s in a normal direction against a vertical surface of area 0.010 m² at the rate of 200 per second. Assuming they rebound without change of speed, find the average pressure over the area.

Change in momentum suffered by each ball

$$= 2mu$$

$$= 2 \times 0.002 \times 3$$

$$= 0.012 \text{ kg m/s or N s}$$

total change in momentum for 200 balls

$$= 0.012 \times 200 \text{ kg m/s}$$

total force over area = rate of change of momentum

pressure over area $\quad = \dfrac{\text{force}}{\text{area}}$

$$= \dfrac{0.012 \times 200}{0.010}$$

$$= 240 \text{ N/m}^2$$

The pressure is 240 N/m² or Pa.

Example 2

A jet of water, 1.5 cm² in cross-section, is projected with a speed of 1.2 m/s in a direction at right angles to a vertical wall. Find the force on the wall.

Assuming the water does not rebound, but drops vertically downwards after its horizontal speed is reduced to zero,

force = change in momentum per second

$\quad$ = mass of water striking wall per second × velocity

$$= \dfrac{1.5}{10\,000} \times 1.2 \times 1000 \times 1.2$$

(using density of water = 1000 kg/m³)

$$= 2.16 \times 10^{-1} \text{ N}$$

The force is 2.16×10^{-1} N.

Atmospheric pressure

Consider a molecule in the small volume $ABCDEFGH$ within an enclosed gas, Fig. 12.5. If $ABCD$ is on the side of the vessel and a molecule strikes this surface normally, with a velocity u,

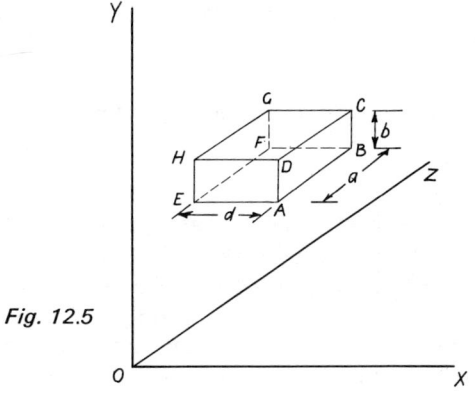

Fig. 12.5

it will rebound with a velocity $-u$, i.e. u in the opposite direction.

When a molecule with velocity u, normal to *EFGH*, passes through this boundary, another molecule passes through *EFGH* in the opposite direction, thus ensuring that the pressure in *ABCDEFGH* remains constant. It follows, therefore, that the volume *ABCDEFGH* always contains the same number of molecules. The particular molecules do not remain in it as there is constant change, but for every molecule travelling with a certain speed in a certain direction which leaves the volume, another one travelling with the same speed in the opposite direction will enter the volume.

This principle provides a method of estimating atmospheric pressure. Note at the outset, however, that the method is very approximate.

We can consider the pressure as being due to molecules falling from the upper layer of the atmosphere to the ground below. The molecule which we consider to start the fall will not be the one which finishes the fall at the ground. The molecules will rebound and reach a height equal to the dropping height. Of course, the atmosphere gets rarer as we go upwards but, if it were the same density throughout, it would have a thickness around the earth of $\dfrac{0.76 \times 13.6 \times 10^3}{1.29}$ m, which is about 9 km.

This is the height of a column of air of density 1.29 kg/m³ which exerts a pressure equal to about 76 cm of mercury (density of mercury is 13.6×10^3 kg/m³).

Consider a molecule falling from the upper region of this layer of air.

If the velocity on reaching the earth's surface is u, then

$$u^2 = 2gs$$

$$u^2 = 2 \times 9.8 \times 9000$$

$$u = 4.2 \times 10^2$$

Thus an estimate of molecular speed of the molecules in air is 400 m/s.

The relationship between pressure and volume—Boyle's law

By employing the principles outlined above, we can derive an expression for the product p and V. That this product is constant under certain conditions is known as Boyle's law. We shall mention the law again in Chapter 16 in conjunction with the other gas laws.

Consider again a molecule in an enclosure *ABCDEFGH* (Fig. 12.5). Let it have a speed u normal to face *ABCD*, i.e. along the x axis. There will be an effective change in momentum of $2mu$ every time it passes through the face for, as we have seen above, another molecule travelling with speed u will enter *ABCD* as the first molecule passes out.

Let $AB = a$: $BC = b$: $AE = d$.

The resulting effect will be as if the molecule passes to and fro along a path of length d. The molecule will make $\dfrac{u}{2d}$ impacts on *ABCD* in unit time. Hence,

rate of change of momentum, i.e. change of momentum in unit time = force

$$= 2mu \times \frac{u}{2d}$$

$$= \frac{mu^2}{d}$$

Let us now consider a molecule travelling with velocity c as shown in Fig. 12.6; c can be resolved

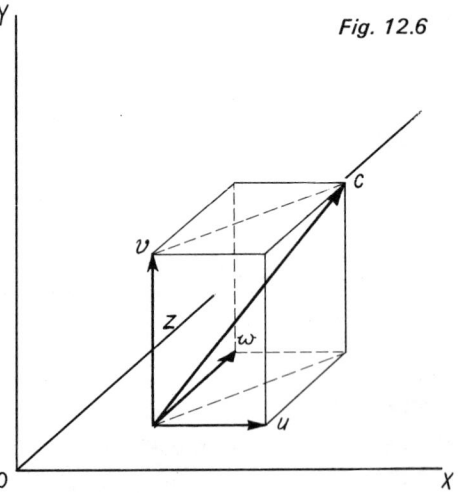

Fig. 12.6

into components u, v and w in directions *OX*, *OY*, *OZ*. But by applying Pythagoras' theorem, we see $c^2 = u^2 + v^2 + w^2$. Now, as the velocities of the molecules are random, just as many molecules will be going in one direction as any other: the sum of all the squares of the velocity components in the *OX*, *OY* and *OZ* directions must be equal,

i.e.
$$\overline{u^2}=\overline{v^2}=\overline{w^2}$$
and
$$\overline{c^2}=\overline{u^2}+\overline{v^2}+\overline{w^2}$$
$$=3\overline{u^2}$$

$\overline{c^2}$ is called the *mean square velocity*. It is the mean value of the (velocity)^2 for all the molecules.

$$\overline{c^2}=\frac{c_1{}^2+c_2{}^2+\cdots c_n{}^2}{n}$$

where c_1, c_2 etc. are the individual velocities for n molecules.

The force over face $ABCD$ due to one molecule $=\dfrac{mu^2}{d}$

area of face $ABCD$ $=ab$

volume of enclosure $=abd$

Let there be N molecules per unit volume,

number of molecules in enclosure $=Nabd$

force over $ABCD$ $=\dfrac{1}{3}\cdot\dfrac{m\overline{c^2}}{d}\cdot Nabd$

pressure over $ABCD$ $=\dfrac{1}{3}\cdot\dfrac{m\overline{c^2}}{abd}Nabd$

$$=\frac{Nm\overline{c^2}}{3}$$

But $Nm=\rho$ (density)

∴ $$p=\rho\frac{\overline{c^2}}{3}$$

$1/\rho$ is the volume V occupied by unit mass

∴ $$pV=\frac{\overline{c^2}}{3}$$

If V_M is the volume occupied by one mole M of the gas, then

$$pV_M=\frac{M c^2}{3}$$
$$=\frac{mL\overline{c^2}}{3}$$

where L is Avogadro's number, i.e. the number of molecules in one mole of gas.

Now $\dfrac{m\overline{c^2}}{2}$ is the average kinetic energy of all the molecules. The temperature is a measure of the average kinetic energy of the molecules,

hence $$pV_M=RT$$

where T is the absolute temperature and R a constant. But V_M is the same for *all* gases and, hence, R is the same for *all* gases. R is called the gas constant per mole and is a universal constant (*see also* Chapter 16).

Molecular speeds

From above, we have
$$p=\frac{\rho\overline{c^2}}{3}.$$

At s.t.p. $p=0.76$ m of mercury. This must be converted into absolute units of pressure.

$$p_a=h\rho\times9.8$$

p_a will be in N/m^2 or Pa when h is in metres, ρ in kg/m^3, and 9.8 in N/kg.

$$p_a=0.76\times13\,600\times9.8$$
$$=101\,300\text{ N/m}^2$$
$$\overline{c^2}=\frac{3\times1.013\times10^5}{1.29}$$
$$=2.3\times10^5\text{ m}^2/\text{s}^2$$
$$\sqrt{\overline{c^2}}\simeq500\text{ m/s}$$

This shows that the average speed of the molecules at S.T.P. is about 500 m/s.*

It will be seen that the mean speed of the molecules of a heavier gas is less than this value. Hydrogen molecules, being the lightest, have the greatest average speed, which is nearly four times that for air molecules.

This very high average molecular speed, which is greater than the speed of sound, can be measured using elaborate timing apparatus. It can be demonstrated with simple apparatus. A coloured vapour is necessary and bromine is suitable, but great care has to be exercised in its use as it is very

* Note that $\sqrt{\overline{c^2}}$ is not the same as $\bar{c}$, but this calculation gives a useful approximate value for the average speed of the molecules.

dangerous to the skin and throat. The experiment simply consists of breaking a glass phial, containing bromine liquid, inside an evacuated long glass tube. The vapour is seen to expand and fill the tube in a very short interval of time. When the experiment is repeated with air at atmospheric pressure in the tube, the coloured vapour can be seen spreading along the tube at a comparatively slow rate. The bromine molecules are impeded by numerous collisions with air molecules.

Estimation of molecular size

There are several methods of determining the size of a molecule which are involved and advanced, but it is possible to arrive at a value by comparatively simple methods. The result has a large percentage error, but it gives us the order, apart from affording an interesting piece of deduction.

The starting point is diffusion. The rate at which a coloured gas, such as bromine, diffuses through a colourless gas, such as air, can be determined by direct observation. The rate depends upon the effectiveness of the barrier of air molecules. This, in turn, depends upon the size and the distance apart of the molecules forming the barrier.

Mean free path

The progress of one molecule through others is, of course, not along a straight path, but is zig-zagged by collisions. The 'free path' between collisions varies over a large range according to the laws of chance governing the random motion. The mean value of this is called the 'mean free path'. It can be estimated by using the idea of a 'random displacement' or 'random walk'. A person decides to move a fixed distance in any direction and then to continue for the same distance in another direction. If this procedure is continued, an equal distance is moved in each stage and the change in direction is randomly chosen. To perform this exercise practically, the positions can be plotted on graph paper. The directions can be chosen by use of dice or by any number of cards. The points of the compass can be represented by numbers. For example, by throwing two dice a selection of 12 directions (30° separation) can be employed. After N throws, the total displacement from the starting spot will be *on average* $\sqrt{N} \times$ each individual distance chosen. Note that this result can only be achieved

by performing the exercise very many times and finding the resultant displacement.

The result is simple to deduce. Refer to Fig. 12.7.

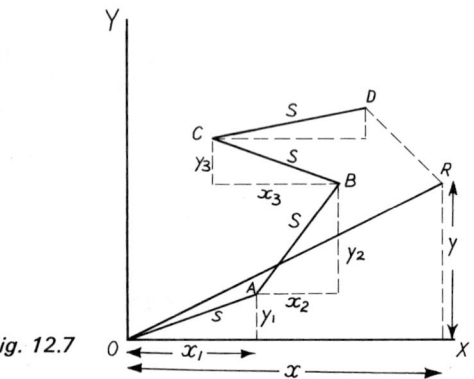

Fig. 12.7

If the fixed displacement is s, then the first displacement is OA, the second AB etc., up to R.

By geometry, $OR^2 = y^2 + x^2$
but y is the algebraic sum of $y_1 + y_2 + y_3 +$ etc.
and likewise $x = x_1 + x_2 + x_3 +$ etc.

Substituting,
$$OR^2 = (x_1 + x_2 + x_3 + \ldots)^2 + (y_1 + y_2 + y_3 + \ldots)^2$$
$$= (x_1^2 + x_2^2 + \ldots + 2x_1x_2 + 2x_1x_3 + \ldots)$$
$$+ (y_1^2 + y_2^2 + \ldots + 2y_1y_2 + 2y_1y_3 + \ldots)$$

The terms such as x_1x_2 and y_1y_2 when summed will equal zero for, with random displacements, the values of x and y will be just as often negative as positive.

$$\therefore \quad OR^2 = (x_1^2 + y_1^2) + (x_2^2 + y_2^2) + \ldots$$
$$= s^2 + s^2 + \ldots$$
$$= Ns^2$$

$\underline{\text{Hence } OR = \sqrt{N} \cdot s.}$

If the distance travelled on each occasion is 10 units and the 'walk' is continued 25 times, the average resultant displacement from the starting position when a large number of walks is carried out is $\sqrt{25} \times 10 = 50$ units.

This treatment is only for movement in two dimensions. It is easily extended to the three dimensions of space by taking three axes OX, OY, OZ. In this case $s^2 = x^2 + y^2 + z^2$.

The resultant displacement OR will be the

distance the bromine molecules travel through the air molecules. s is the average distance between the collisions of a bromine molecule with air molecules. This will be approximately the same as the mean free path of air molecules.

Bromine diffusion through air

This experiment has been mentioned above. For the purpose of estimating molecular size the rate of diffusion has to be measured. The distance the brown coloration travels in a convenient interval of time is measured. As the molecules travel with varying speeds, the boundary of the brown coloration is not clearly defined. The distance is measured to the mean position between *clear* and *maximum brown*. The tube should be long enough to make the time interval several minutes.

The average speed of bromine molecules is less than that of air molecules because they are much heavier. This arises from the fact that the mean kinetic energy of the molecules for any gas is the same, hence (*mean velocity*)^2 varies inversely as *mass*. The mean velocity $\bar{c}$ of bromine molecules is about 200 m/s.

If the path of the bromine molecules were straight the distance travelled would be $\bar{c} \times t$. But the distance actually travelled by the molecules along the tube, and which is measured, is S in time t. We then have $S = \sqrt{N}\lambda$ where λ is the mean free path and N the number of collisions in time t.

But
$$\lambda = \frac{\bar{c}t}{N}$$

hence
$$\lambda = \frac{\bar{c}t}{S^2/\lambda^2}$$

or
$$\lambda = \frac{S^2}{\bar{c}t}$$

S is the average distance the bromine molecules progress through the air molecules in time t. S/t is the mean speed of diffusion.

Experiment shows that when t in seconds is 500, c in m/s is 200, and distance in metres is 0.1,

$$\lambda \text{ in metres} = \frac{0.10^2}{200 \times 500} = 10^{-7}$$

This approximate result gives us a good idea of the average distance travelled between collisions, that is the mean free path of air molecules. It agrees with the value found by other methods.

From above $\quad N = 10^{12}$
Hence,
number of collisions in one second is $\dfrac{10^{12}}{500} = 2 \times 10^9$

Distribution of velocities

We have often mentioned that the speeds of the molecules vary from small to large and are distributed randomly. When dealing with very large numbers, such a chance distribution can be determined precisely by mathematical methods. These methods are used by the actuary who calculates the premiums to cover insurance risks. In his case, however, he has to allow for the fact that he is not dealing with a very large number of cases. He may, for example, have to determine the premium to cover damage to a motor-car by accident. The number of vehicles insured with his company may not be as many as a million (10^6). As the number of molecules in 1 m^3 of gas at s.t.p. is 2.7×10^{25} (27 million million million million) the laws of chance are accurately obeyed. Fig. 12.8 shows the form of the curve giving the

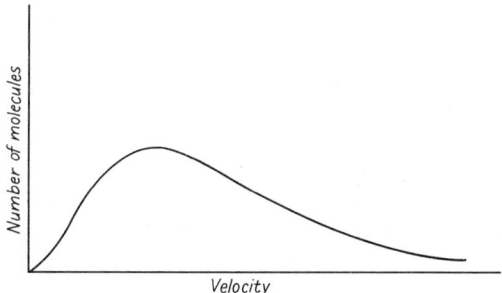

Fig. 12.8 Distribution of velocities among molecules of gas

number of molecules with particular velocities (the distribution of velocities).

The graph shows that there are some molecules with small speeds and some with very large speeds. The majority of the molecules have a speed near to the average value. There is a movement of the maximum when the temperature is increased, for this means an increase in the total kinetic energy of the molecules.

Estimation of the size of an air molecule

We visualize the molecules in the air as possessing free motion and being affected very little by their neighbours. The collisions are frequent but very quick, so that two molecules during collision are close together for a very brief interval.

Consider a molecule X moving among others (Fig. 12.9). It sweeps a path of cross-sectional area πr^2 where r is its radius.* Where this path

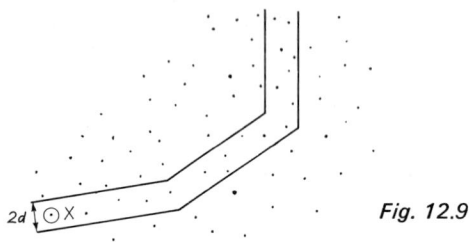

2d $\odot$X

Fig. 12.9

touches another circle a collision occurs. We can study these collisions equally well by representing all the molecules as dots and increasing the cross-sectional area of the path to πd^2, where d is the diameter of an air molecule. We can also consider the path of the one molecule assuming the others to be fixed. This is permissible since the distribution of the molecules at *any* instant will be random.

The average length of this swept region between collisions is the mean free path, which we have found for air to be about 10^{-7} m. Thus the volume in m³ associated with each molecule is $\pi d^2 \times 10^{-7}$.

Now if the molecules in a liquid are packed snugly, each will occupy a cube of side d. When 1 cm³ of liquid air is allowed to evaporate it forms about 750 cm³ of gas at s.t.p. Therefore, the volume containing one air molecule is $750d^3$.

We can combine these two relationships

$$\pi d^2 \times 10^{-7} = 750d^3$$

hence, d in metres is approximately 4×10^{-10}.

We have called this the size of the 'air' molecule. As we are aware, air is not a compound but a mixture of oxygen and nitrogen molecules. These, however, are both about the same size to the degree of accuracy used in the estimation.

This result agrees favourably with the estimate of molecular size obtained by the monolayer

---

* The radius of an atom or of a molecule is really a very indefinite quantity, for we know its outer boundary consists of orbiting electrons. When a molecule is formed, its constituent atoms may be so tightly packed that its atoms are squeezed together.

method when the oil molecule is usually found to have a length of about 20×10^{-10} m.

Avogadro's constant

This is the number of molecules occupying 0.224 m³ of any gas at s.t.p. It is now numerically equal to the seventh SI unit, the *mole*. An approximate value for Avogadro's constant, using the above value for d, is

$$L = \frac{0.0224}{(4 \times 10^{-10})^3 \times 750}$$

$$\simeq 5 \times 10^{23}/\text{mol}.$$

Summary

We have in this chapter built up an important picture of molecules. It will be useful to collect the facts.

Molecules, which constitute all material (matter) are always in motion. In a solid they vibrate, but are not free to move about because of the strong attractive forces between them (cohesion). To gain some extra freedom which enables them to move about while still closely packed together, they require energy (potential energy). This is the *latent heat*. Cohesion still prevents complete freedom of movement, but some molecules, by the process of evaporation, gain complete freedom. For this, *latent energy* is again required and this takes more than what was required in the solid–liquid change. The molecular speeds are randomly distributed. The 'heat' possessed by a body is the sum of the kinetic energies of all the molecules. Temperature is a measure of the average kinetic energies of the molecules.

In gas the molecules move freely about and suffer collisions with each other and with the walls of the containing vessel. These collisions of gaseous molecules are perfectly elastic collisions. There is no loss of energy.

From the volume change from liquid to gaseous air, which is 750 times, the air molecules are about 10 times farther apart in the gaseous state than in the liquid state. The speed of the air molecules is on average 500 m/s. This speed is greater for lighter molecules. The average distance between collisions, that is the mean free path, for air molecules is 10^{-7} m. The number of collisions is 5×10^9 in 1 s. The diameter of air molecules is about 4×10^{-10} m. The number of molecules in one mole is 6×10^{23} (Avogadro's number).

QUESTIONS

1. Describe briefly the forces which act between the molecules in a solid causing them to retain their mean positions.

2. How does the kinetic theory explain the pressure exerted by a gas?

3. How can the small thermal expansion of a solid be explained by a consideration of the internal molecular forces?

4. Describe a simple experimental method of estimating the size of a large molecule.

5. Describe the mechanism causing a change of state (a) from solid to liquid, (b) from liquid to gas.

6. Energy in the form of latent heat is required to convert a liquid into a gas. How is this energy used?

7. The molecules of a gas are said to have random motion. What does this mean?

8. The collisions between molecules are perfectly elastic. What does this mean? What experimental evidence supports this statement?

9. In establishing a simple kinetic theory for gases, we make certain assumptions. Enumerate these assumptions.

10. Assuming that the atmosphere consists of a layer of gas of constant density and 9 km in thickness, calculate the approximate speed of the air molecules at the earth's surface.

11. How does the mean speed of the molecules vary for different gases assuming the pressure and temperature are kept constant?

12. (a) Discuss briefly *two* pieces of evidence which indicate that there are attractive forces between molecules.

(b) Describe an experiment by means of which an estimate may be made of molecular size using a suitable liquid.

(O & C)

13. Explain how the simple kinetic theory accounts for *three* of the following:

(a) The Brownian movement.

(b) The doubling of pressure when the volume of a gas is halved at constant temperature.

(c) The heat produced when a bullet is stopped suddenly by a target.

(d) The rise in temperature when air is compressed in a bicycle pump.

(O & C

14. Give a brief outline of the kinetic theory of gases. Explain from the standpoint of the kinetic theory:

(a) the heating of the barrel of a bicycle pump which is being used to inflate a tyre;

(b) the way heat is transferred by the molecules in the 'vacuum' between the walls of a Dewar vessel containing a hot liquid when the pressure is so low that a gas molecule can go from one wall to another without meeting another molecule.

(O & C)

15. The word 'gas' is supposed to have been based on a Greek word meaning chaos (i.e. disorder). Comment briefly on the aptness of this choice from the viewpoint of the kinetic theory.

(O & C)

16. Diffusion in liquids can be detected after a long time, a matter of days; in gases it is a matter of minutes. State how these facts are related to the kinetic theory of the structure of matter.

(C)

13 Molecules and the forces between them; properties of fluids and solids

Diffusion

The freedom of motion possessed by molecules in liquids is illustrated by the following experiment. A few crystals of a coloured salt, such as copper sulphate, are placed at the bottom of a tall gas-jar filled with water or, alternatively, concentrated copper sulphate solution is passed carefully down a thistle funnel to the bottom of a gas-jar containing water. The coloured liquid, being denser than the water, resides at the bottom of the jar. If the arrangement is left undisturbed for a few weeks, the blue coloration will penetrate upwards into the clear water. This indicatates that some of the copper sulphate molecules have passed into the water. This mixing due to molecular motion is called *diffusion*. It can also take place in a gas which we have already considered in the previous chapter. We mentioned that diffusion involved the molecules pushing their way among other molecules. We saw that bromine molecules, being heavy and also having comparatively slow speeds (about 0.2×10^{-3} m/s), do not pass quickly amongst air molecules. On the other hand, free-moving bromine molecules, as observed when bromine liquid evaporates into a vacuum, move very quickly (200 m/s).

The diffusion of one gas into another can be studied by measuring the rates at which the molecules of one gas pass through a porous parti-tion into another. The demonstration apparatus is shown in Fig. 13.1. It consists of a porous pot provided with a bung and a tube dipping into a beaker of water. When a jar of hydrogen is held over the porous pot, bubbles of air escape from the end of the tube dipping into the water. The moving gaseous molecules can pass through the perfora-

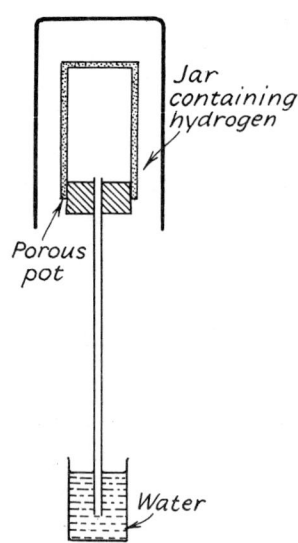

Fig. 13.1 Apparatus for the demonstration of the diffusion of a gas lighter than air

tions of the porous pot. The experiment shows that the hydrogen molecules pass more rapidly into the porous pot than the air molecules pass out of it. The less dense gas always diffuses the faster. This is due to the lighter molecules possessing a mean velocity greater than that of the heavier molecules.

The apparatus is slightly modified in order to compare the rate of diffusion of air with that of a heavier gas. This is shown in Fig. 13.2. If carbon dioxide is in the beaker, the air diffuses faster into the carbon dioxide than the carbon dioxide into

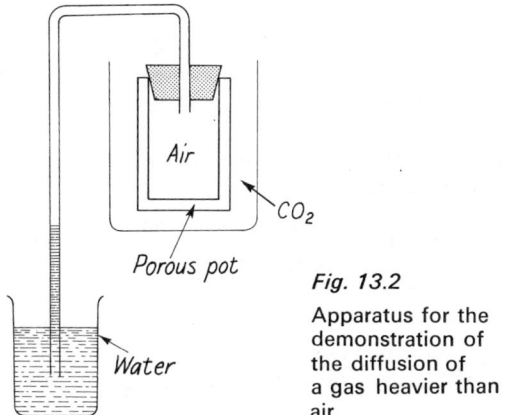

Fig. 13.2

Apparatus for the demonstration of the diffusion of a gas heavier than air

the air. The pressure in the apparatus is reduced, which is shown by a rise in the level of the water in the tube from the beaker.

The difference in diffusion rates is employed as a method of separating mixtures of gases of different atomic masses. It is used extensively in separating the isotopes* 235 and 238 of uranium.

Diffusion between solids and between a solid and a liquid does occur. Silver will slowly pass into other metals, particularly into copper, if the metal and silver are kept close together. Mercury will diffuse into a metal when left on the metal's surface. This is particularly so with the metals copper and silver.

Osmosis

Molecular activity between liquids can take place through separating partitions which are impervious to larger particles. The passage of molecules may take place through parchment, sausage skin,

* *See* Chapter 41.

cellophane, etc. The effect, which is called *osmosis,* can be demonstrated using the apparatus shown in Fig. 13.3. The inverted thistle funnel, to which

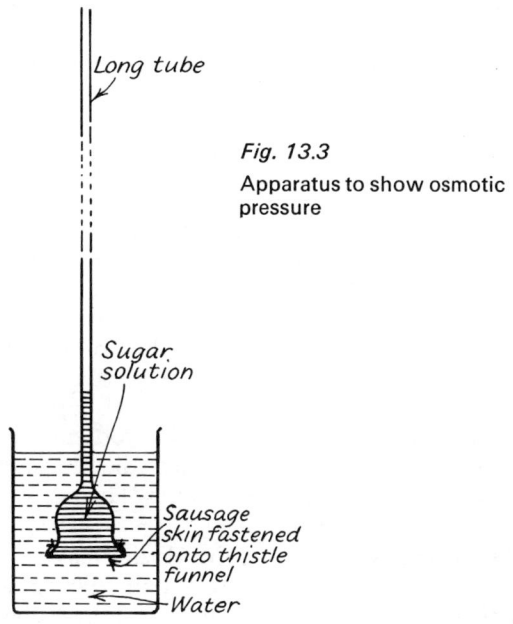

Fig. 13.3

Apparatus to show osmotic pressure

a long tube has been attached, has its end covered with sausage skin (a semi-permeable membrane). A strong sugar solution is put into the funnel and water into the beaker. After a little time the level of liquid in the tube rises, showing that water is passing through the membrane into the tube faster than the sugar solution is passing out.

The pressure difference created by the molecules on the two sides of the membrane is called *osmotic pressure* of the solution and is measured by the head of liquid in the tube when the level is steady. It depends upon the solution, its concentration and its temperature, and may be very considerable. This effect is of great biological importance. It is responsible for the passage of liquids through the walls of plant and animal cells. It contributes to the supply of water from the root system to the top of a high tree.

Viscosity

The introduction of a layer of suitable liquid between two rough surfaces considerably reduces the friction between the surfaces—an effect we call lubrication (*see* Chapter 5). The two rough surfaces are separated by the liquid layer. But there is some friction between the successive layers of liquid as one layer slides over a neighbouring layer. This is shown by the motion of the tea after stirring in a cup. The tea in contact with the cup is at rest and this layer pulls on the next layer and so on. The friction effect between the layers of a moving liquid is called *viscosity*. The variation in viscosity from liquid to liquid is shown in the difference in the times taken to pour out the liquids. Treacle flows from a vessel more slowly than water, but the treacle will drain completely if left for a considerable time. A change in temperature affects the rate of flow of a liquid, because viscosity changes greatly with temperature.

The viscosities of liquids can be compared by observing the rates at which they run through vertical tubes of equal diameters. The shape of the surface of a liquid flowing down a vertical tube is convex downwards, for the sides of the tube have a dragging effect upon the outer layers of the liquid.

The viscosity of a fluid controls the velocity of a body moving through it. If a ball-bearing or a plastic sphere is dropped into a long tube of glycerine supported vertically, it will be observed to attain a constant velocity, called the *terminal* velocity. Repeating the experiment with similar spheres of different diameters establishes the fact that the terminal velocity depends upon the diameter.

Gases also show a viscous effect and a body, which may be a raindrop or a missile, falling through the air attains a terminal velocity.

When a body moves through a fluid with its terminal velocity, the viscous force is equal to the propelling force, which is its weight when falling freely under gravity. The terminal velocity of the very small raindrops which form a mist is very low. For average raindrops it is 2.5 m/s approximately and for the large drops of thunder rain, 5 m/s. But for this viscous effect of the air, the velocity of raindrops from a cloud at the modest height of 500 m would be about 100 m/s or 360 kilometres per hour at ground level.

Fluids in motion

For a liquid to flow through a tube, there must be a pressure difference across the ends of the tube. This may be provided by a pump or by other

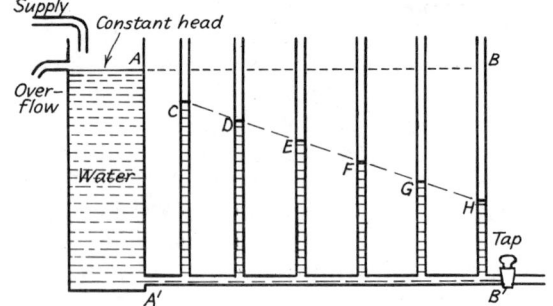

Fig. 13.4 Flow of a liquid through a tube

means such as a head of liquid as shown in Fig. 13.4. In this apparatus the pressure is measured at various points along the tube. When the tap at B' is closed, all columns stand at the common level AB. When the tap is open the levels are at $CDEFGH$. This shows,

(*a*) the pressure drop along the tube is constant —$CDEFGH$ is a straight line,
(*b*) the change in energy from potential to kinetic as the liquid begins to move along the horizontal tube at A'—head AC,
(*c*) the pressure drop HB' created by the tap.

The pressure drop CH along the tube is known as the *friction head*. The rate of flow of liquid is proportional to the pressure difference up to reasonable values, but if this is excessive, the flow becomes irregular. When the flow is regular or *streamlined* all the particles of the fluid are moving in neighbouring paths and not crossing. Irregular flow or *turbulence* applies to a fluid whose particles are moving in a complex manner. In turbulence, secondary motions are created which break up the general flow. Streamlined flow in a pipe necessitates all the particles to be moving in parallel directions. The fluid at the tube walls is stationary, but its velocity increases to a maximum along the central axis.

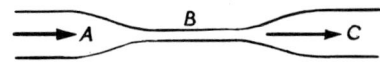

Fig. 13.5 The Venturi tube

Let us now consider the flow of a fluid in a pipe having a narrow section such as shown in Fig. 13.5. The rate of flow (m³/s) must be the same at A, B and C otherwise there would be accumulation of the fluid at some point. To satisfy this condition it is necessary for the speed to be greater at *B* than at *A* and *C*. This difference in speed gives rise to a

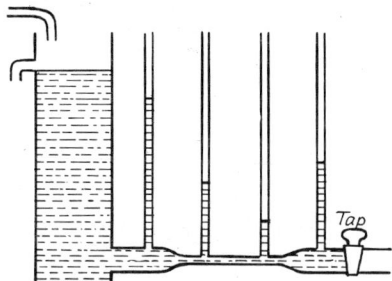

Fig. 13.6 Pressure change due to constriction in pipe

pressure difference and, when tested, the pressure at *B* is found to be less than that at *A* and *C* (*see* Fig. 13.6).

The tube with a constriction is known as a *venturi tube*. The physical principle involved is expressed by Bernoulli's law, which states that,

for streamlined flow of a fluid, an increase in speed is accompanied by a decrease in pressure.

The principle of the venturi tube is used in many devices, from the air-intake on a car to a scent spray. In this latter application it is usual to blow a stream of air over a tube dipping into a liquid. The reduced pressure causes the liquid to be forced up the tube by atmospheric pressure and into the rapid airflow, which breaks it up into a spray (*see* Fig. 13.7).

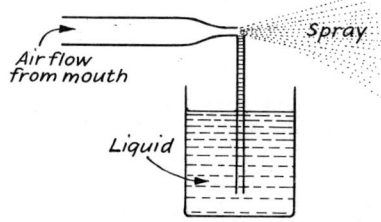

Fig. 13.7 The spray of atomizer

Bernoulli's law has some interesting applications. The swerve in the path of a moving ball can be caused by giving the ball spin. The motion of the ball at *A* (Fig. 13.8) opposes the motion of the air, so reducing the air speed. At *B* the opposite effect obtains, to that a reduction in pressure at *B* and an

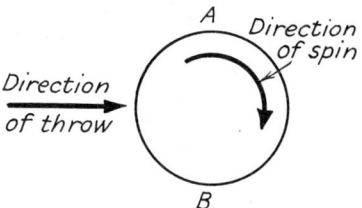

Fig. 13.8 The effect of spin on a pitched ball (looking down from above)

increase at *A*. The ball is thus acted upon by a force from *A* to *B* at right angles to the direction of projection, resulting in a sideways deviation. Likewise top spin causes the ball to fall rapidly and to keep low after bouncing. The tennis player imparts top spin to the ball for, by doing so, he can hit it extremely hard without driving it out of court.

Many simple experiments illustrate the effect. Blowing over the upper surface of a sheet of paper held near to the mouth causes the far edge of the paper to rise as in Fig. 13.9.

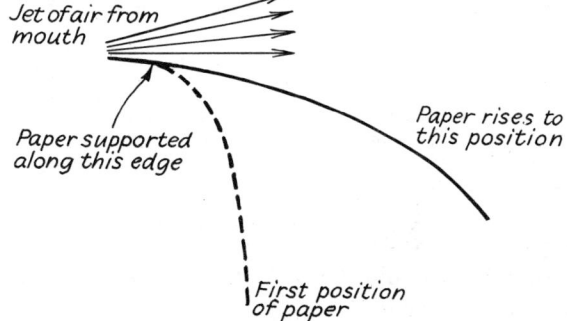

Fig. 13.9 Lift produced by jet of air

A jet of air sent between two suspended ping-pong balls, separated by about 25 mm, causes them to move together (Fig. 13.10).

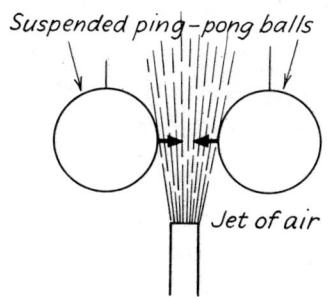

Fig. 13.10 Balls forced together by jet of air

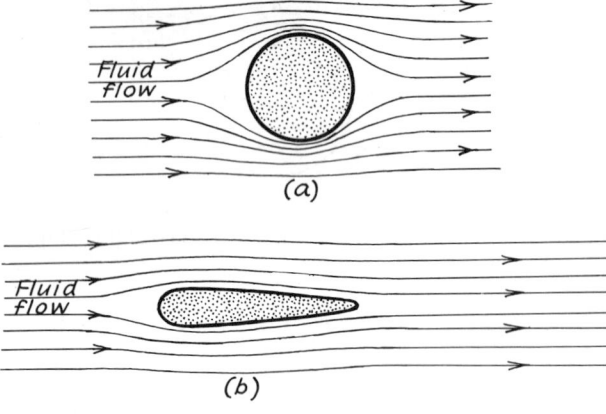

(a)

(b)

Fig. 13.11 Streamlined flow past suitably shaped obstacles

An object placed in a moving fluid offers resistance to the flow. The magnitude of this resistance depends upon the shape of the object and is smallest when the lines of liquid flow are least disturbed by the presence of the object. Fig. 13.11 shows how

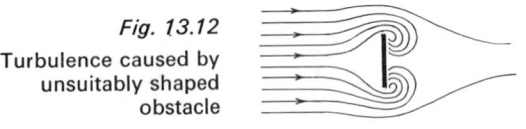

Fig. 13.12

Turbulence caused by unsuitably shaped obstacle

objects of certain shapes do not upset the general flow of the liquid. On the other hand, turbulence

may be produced by an object of a different shape such as a disc in Fig. 13.12.

Streamlined flow, such as that illustrated in Fig. 13.11, may become turbulent if the speed of the fluid is increased beyond a definite value.

These effects just described apply in the case of the passage of an aerofoil through the air and form the basis of the theory of flight.

Fig. 13.14 shows, in section, an aerofoil moving through the air. The inclination θ of the axis of the aerofoil to the direction of motion is called the

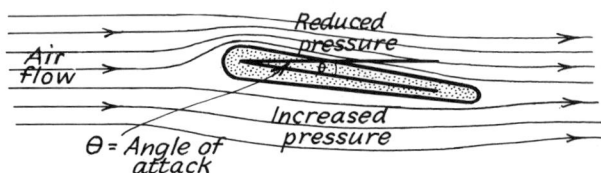

Fig. 13.14 Pressure difference created on the two sides of an aerofoil

angle of attack. For small values of θ, the flow is streamlined as shown. The closing-in of the lines of flow is the same as occurs in a venturi tube and causes a reduction in pressure. Below the aerofoil, the opposite occurs. The lines of flow open out, which means an increase in pressure. Thus the thrust over the upper surface is less than that acting over the lower surface. The resultant thrust (Fig. 13.15) can now be resolved into two components, which are called *lift* and *drag*. If the angle of attack is increased too much, turbulence sets in, the lift quickly decreases and the plane stalls.

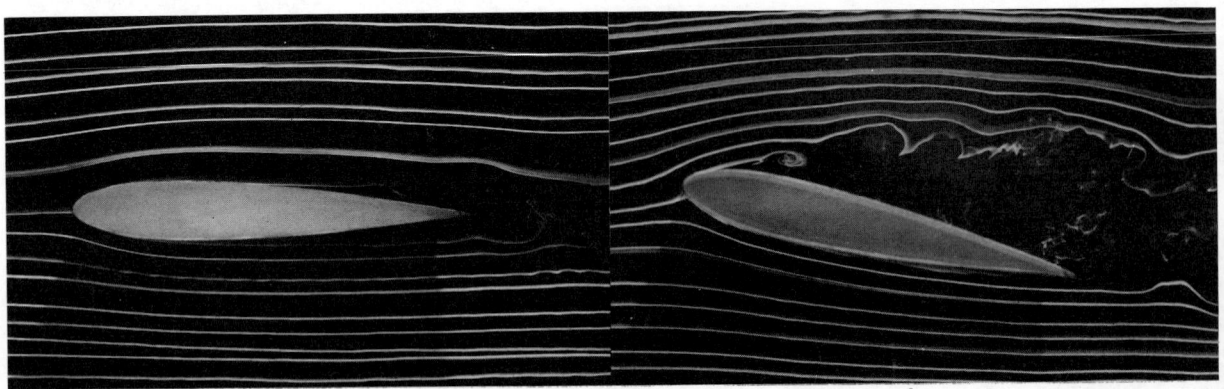

Fig. 13.13 Streamline flow

These photographs show streams of hot air flowing past a model aerofoil. The streams are revealed by the Schlieren effect
Left: perfect flow at zero incidence
Right: slight turbulence at a moderate angle of incidence

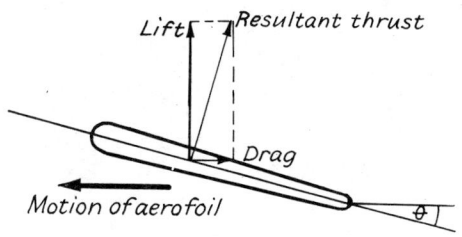

Fig. 13.15 Lift and drag on aerofoil

Surface tension

The well-known experiment of floating a needle or razor-blade on the surface of water is interesting, but it seems to conflict with our ideas of flotation (Fig. 13.16). In performing the experiment, we know we have to lay the needle carefully on the surface without letting either end penetrate through it. This really affords the clue, for the upward force necessary to support the needle is provided by the surface. The surface of any liquid behaves as if it possessed a skin. There is, of course, no such skin in reality. The effect, which is due to forces acting

Fig. 13.16 Razor-blade supported on the surface of water

at the surface, is called *surface tension*. It is largely responsible for the smooth surfaces of glass and sealing-wax, which are formed as they solidify.

For an explanation, we consider the molecular activity. We have already discussed, in connection with evaporation, what happens at the surface of a liquid. The forces of cohesion between the molecules, which prevent their escape, are responsible for the surface tension effects and also for the

surface energy. These effects are caused by the first few layers of molecules: soap films can have a thickness very small compared with the wavelength of light.

A molecule which is moving upwards at the surface experiences a downward pull. This condition can be likened to a chain supported by its ends from two points at the same horizontal level. Gravity acting downwards on all the links of the chain produces a tension in the chain. Or in the case of an inflated balloon, the excess pressure inside produces a force normal to the surface which causes a tension in the rubber.

The cohesion between the molecules of a solid holds the solid together, and is greater than that in a liquid. Many liquids wet glass. This is due to the fact that the molecules of the solid exert a greater force of attraction on the molecules of the liquid near them, than do the molecules of the liquid on each other. This force between molecules of solids and liquids is called adhesion, and the term 'adhesive' is applied to a substance such as gum or glue in which the effect is large.

The cohesive forces in liquids are overwhelmed by gravity, and therefore, a liquid will not support itself and has to be contained in a vessel. In very small drops, however, the cohesive forces assert themselves, and the drops take on a spherical form. As the resultant cohesive forces act inwards on the molecules near the surface, a quantity of liquid always tries to shrink and in so doing makes its surface area a minimum and thus its surface energy also tends to a minimum. Small drops of a liquid are spherical, but larger drops are distorted by

Fig. 13.17 Spherical drop of aniline suspended in brine

gravity. Large spherical drops of one liquid in another of the same density can be formed provided the liquids do not mix. Large aniline drops of true spherical form can be suspended in brine if its density is adjusted to be equal to that of aniline (Fig. 13.17).

Adhesion accounts for the distortion of a liquid surface where it touches the containing vessel. In a glass tube, the surface of a water column is concave, but that of a mercury column, convex. This is because,

cohesion between water molecules
>   < adhesion between water and glass,

cohesion between mercury molecules
>   > adhesion between mercury and glass.

Contamination of a solid or liquid surface greatly alters the forces of adhesion and surface tension. Thus, water will not form drops on a clean glass surface, but will readily do so when the surface is dusty or waxed.

A small rectangular box, open at the top, can be made from ordinary iron gauze. It will not hold water. If it is dipped in molten paraffin wax, and shaken to keep the holes open, it can then be filled with water. In this experiment the wax coating on the gauze changes the adhesive forces. If the under-surface of the gauze box be touched water will run through, for the curved surface of the water droplet at each hole will be broken. (It may be found advisable to pour the water on to a piece of paper on the bottom of the box, so that the force of the water falling will not drive it through the holes. The paper can be carefully removed afterwards.)

This principle explains why a tent is rainproof unless it is touched on the inside. Cloth coats are sometimes proofed by rubbing them with wax, the cotton fibres then acting like the wires of the gauze box.

The radical change in the surface tension of water by a contamination of its surface is shown by many experiments. If a light powder is sprinkled over the surface of water and then a drop of alcohol added, the dust is cleared over the area covered by the alcohol.

The expression 'pouring oil on the troubled waters', implying the smoothing out of a dispute, arises from the practice at sea of pouring oil on the water to quell the waves. Sea-water has a high surface tension and the surface of the sea is not easily broken up, so that the large waves are formed.

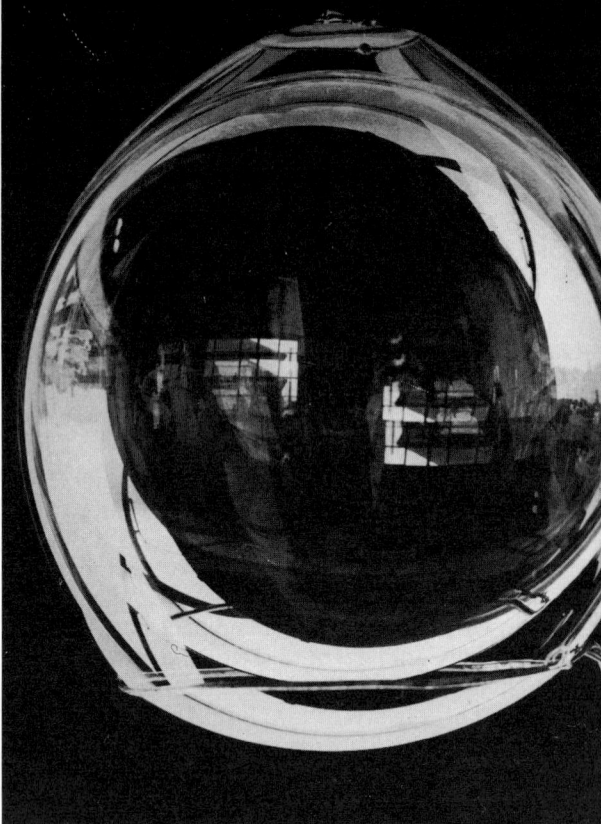

Fig. 13.18 One bubble blown inside another—an arrangement requiring skill to produce

A thin layer of oil reduces the surface tension so much that the surface is easily broken and large waves are not produced.*

The mosquito larva, which is present in stagnant pools, comes to the surface to breathe. It attaches itself to the underside of the water surface, through which it probes its small breathing-tube. We know that the mosquito is the carrier of infectious germs such as those of malaria and yellow fever, and so its extermination is vital in many foreign countries. An effective way of dealing with it is by spraying an oil containing a toxic constituent over the surface of infested ponds: the oil must have suitable spreading properties. The larvae absorb some of the oil and die.†

Small pieces of camphor when placed on a clean water surface move about it in irregular paths. This

---

* There are other explanations of this effect, but all are based on the change in the surface tension produced by the oil.

† The explanation that the oil reduces the surface tension and so prevents the larvae adhering to the under-surface has been proved incorrect. The action of the oil is purely toxic.

is due to the camphor dissolving slightly and reducing the surface tension. The camphor does not dissolve equally all round each fragment, and so the alteration in surface tension is not uniform. The unequal forces acting around a fragment cause it to move. Celluloid boats or ducks can be made to sail by attaching small pieces of camphor to them. A small amount of grease on the water reduces the surface tension so much that motion ceases.

Adding detergent to water reduces its surface tension and so increases its spreading or wetting power. This enhanced ability of spreading and penetrating into tiny crevices accounts for its cleansing action.

The surface tension of soap solution permits the blowing of soap bubbles, which have a small excess pressure inside.

A liquid film always tends to reduce its area. This can be shown in the following way. Make a wire loop, and form a soap film over it by slowly bring-it through the surface of soap solution. If a loop of cotton is floating on the solution, this may be brought away in the soap film. Now break the soap film inside the cotton loop and it will become circular, showing that the forces of surface tension act equally over the surface in all directions (Figs. 13.19, 13.20).

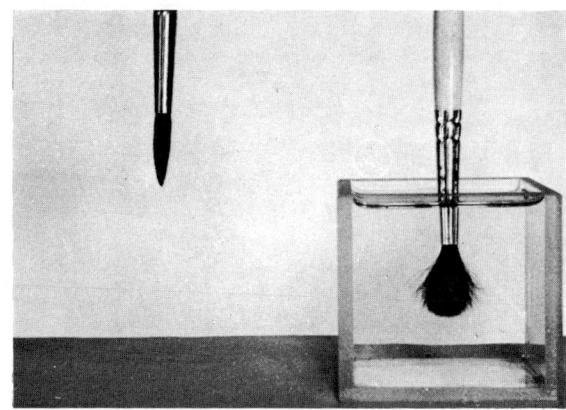

Fig. 13.20 Effect of surface tension on wet brush
When brush is removed from water, the water surface contracts and pulls the bristles together

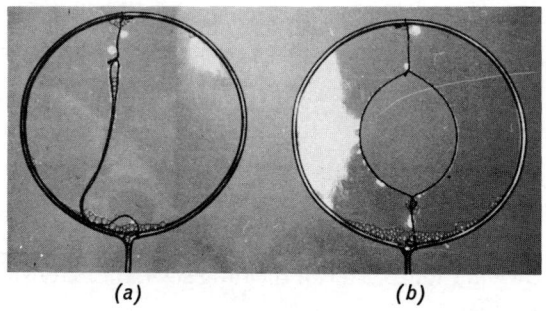

(a)   (b)

Fig. 13.19 A soap film tends to reduce its area
(b) shows effect of breaking film inside loop of cotton in (a)

A molecule on the surface of a liquid has a different potential energy from one below the surface. Thus a liquid surface has a 'surface energy'. Work has to be done to increase the surface area and there is always a tendency for it to decrease. Most of the work done in producing a cloud of tiny water droplets with a spray goes to increasing the surface area of the water.

Capillarity

If one end of a length of glass tubing of very small bore (called capillary tubing) be dipped into water, the level inside the tube is above that in the vessel. The smaller the bore, the greater is this difference (Figs. 13.21, 13.22). The effect is called capillarity and is due to adhesion being greater than the cohesion between the water molecules. With mercury, in which the cohesion is great, there is a depression of the mercury in the tube below the level in the vessel.

Capillarity is responsible for the 'drying' effect of blotting-paper. The spaces between the fibres of the paper act as capillaries and the ink is drawn into them. In a similar way, oil rises in a lamp wick,

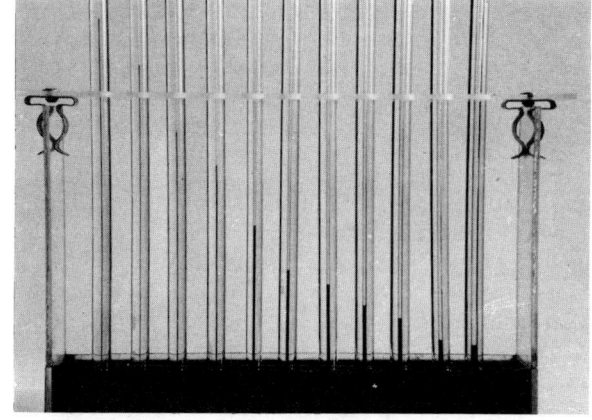

Fig. 13.21 Rise of liquid in capillary tubes of different bore

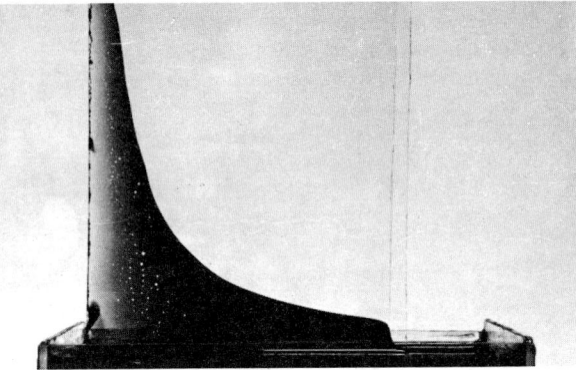

Fig. 13.22 Rise of liquid in wedge-shaped gap between glass plates (capillary rise)

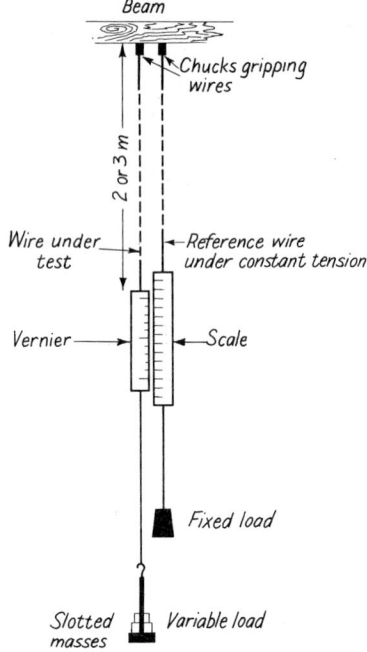

Fig. 13.23 Apparatus for investigation of the stretching of a wire

and water is drawn from the wet subsoil into the top layers of soil during a hot period. Sandy soil is dry because the sand particles are large and the interspaces large, hence capillarity is not so effective in transporting the water from below. On the other hand, clay particles are very small, and consequently the fine capillaries in a clay soil enable it to keep wet.

Properties of solids

The molecular attractions and molecular arrangements in solids give them their special properties. Copper has a high ductility: it can be drawn out into very fine wires. Lead and gold are malleable. This property of gold enables it to be hammered out into sheets only a millionth of a centimetre thick.

Elasticity

Elasticity is the tendency of a substance to return to its original shape after being distorted, e.g. compressed, stretched, or twisted. It is this property which causes a ball to bounce. In a football, the elastic substance is air: in a golf ball, it is rubber.

Many metals are elastic up to definite limits. The investigation of the relation between the extension of a wire and the load can be carried out using the arrangement shown in Fig. 13.23. Two similar wires 2–3 m long are suspended so that they hang side by side from a high support. One is kept taut by a constant load and the other is stretched by the application of a varying load. A scale is attached to the first wire and a vernier to the second, which

is so arranged that it can move alongside the scale. With this arrangement, the extension is measured for various loads during loading and unloading. The mean *extension* is plotted against the *load*. This is found to be a straight line for a limited range of loads.

Hooke, in 1678, published an account of the experiments he had carried out on the extension of a spring, from which he concluded that *extension* was proportional to *load*.

If the load on a wire is progressively increased until it breaks, the graph of *load/extension* is as shown in Fig. 13.24. For the range OA, the extension

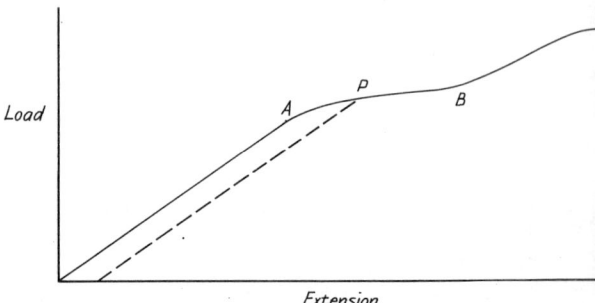

Fig. 13.24 Graph of load/extension of a stretched wire

is proportional to the load. If at any point along *OA* the load is removed, the extension falls back to *O*. For the region *A* to *B* removal of the load does not cause a return to condition *O*, but instead to some point such as *O'*. The wire has been permanently distorted and *OO'* is called the permanent set. The point *A* is called the *elastic limit*. At *B* a small increase in the load produces a large extension —the wire changes its properties. From *B*, which is called the *yield point*, the behaviour of the wire is irregular and depends much on the material from which it is made. At some point such as *C*, the wire breaks.

We have seen in the previous chapter that if the distance between atoms becomes less than normal, a force of repulsion results, which increases very rapidly as the distance decreases. This causes the elasticity of the substance. A similar, but opposite, effect occurs when a substance is stretched. These effects produce the tendency of a solid to resist deformation and, further, to return to its original form. The effect, however, is not simple because the atoms in a solid are not randomly arranged. Most substances have a crystalline structure, and so too much distortion causes a permanent rearrangement of the atoms. This causes a radical change in the elastic properties, which accounts for the yield point. Here atomic attraction is at its limit and a further increase in the pull causes the atoms to slide over each other. If a copper wire is stretched by a strong quick pull, it becomes springy and much less malleable.

Energy change in stretching

Work is done in stretching a wire. For the range *OA*, the energy is stored and is recovered when the load is removed. When the stretching exceeds the yield point, the work is not all recoverable: the permanent distortion has absorbed some energy which is retained in the form of internal potential energy. The work is measured by the product of the extension and the mean load. It is represented by the area under the *load/extension* graph.

Strength of materials: tenacity

The engineer must know the strength of his materials when he designs a bridge or other structure. The *tenacity* of a substance is its ability to withstand a straight pull and is measured by the *tensile strength*. This is the force required to break a uniform rod of unit cross-sectional area by a straight pull. It is measured in N/m^2. A ductile substance (e.g. steel) stretches before it breaks, whereas a brittle substance (e.g. cast-iron) yields with little stretch. The tensile strength of fibres for the manufacture of various cloths is of importance in deciding the wearing properties of the materials.

A liquid shows tenacity: work has to be done to break it up. This we mentioned in connection with the siphon in Chapter 7.

TENSILE STRENGTHS / N/m^2	
Aluminium (cast)	$0.9–1.0 \times 10^8$
Brass (cast)	$1.5–1.9 \times 10^8$
Copper (cast)	$1.2–1.7 \times 10^8$
Iron (wrought)	$2.9–4.5 \times 10^8$
Iron (cast)	$1.0–2.3 \times 10^8$
Steel (cast)	$4.0–6.0 \times 10^8$
Steel (nickel-chromium)	$10.0–15.0 \times 10^8$
Carbon steel	$9.3–10.8 \times 10^8$
Tin (cast)	$0.2–0.35 \times 10^8$

QUESTIONS

1. What is diffusion and how do we explain it?
   How can we account for different gases having different rates of diffusion?

2. Describe an experiment to illustrate diffusion (*a*) in liquids, (*b*) in gases. (One experiment in each case.)

3. Describe the phenomenon of osmosis and how it is explained in terms of molecular motion.

4. Explain (*a*) the rise of sap up a plant, (*b*) the swelling of dried peas and beans, when soaked in water.

5. State Bernoulli's theorem and apply it to explain the curved flight of a spinning cricket ball.

6. Discuss the effect of top spin on the flight of a tennis ball.

7. Suggest a design for a chimney-cowl to prevent the chimney's 'smoking'. Explain how it functions in accordance with Bernoulli's theorem.

8. Describe and explain how a water-jet directed vertically upwards can support a table-tennis ball.

9. Describe two practical applications of Bernoulli's theorem.

10. What is meant by streamlined flow? What happens when the flow is not streamlined? Give two illustrations for (*a*) a liquid, (*b*) a gas.

11. Explain the production of lift and drag on an aerofoil.

A large liner steered parallel and close to the side of a dock may be pulled into the dock side. Explain how this may happen.

12. Describe experiments, one in each case, you would perform to demonstrate (*a*) convection in a liquid, (*b*) diffusion in a liquid. What precautions would you take, in (*b*), to ensure that no convection occurs? Suggest reasons why convection is likely to be more rapid in liquids with a large coefficient of expansion, and diffusion more rapid at higher temperatures. (C)

13. Water rises in a capillary tube. If the tube is cut off very short, explain why the water does not run out at the top.

14. Explain the following:
    (*a*) liquids are absorbed by blotting paper;
    (*b*) a very dry cloth will not easily dry dishes and a dry sponge will not readily absorb water;
    (*c*) the sharp end of a piece of glass tubing is smoothed by heating.

15. A drop of oil, slightly denser than water, can float on water and it is then a *lens-shaped* drop. A drop of the same oil formed under water has a nearly *spherical shape* and it *sinks*, *uniformly* and *slowly*, to the bottom of the water. Account, in terms of surface tension, density, and viscosity, for the features printed in *italics*. (C)

16. You are provided with some water and with a certain mixture of alcohol and glycerine which has the same density as the water. How would you use a capillary tube (or any other apparatus) to show that the water has the greater surface tension and the lesser viscosity? State how you deduce these facts from the experiments performed. (C)

17. A small quantity of mercury placed on a horizontal glass surface forms approximately spherical drops, but water used instead of mercury spreads over the surface of the glass. Account for these effects in terms of the forces between molecules.

In certain circumstances water will form spherical drops. Give any one example of this. (C)

18. A small bubble of air is at rest in contact with the base of the vessel containing motor oil. State the forces, acting on the bubble, which keep the bubble in equilibrium. When the bubble is dislodged it rises through the oil with uniform speed. State the forces which are then acting on the bubble and state, giving your reason, whether these forces are in equilibrium. (C)

19. Explain what is meant by the *surface tension* of a liquid. Describe experiments to demonstrate its existence in (*a*) water, (*b*) mercury. (L)

20. State Hooke's law. Describe and explain any *one* method by which you could test the validity of this law experimentally.

A 100 g brass weight is suspended in air by a perfectly elastic cord, and produces an extension of 10.3 cm in the length of the cord. When the weight is arranged to hang suspended in water, the extension is 9.1 cm. When it is suspended in alcohol, the extension is 9.3 cm. Calculate the relative density of (*a*) brass, (*b*) alcohol. (C)

21. Describe briefly an experiment to investigate how the stretching force affects the length of a spiral spring.

What result would you expect?

When a piece of aluminium is suspended in air from the lower end of a vertical spiral spring, of unstretched length 30.6 cm, its length is 35.0 cm. On surrounding the aluminium with water the length of the spring becomes 33.0 cm. Find the relative density of the aluminium. (L)

22. State Hooke's law and illustrate your statement by a sketch graph.

A light spiral spring, one end of which is attached to a fixed point, hangs vertically and is extended by 10 cm when a mass of 1 kg is hung from its lower end. Find the additional force which must be applied to the mass to pull it down a further 5 cm.

What happens to the work that has to be done in extending the spring 5 cm?

Explain why the mass oscillates vertically when the force is removed, and describe the energy changes which take place during one complete cycle of the oscillation. (O)

23. A pan is suspended from the lower end of a supported spiral spring. A small load is added to the pan which just extends the spring to a length of 10.3 cm. Additional loads are added to the pan and the corresponding length of the spring is given in the table below:

additional load/g	100	200	300	400	500	600	700	800
length of spring/cm	11.1	11.9	12.7	13.5	14.3	15.3	16.7	19.4

Plot a suitable graph showing how the extension varies with the load. What conclusions can you obtain from the graph?

If a spring balance were to be constructed using a similar spring, what would be the most suitable working range and how long would the scale have to be? (JMB)

Part Six HEAT, THERMAL ENERGY

Model of a wingless manœuvrable space vehicle on test in the high-speed wind tunnel at the Ames Research Center, California. The air flow at 4300 m/s produces surface heating to 5000 °C at the nose. The heat-transfer distribution is being investigated

14 Heat

Historical introduction

In our everyday lives we recognize heat as something which can be produced by a variety of methods, such as by fire, by friction, by hammering, or by chemical action. We are also familiar with its production in the sun. We recognize, too, many ways by which it can be detected.

The ancients, although unaware of the nature of heat, were able to appreciate the effects which it can produce. Some of these can be represented diagrammatically.

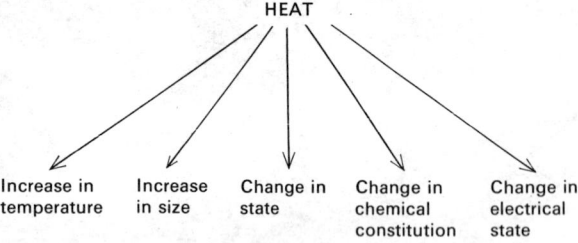

By studying these effects and inventing means of measuring them, many important developments were effected without even knowing what heat was.

Heat and temperature

We must first distinguish quite clearly in our minds the difference between *heat* and *temperature*—two words that are often misused in general conversation.

As we see above, a rise in temperature is one of the effects that may be produced when heat is given to a body. A kettle of water is supplied with heat from a gas jet: the temperature of the water rises. *Heat* is the *cause* and a *rise in temperature* the *effect*.

Temperature is the degree of hotness of a body. A body may be at a very high temperature and yet possess very little heat. A spark from a 'sparkler' firework is quite harmless, although it is a white-hot fragment of iron. When it falls on the hand it does not give out sufficient heat in cooling to burn the skin. On the other hand, a teaspoonful of boiling water, which is at a comparatively low temperature, will produce a nasty burn.

When two bodies at different temperatures are placed in contact, heat flows from the one at the higher, to the one at the lower temperature. The size of the bodies does not matter. The flow of heat continues until the temperatures are the same.

Theories of heat

The fluid theory

Until the end of the eighteenth century, heat was thought to be an elusive fluid, termed 'caloric', whose presence could not be detected by weighing, but by the effects given above. Hammering a body caused caloric to flow out of it; thus continued hammering required an inexhaustible supply of caloric!

In the early experiments performed in the latter half of the eighteenth century, the rise in temperature was used to measure the quantity of heat

122

involved and units of heat were devised, based upon the rise in temperature of unit mass of water through one degree. The metric unit was called the *calorie* (*see* below and also Chapter 17).

There were criticisms of the fluid theory, but most of these were conveniently ignored in the eighteenth century. The production of heat by friction and by hammering certainly raised doubts, but explanations, such as there being a radical change in the heat properties of a substance when in the form of small particles abraded by friction, were freely accepted.

The dynamic theory of heat

The fluid theory of heat was first seriously challenged by Rumford* in 1798. It was while superintending the boring of cannon at the Munich arsenal that Rumford noticed the high temperature of the fragments of metal from the drill and investigated the effect by using a blunt borer. He found that the amount of heat generated depended upon the number of revolutions of the drill and the force with which it pressed against the metal. He concluded that heat was associated with motion and was a form of energy.

The researches of Joule

Although supported by Sir Humphry Davy, the new dynamic theory of heat gained little support until some forty years later. Joule, a private pupil of John Dalton, became interested in energy changes,† particularly those involving heat. His experiments led him to an investigation of the heat developed by the expenditure of mechanical energy. He considered that the new theory of heat could be unquestionably established by showing that there exists a constant relation between the amount of heat produced and the amount of mechanical energy expended.

Joule's researches on this topic covered the period 1840–60. They are famous, not only because of their importance in contributing support to the

theory, but because they are models of careful research. Opposition to Joule's early work was due to its dependence on a very small change in temperature which, it was maintained, could not be accurately measured. When Joule first reported his work to a British Association meeting in 1843, he received little encouragement. At a later meeting, it was only by the support of Lord Kelvin that the true value of his work was appreciated.

The purpose of Joule's investigations was to show that the expenditure of a certain amount of energy always produced the same amount of heat. This meant finding the number of joules of work which would raise the temperature of a known mass of water by a certain amount, in other words, finding the specific heat capacity of water in J/kg K (or J/kg °C: *see* Chapter 1).

Joule devoted much of his life to this theme—the relation between heat and other forms of energy. His name is associated with the heat developed in a resistor when an electric current is passed through it (*see* Chapter 35). He investigated the work done and the consequent rise in temperature when water is forced through a capillary tube. Kelvin relates how he came upon Joule by chance in the Valley of Chamonix. He was carrying a large thermometer with which to measure the difference in the temperatures at the top and bottom of the high waterfalls in the district. He had deduced that the loss of potential energy should cause a rise in temperature. By showing this relationship to be the same for all energy conversions, he definitely proved the theory and so convinced many unbelievers of the day.

His most famous experiment was one in which water contained in a calorimeter was churned by paddles rotated by falling weights. Both the heat and the work were easily measured, but the accuracy of the result depended solely upon the skill with which the experiment was executed.

The calorimeter, shown diagrammatically in Fig. 14.1, was provided with baffle plates to prevent the whirling of the water and to churn it thoroughly. The paddles were rotated by the falling of the heavy masses (two masses were used, simply to balance the apparatus). These masses could be wound up, again and again, so that many falls could be made.

Joule found that the following ratio was a constant in all his experiments:

$$\frac{\text{energy expended}}{\text{heat developed}}$$

---

* Rumford (1753–1814)—an international courtier of the period. He was adviser to many countries on engineering, armaments and strategies. He is known in England as a founder of the Royal Institution. It was he who designed the building occupied by the Institution which contains the lecture theatre made famous by Michael Faraday.

† His researches included the work done by electric motors and batteries, and the heat developed by an electric current.

The result can be expressed:

> **Whenever work is totally converted into heat, or heat is totally converted into work, the number of work units equivalent to one heat unit is always the same.**

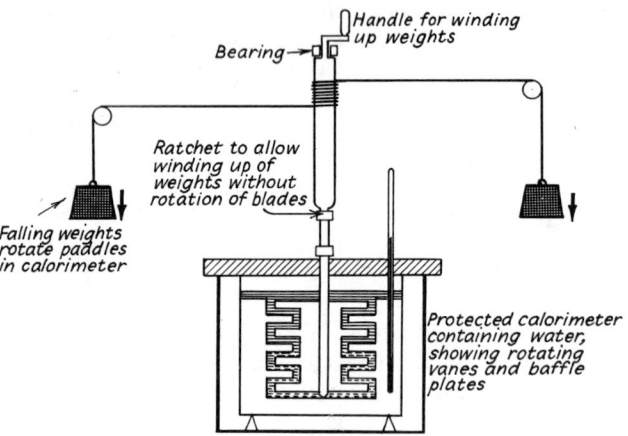

Fig. 14.1 Joule's apparatus

The unit of heat referred to here is the calorie, and this relation between the units of energy and heat is called the *mechanical equivalent of heat* (denoted by J): $J=4.2$ J/cal.

Now that the specific heat capacity of water is expressed as 4.2×10^3 J/kg K, there is no further use for J, the mechanical equivalent.

Historical note

It must be appreciated that the above researches of Joule and others were carried out to prove that heat is a form of energy. This having been decided, there was no longer any need for perpetuating the units of heat dependent upon the performance of water. Unfortunately, the discarding of these units has been delayed until recently. It will take some years, however, before their use will be entirely ended.

The SI unit of energy, the joule, is now used as the unit of heat. Hence, the heat required to raise the temperature of one gramme of water from 15 °C to 16 °C, which was called the calorie, is equal to 4.1855 joules. The specific heat capacity of water is 4.1885×10^3 J/kg K.

The measurement of quantity of heat and the determination of specific heat capacities are dealt with in Chapter 17.

Engines

In all experiments like those of Joule described above, all the mechanical energy is changed into heat energy.*

The transfer of energy the other way round, i.e. from heat energy to mechanical energy, cannot be carried out without an appreciable amount of energy being lost, i.e. not being transferred to mechanical energy. A machine performing this task is called an engine and, of the heat energy supplied, a good proportion of it—in many cases more than 50 per cent is not converted into useful work.

The steam engine

In a steam engine, steam from a boiler enters a cylinder under pressure and drives a piston along the cylinder. By means of valves this first supply is cut off and steam is directed into the cylinder at the other end, so driving the piston back along the cylinder (Fig. 14.2).

Fig. 14.2 Cut-away model of steam-engine

The steam turbine

The steam turbine has now largely replaced the piston engine. The principle of the turbine is simple. Steam at high pressure is injected into one end of a cylinder which contains a central shaft provided with a system of vanes. The fast-flowing steam rotates the vanes, like the sails of a windmill. The shape, size and number of vanes—usually called 'blades'—are carefully worked out in designing a turbine.

---

* Apart from relatively small amounts of energy lost in such ways as friction and as sound energy in the strings supporting the masses in Joule's experiment.

The petrol engine

In petrol and heavy-oil engines the fuel is burned inside cylinders, which are closed by movable pistons. The fuel is drawn in and exploded either by a spark, as in the petrol engine, or by using the heat produced by compressing the vaporized fuel, as in a diesel engine. The explosion forces the piston along the cylinder and the movement is connected to the driving shaft by a rod, one end of which is attached to the piston and the other to a

The fundamental principle of the jet engine is contained in Newton's third law—*to every action there is an equal and opposite reaction.* Air passes into the engine at a front intake duct and promotes the burning of the liquid fuel. The hot gases are ejected at the back and provide the thrust. This thrust depends upon the mass of the gas and its speed of ejection.

The rocket is fundamentally different from the jet, for the former carries its own fuel to eject

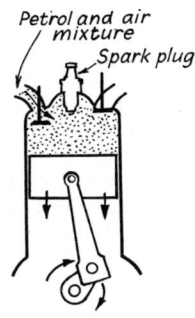
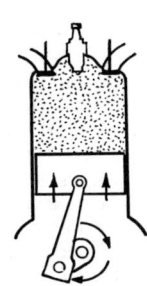

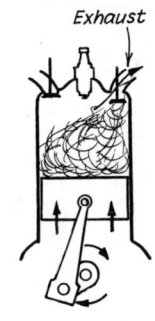

Fig. 14.3 The four-stroke petrol engine

1. *Inlet stroke*
Inlet valve open. Petrol and air mixture taken in

2. *Compression stroke*
Valves closed. Piston compresses mixture

3. *Explosion stroke*
Valves closed. Spark explodes mixture

4. *Exhaust stroke*
Exhaust valve open. Waste gas driven out

crank on the mainshaft. The modern petrol engine is most complex. It is a four-stroke engine, i.e. the piston moves four times along the cylinder for each time the gas is exploded. Details of each stroke are given in Fig. 14.3. Valves are fitted to the cylinders for the intake and exhaust of the fuel and waste gases. The coupling of the drive from the piston rods is done through a complicated system of gears.* Lubrication of the engine is important because of the high speed of the pistons. Oil is fed to the essential points under pressure. Not all the heat generated in the combustion is converted into useful work and the unused heat has to be removed to keep the temperature of the engine reasonably low. For this purpose water is circulated around the cylinders. The heated water passes through a 'radiator' where it is cooled by cold air flowing around it when the car is in motion.

The jet engine

In the last three decades a new type of engine, the *jet*, has been developed and is proving very satisfactory, particularly for aircraft.

whereas the latter requires an external supply of air in which to burn the fuel.

It must be appreciated that the jet does not push *against* the air, but uses it simply to promote the combustion of the fuel it carries. Jet and rocket propulsion can be likened to the propulsion of a trolley by throwing away bricks at the rear. The greater the speed at which the bricks are thrown, the greater the forward thrust. The energy derived from the burning fuel enables the heated gases to be ejected with a very high speed and so to produce

* *See* page 39.

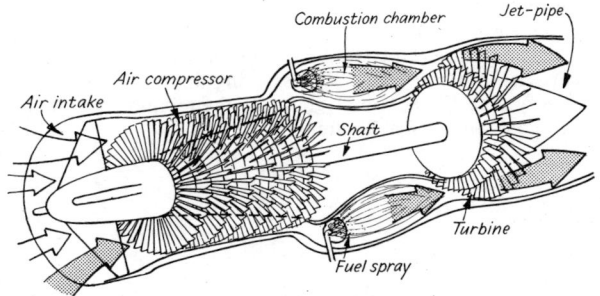

Fig. 14.4 The turbo-jet

a large thrust. The rocket functions in space where there is no air.

The *turbo-jet* (Fig. 14.4) works somewhat differently from the pure jet engine. In such an engine, the intake air is compressed and fuel injected into it. This is then burned and the expelled hot gases provide the thrust. Before ejection, the hot gases rotate the blades of a turbine which drives the compressor. This takes only a part of the power avail-

able, the majority of the power being used in forward propulsion.

The *turbo-prop* is another modification of the jet engine. It closely resembles the turbo-jet, but in this case the turbine is used, not only to work the air compressor, but also to drive a propeller. This type of engine is economical of fuel, but is not, however, as efficient as is the pure jet engine at high speeds.

QUESTIONS

(It will be advisable to gain some additional knowledge of specific latent heat and specific heat capacity, Chapter 17, before attempting to answer some of the following numerical examples.)
[Take the specific heat capacity of water as 4.2×10^3 J/kg K and *g* as 10 m/s^2.]

1. Describe Joule's contribution to the dynamic theory of heat.

2. Explain the energy changes which take place when a body, e.g. a piece of lead, is dropped to the ground from the top of a high building.

3. Give an account of the experiments of Rumford and Joule in support of the idea that heat is a form of energy.

4. A piece of lead falls 3 m from rest, coming to rest again on the ground. Calculate the rise in its temperature, the specific heat capacity of lead being 0.134×10^3 J/kg K. State the assumptions you make in the calculation.   (L)

5. A waterfall is 200 m high. How much warmer will the water be at the bottom than the top? State, giving your reasons, whether you would expect to observe the rise in temperature you have calculated.   (D)

6. A lead bullet travelling at a speed of 250 m/s is brought to rest in a heavy target. Assuming that all its kinetic energy is transformed into heat, find its rise in temperature.
[Specific heat capacity of lead=0.134×10^3 J/kg K.]

7. Describe a simple experiment to *illustrate* that work can be transformed into heat and that the heat produced depends on the work done.
An electric cake mixer, of power 40 W, is used to beat up a cake mixture of mass 0.80 kg and of average specific heat capacity 2.10×10^3 J/kg K. Assuming that only 50% of the energy used by the motor goes into the mixture, find its rise in temperature after 2 minutes working.   (JMB)

8. Describe an experiment to show that when mechanical energy is transformed into heat energy the ratio of the work done to the heat generated, is a constant.
A 20 W electric heater immersed in water in an open Dewar flask just keeps the water boiling. If no heat is lost by conduction, convection and radiation, find how many grammes of water boil away in 20 minutes.
Explain briefly what has happened to the electrical energy supplied to the heater.
[Specific latent heat of vaporization of water=226×10^3 J/kg.]   (O & C)

9. Give an account of the experimental work which led to the conception of heat as a form of energy.
State clearly the sources of the heat energy produced (*a*) when a bicycle pump is used to pump up a tyre, (*b*) when a mass of lead shot is shaken up vigorously inside a cardboard tube.   (O & C)

10. A lead bullet of mass 25 g at a temperature of 17 °C is brought to rest by hitting a target, and is just melted by the impact. Assuming that the kinetic energy of the bullet is all converted into heat in the bullet itself, calculate (*a*) the kinetic energy of the bullet just before impact, (*b*) the velocity of the bullet just before impact.
[Melting point of lead=327 °C. Specific heat capacity of lead=0.126×10^3 J/kg K. Specific latent heat of fusion of lead=21×10^3 J/kg.]   (O & C)

11. A lead shot at a temperature of 37 °C is fired at a speed of 3×10^2 m/s into a resisting target. Assuming that all the heat remains in the shot what fraction of the shot melts?
[Specific heat capacity of lead=0.126×10^3 J/kg K; specific latent heat of lead=25.2×10^3 J/kg; melting-point of lead =327 °C.]   (O & C)

12. State the principle on which the action of jets and rockets depends.
Describe the action of either a jet engine or a rocket engine, explaining how the principle is applied.   (JMB)

15 Heat and temperature

In the previous chapter we have established that heat is a form of energy and in Chapter 2 we learned that molecules are always in motion. Heat is the energy of motion or kinetic energy of the molecules. When a body is heated its molecules move faster. The hotness or temperature of a body is determined by the kinetic energy of the molecules. But the molecules have different kinetic energies because they possess different speeds. It is the *average* kinetic energy which indicates the temperature of a body.

> **Heat is the kinetic energy of molecules. The total heat contained in a body is the sum of the kinetic energies of all the molecules.**
> **Temperature is a measure of the average kinetic energy of the molecules.**

Measurement of temperature

In order to measure temperature we can use any of the effects of heat which produce regular changes as the temperature varies. The suitable changes are those of size and electrical resistance. The latter is usually reserved for high-temperature measurement.

The sense of touch can be used to estimate temperatures very roughly, and the eye is actually employed to judge high temperatures. The colour of a luminous body is an indication of its temperature, e.g. a body glowing bright red is at about 800 °C (1100 K), flesh colour about 1000 °C, lemon 1200 °C, and white 1300 °C. A blacksmith judges when his iron is hot enough to work by its colour.

An instrument designed to measure temperature is called a thermometer. Those designed specially for high-temperature measurement are called pyrometers.

The most common type of thermometer uses the change in volume of mercury, which is enclosed in a bulb with a capillary tube attached. The most accurate thermometer, which is adopted as a standard, is a gas thermometer containing hydrogen.

History of the thermometer

The first thermometers were made at the end of the sixteenth century by Sanctorius of Padua (1590) and Galileo (1602). These early types depended on the expansion of air, and so were influenced by changes in atmospheric pressure.

In the middle of the seventeenth century liquid thermometers were made. Although the expansion of the liquid was less than that of air, very fine capillary tubes could be used and, further, such thermometers were uninfluenced by atmospheric pressure.

In 1701, Newton suggested a method of calibration which could be made universal.

Temperature scales

To establish a scale of temperature, two fixed temperatures have to be chosen and the range between them divided into a definite number of parts. Newton chose the freezing point of water and the temperature of the body for his two fixed temperatures and divided the interval into twelve parts.

The next important improvement was due to Halley, the famous Astronomer Royal. He was the first to suggest the use of mercury, and its superiority was revealed by Fahrenheit in 1714.

Fahrenheit scale

Fahrenheit devised a new scale, using a mixture of ice and salt in definite proportions to give the zero of the scale, and the temperature of the body as the upper fixed point, which he finally called 96°. This scale of temperature was fixed by putting the freezing and boiling points of water at 32° and 212° respectively, but is now no longer used in scientific circles in England.

Celsius scale

The commonest scale was devised by Celsius and modified by Christen about the middle of the eighteenth century. It is called the Celsius or centigrade scale. The fixed levels are marked 0 °C and 100 °C.

The Kelvin or absolute scale (SI unit)

About 100 years ago Kelvin devised a scale of temperature based upon the performance of an ideal gas and so quite independent of the performance of a particular substance. Its zero is the absolute zero of temperature (page 140), i.e. the temperature of a substance when all its heat has been extracted from it. Temperatures within a very small fraction of a degree of absolute zero have been attained.

The size of a kelvin degree is defined so that it is the same as that of the Celsius degree, but 0 °C is 273 K approximately and 100 °C is 373 K. Note that no ° is necessary and K denotes both an actual temperature on the kelvin scale and also a change in temperature, i.e.

$$t \,°\text{C} = (t + 273) \text{ K}$$

but a change in temperature of t °C (i.e. t degrees C) equals a change in temperature of t K.

Standard temperatures

0 °C—the melting point of ice—is the equilibrium temperature between ice and water under a pressure of one standard atmosphere.
100 °C—the boiling point of water—is the equilibrium temperature between water and its vapour under a pressure of one standard atmosphere.

Graduation of a thermometer

A thermometer is calibrated by first putting it in melting ice and later in steam at normal atmospheric pressure. The readings of the thermometer at the two temperatures are recorded. These may be two electrical resistances or two gas volumes, etc. The difference in the two readings is divided into 100 equal parts in order to obtain Celsius degrees. The temperature is then given by,

$$t = \frac{R_t - R_o}{R_{100} - R_o} \times 100$$

where, R_o is the value of some constantly varying property at the ice point,

R_{100} is the value of the same property at the steam point,

R_t is the value of the same property at the temperature t °C.

In this calibration it is assumed that the property measured varies regularly with temperature, but there is no direct means of verifying this. Kelvin discovered a thermodynamic scale of temperature which is independent of any particular property. It has been found that a scale of temperature based on the variation of the pressure of a gas with temperature is in very close agreement with this absolute scale. A gas thermometer is used as the standard.

Graduation of a mercury thermometer

The graduation of a mercury thermometer consists of marking the levels of mercury in the tube at the two fixed temperatures and dividing the length of the tube between them into the necessary number of divisions. For the lower fixed temperature, the thermometer is placed with its bulb in melting ice and left for about ten minutes. The ice shavings can be contained in a filter funnel so as to allow the excess water to drain away: the thermometer bulb should be completely immersed. In order to mark the upper level, all the mercury should be at the temperature of steam from boiling water. To do this, the stem, as well as the bulb, has to be placed in a steam-bath. The instrument shown in Fig. 15.1, called a hypsometer, is used. It consists simply of a double-walled steam-bath. The steam passes up the inside tube and then round the outer tube, thus making sure that the steam in the inner tube is at the boiling point of the water.

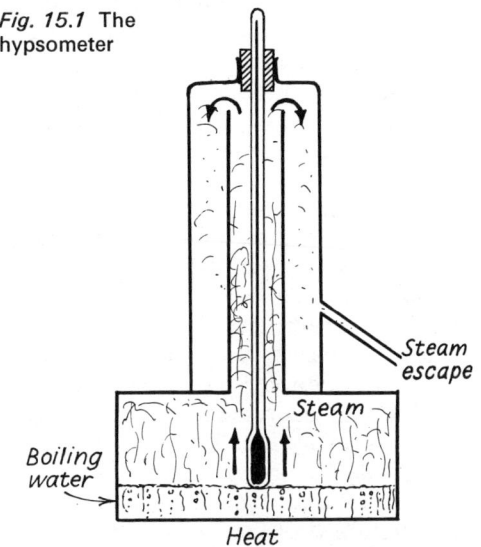

Fig. 15.1 The hypsometer

Steam escape

Steam

Boiling water

Heat

If the height of the barometer is not 76 cm, allowance must be made when marking the upper fixed point.

After marking the two levels of the mercury in the tube, the length between is divided into 100 equal parts for a Celsius thermometer.

Special thermometers

It is often necessary to record the highest and the lowest temperatures attained over a period of time, and for this purpose maximum and minimum thermometers are used.

Clinical thermometer

The clinical thermometer (Fig. 15.2) (used by the doctor for taking the temperature of the body) is a maximum thermometer. It is an ordinary mercury-in-glass thermometer designed to cover a small range from about 35 °C to 42 °C (normal body temperature is 36.9 °C).* There is a constriction in the tube just above the bulb. When the temperature rises, the mercury expands and in doing so forces its way through the constriction and into the tube. On cooling, however, there is no force acting to eject the mercury from the tube through the constriction, and so the column remains in the tube. This enables the maximum temperature to

* Most clinical thermometers are still calibrated in °F. Normal body temperature is 98.4 °F.

be read at leisure. It is necessary to shake the thermometer to restore the mercury into the bulb.

Six's thermometer

Six's thermometer (Fig. 15.3) is popular for recording both maximum and minimum temperatures. It is much used for daily records of air temperature. The tube is in the form of a U with a large bulb, *B*, containing alcohol on one end, and a small bulb, *A*, containing alcohol and alcohol vapour on the other. In the U-tube is a column of mercury. In each limb, above the mercury, is a small, light steel index to which is attached a spring just strong enough to support it and yet readily permit its movement. When the tempera-

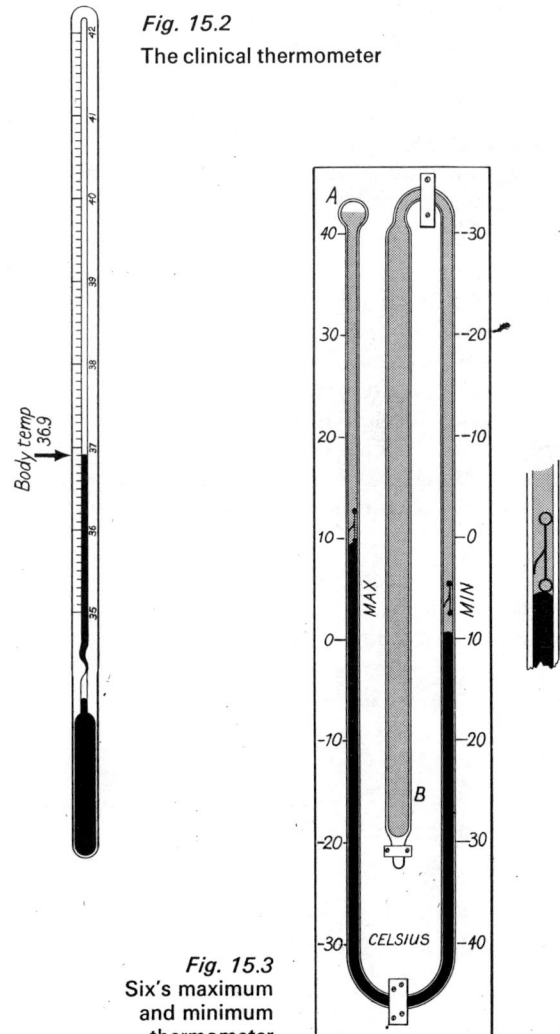

Fig. 15.2
The clinical thermometer

Body temp 36.9

Fig. 15.3
Six's maximum and minimum thermometer

ture rises, the alcohol in bulb *B* expands and moves the mercury column round the U-tube, pushing up one index and leaving the other behind. When the temperature falls, the mercury recedes, leaving the first index to record the maximum temperature, and pushes back the second index towards bulb *B* to record the minimum temperature in the other limb. The alcohol vapour in the small bulb exerts a pressure so keeping the mercury column joined to the alcohol in the bulb.

After reading the thermometer, it is re-set by bringing the indexes into contact with the mercury. This is done with the aid of a magnet.

Electrical thermometers

Electrical thermometers are largely used in industry and research. They possess certain advantages over ordinary mercury-in-glass thermometers.

(*a*) They can be used for measuring high temperatures.

(*b*) The indicating dial can be far removed from the actual thermometer.

(*c*) They can be made to record automatically.

Resistance thermometer. The electrical resistance* of metals increases with a rise in temperature. This change in resistance reduces the current in a simple circuit, and the change is recorded by a suitable current meter. A platinum coil, contained in a silica tube, is usually employed. This is connected in a circuit including a current meter. The latter, which may be in any convenient and remote position, is graduated to read the temperature directly.

The thermocouple. This is another, and different type of electrical thermometer. When a loop is made of two different metals (Fig. 15.4), and the

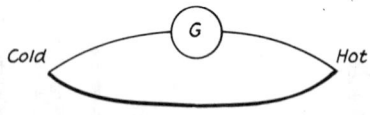

Fig. 15.4 Thermo-junction

two junctions are maintained at different temperatures, an electromotive force (e.m.f.) is created

---

* *See* Chapter 34.

in the loop. This e.m.f. can be measured by a potentiometer method, or by inserting a low-resistance galvanometer in the loop. The thermo-e.m.f. is of the order of only microvolts per degree temperature difference.

The greatest effect is produced by using antimony and bismuth, but any two metals can be used. The e.m.f. is not strictly proportional to the temperature difference of the junctions, but pairs of metals can be chosen which give an e.m.f. proportional to the temperature difference over a wide range of temperatures. Such an arrangement, which usually consists simply of two fine wires, is called a thermocouple and it is used extensively for temperature measurement. Its great advantage is that it can be very small and so have a small thermal capacity. In use, one junction is kept at a fixed temperature, e.g. in melting ice, while the other junction is placed where the temperature is required. It has great value in determining local temperatures such as are met with in engineering and in medicine. One recording instrument is used in conjunction with a dozen or more thermocouples, which may be advantageously placed at particular points.

Pyrometers

A completely different method of measurement has to be employed for very high temperatures, such as that of a furnace. The radiation emitted from an aperture in the wall of the furnace is focused by means of a lens or concave mirror on to a thermocouple which is heated and the effect, as outlined in the preceding section, is recorded using a galvanometer. In another type the colour of the radiation is matched against that of an electrically heated filament. The glow of the filament is adjusted with a rheostat and the current measured.

The calibration of these pyrometers is too complicated for inclusion here, but in practice, the following accurately known temperatures are used:

boiling point of oxygen	−182.97 °C
boiling point of sulphur	444.6
freezing point of silver	960.8
freezing point of gold	1063

The gas thermometer

This is described in the next chapter.

QUESTIONS

1. Define the fixed points 0 and 100 on the centigrade scale. How would you check these readings on a mercury thermometer?

2. A mercury-in-glass thermometer is graduated from $-10\,°C$ to $+110\,°C$. How would you test experimentally whether the 0 and 100 marks were correct? If you found them to be correct, how would you then check the accuracy of the 50 mark?

The thermometer, with its bulb in a tank of hot liquid and with its stem in the air, reads 109.0 °C. When totally submerged, the thermometer gives a different reading. How may each of the following factors have affected this difference? Give a reason for each answer.

(a) Pressure in liquids increases with depth.
(b) More of the mercury has been heated.
(c) More of the glass has been heated. (C)

3. What do you understand by (a) the fixed points of a thermometer and (b) a scale of temperature? Tabulate the advantages and disadvantages of mercury and alcohol as suitable liquids for use in thermometers.

Changes of pressure affect (a) the boiling-point of water and (b) the melting-point of ice. Describe one experiment to illustrate each of these effects. Indicate the kind of result you would expect to obtain. (L)

4. Describe the important features of a mercury thermometer and the way in which you would calibrate it for use over a temperature range of $-5\,°C$ to $105\,°C$.

Explain why in the determination of the upper fixed point, (a) the bulb of the thermometer should not be in contact with the water, (b) the atmospheric pressure should be read.

State, with reasons, the effect on the spacing of the degree markings of (i) halving the volume of the bulb, (ii) halving the diameter of the capillary. (O & C)

5. Explain as far as you can the difference between *heat* and *temperature*, and state *two* properties of matter which would enable it to be used for measuring temperatures.

Give the reasons for the following features of a clinical thermometer: (a) the bulb is small, (b) it is a long cylinder and not spherical, (c) the bulb is made of thin glass, (d) the capillary is of very small cross-sectional area, (e) there is a constriction in the capillary. (O & C)

6. Explain the following terms: *thermometer, fixed points, scale of temperature*.

Tabulate the advantages and disadvantages of mercury and alcohol thermometers and give reasons why water is unsuitable as a thermometric liquid. (O & C)

7. What is meant by the *fixed points* on a temperature scale? Describe briefly how one of them is obtained in practice.

In looking for a good liquid to use in a thermometer would you look for one with a large or small thermal conductivity and coefficient of expansion? Would the specific heat capacity and density of the liquid used in a clinical thermometer have any effect on its behaviour?

What property of a perfect gas has led to the use of the perfect gas scale of temperature? (O & C)

8. Enumerate the advantages of a platinum resistance thermometer.

9. Give reasons why (a) a platinum resistance thermometer is used for measuring the temperature of a furnace, and (b) a thermocouple is used for measuring a local temperature in the body such as that just below the skin.

10. Describe the structure and action of Either a maximum Or a minimum thermometer.

State the properties which a liquid should possess to make it suitable for use in a thermometer.

A liquid-in-glass thermometer contains $0.5\,cm^3$ of mercury at $0\,°C$ and has a bore of cross-section $0.03\,mm^2$. Find the distance between the 0 degree and the 10 degree marks.

[Coefficient of real expansion of mercury$=0.00018/°C$. Coefficient of volume expansion of glass$=0.00003/°C$.]

(JMB)

16 Expansion

Expansion is one effect resulting from adding heat to a body and, in the previous chapter, we have already used this effect to measure temperature. Solids, liquids and gases all expand when heated. A few substances, such as wood, contract,* but this is due to certain changes, mostly chemical, which take place on heating.

The expansion of a gas is shown by use of the simple apparatus shown in Fig. 16.1. Warming the flask by the hands is sufficient to show the expansion. Fig. 16.2 shows the apparatus for demonstrating the expansion of a liquid. When heat is first applied, the level of the liquid goes down a little in the tube before it begins to rise. Why is this?

These experiments show that a gas is more expansible than a liquid. The second one also shows that a liquid is more expansible than a solid, for, if the glass flask had expanded more

than the liquid in it, the liquid level would have fallen in the tube.

The expansion of a solid is very small in comparison with that of a liquid and a gas, but it can be demonstrated by many well-known experiments, e.g. Figs. 16.3 and 16.4.

Expansion is a hindrance to the engineer. If any machine or structure is subjected to temperature changes, allowance has to be made for expansion. Bridges with long spans have to be specially designed to allow for the seasonal changes in length (Figs. 16.5 and 16.6): a special coupling unit is necessary where a hot part of a machine connects with a cold part.

If allowance is not made for expansion and contraction a tremendous force is called into play. This force is demonstrated with the apparatus in Fig. 16.7. A strong iron bar is heated and then clamped tightly between strong pillars as shown. A cast-iron rod, placed through a hole in the bar before heating, is broken when the bar contracts on cooling.

---

* A pure contraction on heating occurs with water between 0 °C and 4 °C (*see* page 138).

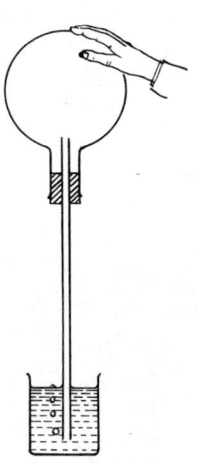

Fig. 16.1 Apparatus to show the expansion of air

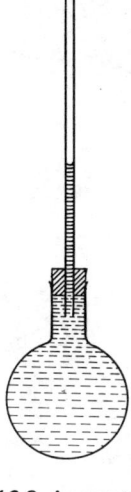

Fig. 16.2 Apparatus to show the expansion of a liquid

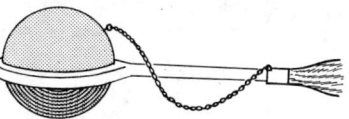

Fig. 16.3 Ball and ring

Ball passes through ring when cold but not when hot

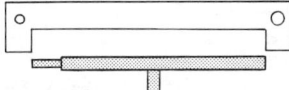

Fig. 16.4 Rod and gauge

Rod fits gauge when cold but not when hot. Circular holes in gauge show increase in diameter of rod

Fig. 16.5 Expansion joint in roadway deck and kerb on Thelwall Bridge on Preston–Birmingham M6 motorway

Fig. 16.6 Expansion link bearing at one end of Thelwall Bridge on the Preston–Birmingham M6 motorway

The roadway deck shown in Fig. 16.5 has to be supported in such a way as to permit a change in length. Fig. 16.6 shows a support at one end which, with its two pivots, enables movement to take place

Fig. 16.7 Apparatus to show the large force exerted on contraction

Fig. 16.8 Buckling of railway lines due to thermal expansion

Fig. 16.9 Heating continuously welded railway lines before fixing

In order to provide smooth travel, the number of joints between the rails is reduced by welding together several lengths of rail. So that the long rail will not buckle under the strain resulting from a normal temperature variation, it is heated during fixing and pegged down while at a temperature of about 21°C. The plate shows the heating of the rail to this temperature with a specially designed burner.

Fig. 16.10 Riveting

The photograph shows a white-hot rivet *A* about to be inserted in the steel structure at *C* of the United Nations building in New York

This force is also illustrated in the bi-metal strip (*see* Fig. 16.12). This consists of two thin metal plates of different metals bonded together. When heated, the more expansible metal must expand more than the less expansible; consequently, forces are set up which bend the strip with the more expansible metal on the outside of the arc. Bending the other way occurs when the strip is cooled.

(d) Temperature control
Many devices for the automatic control of temperature depend upon expansion. Such devices are called thermostats. The Regulo attachment fitted to many gas cookers is one of these. Fig. 16.11 shows the construction of this unit.

Many electrical thermostats depend simply upon the bending of a bi-metal strip when heated to switch the current on or off (Fig. 16.13).

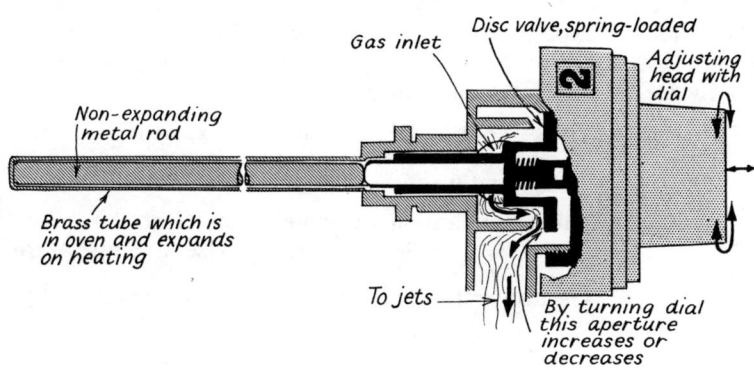

Fig. 16.11 'Regulo' gas thermostat

The brass tube is contained in the oven. On heating, the tube expands, but not the rod inside it, hence the gas supply is gradually closed by the disc valve

Uses of expansion

(a) Thermometers
We have already seen that we use the expansion of a substance as a means of measuring temperature. The commonest thermometers are the mercury-in-glass type, which depend on the expansion of mercury.

The expansion of a short rod can be magnified by a system of levers and so made to give a rough estimate of temperature.

(b) Riveting
Riveting of ships' plates and steel structures is carried out by strongly heating the rivet before fixing it. On cooling, it draws the two sections tightly together (Fig. 16.10).

(c) Fitting a collar on a cylinder
A metal collar can be fitted tightly round a cylinder by making it slightly too small and then heating it to allow it to be fitted. On cooling, a tight fit is obtained by the force of contraction.

The same principle in reverse is employed in fixing a car-engine cylinder liner. It is made slightly larger and is then shrunk by immersion in liquid oxygen at −180 °C. This enables it to be fitted, and when it attains normal temperature it grips tightly.

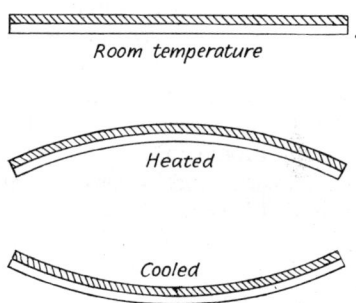

Fig. 16.12 Bi-metal strip, heated and cooled

Metal with the greater coefficient of expansion is shown shaded

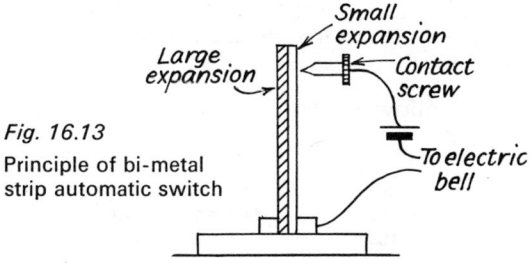

Fig. 16.13

Principle of bi-metal strip automatic switch

When heated, strip bends and makes contact with screw, thus completing circuit

(e) Switches

The automatic cut-off incorporated in the base of an electric kettle consists of a metal rod, which, when heated to a sufficiently high temperature, releases a spring that switches off the mains current.

Substances with very small expansions

The breaking of glass vessels on heating is due to local expansion of the heated parts. Glass, being a poor conductor of heat, does not allow the quick distribution of the heat which is necessary to produce uniform expansion. 'Pyrex' glassware can be used for cooking, as its expansibility is very much less than that of ordinary glass.

Balance wheels in watches and pendulums in clocks are now often made of invar, a metal with a very small expansibility, instead of being of special design to compensate for expansion.

Measurement of expansion

Expansion is due principally to the molecules vibrating more vigorously when heated, and appropriating more space. Whatever may be the complete explanation of all the factors involved, it is important in science and in many branches of industry to be able to calculate accurately the size of a body at any temperature.

Let us first consider the increase in length of a bar when heated. If a bar of length 1 m increases in length by a m when heated through 1 °C, a similar bar of length 10 m will expand $10a$ m. The fractional expansion in each case is a. Thus, a bar of length l units will expand la units when heated through 1 °C; a is called the coefficient of linear expansion and is defined in the following way.

The coefficient of linear expansion is the increase in length of unit length when heated through 1 °C, i.e. 1 K.

Or the following wording is better if it can be understood.

The coefficient of linear expansion is the fractional increase in length per degree Celsius (kelvin) rise in temperature.

A few coefficients are given in the table at the top of page 138.

If a bar of length l m expands la m per degree rise in temperature, its expansion for t degrees rise will be lat m.

Therefore,

expansion = length × coefficient × rise in temperature

This simple relationship is all that is required for finding the expansion of a solid.

A metal plate expands in all directions and so increases its area.

A solid increases its volume. A hollow vessel expands as much as would a solid of the same size.

For the volume expansion of a solid, we use the cubical coefficient γ.

The coefficient of cubical expansion is the fractional increase in volume per degree rise in temperature.

Volume expansion = volume × coefficient of cubical expansion × rise in temperature

The above theory can be expressed mathematically.

Let l_1 = length of rod at $t_1°$
l_2 = length of rod at $t_2°$
a = coefficient of linear expansion.

From above,

expansion = length × coefficient × rise in temperature.

If the rod is heated from t_1 to t_2,

$$l_2 - l_1 = l_1 a(t_2 - t_1)$$

hence
$$l_2 = l_1\{1 + a(t_2 - t_1)\} \qquad (1a)$$

This is a formula for the length at $t_2°$ when the length at $t_1°$ is known.

A similar formula for volume is:
$$V_2 = V_1\{1 + \gamma(t_2 - t_1)\} \qquad (1b)$$

where γ is the coefficient of cubical expansion.

Determination of the coefficient of linear expansion

The expansion of a metal bar can be measured using a device which measures a small change in length, such as a micrometer screw gauge. The bar is enclosed in a jacket through which cold water and then steam can be circulated. Alternatively, if the metal is in the form of a tube, the cold water and steam can be passed through it

(Fig. 16.14*a*). In this case an asbestos tube around the metal tube prevents heat loss. Keeping one end of the tube fixed, the small expansion is measured by applying a micrometer across the

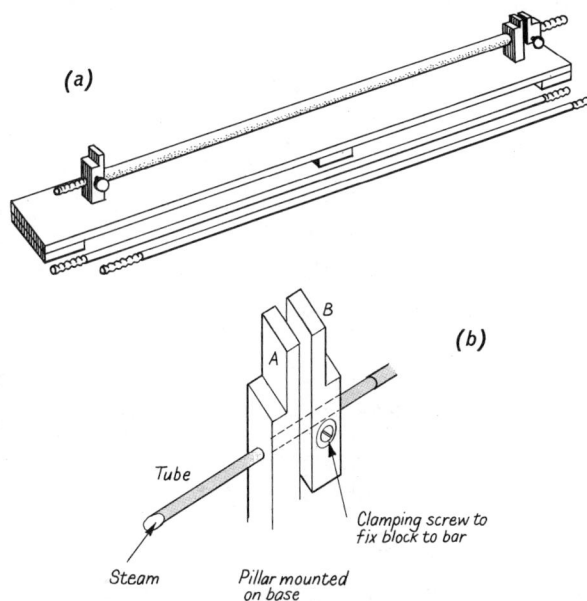

Fig. 16.14 Apparatus for measuring the coefficient of expansion of a metal in the form of a tube

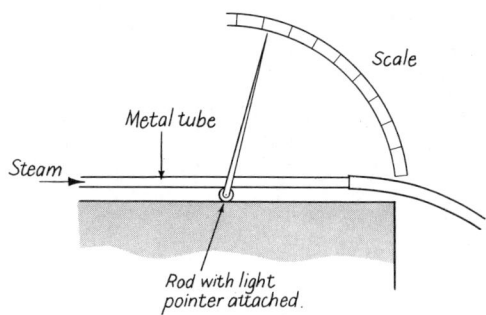

Fig. 16.15 Apparatus for measuring expansion

faces *A* and *B* of the blocks mounted as shown in Fig. 16.14*b*. The following are typical readings taken with this apparatus.

Working length of copper tube   =   0.496 m
Initial temperature of tube   =   16 °C
Final temperature of tube   =100 °C
Initial reading of micrometer   =   4.62 mm
Final reading of micrometer   =   5.34 mm
Expansion   =   0.72 mm

$$\text{Expansion}=\text{length} \times \text{linear coefficient} \times \text{rise in temperature}$$

$$0.72 = 496 \times a \times 84$$

$$\underline{a = 0.000\ 017/K}$$

The coefficient of expansion of copper is 0.000 017/K

A very simple method which gives a reasonable value for the coefficient is illustrated in Fig. 16.15. The jacketed tube has one end resting on a rod to which is attached a long light pointer. The other end is wedged against a heavy stop which prevents

it expanding in that direction. A small movement of the supported end causes an appreciable movement of the pointer round the scale. From the diameter of the rod and the length and movement of the pointer, the expansion is easily calculated.

Expansion of liquids

When a liquid is heated, the vessel containing it expands also, and the observed expansion therefore is not as great as it should be. The coefficient determined when the expansion of the containing vessel is ignored, is called the coefficient of apparent expansion.

The coefficient of absolute expansion of a

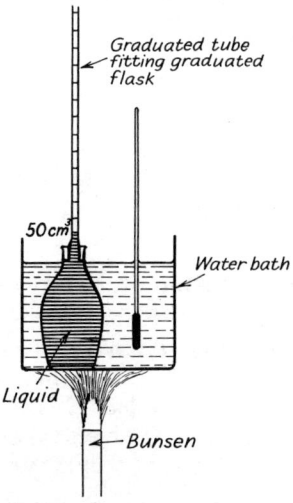

Fig. 16.16 Apparatus for measuring the coefficient of apparent expansion of a liquid

COEFFICIENTS OF LINEAR EXPANSION/K	
Aluminium	0.000 025 5
Copper	0.000 016 7
Iron (wrought)	0.000 012
Steel	0.000 011
Platinum	0.000 008 9
Glass	0.000 009
Glass (Pyrex)	0.000 003

COEFFICIENTS OF CUBICAL EXPANSION/K AT 20 °C	
Water	0.000 21
Mercury	0.000 18
Paraffin oil	0.000 9
Turpentine	0.000 96

liquid is approximately equal to the sum of the coefficient of apparent expansion and the coefficient of cubical expansion of the material of the vessel. The coefficient of apparent expansion can be determined using a flask with a graduated stem (Fig. 16.16). The flask, filled with the liquid, is first put in cold water, which is heated up and the expansion read off on the graduated tube. Substitution of the readings in the formula for volume expansion gives the coefficient of apparent expansion.

Relation between the density and the coefficient of expansion

The expansion of a substance is accompanied by a decrease in the density. As mass remains constant, we have:

$$\text{mass} = \rho_1 V_1 = \rho_2 V_2$$

where ρ_1 and ρ_2, V_1 and V_2 are the densities and volumes respectively at temperatures t_1 and t_2.

$$\therefore \qquad \frac{\rho_1}{\rho_2} = \frac{V_2}{V_1} = 1 + \gamma(t_2 - t_1)$$

This relationship is useful in finding the coefficient of expansion of a liquid: determinations of the density of the liquid at two temperatures are all that is required.

Peculiar expansion of water*

Most substances expand when heated, and contract when cooled, but water is an exception to this rule. If the volume of water in a graduated flask is observed as it is cooled, it is found that cooling between 4 °C and 0 °C causes expansion (Fig. 16.17). Thus, at 4 °C water occupies its smallest volume,

---

* This phenomenon is attributed to a change in the form of the water molecule occurring between 0 °C and 4 °C.

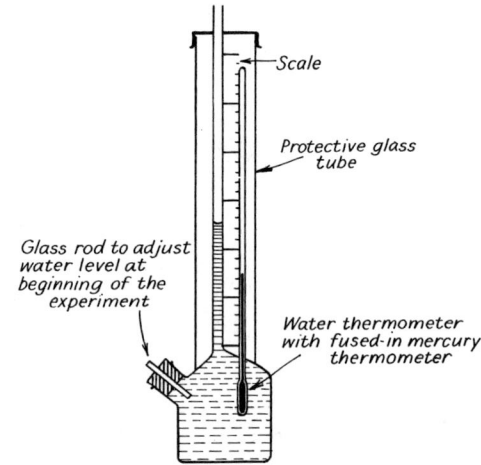

Fig. 16.17 Apparatus for showing volume change of water from 0 °C to 10 °C

The apparatus is placed in melting ice. Allowance for expansion of the glass can be made by putting a specific volume of mercury in the bulb

that is, at this temperature its density is a maximum.

This simple fact has far-reaching consequences. It is due to this that ice forms first on the surface of water. The cold air above the surface of a pond cools the water, which then becomes denser and forces the warmer, less dense water to the surface. This convection current* continues until all the water in the pond is cooled at 4 °C. After this, further cooling of the surface water causes its density to fall and so it remains on the surface, eventually becoming ice (*see* Fig. 16.18). As ice is a bad conductor† of heat, the layer of ice on the surface increases in thickness very slowly. In the severest winter in England, any but a very shallow pond has water under the layer of ice, so enabling the pond life to survive. If water did not attain its

---

* *See* page 168.
† *See* Chapter 19.

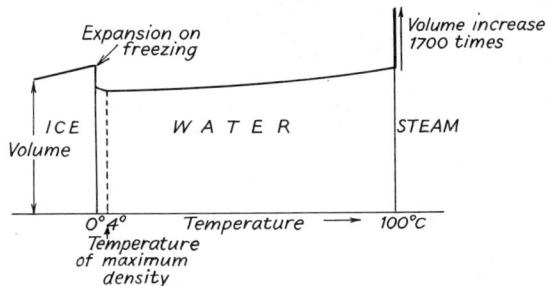

Fig. 16.18 Graph showing volume change of water with temperature (not to scale)

maximum density at a temperature above 0 °C, convection would continue until 0 °C, when the pond would become completely frozen.

Expansion of gases

The restrictions put upon the movements of the molecules in solids and liquids due to the mutual forces of attraction between them (cohesion) prevent the molecules exerting an outward pressure. As gaseous molecules, at pressures not far from normal, are free from the effects of their neighbours, a gas always fills the containing vessel and exerts a pressure on the walls. This pressure is created by the impacts of the moving molecules, i.e. by the molecules' change in momentum. It depends upon the mass and upon the speed of the molecules and also upon the number of molecules

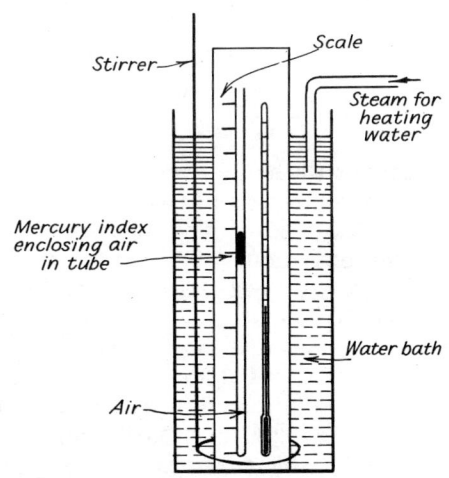

Fig. 16.19 Apparatus for measuring the expansion of air at constant pressure

per unit volume. This amounts to saying that the pressure depends upon the volume and the temperature of the gas.

In considering the thermal expansion of a gas, we must take into account the pressure, and hence we have three variables—

pressure, volume, and temperature.

Pressure has little effect upon the volumes of solids and liquids, so we have not troubled about it previously.

To study the variations of three quantities which may all be changing at the same time, we keep each one constant, in turn, and vary the other two.

1. Pressure constant: variation of volume with temperature

Enclose a volume of air in a narrow glass tube, sealed at one end, by a short length of mercury (Fig. 16.19). Mount this along a scale with a thermometer attached. Place the complete apparatus in a water-bath and note the length of the air column at different temperatures of the bath. It is advisable to stop heating, and to stir well before taking a reading, so as to make certain that the air has attained the temperature of the water. (The pressure is constant throughout the experiment, for it is equal to atmospheric pressure plus the pressure of the short mercury column.) Tabulate the results and plot a graph of the volume against the temperature. This graph is a straight line as shown in Fig. 16.20. From it the coefficient can most easily be determined by reading off values for V_0 and V_{100}, and substituting in

$$V_{100} = V_0(1 + 100\gamma)$$

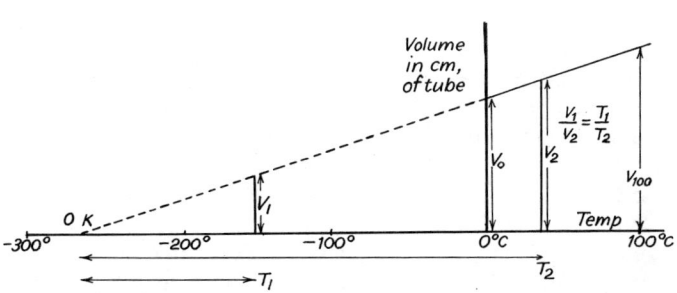

Fig. 16.20 Graph of volume against temperature of a gas at constant pressure

γ comes out to be $\dfrac{1}{273}$ per °C for *all* gases and the result is expressed in Charles's law.

> **The volume of a fixed mass of gas at constant pressure increases by 1/273 of its volume at 0 °C for each degree rise in temperature.**

2. Volume constant: variation of pressure with temperature

In this case we have to find the variation in the pressure of a gas when the temperature is changed, and we therefore require some form of pressure gauge attached to the apparatus. An open mercury manometer is simple to use. The complete apparatus is shown in Fig. 16.21. The dry gas is enclosed

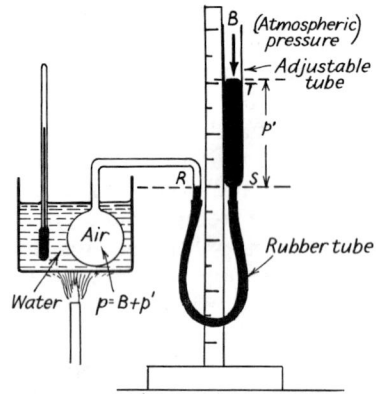

Fig. 16.21 Apparatus for measuring the change in pressure of a gas at constant volume

Mercury is always brought to level *R* by alteration of the tube on right.

Pressure at *R* = pressure at *S*
= atmosphere pressure + pressure of column *ST*

in the bulb, which is connected by a narrow tube to the mercury manometer. Each time, before taking a reading of the manometer, the mercury must be brought back to a fixed point *R*, in order to satisfy the condition of constant volume. This is done by altering the height of the tube *ST*, which is adjustable. The pressure in the bulb is equal to that at *R*, which is atmospheric pressure plus the pressure due to the column *ST*. The gas in the bulb is heated by means of a water-bath. Sufficient time should be given in taking each reading for the gas inside the bulb to attain the temperature of the water in the bath.

The readings should be tabulated and the graph

of pressure against temperature plotted (Fig. 16.22). This graph is a straight line and from it we can calculate the coefficient of increase of pressure.

> **The coefficient of increase of pressure is the fractional increase in pressure of the pressure at 0 °C per degree rise in temperature.**

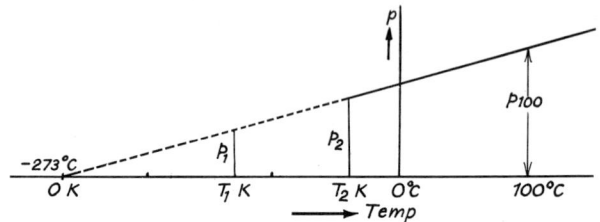

Fig. 16.22 Graph of pressure against temperature of a gas at constant volume

It is calculated from the graph by reading off p_0 and p_{100}, and substituting in

$$p_{100} = p_0(1 + 100\gamma')$$

γ' is found to be the same for all gases and equal to the coefficient of expansion, viz. $\dfrac{1}{273}$ per °C. The result is stated in the form of a gas law.

> **The pressure of a fixed mass of gas at constant volume increases by 1/273 of the pressure at 0 °C for each degree rise in temperature.**

Absolute zero

In the last chapter it was recorded that Kelvin devised an absolute scale of temperature, i.e. a scale independent of the performance of any working substance, which is based on thermodynamic principles. The zero of this scale is −273.16 °C.

$$K = 273 + °C$$
$$T = 273 + t$$

This corresponds to the very rock-bottom of temperature, i.e. the temperature at which the molecules have lost their kinetic energy. We cannot make simple conclusions about this temperature for, at such low levels of energy, the molecules have unusual properties and are not in the gaseous state. Because it so happens that the graphs of the variations of the pressure and volume with temperature over a large range, when produced back, meet the axis at about −273 °C (Figs. 16.20 and 16.22) it is often stated that the pressure and volume of a gas vary as the 'absolute' temperature. When the expansions of solids and liquids are so treated, the intercepts on the temperature axis are very much lower.

3. Temperature constant: variation of volume with pressure

The relationship between volume and pressure can be studied by compressing a gas enclosed in a strong glass tube by a mercury column, and

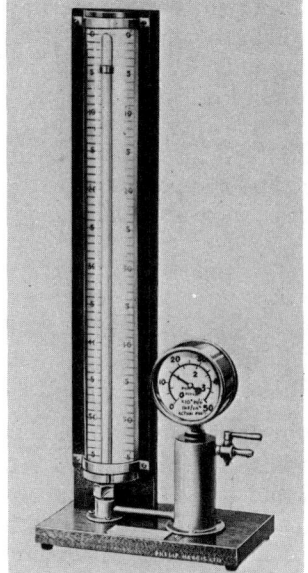

Fig. 16.23

Apparatus for measuring variation of volume of a gas with pressure at constant temperature (Boyle's law)

measuring the pressure with a direct-reading pressure gauge (Fig. 16.23). This apparatus is not so popular as that shown in Fig. 16.24, in which the pressure of the gas enclosed in the tube R is

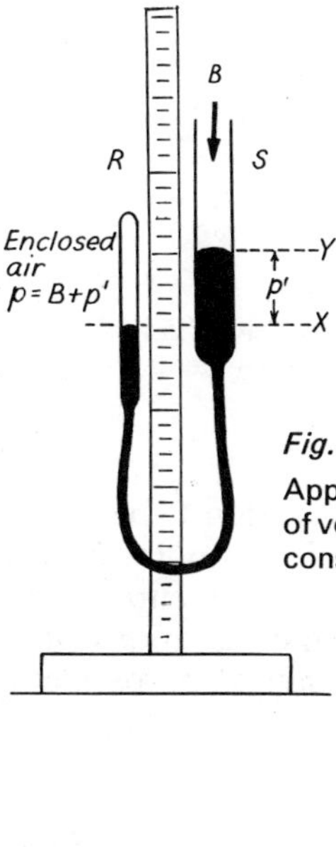

Fig. 16.24

Apparatus for measuring variation of volume of a gas with pressure at constant temperature (Boyle's law)

measured by an open manometer S. The volume of the gas is measured in centimetres of tube, and the pressure in the gas is equal to that at X, which is atmospheric pressure B plus the pressure p' due to the column XY. The pressure can be varied by altering the position of the adjustable tube S. Some negative values of p can be included, which are obtained by lowering Y below X. The results should be tabulated as shown.

Atmospheric Pressure $B=$ cm of mercury.

Volume V /cm of tube	p' /cm of mercury	$p=B+p'$ /cm of mercury	$p.V$

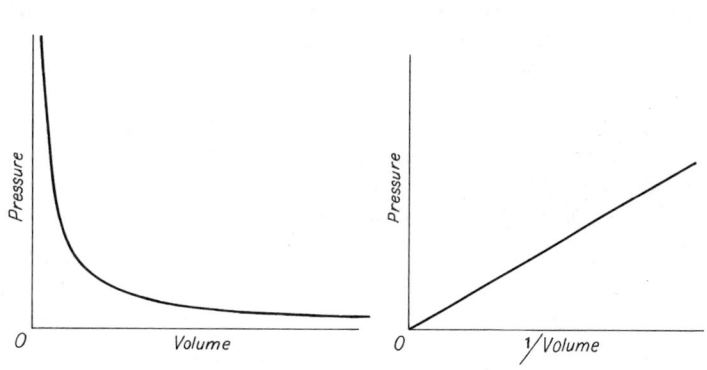

Fig. 16.25 Graphs of pressure against volume, and pressure against 1/volume for a gas at constant temperature

The product $p.V$ is a constant which shows that the volume and pressure are inversely proportional to each other at constant temperature (*see* Fig. 16.25).

Summary of the gas laws

1. Provided the pressure of a fixed mass of gas is kept constant, the volume increases by 1/273 of its volume at 0 °C for each degree rise in temperature.

$$V \propto 273 + t$$

2. Provided the volume of a fixed mass of gas is kept constant, the pressure increases by 1/273 of its pressure at 0 °C for each degree rise in temperature.

$$P \propto 273 + t$$

3. Provided the temperature of a fixed mass of gas is kept constant, the volume varies inversely as the pressure.

$$p_1V_1 = p_2V_2$$

Combination of the gas laws

$$\frac{p_1V_1}{273 + t_1} = \frac{p_2V_2}{273 + t_2}$$

For a perfect gas we can deduce

$$pV = RT$$

where T is the temperature on the kelvin scale.

This is known as the gas equation, and R is called the *gas constant*. If V is the volume of a mole of gas, then R is the gas constant per mole and is the same for all gases (8.3 J/mol K).

Normal temperature and pressure

As the volume of a gas is dependent upon its temperature and pressure, a standard method of quoting its volume is to state that which it would occupy at a pressure of 76 cm of mercury and a temperature of 0 °C. This is called normal or standard temperature and pressure (n.t.p. or s.t.p.).

Reduction of gaseous volumes to n.t.p. is very important in chemistry (*see* Example).

Example

In an experiment, a quantity of hydrogen occupied a volume of 2×10^{-3} m³ when liberated at a pressure of 95 cm of mercury and a temperature of 13 °C. Find what volume it would occupy at n.t.p.

$p_1 = 95$ cm of mercury	$p_2 = 76$ cm of mercury
$V_1 = 2 \times 10^{-3}$ m³	$V_2 = V$ m³
$T_1 = 286$ K	$T_2 = 273$ K

Note the units for the pressures and the volumes must always be the same and the temperatures must always be in ° Absolute.
Substituting in

$$\frac{p_1V_1}{T_1} = \frac{p_2V_2}{T_2}$$

We have

$$\frac{95 \times 2 \times 10^{-3}}{286} = \frac{76 \times V}{273}$$

Hence

$$V = 2.39$$

The volume of the hydrogen at n.t.p. is 2.39 m³.

The gas thermometer

The variations of the pressure and volume of a gas with temperature have been described above using a mercury thermometer to measure the temperature. As mentioned in the previous chapter these variations are very close to an absolute standard. On the absolute scale the value of a property at 50 °C is exactly the mean of the values of the property at 0 °C and 100 °C. The standard thermometer at the International Bureau is a gas thermometer using hydrogen, nitrogen, or helium. It is a constant volume thermometer. In principle it is equivalent to the apparatus in Fig. 16.24. The pressure of the gas is measured first with the bulb in ice shavings and then in a double-walled steam bath. From the two readings, p_0 and p_{100}, of the pressure any temperature can then be measured using,

$$t = \frac{p_t - p_0}{p_{100} - p_0} \times 100$$

Molecular explanation of expansion

We have seen that the atoms in a solid are packed quite closely and they vibrate about their positions of stability. Thus an atom may be vibrating between limits corresponding to A and B (Fig. 16.26), on

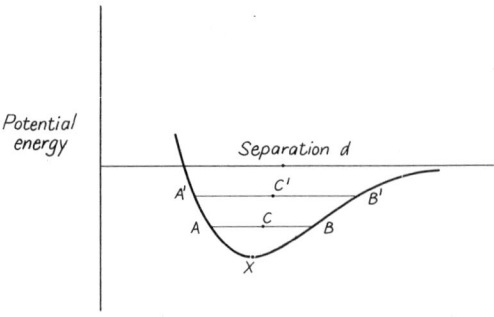

Fig. 16.26 Graph of potential energy of a molecule against distance from an influencing molecule

its energy curve. The mean position C between A and B will be near to the minimum point X of the curve. If the temperature of the substance is increased, the amplitude of the vibrations of the atoms will increase and the limits of the oscillation of the atom considered will move to A' and B'. The mean C' of this new vibration will have a larger value of the separation d than that for AB, due to the curve not being symmetrical. Hence the atom moves farther away from its neighbour and we get expansion. The effect will be different for different substances and so coefficients of expansion will be different. The case for liquids will be something similar to that for solids but the movement will be larger.

The case for gases is very different, for here there is no cohesion at ordinary pressures. The increase in temperature means an increase in the kinetic energy of the molecules. This will cause changes in both the pressure and the volume. These changes are the same for all gases.

QUESTIONS

1. How would you show experimentally that, when heated (a) a liquid expands more than a solid, and (b) a gas expands more than a solid?

2. Explain the following:
(a) warming the neck of a glass bottle often loosens the stopper;
(b) a platinum wire can be sealed through the wall of a glass tube.

3. Calculate by how much the length of the standard metre bar would expand if its temperature were increased by 50 K.
[Coefficient of expansion of platinum $= 8.9 \times 10^{-6}/\text{K}$.]

4. Describe an experiment which demonstrates that different metals possess different coefficients of expansion and explain with a diagram *one* everyday application of this fact.
An iron rod is 1 m long at 0 °C. What must be the length at 0 °C of a copper rod if the difference between its length and that of the iron rod is not to vary with temperature?
[Coefficient of expansion of copper $= 17.0 \times 10^{-6}/°\text{C}$; coefficient of expansion of iron $= 11.9 \times 10^{-6}/°\text{C}$.] (O & C)

5. Define *coefficient of linear expansion* and describe in detail how you would measure its value for copper.
An aluminium piston slides in a steel cylinder of diameter 3.0 cm at 15 °C. If the difference in diameter of piston and cylinder is 0.005 cm at this temperature, find at what temperature the gap would become zero.
[Coefficients of linear expansion: steel, $1.1 \times 10^{-5}/°\text{C}$; aluminium, $2.6 \times 10^{-5}/°\text{C}$.] (O & C)

6. Describe a method of measuring the coefficient of linear expansion of copper, supplied as a rod or tube. Draw a diagram of the apparatus including the measuring device, and explain how the result is calculated.
Give *two* examples, with short explanations, of structures where precautions to minimize the effects of thermal expansion are essential. (O & C)

7. What is meant by the statement 'the coefficient of linear expansion of zinc is 0.0000260/°C'?
A zinc rod, exactly 1 m long at 10 °C, is 2.30 mm shorter than a copper rod at the same temperature. When both rods are at 260 °C their lengths are equal. Calculate the coefficient of linear expansion of copper. (C)

8. Distinguish between the real and apparent coefficients of expansion of a liquid, and describe a simple experiment to demonstrate the part played by the expansion of the vessel.

A steel can, of volume 2000 cm³ is filled to the brim with paraffin at 15 °C. Find the volume that overflows if the temperature rises to 25 °C.
[The real coefficient of expansion of paraffin is 0.00097/°C, and the coefficient of *linear* expansion of steel is 0.00001/°C.] (O)

9. How would you demonstrate the difference between the real and the apparent expansion of a liquid?
Describe an experiment to measure the apparent coefficient of expansion of a liquid.
A glass vessel of volume 25.00 cm³ contains 23.50 cm³ of paraffin at 15 °C. At what temperature does the liquid just fill the vessel?
[Real coefficient of expansion of paraffin is 0.000924, and coefficient of linear expansion of glass is 0.000008/°C.] (O)

10. Define *coefficient of expansion*. Distinguish between the coefficients of *real* and *apparent* expansion of a liquid, and state a relation between them.
A glass bottle of volume 10.0 cm³ at 0 °C contains 132.6 g of mercury of density 13.60×10^3 kg/m³ at 0 °C. What is the volume of this mercury at 0 °C? Assuming the coefficients of volume expansion of mercury and glass to be $18 \times 10^{-5}/°\text{C}$ and $3 \times 10^{-5}/°\text{C}$ respectively find the temperature at which the mercury just fills the bottle. (L)

11. Define *coefficient of volume expansion*. Distinguish between the coefficients of *real* and *apparent* expansion of a liquid in a glass vessel.
A mercury-in-glass thermometer has a bulb of internal volume 0.30 cm³ and a tube of internal cross-sectional area 2.5×10^{-4} cm². If the 0 °C division on the tube is just above the bulb find (a) the apparent increase in volume of the mercury when the temperature of the thermometer rises from 0 °C to 100 °C, (b) the distance between the 0 °C and 100 °C divisions.
[The coefficient of apparent expansion of mercury in glass may be taken as $16 \times 10^{-5}/°\text{C}$.] (L)

12. Define *coefficient of linear expansion* and describe how you would determine its value for brass in the form of a rod or tube.
An iron tyre of diameter 50 cm at 15 °C is to be shrunk on to a wheel of diameter 50.35 cm. To what temperature must the tyre be heated so that it will slip over the wheel with a radial gap of 0.5 mm?
[Coefficient of linear expansion of iron, 0.000012/°C.] (O & C)

13. Explain why, in the case of a liquid, we distinguish between the coefficients of apparent and real expansion, whereas the distinction is not usually necessary in the case of gases.

Describe an experiment to determine the coefficient of expansion of air at constant pressure. (JMB)

14. Describe an experiment to find the coefficient of increase of pressure of a gas at constant volume. The pressure of the air in a motor tyre is 2×10^5 N/m^2 when its temperture is 17 °C. During the running of the car the temperature rises to 46 °C. Find the new pressure of the air, assuming that the tyre does not expand. (JMB)

15. State Boyle's law. Draw a labelled diagram of the apparatus you would use to verify the law. Do not describe the experiment, but show by headed columns how you would tabulate your readings. Show how you would use your readings to obtain the verification.

The barrel of a pump has an effective volume of 100 cm^3. The pump is used to drive air into a cylinder of capacity 1000 cm^3. If the pressure of the air in the cylinder is the barometric pressure (76 cmHg) calculate the pressure in it after the first stroke of the pump, assuming that there is no temperature change. (JMB)

16. The following results were obtained, using a constant volume air thermometer, in an experiment to find the relation between the pressure and the temperature of a mass of air at constant volume.

Temperature (°C)	8	32	54	78	99
Pressure (cmHg)	78.3	84.9	91.1	97.7	103.7

Draw a graph to show the relation between the pressure and the temperature of the air and from it find the pressures of the air at 0 °C and 100 °C. Use these two values of the pressure to calculate the coefficient of increase of pressure of air at constant volume.

Why is a *capillary* tube used in the apparatus to connect the bulb containing the air to the U-tube containing the mercury?

What effect has the expansion of the bulb upon the value of the coefficient? (JMB)

17. State Boyle's law, and describe an experiment which tests its validity for a given gas.

Air is compressed into a 3.5 dm^3 vessel by means of a pump. Each stroke results in the intake of 100 cm^3 of atmospheric air and its transference to the vessel. Inside the vessel the pressure is one atmosphere at the start; find the pressure (in atmospheres) after 30 strokes of the pump. (O)

18. Describe the constant volume air thermometer, and explain how you would use it to find the temperature of a water bath.

An electric light bulb contains argon at a pressure of 20 cm of mercury at 15 °C. Assuming that, when the lamp is alight, the average temperature of the argon is 150 °C, calculate the pressure inside the bulb when the lamp is in use. (O)

19. State Charles's law and describe how you would proceed to verify it experimentally.

Explain how measurements on the expansion of gases can lead to the idea of absolute temperature.

A certain motor-car engine ejects 25 dm^3/s of gas at a temperature of 303 °C and a pressure of 380 cmHg. Assuming that the gas laws are obeyed, calculate the volume ejected per second when measured at 15 °C and 1 atmosphere pressure. (O & C)

20. Describe how you would investigate the way in which the volume of a fixed mass of gas heated at constant pressure depends on the temperature.

The density of air at s.t.p. is 1.29 kg/m^3. Find the volume occupied by 1 g of air at a temperature of 100 °C and a pressure of 76 cmHg. (O)

21. State Boyle's law and describe how you would carry out an experiment to test its validity.

A tank of volume 10 dm^3 is being filled with compressed air by a simple pump. At each stroke the pump takes 500 cm^3 of air at atmospheric pressure and, after compression, forces it all into the tank. Assuming that the temperature remains constant throughout, calculate how many strokes of the pump are required to raise the pressure in the tank from 1 atmosphere to 2.5 atmospheres. (O & C)

22. What is understood by (i) s.t.p., (ii) absolute gas scale of temperature?

A fixed mass of air is heated, (*a*) without altering the pressure, and (*b*) without altering the volume. State in each case the law governing the change which takes place.

3.0 dm^3 of air at 0 °C and a pressure of 1.0 atmosphere are heated in such a way as to keep its volume unchanged. Calculate the temperature of the air when its pressure becomes 5.0 atmospheres.

The temperature is now kept constant and the pressure reduced again to 1.0 atmosphere. What will be the resulting volume?

If, finally, this last pressure be maintained, calculate the temperature for which the volume will be reduced to 2.0 dm^3. (L)

23. State the law which indicates the way in which the pressure of a fixed mass of gas varies with its temperature as recorded with a mercury thermometer, when its volume remains constant.

Describe an experiment to verify the law.

A vessel used for storing compressed air is fitted with a safety valve which lifts at a pressure of 1.0×10^4 N/m^2. It contains air at 17 °C and 8.75×10^3 N/m^2. If it is heated, at what temperature will the valve lift? (L)

24. State the law governing the change of pressure which takes place when a fixed mass of gas is heated in a closed vessel whose volume remains constant. Describe an experiment by which your statement could be verified.

A tin, closed by an air-tight lid, contained air at a pressure of 1.0×10^5 N/m^2 and at a temperature of 12 °C. It was heated in a water bath until the lid blew off. Calculate the pressure required to do this if the temperature of the bath was then 88 °C. (L)

25. A fixed mass of gas is kept at a constant temperature. How does (*a*) the volume, (*b*) the density, vary with the pressure?

Describe, carefully, an experiment to show how the volume varies with the pressure.

A vessel of capacity 1.0 m³ contains air at a pressure of 10 atmospheres. If there is a small leak in the vessel, what mass of air eventually escapes into the atmosphere?

[Take the density of air at atmospheric pressure as 1.3 kg/m³.] (S)

26. State Boyle's law. Describe how you would test the law experimentally for air, over a pressure range of about 0.5 to about 1.5 atmospheres.

A glass tube, of uniform cross-section 0.75 cm², is corked at one end and contains 40.0 cm² of air at atmospheric pressure (75.0 cmHg). The tube, held vertically with its open end downwards, is pushed below the surface of mercury. Calculate the change in atmospheric pressure which would be necessary to equalize the mercury levels inside and outside the tube, the open end of the tube being then 2.0 cm below the surface. (Assume the temperature to remain constant.) (C)

27. A gas container of volume 10 dm³ contains gas at a pressure of 120 atmospheres.

(a) All the gas in the container is used to fill a balloon of volume 240 dm³. Calculate the pressure in the balloon.

(b) Calculate the volume of the gas when all of it is allowed to escape into the atmosphere. (C)

28. Make a diagram of an apparatus which enables you to use the expansion of a gas as a means of measuring temperature.

Explain how you would calibrate it to enable you to do this. Mention one advantage of using a gas instead of a liquid as the expanding substance.

The gas in a gas thermometer occupies 54.0 cm³ at 0 °C and 74.0 cm³ at 100 °C. What are the temperatures when it occupies (a) 62.0 cm³, (b) 46.0 cm³, if its pressure is constant throughout? (O & C)

29. State Charles's law and Boyle's law.

A cylinder 8.0 cm long and 3.0 cm² cross-sectional area is closed at both ends. Midway between the ends is a thin frictionless air-tight piston. Each half of the cylinder contains air at a pressure of 1.5×10^5 N/m² and a temperature of 27 °C. The piston is now moved 2.0 cm along the cylinder. Calculate the force needed to hold it in the new position with the temperature unchanged.

To what temperature must the air in the larger compartment be heated in order to make the pressure there equal to the pressure in the smaller (unheated) compartment? (O & C)

30. A mass of gas is contained in a cylinder fitted with an air-tight frictionless piston. Describe the change in volume of the gas as the temperature is increased from −50 °C to 150 °C, the pressure on the gas remaining constant, and draw a graph to represent this change in volume with temperature. State how the volume of the gas will vary, at constant temperature, when the pressure is changed. Give the gas equation, making clear the meaning of the symbols used.

Describe how you would calibrate a constant volume gas thermometer to measure temperatures of the order of 50 °C (no other thermometer being available). Discuss one advantage possessed by the thermocouple over the constant volume gas thermometer when used to measure the same temperatures. (C)

31. Describe as fully as you can the kinetic picture of a gas, and show how it accounts for the change in pressure when the volume of a gas is increased, as in a Boyle's law experiment.

What explanation can be given, in terms of kinetic theory, of the rise in temperature of a gas when it is compressed, as in a bicycle pump?

The pressure inside a bicycle tyre can be increased either (a) by pumping, or (b) by heating alone. What explanation can you give to account for the rise in pressure in each case? (O & C)

17 Quantity of heat

In Chapter 14 we established that heat is a form of energy and so is measured in joules, the SI unit of energy. But measurements of heat quantities were undertaken in the eighteenth century before the equivalence of heat and energy was established.

At this time heat was measured in an arbitrary unit based on the performance of water. This unit, called the *calorie*,* was defined:

the quantity of heat required to raise the temperature of one gramme of water from 15 °C to 16 °C is one calorie.

The temperature range is given because the amount of heat varies slightly according to the particular range taken.

The calorie is still used as a unit of heat in some branches of science.

$$1 \text{ calorie} = 4.1855 \text{ joules}$$

---

* The kilogramme or great calorie (written sometimes Calorie) is 1000 calories. *See* Chapter 14.

The corresponding British unit is the British thermal unit, Btu, and is equal to the amount of heat required to raise the temperature of 1 lb of water through 1 deg. F.

$$1 \text{ therm} = 100\,000 \text{ Btu}$$

These units are obsolescent.

Measurement of heat

The measurement of heat quantities is important in industry as well as in scientific research. Thus it is important to be able to measure how much heat is given out when a quantity of a substance is burnt and how much heat is required to raise the temperature of a body.

Fuels

Many fuels are bought according to the amount of energy which they produce when burnt. For example, gas is charged at so much per unit of energy.*

The calorific value of a fuel is the amount of energy liberated when unit mass is burnt (unit: J/kg).

Sometimes unit volume is used in place of unit mass for a liquid fuel (unit: J/m^3).

The calorific value of coal or fuel oil is very important to the large consumer. At a generating station, constant check has to be kept on the

---

* The therm is the unit in present use, but this is to be replaced by the MJ.

quality of the coal by determining the calorific value of each consignment.

In biology too, calorific values are important. Energy is supplied to the body from food which is slowly oxidized or burned in the process of digestion. It is estimated that the body requires about 12×10^6 J each day for this purpose. Sufficient food of the correct type should be eaten to make up this amount of heat. Some athletes recommend eating sugar during a strenuous game to provide quickly the energy necessary. Below is a table of the calorific values of several foods.

FOOD	CALORIFIC VALUE
	J/kg
Carbohydrate	17×10^6
Fat	38×10^6
Protein	17×10^6
Milk	2.9×10^6
Cheese (Stilton)	21×10^6
Egg	6.7×10^6
Flour	15×10^6
Brazil nut	31×10^6

Heat capacity

Two kilogrammes of water (or two kilogrammes of iron) require twice as much heat as does one kilogramme of water (or one kilogramme of iron) to raise their temperature by the same amount.

The thermal capacity of a body is the quantity of heat required to change its temperature by one degree.

Obviously the heat capacity will depend upon the mass, also upon the substance. The heat capacity of one kilogramme of water is 4.19 kJ/K and of one kilogramme of copper 0.4 kJ/K. It is necessary to know, for many reasons, the heat capacity of unit mass of a substance, and this quantity is called the *specific heat capacity*.

Specific heat capacity

The specific heat capacity of a substance is the heat required to change the temperature of unit mass of the substance by one degree.

It is the heat capacity of unit mass. (Unit: J/kg K.)

If c is the specific heat capacity of a substance,

the value of c is the number of joules required to raise the temperature of 1 kilogramme through 1 °C (or 1K).
mc joules are required to raise the temperature of m kilogrammes through 1 °C.
mct joules are required to raise the temperature of m kilogrammes through t °C.

i.e. **heat required to warm a substance = mass × sp. ht. cap. × rise in temperature**

Similarly,

heat given out when a substance cools = mass × sp. ht. cap. × fall in temperature

Not only in pure science, but in industry also, are specific heat capacities very important.

Determination of specific heat capacity

The direct determination of specific heat capacity has to be carried out by supplying a known amount of energy to a body and determining the rise in temperature produced. A method must be adopted which allows all the energy supplied to be converted into heat. In practice, however, some heat is always lost, but precautions are taken to keep this to a minimum.

If the specific heat capacity of one substance is found, that of other substances can be found by comparison.

Mechanical methods

Joule's churning experiment, described in Chapter 14, is a method of determining the specific heat capacity by finding the rise in temperature of a known mass of water produced by a known amount of mechanical energy.

A simplified form of the apparatus devised by Callendar in this study of the conversion of mechanical energy into heat is shown in Fig. 17.1. It is now used to find the specific heat capacity of a good conductor of heat such as a metal. The metal is made in the form of a cylinder around which a cord passes. Tension is maintained by a load (5 kg in the diagram) at the free end of the cord and a spring balance at the other. When the cylinder is rotated at a suitable speed, the mass becomes suspended. The work done against friction between cord and cylinder is changed into heat,

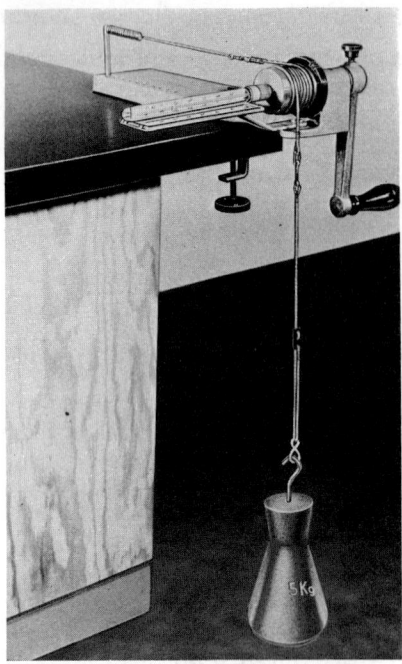

Fig. 17.1 Measurement of heat generated by mechanical work

causing a rise in temperature of the metal cylinder. The temperature rise is measured directly with the thermometer inserted in one end of the cylinder. The specific heat capacity is calculated in the following way.

Work done in rotating cylinder
$$= (\text{load} \times 9.8* - \text{reading of balance}) \times \text{circumference of cylinder} \times \text{no. of turns}$$

Heat generated
$$= \text{mass of cylinder} \times \text{sp. ht. cap.} \times \text{rise in temp.}$$

Using SI units both these energies are expressed in joules and by equating them (assuming no heat losses) the specific heat capacity is found in J/kg K.

An approximate value for the specific heat capacity of lead shot can be obtained by allowing a quantity of shot to fall through a measured distance. The initial temperature of a mass of lead shot is first found. The shot is then transferred to a thick cardboard tube about 1 m long and 4 cm diameter. The ends are closed with rubber

---

* This factor 9.8 was explained in Chapter 3. It is the strength of the earth's gravitational field and is the force in newtons acting on a mass of 1 kg on the earth's surface.

bungs. The shot is then allowed to fall about 100 times from one end of the tube to the other by holding the tube vertical for one fall and then quickly inverting it for each successive fall. The final temperature of the lead shot is found—the shot can be emptied into a beaker for this measurement. We have then:

energy, in joules, lost by shot in falling
$$= \text{mass in kg} \times 9.8 \times \text{distance fallen in m} \times \text{number of falls}$$

increase in heat energy of shot
$$= \text{mass} \times \text{sp. ht. cap.} \times \text{rise in temp.}$$

By equating these quantities, the specific heat capacity in J/kg K is found.

Electrical methods

The specific heat capacity of a liquid can be determined very accurately by supplying the energy electrically.* The electrical energy consumed in a wire is *all* converted into heat. A coil of wire (Fig. 17.2) is contained in a known mass of liquid in a vacuum flask. (The flask reduces the heat losses. *See* Chapter 19.) A current is passed

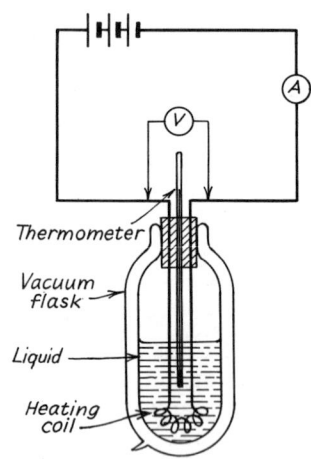

Fig. 17.2 Heating effect of a current

through the coil for a measured time t and the rise in temperature noted. When the current I through the coil and the potential difference V across it are known, the energy supplied is VIt. This can be equated to the product of the mass of the water,

---

* *See also* Chapter 35.

the rise in temperature and the specific heat capacity.

A modification of this experiment was designed by Callendar and Barnes in which a steady stream of water was passed through a tube containing the heating coil. The temperatures of the water on entering and leaving were measured when they were constant. By correcting for heat losses and measuring the current and p.d. carefully, a very accurate value for the specific heat capacity of water was found.

A laboratory version of Gaede's apparatus* for determining the specific heat of a metal is shown in Fig. 17.3. The metal is in the form of a large block of known mass, which is prepared with two holes, one to contain a small heating coil and the other a thermometer. The block is mounted within a wooden box, and the space around the block is filled with cork chippings to minimize heat losses by conduction. After taking the initial temperature, the current is switched on and the temperature taken every half minute. After passing the current for a few minutes, so as to obtain a rise in temperature of about 10 degrees, the current is switched off, but the temperature readings are continued for a further few minutes.

---

* Suitable only for good conductors.

Readings

Reading of ammeter	$=1.46$ A
Reading of voltmeter	$=12.1$ V
Time current flowed	$=300$ s
Mass of copper block	$=1.56$ kg

Temperature readings:

Time /min	Temp. /°C	Time /min	Temp. /°C
0	17.15	8	25.3
1	18.4	9	25.25
2	19.85	10	25.2
3	21.5	11	25.1
4	23.2	12	25.1
5	24.85	13	25.0
6	25.4	14	25.0
7	25.4	15	24.9

Cooling correction. This experiment provides a good example of a method of allowing for heat losses. The temperature/time graph is plotted, and from the cooling section of the graph—which is approximately a straight line—the rate of fall in temperature at the final temperature is found. But as there is no cooling at the initial temperature, the average rate of fall is half that at the final temperature. Thus, by multiplying this average

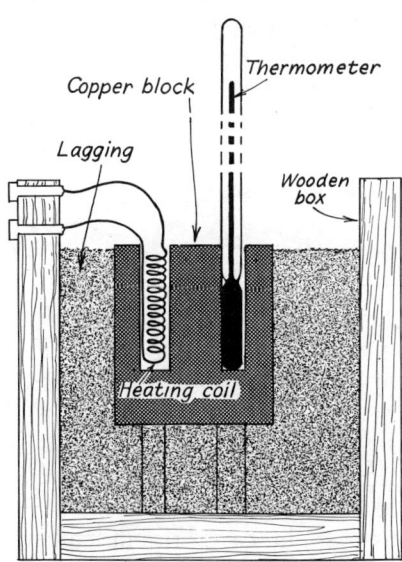

Fig. 17.3 Gaede's apparatus

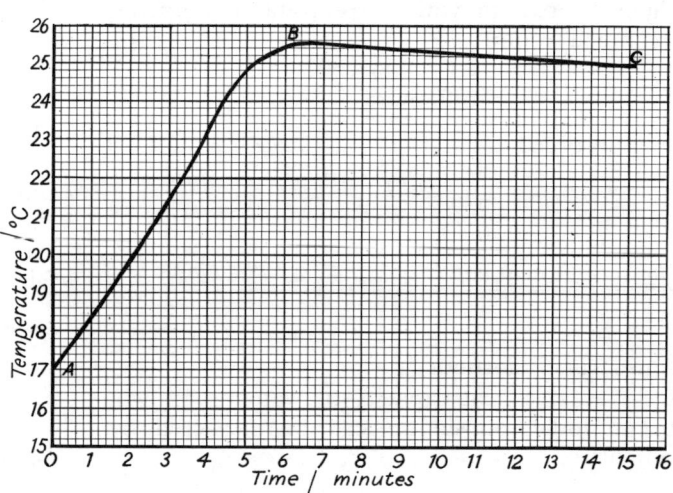

Fig. 17.4 Gaede's experiment-graph of temperature/ time

rate of fall by the time taken to reach the maximum temperature, the fall in temperature due to cooling is found, and the corrected final temperature obtained by adding it to the maximum temperature recorded.

Taking the readings from the graph (Fig. 17.4) we have:

Slope of BC = rate of cooling at temperature B
$$= 0.5 \text{ K in 8 min}$$
$$= 0.062 \text{ K/min}$$

As rate of cooling at temperature at A
$$= 0$$

Average rate of cooling over range AB
$$= 0.031 \text{ K/min}$$

Cooling correction over range AB
$$= \text{average rate of cooling} \times \text{time}$$
$$= 0.031 \times 6.5$$
$$= 0.20 \text{ K}$$

Corrected final temperature
$$= \text{maximum temp. observed}$$
$$+ \text{loss in temp.}$$
$$= 25.45 + 0.20$$
$$= \underline{25.65 \text{ °C}}$$

Heat absorbed by copper block
$$= \text{mass} \times \text{sp. ht. cap.} \times \text{rise in temp.}$$
$$= 1.56 \times c \times 8.5$$
$$= \underline{13.27c \text{ joules}}$$

Heat generated electrically*
$$= VIt \text{ joules}$$

where V = p.d. in volts, I = current in amperes, t = time in seconds
$$= 12.1 \times 1.46 \times 300$$
$$= \underline{5300 \text{ joules}}$$

Heat absorbed
$$= \text{heat generated}$$
$$13.270c = 5300$$
$$c = 0.40 \times 10^3$$

The specific heat capacity is 0.40×10^3 J/kg K.

Comparison methods
If the specific heat capacity of water is known, that of other substances can be measured by comparison.

---

* *See also* Chapter 35.

Electrical method. A heating coil (immersion heater) is left switched on in a known mass of water for a known time and the rise in temperature measured. The coil is then immersed in a known mass of another liquid and left switched on for an equal time. As the times are equal, the quantities of heat supplied are equal. Hence,

mass of water $\times 4.2 \times 10^3 \times$ temp. rise of water
= mass of liquid $\times$ sp. ht. cap. $\times$ temp. rise of liquid.

This gives the specific heat capacity in J/kg K.

Determination of the specific heat capacity of a metal
The specific heat capacity of a metal can be found very easily if it is in the form of a heavy vessel.

The mass of the heavy vessel is first determined, using a spring balance. (It is good enough to measure this mass to the nearest 5 g, if it is about 1000 g.) The initial temperature is found by placing a thermometer in it, and leaving it for a few minutes. 100 g of water, which can be measured conveniently with a pipette, is then heated to 80 or 90 °C and poured into the vessel. As the water gives out heat to the vessel, the temperature falls rapidly, but when the temperatures are equalized, further reduction is slow. The value of the steady temperature before the slow fall begins, is noted.

The heat given out by the water in cooling is given to the vessel, and so is equal to that absorbed by the vessel in warming. Of course, some heat will be lost to the bench and surroundings during the experiment, but this can be reduced considerably by enclosing the vessel in a felt jacket.

The specific heat is calculated as follows:

Mass of copper vessel	= 0.90 kg
Initial temperature of vessel	= 15 °C
Mass of water	= 0.100 kg
Initial temperature of water	= 80 °C
Final temperature of mixture	= 49 °C
Fall in temperature of water	= 31 K
Rise in temperature of vessel	= 34 K

Heat given out by water in joules
$$= \text{mass} \times \text{sp. ht. cap.} \times \text{fall in temperature}$$
$$= 0.100 \times 4.2 \times 10^3 \times 31$$
$$= \underline{13\,000.}$$

Heat absorbed by vessel in joules
$$= \text{mass} \times \text{sp. ht. cap.} \times \text{rise in temperature}$$
$$= 0.90 \times c \times 34$$
$$= \underline{30.600 \, c.}$$

Heat absorbed
          = heat given out
30 600 c = 13 000.
Therefore
          $c = 0.42 \times 10^3$.

The specific heat capacity of the metal is 0.42 ×
10^3 J/kg K.

Specific heat capacity of a liquid

Having determined the specific heat capacity of
the metal of the heavy vessel, it can now be used
to find the specific heat capacity of a liquid. A
similar experiment to that described above is
carried out, using the liquid of unknown specific
heat capacity instead of water. The value of the
specific heat capacity of the metal of the vessel
is substituted, and the only unknown quantity in
the final equation is the specific heat capacity of
the liquid.

SPECIFIC HEAT CAPACITIES	
	J/kg K × 10^{-3}
Water	4.1855
Aluminium	0.88
Copper	0.39
Iron	0.46
Silver	0.234
Zinc	0.39
Brass	0.37
Glass	0.67
Ice	2.1
Alcohol	2.3
Glycerine	2.4
Paraffin oil	2.2
Mercury	0.139
Air (at constant volume)	0.72
Nitrogen (at constant volume)	0.73
Hydrogen (at constant volume)	10.1

Latent heat

We are often told that we waste fuel by leaving the
gas or electric cooker on full after vegetables
have been brought to the boil. This is quite simple
to understand, for the temperature of boiling water

is the same, whether it is boiling rapidly or slowly.
Rapid boiling means rapid conversion of water
into steam. Thus, it appears, boiling water requires
heat to turn it into steam. This is a change from
liquid to vapour. The change from solid to liquid
also requires a supply of heat. Heat is supplied
simply to perform the change from solid to liquid
and its effect cannot be detected with a thermometer. The heat is called *latent* or *hidden heat*.

In Chapter 12 we have discussed at length the
behaviour of molecules in the solid, liquid and
gaseous states and also the conditions affecting
the changes between the states. It will recalled
that the extra freedom of molecules in the liquid
state has had to be bought by supplying them,
when in the solid state, with extra energy. This
energy is the latent heat. In simple terms, it appears
that the change from complete domination of the
molecules in a solid to their total freedom in a
gas, has to be effected in two stages. Partial freedom
is obtained in the liquid state. The greater step
is that between liquid and gas and so more energy
is involved. The latent heat is all transferred to
potential energy of the molecules; there is no
change in temperature. In quoting the energy
involved we do so for unit mass (1 kg) which is then
called the *specific latent heat*.

The specific latent heat of $\begin{Bmatrix} \text{fusion} \\ \text{vaporization} \end{Bmatrix}$ **is the amount of heat required to convert unit mass of a** $\begin{Bmatrix} \text{solid} \\ \text{liquid} \end{Bmatrix}$ **into a** $\begin{Bmatrix} \text{liquid} \\ \text{vapour} \end{Bmatrix}$ **without a change in temperature.**

Fusion is the change from solid to liquid.
Vaporization is the change from liquid to vapour.

Determination of the specific latent heat of vaporization of water

This determination is conveniently carried out
using a heavy vessel (Fig. 17.5). First weigh the
vessel on a spring balance: put it in the felt jacket
and place a thermometer in it. After a few minutes,
read the temperature and insert the large bung
provided with tubes, as shown in the figure. Invert
the vessel and connect it to a can containing
boiling water. Steam now enters the cold vessel and
is there condensed. Condensation continues until
the vessel is brought to 100 °C. After reaching
this state, the steam issues from the side tube.
The condensed steam collects above the bung in
the vessel. (Note the glass tubes must project
about 3 cm into the vessel.)

SPECIFIC LATENT HEATS

	Melting point/°C	Specific latent heat of fusion /J/kg × 10⁻³		Boiling point/°C	Specific latent heat of vaporization at normal boiling point /J/kg × 10⁻³
Water	0	335			
Aluminium	660	404	Water	100	2260
Copper	1083	205	Mercury	357	285
Platinum	1769	113	Turpentine	159	293
Lead	327	25	Benzene	80	389
Mercury	−39	13	Nitrogen	−196	201
Nitrogen	−210	25.5	Oxygen	−183	214
Oxygen	−219	13.8			

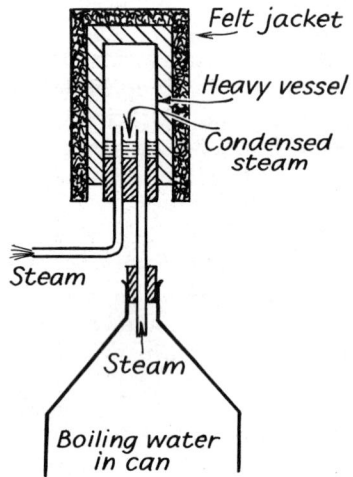

Fig. 17.5 Apparatus for determining the specific latent heat of steam

After steam has been issuing freely from the side tube for two or three minutes, disconnect the vessel and allow the water in it to cool a little. Next determine the mass of this water by measuring its volume, using either a burette or a fine measuring cylinder.

Arrange the reading as follows:

Mass of heavy vessel (copper) = 0.940 kg
Sp. ht. cap. of vessel (given) = 0.40×10^3 J/kg
Initial temperature of vessel = 15 °C
Final temperature of vessel = 100 °C
Rise in temperature of vessel = 85 K
Mass of steam condensed = 14.8 g

Heat in joules given out by steam in condensing

$$= \text{mass} \times \text{sp. lat. ht.}$$
$$= \underline{14.8 \times 10^{-3} l}$$

Heat in joules absorbed by vessel

$$= \text{mass} \times \text{sp. ht. cap.} \times \text{rise in temp.}$$
$$= 0.94 \times 0.40 \times 10^3 \times 85$$
$$= \underline{3.2 \times 10^4}$$

Heat given out = heat absorbed
$$14.8 \times 10^{-3} l = 3.2 \times 10^4$$
$$l = \underline{2.2 \times 10^6}$$

The specific latent heat of vaporization of water is 2.2×10^6 *J/kg.*

Electrical methods

Modern methods of determining the specific latent heat of vaporization incorporate electrical heating of the liquid. If heat losses are eliminated, all the heat supplied to a liquid at its boiling point is used to change the state of a liquid into vapour. A simplified form of apparatus for this determination is shown in Fig. 17.6. It is called a 'self-jacketing vaporizer'.

The liquid under investigation is placed in *A*, a sealed inner container, and is boiled by means of a heating coil. The vapour given off passes through holes *H* at the top of *A* into an enclosure *B* surrounding it. As the temperature of *B* is always the same as that of *A*, no heat passes from *A*, hence all the heat generated electrically provides latent heat to vaporize the liquid. From *B* the vapour passes down a condenser, and the liquid formed is collected in a fine measuring cylinder. Some time is allowed for conditions to become steady, and then the liquid drips from the condenser at a steady rate into the cylinder.

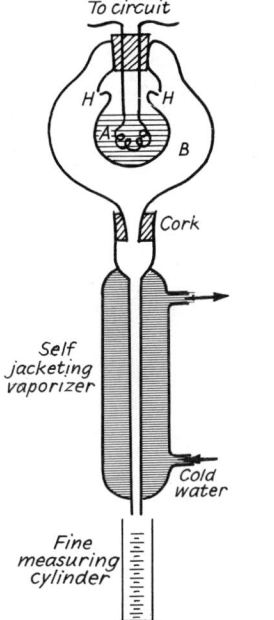

To circuit

H   *H*

A   *B*

Cork

Self jacketing vaporizer

Cold water

Fine measuring cylinder

Fig. 17.6
Self-jacketing vaporizer

One important feature of the method is that there is no cooling correction to apply as the liquid is maintained throughout at the same temperature.

Determination of the specific latent heat of fusion of ice

First weigh a calorimeter* empty and then half-full of warm water. As the experiment takes some time to carry out, it is advisable to start with the water in the calorimeter at a temperature approximately as much above room temperature as it will finish below room temperature.

Determine the initial temperature and then add several small pieces of ice which have been dried on blotting paper. This drying is essential, for the surface water has already given up its latent heat.

Stir well, and note the final temperature when sufficient ice has been added to bring the temperature down to the planned value.

Re-weigh the calorimeter and so find the mass of ice added.

Specimen readings are given below.

Room temperature	$=20\ ^\circ C$
Mass of calorimeter	$=51.6\ g$

Let mass m kg of liquid be collected in time t.

Heat supplied electrically $= VIt$.

Heat absorbed by liquid in changing into solid
$$= \text{mass} \times \text{sp. lat. ht.}$$
hence, $ml = VIt$

$$l \text{ in J/kg} = \frac{VIt}{m}$$

_____

* A vessel made of any good conductor of heat, usually copper, used for heat exchange experiments. As its temperature is that of its contents, due allowance for the heat it absorbs in an experiment can be made.

Example

In an experiment to determine the specific latent heat of vaporization of water, steam was bubbled into a copper calorimeter weighing 50 g and containing 100 g of water at 4.7 °C. The final mass of the calorimeter and contents was 104.0 g and the final temperature was 28 °C. From these readings calculate the specific latent heat of vaporization in J/kg. (Specific heat capacity of water $= 4.19 \times 10^3$ J/kg K and specific heat capacity of copper $= 0.4 \times 10^3$ J/kg K.)

Heat given out (in joules):

(a) by steam in condensing
$$= \text{mass} \times \text{sp. lat. ht.}$$
$$= 4.0 \times 10^{-3}\ l$$

(b) by condensed steam in cooling
$$= \text{mass} \times \text{sp. ht. cap.} \times \text{fall in temp.}$$
$$= 4.0 \times 10^{-3} \times 4.19 \times 10^3 \times 72$$
$$= 1207$$

Heat absorbed (in joules):

(a) by water
$$= \text{mass} \times \text{sp. ht. cap.} \times \text{rise in temp.}$$
$$= 0.100 \times 4.19 \times 10^3 \times 23.3$$
$$= 9763$$

(b) by calorimeter
$$= \text{mass} \times \text{sp. ht. cap.} \times \text{rise in temp.}$$
$$= 50 \times 10^{-3} \times 0.4 \times 10^3 \times 23.3$$
$$= 466$$

Heat given out = heat absorbed
$$4.0 \times 10^{-3}\ l + 1207 = 9763 + 466$$
$$4l = 9022 \times 10^3$$
$$l = 2.26 \times 10^6$$

The specific latent heat of vaporization is 2.26×10^6 J/kg.

Mass of calorimeter and water $= 152.8$ g
Mass of water $= 101.2$ g
Mass of calorimeter, water and ice $= 183.2$ g
Mass of ice added $= 30.4$ g
Initial temperature of calorimeter
and water $= 33.1$ °C
Final temperature of mixture $= 8.2$ °C

Heat absorbed in joules

(*a*) by ice in melting

$$= \text{mass} \times \text{sp. lat. ht.}$$
$$= 30.4 \times 10^{-3} \, l$$

(*b*) by melted ice in warming

$$= \text{mass} \times \text{sp. ht. cap.} \times \text{rise in temp.}$$
$$= 30.4 \times 10^{-3} \times 4.2 \times 10^3 \times 8.2$$
$$= \underline{946}$$

Heat given out in joules

(*a*) by calorimeter $= \text{mass} \times \text{sp. ht. cap.} \times \text{fall in temp.}$

$$= 51.6 \times 10^{-3} \times 0.4 \times 10^3 \times 24.9$$
$$= \underline{512}$$

(*b*) by water $= \text{mass} \times \text{sp. ht. cap.} \times \text{fall in temp.}$

$$= 101.2 \times 10^{-3} \times 4.2 \times 10^3 \times 24.9$$
$$= \underline{11\ 080}$$

Heat absorbed $=$ heat given out
$$30.4 \times 10^{-3} l + 946 = 512 + 11\ 080$$
$$l = \underline{333 \times 10^3}$$

The latent heat of fusion of ice is 333×10^3 J/kg.

Cooling curves

If the temperature of hot water contained in a boiling tube is recorded at regular intervals as it cools, and the readings are plotted on a temperature/time graph, a cooling curve as shown below is obtained. The cooling curve for a substance which changes its state during cooling shows an irregularity at a particular temperature.

A cooling curve for a substance such as naphthalene should be obtained experimentally (Fig. 17.7). To do this, a small quantity of naphthalene contained in a test-tube is heated in hot water. A thermometer is inserted and, when the temperature is about 95 °C, the tube is removed and the naphthalene allowed to cool. Readings are

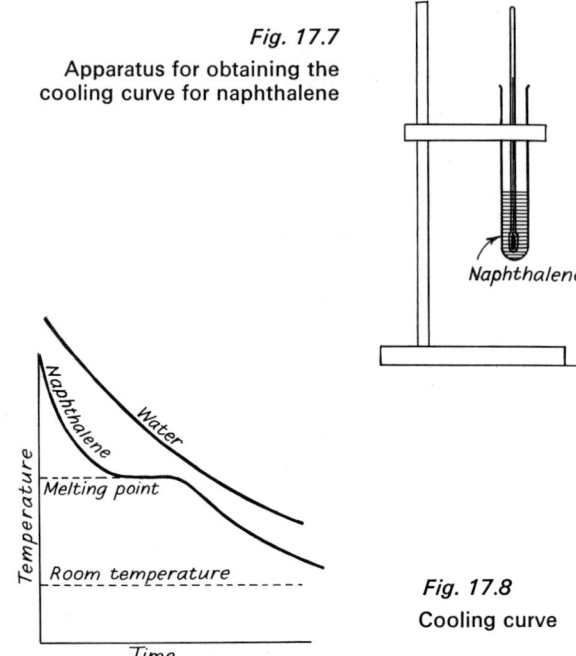

Fig. 17.7

Apparatus for obtaining the cooling curve for naphthalene

Fig. 17.8

Cooling curve

taken every half-minute and plotted on a graph (Fig. 17.8).

At some particular temperature the graph becomes horizontal. Although the naphthalene is above room temperature, its temperature does not fall for a few minutes. It is at this temperature that it solidifies. In doing so, it continues to give out heat, but does not cool because it is giving out its latent heat.

Sometimes, when a graph such as the above is plotted, it is found that the temperature falls below the melting point and then suddenly rises to the melting point (Fig. 17.9). This is due to super-cooling. Or again, the graphs of some substances

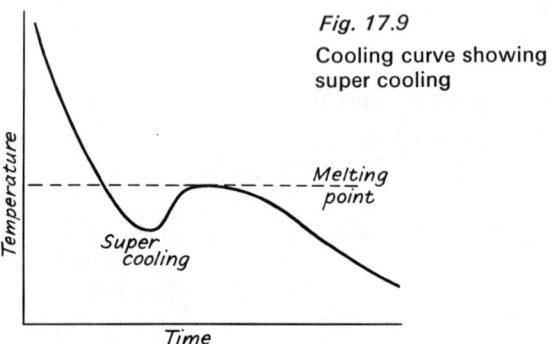

Fig. 17.9

Cooling curve showing super cooling

such as paraffin-wax show more than one horizontal section. This is caused by different constituents of a mixture solidifying at different temperatures.

Several substances melt or solidify over a range of temperature, e.g. butter, glass—this is most useful for it enables butter to be easily spread and makes glass blowing possible.

Change in volume on change of state

(a) Solid to liquid

On solidification a liquid may either expand or contract. Water expands when it becomes ice, increasing its volume by about 10 per cent (see Fig. 16.18). This expansion accounts for the bursting of pipes, and is of great importance in refrigeration. In general, food in cold storage is kept at a temperature just above freezing point so that the cells are not burst by freezing. The quality of the food is thus maintained and, at the same time, multiplication of bacteria prevented. New methods of 'quick-freeze' have been developed by which lower temperatures can be safely used.

The crumbling of rocks is largely due to the freezing of water in the fissures, which causes further splitting. In a similar manner large lumps of soil are broken into small pieces during frosty weather.*

Most substances, unlike water, contract on solidification. In making a casting, due allowance has to be made for the contraction of the molten metal on solidification. Type metal is an exception to this rule, for it is a special alloy which expands on solidifying, thereby producing a very sharp cast.

The contraction of fats and waxes is shown by the depressed surfaces formed as they change to the solid state.

A solid which contracts on melting, floats in its own liquid, e.g. ice floats on water, but solids which expand on melting sink in their own liquids.

---

* '. . . the thermometer had dropped from 10 above zero, with probably a fairly high degree of humidity, to 66 below zero in about 14 hours. The sudden expansion of the trees, rocks, lakes and atmosphere produced a vivid and fantastic effect. On this day, instead of traversing a silent brilliant land, we found ourselves in the midst of a volley of explosions and gunlike reports. The lakes, woods and rocks, suddenly being forced to freezing point, and filled as they were with a certain amount of moisture, simply were all bursting at once.' From *Grass Beyond the Mountains*, R. P. Hobson (G. Bell and Sons), by courtesy of author and publishers.

(b) Liquid to vapour

When a liquid boils, it produces vapour which, at atmospheric pressure, occupies a volume greater than that of the liquid: e.g. steam at atmospheric pressure has a volume 1700 times that of the water from which it is formed.

The change in volume when a liquid is vaporized can be directly determined by allowing a known volume of liquid to evaporate into an enclosure. The experiment can be carried out for water* by injecting a very small volume of water, using a hypodermic syringe, into a large syringe. If this is then heated in a suitable bath of oil or brine, the water will be converted into steam which will push the plunger up the cylinder of the syringe.

An interesting determination can be performed with liquid air.† If a small volume (1 or 2 cm^3) of liquid air is quickly poured into a small glass bottle to which a tube can be quickly attached, the gas evolved can be collected over water in jars of known volume. In this simple way, the volume occupied by the air formed from 1 cm^3 of liquid air can be found.

A similar experiment can be performed on a small cube of carbon dioxide‡ of known size.

Effect of pressure on change of state

(a) Solid to liquid

The freezing point of water is lowered by increase of pressure. It is said that we skate on water rather than on ice, as the blade of the skate exerts a high pressure on the ice, lowering its freezing point below its own temperature and so causing it to melt. When the pressure is removed the water freezes again.

This can be shown by an interesting experiment (Fig. 17.10). Support a large block of ice across two laboratory stools and pass over it a copper wire with a large mass (e.g. 10 kg) attached. The copper wire will pass through the block of ice and leave it in one piece. The effect is known as *regelation*.

The wire exerts a high pressure which lowers the freezing point so that the ice beneath it melts. The water so formed passes above the wire, allowing it

---

* Nuffield *G. to E.*, IV, No. 91 (*d*).
† Ibid., IV, No. 91 (*a*).
‡ Ibid., IV, No. 91 (*c*).

to sink into the ice. The pressure on it having been released, the water freezes again, and the latent heat it gives up passes through the good conducting

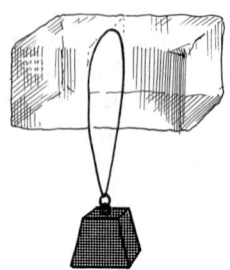

Fig. 17.10 Regelation
Block of ice supported across two stools. Heavy mass is hung on wire which slowly passes through block. The ice melts to pass the wire, then re-freezes

wire to melt more ice underneath. The experiment will work only slowly if string is used instead of wire, because the string is a bad conductor of heat.

Other substances which expand on solidification also have their freezing points lowered by pressure, but the effect is reversed for substances which contract on solidification, i.e. their freezing points are raised by increase in pressure.

We can explain the effect by considering pressure to aid a decrease in volume. Thus, it will help a substance which contracts on freezing to change its state and to raise the freezing point.

(b) Liquid to vapour

The boiling point of a liquid is raised by increasing the pressure and reduced by lowering the pressure. The temperature of the steam used in a turbine or · a steam piston engine is much above 100 °C, which means that it is at a pressure well above atmospheric.

Many chemical processes require a temperature just over 100 °C. This is often attained by using steam at high pressure. For this purpose an autoclave is used. Vulcanization of rubber is carried out in such a machine.

The lowering of the boiling point of a liquid when the pressure is reduced can be demonstrated easily for water, using the apparatus shown in Fig. 17.11. Water in the round-bottom flask is first brought to the boil. The bunsen is then removed and the bung, with thermometer, inserted. The apparatus is

inverted and cold water poured over the bottom of the flask. This condenses the water vapour filling the space above the water and reduces the pressure

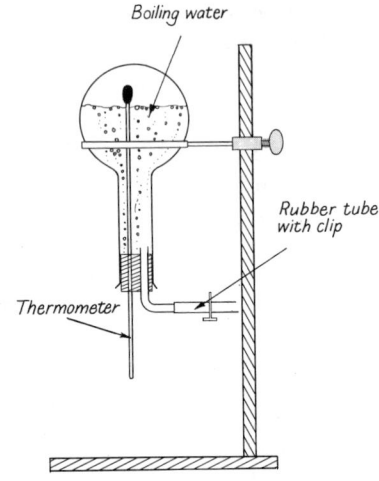

Fig. 17.11 Water boiling under reduced pressure

in the flask. The water will be seen to boil vigorously although the thermometer will be reading possibly less than 50 °C.

A modern method of cooking is to put vegetables in wire trays which fit into a strong vessel containing a small amount of water. The lid is fastened down securely and the water boiled. A valve in the lid does not let out the steam until the pressure inside is well above atmospheric pressure. In this way the food is cooked by steam at a temperature above 100 °C, which completes the process in a shorter time, and does not remove many valuable constituents extracted from food cooked in boiling water.

As the boiling point of a liquid depends on the pressure, we always quote the normal boiling point, i.e. the temperature at which the liquid boils under standard pressure, 76 cm of mercury.

Effect of dissolved substances

The effect of dissolved substances is to lower the freezing point and raise the boiling point of the solvent. It was for this reason that the use of brine as a bath for boiling water was suggested earlier in this chapter.

QUESTIONS

In the examples below, the specific heat capacity of water can be taken as 4.2×10^3 J/kg K. Other specific heat capacitances, when required, can be found in the table in this chapter. Specific latent heats also can be found in the tables given above, e.g. specific latent heat, ice→water = 335×10^3 J/kg; specific latent heat, water→steam = 2260×10^3 J/kg.

1. Calculate the heat evolved when:
   (a) 100 g of copper are cooled from 90 °C to 10 °C,
   (b) 50 g of aluminium are cooled from 150 °C to −100 °C.

2. 100 g of boiling water are poured into a metal vessel weighing 800 g and at a temperature of 20 °C. If the final temperature is 50 °C, what is the specific heat capacity of the metal?

3. Describe how you would find the specific heat capacity of a solid by the method of mixtures.
   Mention the precautions you would take to reduce heat losses during the experiment.
   A 100 g brass weight is quickly transferred from a heating bath at 100 °C to a calorimeter, of water equivalent 8 g, containing 62 g of water at 15 °C.
   If the specific heat capacity of brass is 0.37×10^3 J/kg K, what is the final temperature?   (O)

4. Define *thermal capacity* and *specific heat capacity*.
   Describe how, given a block of copper and a copper calorimeter, you would determine the specific heat capacity of copper.
   A metal block of mass 120 g is heated in an oven and then quickly transferred to a calorimeter of thermal capacity 84 J/K containing 200 g of oil, of specific heat capacity 2.1×10^3 J/kg K, at a temperature of 15 °C. If the temperature of the calorimeter and contents rises to 55 °C estimate the temperature of the oven. Assume the specific heat capacity of the material of the block to be 0.42×10^3 J/kg K.   (L)

5. A copper calorimeter of mass 80 g contained 120 g of oil at 15 °C. A piece of copper of mass 30 g was transferred quickly from a furnace to the calorimeter. The temperature of the oil rose to 45 °C. Find the temperature of the furnace. State any possible sources of error.
   [Specific heat capacity of copper = 0.42×10^3, and that of the oil = 1.89×10^3, both in J/kg K.]   (L)

6. 150 g of brass is cooled in solid carbon dioxide. It is then dropped into a brass calorimeter of mass 110 g, containing 140 g of water at 23 °C. The temperature of the mixture is 14 °C. Find the temperature of the solid carbon dioxide.
   [Specific heat capacity of brass = 0.378×10^3 J/kg K.]   (C)

7. Describe an experiment you would carry out to determine the specific heat capacity of a metal. Show how you would work out your result. State the precautions you would take to ensure a reliable result.
   A thermostat switches on the heater in a tank of water when the temperature falls 2 °C below the temperature at which it switches off. When switched on the heater supplies 3024 J/s, and it is found that when no water is being drawn from the tank the heater is on for one period of 20 s in every 4 minutes.

Calculate: (a) the heat supplied while the heater is switched on for 20 s; (b) the average rate at which heat is lost from the tank; (c) the mass of water in the tank, if the thermal capacity of the tank and the heater is neglected.   (C)

8. Explain the following:
   (a) Lumps of ice are more effective in cooling a drink than iced water.
   (b) Burns from steam are usually more severe than those from boiling water.
   (c) It takes much longer to boil away a quantity of water than it does to bring it to the boil.

9. How much heat is required to convert 10 g of ice at 0 °C into steam at 100 °C?
   [Use the values of the specific latent heats given in the table.]

10. 20 g of copper at 17 °C are dropped into liquid oxygen at −183 °C, and 5.5 g of oxygen are evaporated. What is its specific latent heat?
    [Take the specific heat capacity of copper over this range as 0.336×10^3 J/kg K.]

11. What is the effect of adding 100 g of ice at 0 °C to 200 g of water at 12 °C?

12. A piece of red-hot metal, specific heat capacity 0.462×10^3 J/kg K, mass 21 g, temperature 710 °C, is dropped into 100 g of water, temperature 0 °C. Some steam is given off and after stirring the temperature of the water is found to be 10 °C. The remaining water and metal together weigh 120 g. How much heat is lost by the metal? How much heat is gained by the remaining water? What result do these figures give for the specific latent heat of steam?   (JMB)

13. A copper calorimeter of mass 120 g contains 200 g of oil at 35 °C. What mass of ice at 0 °C must be added in order to reduce the temperature of the calorimeter and its contents to 5 °C?
    [Neglect heat losses to the surroundings. Take the specific heat capacity of oil to be 2.52×10^3 J/kg K, that of copper to be 0.42×10^3 J/kg K, and the specific latent heat of fusion of ice to be 336×10^3 J/kg.]   (O)

14. Define *specific latent heat*, and describe a method of measuring the specific latent heat of fusion of ice.
    A copper calorimeter, of mass 70 g, contains 100 g of paraffin oil at 20 °C. A lump of ice, at 0 °C and of mass 20 g, is added to the oil. How much ice will be left when the temperature of the calorimeter and its contents has become steady at 0 °C?
    [Take the specific heat capacity of copper to be 0.42×10^3 J/kg K, that of oil to be 2.23×10^3 J/kg K, and the specific latent heat of fusion of ice to be 336×10^3 J/kg.]   (O)

15. Define *specific latent heat of vaporization* of a liquid.
    Describe an experiment for the determination of the specific latent heat of vaporization of water boiling under atmospheric pressure. Draw a diagram of the apparatus used and point out any precautions which should be taken to reduce the experimental errors.
    A vessel of negligible thermal capacity contains a mixture

of 500 g of ice and 1000 g of water at 0 °C. What mass of steam at 100 °C must be condensed in the vessel in order that the temperature of the vessel and its contents may rise to 20 °C? (L)

16. A piece of copper of specific heat capacity 0.40×10^3 J/kg K and mass 140 g is quickly transferred from an oil bath to a cavity in a block of ice at 0 °C. If 20 g of ice are melted, calculate the temperature of the oil bath. Point out two possible sources of error in this experiment. (L)

17. Describe how you would determine the specific latent heat of fusion of ice.
A plastic tray weighing 48 g and containing 200 g of water at 20 °C is put in a refrigerator which abstracts heat at a uniform rate of 2100 J/min. Calculate (a) the time taken for the tray and water to reach 0 °C, (b) the total time taken to freeze all the water to ice at 0 °C.
[Specific heat capacity of the plastic = 1.05×10^3 J/kg K.] (C)

18. A flask containing water at 100 °C is suspended in a can of boiling water (boiling at 100 °C) with the open neck of the flask just above the level of the boiling water. Draw a diagram of the arrangement. When a lump of metal of mass 56.0 g and at 250 °C is dropped into the flask, *boiling occurs in the flask* and then *ceases* when a mass of 3.42 g of water has been boiled away. Explain the actions printed in *italics* and calculate the specific heat of the metal. (C)

19. Describe an experiment to determine the specific latent heat of steam. State the sources of error in the experiment and point out the precautions you would take to reduce them.
A 200 g brass weight is held suspended in liquid oxygen contained in an open vacuum 'Thermos' flask. It is then quickly transferred to a beaker of water at 0 °C. What weight of ice will form on the brass?
[Boiling point of oxygen is −183 °C, specific heat capacity of brass = 0.37×10^3 J/kg K, and specific latent heat of ice = 335×10^3 J/kg.] (JMB)

20. How would you determine the specific latent heat of condensation of steam? State the precautions you would take, and indicate how the probable errors affect the result.
Alcohol vapour at its boiling point of 78 °C is passed into a brass vessel of mass 600 g at 10 °C. What mass of alcohol will be condensed? Heat losses may be neglected.
[Specific heat capacity of brass = 0.37×10^3 J/kg K; specific latent heat of vaporization of alcohol = 860×10^3 J/kg.] (C)

21. A copper cylinder of mass 100 g is cooled to the temperature of liquid air (−185 °C) and is then completely immersed in a large calorimeter of water at a temperature of 0 °C. Calculate the mass of ice formed in the water.
[Specific latent heat of fusion of ice = 335×10^3 J/kg; mean specific heat capacity of copper = 0.29×10^3 J/kg K.] (O & C)

22. An electric kettle is required to heat 1 kg of water from 15 °C to 100 °C in 5 min. Neglecting heat losses, find the power consumption, in watts, of the heater. At what rate does the water vaporize when the boiling-point is reached? (O & C)

23. When heat energy at the steady rate of 100 watts is supplied to the water inside a closed vessel the pressure inside is found to be atmospheric when the temperature is 100 °C. At what rate in g/min would water boil away if a small hole were opened in the lid of the vessel, and simultaneously the rate of supply of energy were increased to 520 watts? (O & C)

24. A 30 W immersion heater just keeps 600 g of a metal molten. The heater is switched off and 6 min later the temperature of the metal starts to fall. What is the specific latent heat of fusion of the metal in J/kg? (O & C)

25. Sunshine falls on a blackened copper plate which is lying on snow at 0 °C. Each cm² of plate surface absorbs energy at the rate of 0.01 W and conducts it to the snow beneath. Take the specific latent heat of snow as 335×10^3 J/kg and the density of snow as 90 kg/m³ and calculate the depth of snow that will be melted in an hour. (O & C)

26. The average kinetic energy of the molecules in boiling water is the same as that in the issuing steam; how does the kinetic theory explain the fact that an appreciable amount of energy is required to convert water at 100 °C to steam at 100 °C? The actual value of the energy in this case is 2260×10^3 J/kg. Describe any *one* experimental method of checking this value (to within about 10%).
Steam from a boiler issues, through a pipe, at 100 °C. What is the pressure of the steam on the wall of the pipe and how is this pressure explained by the kinetic theory? If the exit pipe becomes nearly blocked, how will this affect the temperature and pressure of the steam in the pipe? Give reasons. (C)

27. Explain the distinction between *saturated* and *unsaturated* vapour, and between *evaporation* and *boiling*.
Heat is supplied at a rate of 500 W to a pressure cooker containing water and fitted with a safety-valve. Steam escapes at such a rate that the loss of water is 10.4 g/min. If heat is supplied at the rate of 700 W, 15.6 g of water is lost per minute.
Suggest an explanation of these figures and deduce (a) the specific latent heat of steam in J/kg at the temperature of the cooker, and (b) the rate of loss of heat from the cooker at this temperature by other processes than evaporation. (O & C)

28. Distinguish between *evaporation* and *boiling*.
A jet delivering 0.44 g of dry steam per second, at 100 °C, is directed on to crushed ice at 0.0 °C contained in an unlagged copper can which has a hole in the base. 4.44 g of water at 0.0 °C flow out of the hole per second. How many joules of heat are given out per second by the condensing of the steam and the cooling to 0.0 °C of the water formed? How much heat is taken in per second by the ice which melts? Suggest why these amounts are different.
Take the specific latent heat of steam as 2260×10^3 J/kg, the specific latent heat of fusion of ice as 334×10^3 J/kg, and the specific heat capacity of water as 4.18×10^3 J/kg K.
If the steam jet were replaced by an electric heater buried in the ice, and connected to a 240 V supply, what current would be needed to give the same heat per second to the apparatus? (O & C)

29. An immersion heater is rated at 60 W. When it is inserted in 0.25 kg of water contained in a copper vessel, the temperature of the water and vessel increases from 15 °C to 31 °C in 5 minutes. What is the mass of the copper vessel?

[Specific heat capacity of water is 4.2×10^3 J/kg K; specific heat capacity of copper is 0.42×10^3 J/kg K.]

30. A mains immersion heater rated at 300 W is placed in a vessel containing 0.75 kg of water at 10 °C. Find the rise in temperature after 3.5 minutes, neglecting the heat absorbed by the containing vessel. If the water were replaced by an equal mass of oil, what would be the final temperature?

[Specific heat capacity of water $= 4.2 \times 10^3$ J/kg K; specific heat capacity of oil $= 3.0 \times 10^3$ J/kg K.]

31. What is meant by the terms *change of state* and *latent heat*?

Describe a simple laboratory demonstration to show (*a*) that there is no change of temperature while a change of state is taking place, (*b*) the heat exchange that occurs during a change of state.

Calculate the heat needed to melt completely 0.50 kg of lead initially at 20 °C.

[Melting point of lead $= 327$ °C; specific heat capacity of lead $= 0.134 \times 10^3$ J/kg K; specific latent heat of fusion of lead $= 21 \times 10^3$ J/kg.]   (C)

18 Evaporation, boiling, vapour pressure, humidity

Evaporation

We have seen, in Chapter 13, that evaporation is the escaping of the fastest-moving molecules. This will result in a lowering of the average kinetic energy of those remaining, and will produce cooling. Another way of looking at this phenomenon is to consider that the latent heat necessary to convert the liquid into vapour is absorbed from the liquid itself, thereby causing cooling.

Butter and milk coolers are unglazed vessels which, after being soaked in water, produce cooling by the evaporation of the absorbed water. The practice of moistening the forehead of a sick person with eau-de-cologne provides not only a sweet atmosphere but also cools the sick person's head.

A simple demonstration of this cooling is the freezing of water by the rapid evaporation of ether (Fig. 18.1). A little water is put on a block of wood and a small beaker placed upon it. Some ether is poured into the beaker and rapidly evaporated by blowing air through it by means of a glass tube dipping into the ether and connected to the foot bellows. Very soon the cooling is sufficient to freeze the water and so cause the beaker to stick to the wooden block.

This principle is used in refrigeration. Here, a substance normally in the gaseous state is first liquefied and then allowed to evaporate quickly.

We have seen that increase of pressure raises the boiling point of a liquid and, conversely, increase of pressure helps a gas to liquefy. But we cannot liquefy all gases by pressure alone, for it has been found that every gas must be cooled below a certain temperature, termed the critical temperature, before it can be liquefied by pressure. When the 'gas' is below its critical temperature, i.e. when it can be liquefied by pressure alone, it is termed a vapour. The critical temperature for some gases, e.g. ammonia, carbon dioxide, sulphur dioxide, is above normal air temperature, and so they can be liquefied conveniently by pressure without any cooling.

In a modern refrigerating plant ammonia is used as the refrigerating substance.* First, it is compressed in order to liquefy it. In Fig. 18.2, the pump P compresses the ammonia in the spiral S_1 (valve V_1 is open and V_2 closed). This compression causes

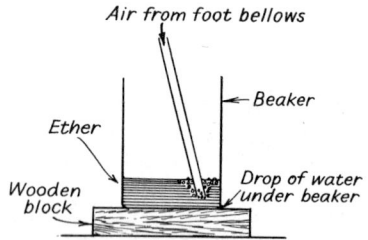

Fig. 18.1 Cooling produced by rapid evaporation of ether

Water freezes and fastens beaker to block

---

* 'Freon' and other specially chosen substances are used in household refrigerators. Suitable vapours are easily liquefied by pressure and have a high specific latent heat.

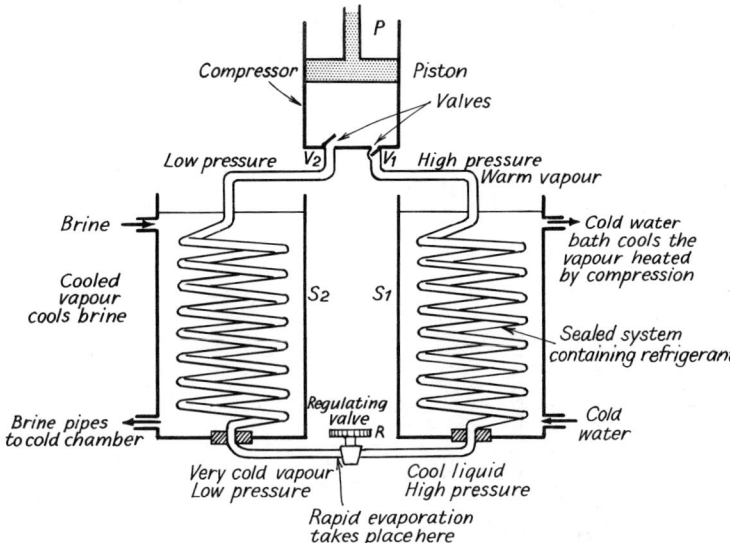

Fig. 18.2
Diagram illustrating
principle of refrigerator

some heating and so the spiral is water cooled and the ammonia liquefied. From the high-pressure side, the liquefied ammonia passes through the regulating valve *R* into the low-pressure side. The low pressure is created by drawing the gas into the pump on the upstroke of the piston (V_2 is open, V_1 closed). In the spiral S_2 the liquid quickly evaporates, and in so doing produces cooling. Around S_2, brine circulates and, when cooled, is led away through pipes which surround the low-temperature storage vaults. In the manufacture of ice, large cast-iron vessels filled with water are lowered into a reservoir of the cooled brine. When the water is frozen, the iron vessels are immersed in warm water to release the ice blocks.

Factors which influence the rate of evaporation

1. *Temperature.* When the temperature of a liquid is increased, more molecules possess sufficiently high speeds to escape and so the rate of evaporation is increased.

2. *Pressure.* When the pressure over a liquid is reduced, the molecules can escape more easily and so the rate of evaporation is increased.

3. *Area of surface.* If the area of the surface of a liquid is increased, more molecules have the opportunity of escaping.

4. *Draught over surface.* A current of air blown over the surface removes the molecules as they escape and so prevents their returning to the liquid.

Boiling

If the temperature of a liquid is increased, evaporation continues at a greater rate and eventually the majority of the molecules have sufficient speed to escape. Those beneath the surface cannot escape into the air and so they evaporate internally, producing a bubble of vapour. This bubble is pushed up to the surface by the surrounding liquid, and increases in size because of the evaporation into it of more molecules, and the reduction in pressure. This process is called boiling.

Boiling is rapid evaporation from all parts of a liquid and takes place at a definite temperature which depends only on the pressure. It differs from evaporation, which is a surface effect taking place at all temperatures.

Vapour pressure

If a liquid is allowed to evaporate in a closed space, the vapour formed exerts a pressure. With sufficient liquid present, evaporation continues until the number of molecules leaving in a certain time is equal to the number re-entering. When this steady state (dynamic equilibrium) is attained, the space is said to be saturated: it contains the maximum number of molecules per cubic centimetre and so exerts the greatest pressure. This is called the maximum or saturated vapour pressure. The simplest way of demonstrating vapour pressure is to introduce some liquid into the vacuum above the mercury in a barometer (Fig. 18.3). This is accomplished by using a bent pipette. If the liquid is added a drop at a time, the first few drops will

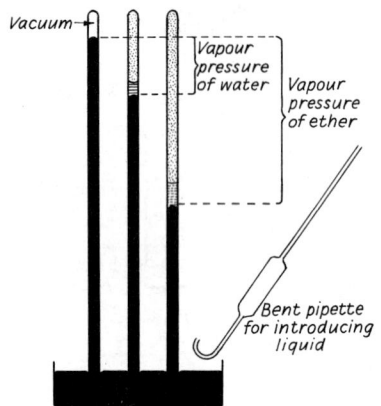

Fig. 18.3 Measurement of saturation vapour pressure

all evaporate, each causing a small depression of the mercury column. When excess liquid is seen on the top of the mercury, further addition will not cause any further depression of the mercury. The total depression is a measure of the maximum vapour pressure at room temperature.

The pressure of a saturated vapour increases with temperature. This increase is due to two causes: (1) at a higher temperature the molecules have greater speeds, and (2) there are more molecules per cm³. At the higher temperature the rate of escape of the molecules from the liquid is increased. Equilibrium is again restored when the

Fig. 18.4

Apparatus to show large vapour pressure of ether

When ether is added to water by opening clip and straightening tube, vapour pressure forces water up glass tube

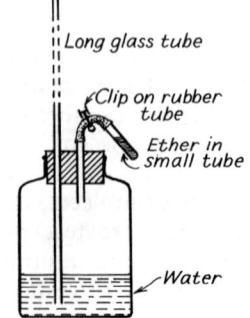

number of molecules re-entering has likewise increased. But the number in the space is also greater. This can be appreciated by considering a juggler who throws three balls per second with one hand, and catches them in the other. If it takes one second for each ball to pass through the air from one hand to the other, there will be three balls in flight at any instant. If he now increases his rate of throwing to four balls per second, he will have to catch them at the rate of four per second, but there will be four balls in the air at any instant.

The effect of increasing the temperature of a saturated vapour can be shown by running the bunsen flame up and down the upper part of the barometer tube containing the vapour. To obtain a quantitative result, the upper end of the barometer tube is enclosed in a water-bath. The bath is heated and kept well stirred, the depression of the mercury being measured at different temperatures. There must always be a little liquid above the mercury, otherwise the space will not be saturated.

When a liquid boils, its saturation vapour pressure is equal to the external pressure, hence, the graph of pressure against boiling point is the same as saturation vapour pressure against temperature.

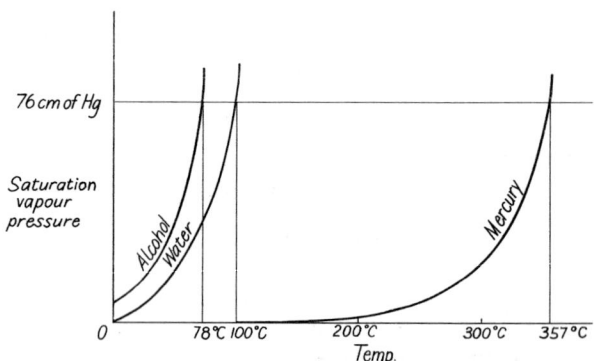

Fig. 18.5 Graph showing SVP against temperature

Fig. 18.5 shows the graphs of water, mercury, and alcohol. At 100 °C the saturation vapour pressure of water is 76 cm of mercury. The curves for other liquids are similar, and all must show a saturation vapour pressure of 76 cm of mercury at their respective boiling points, as this is what we mean by *boiling point* (*see* below).

Change in temperature, then, is the only factor which can influence the pressure of a saturated vapour.

Fig. 18.6 Upward current of moisture-laden air producing a cloud

If such a vapour is cooled, the space cannot hold as many molecules and the excess is delivered as liquid, i.e. condensation occurs.

Decreasing the volume of a saturated vapour does not cause an increase in pressure, for the number of molecules per cm^3 must remain constant. Again, the surplus molecules are returned to the liquid, i.e. condensed.

These two effects constitute the tests for a saturated vapour:

1. Decrease the temperature—condensation occurs.

2. Decrease the volume—condensation occurs.

The pressure of a saturated vapour increases with temperature and becomes equal to atmospheric pressure when the liquid boils.

The normal boiling point of a liquid is the temperature at which its vapour exerts standard atmospheric pressure.

Clouds

Clouds consist of very fine particles of water formed by condensation from air saturated with water vapour which has been cooled by expansion on rising (Fig. 18.6).

Fog and mist

A rapid fall in temperature when the air is damp causes condensation. The vapour condenses most readily on any particle present in the air. In towns where the air is polluted with soot particles, these form suitable nuclei for condensation, and so we have the dark fogs which are familiar in cities.

Humidity of the air

The air always contains some water vapour. Its content is of great importance. For ideal conditions the air must not be too dry nor too humid. If it is too dry, evaporation from the body is rapid: the mouth becomes parched and the eyes dry and overheated. In a very humid atmosphere, evaporation from the body is too slow and perspiration follows.

For ideal working conditions a room should not only be suitably heated and lighted, but it should be supplied with conditioned air, that is, with air containing the correct amount of water vapour. This amount depends on the temperature of the air, for the atmosphere on a hot, dry day in summer may contain more water vapour than on a cold wet day in winter. The deciding factor is the relative amount of water vapour present, that is, the mass actually present compared with the maximum mass which could be present. This ratio is called the *relative humidity*.

If unsaturated air is cooled, it eventually becomes saturated. The temperature at which this occurs is called the *dew point*. The effect is noticed when a cold object, such as a tumbler of cold water, is brought into a warm room. Moisture from the air in the room condenses on the outside of the cold tumbler.

QUESTIONS

1. What is evaporation and what factors influence the rate of evaporation? Distinguish between evaporation and boiling.

2. Explain the following:
   (*a*) Cooling is produced by evaporation.
   (*b*) Clothes dry more quickly on a windy day than on a calm day.
   (*c*) A heavy damp soil is cold.

3. What is meant by *saturated vapour*? How can the maximum pressure exerted by a vapour be determined?

4. A tumbler of cold water brought into a warm room often becomes covered with a film of moisture. Explain this.

5. How would you find the saturated vapour pressure of water in the laboratory at 10 °C above room temperature?
   How would you ensure that the vapour is saturated? (C)

6. What do you understand by a *saturated vapour*?
   Describe a method of measuring the saturated vapour pressure of water at room temperature.
   How is the boiling-point of water affected by (*a*) a change in the external pressure; (*b*) the presence in the water of dissolved salt? (O)

7. Give an account of the process of evaporation, and explain why a vapour exerts a pressure. Distinguish between a saturated and an unsaturated vapour.
   Describe how you would determine experimentally the saturated vapour pressure of water at different temperatures within the range 15 °C to 60 °C. (O)

8. Distinguish between a *saturated vapour* and an *unsaturated vapour*.
   Describe how you would measure the saturated vapour pressure at room temperature of a volatile liquid such as ether.
   Explain briefly, in terms of the kinetic theory, why you would expect the saturated vapour pressure of a liquid to increase as the temperature is raised. (O)

9. Distinguish between *boiling* and *evaporation*.
   State the effect on the boiling-point of a liquid of (*a*) changing the pressure on its surface, (*b*) adding a soluble solid to it. Describe one experiment in each instance to support your answer. (L)

10. Air is blown through some ether in a copper can. Explain (*a*) the effect on the temperature of the ether, (*b*) the formation of mist on the outside of the can. On another day, repetition of the experiment produced the mist at a lower temperature. Explain the reason for this and describe a practical application of these effects. (L)

11. Describe how you would measure the saturated vapour pressure of water at a series of temperatures between about 15 °C and about 60 °C.
   A certain mass of gas, stored in a gas jar over water, occupies a volume of 246 cm³ at 19 °C, the barometric pressure being 755 mmHg. Find the volume which this mass of gas would occupy, after drying, at 0 °C, and 760 mmHg pressure.
   [The saturated vapour pressure of water at 19 °C is 17 mmHg.] (O)

12. Describe, with a diagram of the apparatus used, how you would demonstrate the following phenomena:
   (*a*) The boiling-point of water is reduced by reduction of the pressure on its surface.
   (*b*) The saturation vapour pressure of a substance increases with rise of temperature.
   The height of the mercury column of a simple barometer is 76.9 cm. When some alcohol is put into the barometer tube, a column of alcohol 15.3 cm long stands on top of the mercury; the total height of the alcohol and mercury columns is 76.9 cm. Calculate the saturation vapour pressure of the alcohol under the conditions of the experiment.
   [Density of alcohol $= 0.80 \times 10^3$ kg/m³. Density of mercury $= 13.6 \times 10^3$ kg/m³.] (C)

13. Explain the terms *saturated vapour*, *saturated vapour pressure*.
   State and explain the conditions under which (*a*) evaporation, (*b*) boiling of a liquid can take place.
   Describe a method by which you would determine the saturated vapour pressure of water at temperatures between about 20 °C and about 100 °C. (O & C)

14. What do you understand by the terms *saturated vapour, unsaturated vapour, boiling-point*?
   Describe a method of measuring the saturation vapour pressure of alcohol between room temperature and a few degrees below the boiling-point.
   Sketch roughly the type of curve you would expect to obtain.
   [Saturation vapour pressure at 15 °C = 30 mmHg. Boiling-point of alcohol = 78 °C.] (O & C)

15. What is meant by (*a*) a saturated, (*b*) an unsaturated vapour?
   Describe a method of measuring the saturation vapour pressure of ether at room temperature.

Explain the distinction between *evaporation* and *boiling*.

Suggest reasons why the rate of evaporation from the water in an electric kettle should increase when the boiling-point is reached. (O & C)

16. Explain what is meant by *saturated vapour pressure*. Sketch a graph to show the manner in which the saturated vapour pressure of a substance varies with temperature.

Consider three barometer tubes *A*, *B*, *C*, each about 80 cm long; *A* has a good Torricellian vacuum in the space above the mercury; *B* has some air trapped in this space; *C* has alcohol vapour in the space and a little liquid alcohol floating above the mercury.

Explain what you would expect to happen to the mercury level in each tube, (*a*) if you were to tilt the tube at a small angle with the vertical, (*b*) if you were to warm the space above the mercury. (C)

17. A faulty barometer, containing a little air in the top of the tube, reads 73.5 cm when the correct barometric pressure is 76 cm. The distance from the mercury level in the bowl to the sealed end of the uniform tube is 80 cm. Calculate the correct barometric pressure on a day when this barometer reads 72.5 cm, the temperature remaining constant. (C)

18. What is meant by the statement 'the saturation vapour pressure of water at 60 °C is 149.4 mm of mercury'? Does the correctness of this statement depend upon the presence or absence of air in the space containing the vapour?

Draw a graph of the variation of the s.v.p. of water with temperature using the following table of values:

Temperature (°C)	10	20	30	40	50	60
s.v.p. (mm of mercury)	9.2	17.5	31.8	55.3	92.3	149.4

From your graph deduce, with reasons,

(*a*) the value to which the atmospheric pressure must be reduced to cause water to boil at 55 °C;

(*b*) the lowest temperature at which evaporation can occur when the vapour pressure of water in the atmosphere is 25 mmHg. (O & C)

19. Describe an observation which supports the hypothesis that the molecules of matter possess energy of motion.

A closed vessel contains a small quantity of liquid, and vapour occupies the remaining space. Explain how the hypothesis describes the process of evaporation and how it distinguishes between saturated and unsaturated vapour. (JMB)

19 Transmission of heat

We return again to the idea that heat is a form of energy—the kinetic energy of the molecules. When this molecular movement is handed on from one molecule to another we say the heat is *conducted* through the medium. Imparting kinetic energy to a molecule can be effected by two methods, which can be illustrated by considering how a cork floating on a pond may be given K.E. This can be done by (*a*) direct impact, e.g. by throwing a stone at it, or (*b*) by creating ripples on the pond which, as they pass the cork, cause it to vibrate up and down.

These examples illustrate transmission of heat by (*a*) conduction and (*b*) radiation.

We usually include a third method called *convection*.* In this case it is not the *heat* which is transmitted but the hot *substance*. It is therefore sometimes regarded as not a true method of transmission of heat.

Conduction

Conduction is the handing on of increased energy of vibration from molecule to molecule. If the end of a rod is heated, the molecules at this end vibrate more vigorously and affect the next molecules, causing them to vibrate more quickly, and so on throughout the rod.

---

* Convection needs a gravitational field in which to work.

The effect in solids can be studied by arranging three similar rods of copper, iron and glass on heat-sensitive paper,* so that one end of each

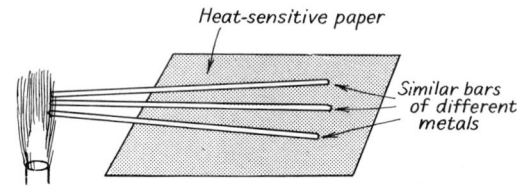

Fig. 19.1 Experiment to show different conductivities of metals

Passage of heat along the bars is observed by coloration of the heat-sensitive paper

may be heated in the same bunsen flame (Fig. 19.1). After a short time the paper will show roughly the relative conductivities of the rods: the heat travels along the bars, and the paper under the rods becomes coloured. The result will show that copper is a good conductor of heat, iron not quite so good, and glass a poor conductor.†

The good conductivity of brass contrasted with

---

* Heat-sensitive paper can be made by soaking white blotting paper in cobalt chloride solution and drying slowly. When heated, it turns green, but loses its coloration on cooling.

† This observation should be made after equilibrium has been attained. In the initial stage, while the rods are warming up, specific heat capacity plays a part.

the poor conductivity of wood is shown by heating a compound bar of brass and wood wrapped with a piece of paper. The paper covering the wood is charred, whereas that covering the brass is unharmed. This is because the brass rapidly conducts away the heat and so keeps the temperature low.

These experiments illustrate that metals are good conductors and non-metals poor conductors. By using different metallic rods in the first experiment it can be shown that, of the common metals, copper is the best conductor.

Liquids

Liquids are bad conductors of heat. A well-known experiment which shows this, is the heating of a tube of liquid at the top (Fig. 19.2). In this

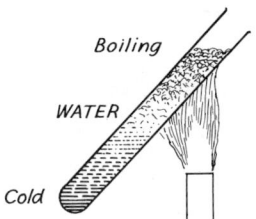

Fig. 19.2

Experiment to show water is a bad conductor of heat

Very little heat is conducted to the bottom of the tube

way water can be boiled at the top of the tube while that at the bottom is cold, as can be shown by holding the tube in the hand, or by putting a small piece of ice, wrapped in gauze to weight it down, at the bottom of the tube.

Gases

Gases have a still lower conductivity than liquids. It is impossible to demonstrate this by a simple experiment because of convection currents.

Uses of good and bad conductors

We can now account for the particular use of many materials in our homes. Metals, being good conductors, are used where heat has to be readily transmitted. Thus, pans, kettles and other cooking utensils which have to be heated directly are metal, but are provided with bad conducting handles of wood or plastic material. We wear 'warm' clothes in winter, which are made of bad conducting material. Such is often manufactured from the coats of animals, e.g. wool and fur. The

Fig. 19.3 Air cavity in outside wall of house to provide heat insulation (it also prevents the passage of moisture)

poor conductivity of many materials is due to their open structure which encloses small air pockets. Thus, their heat insulating property is often due to a layer of air (Fig. 19.3). Open string vests are a good example of the retention of body heat by an air layer. Double-glazed windows, which consist of two sheets of glass separated by an air layer, are becoming popular because they reduce the loss of heat from a room.

Asbestos is a well-known insulator which is much used because it is non-flammable and can be easily worked and artistically decorated. Sometimes an asbestos cement is employed as a covering for boilers and pipes conveying steam. Such covering is termed heat lagging. Many new insulating materials, which are also extremely light, are being developed and used extensively, e.g. for lining the walls of refrigerators.

It is well known that the hard chalk deposits on the inside of kettles result from a hard water supply. This fur, which rapidly accumulates on the inside of boilers in limestone areas, is a bad conductor of heat, and the scaling of the boilers has to be carried out frequently because of the marked reduction in efficiency and the consequent wastage of fuel.

Further notes on thermal conduction

Conduction is somewhat more involved than has been suggested in the short statement above. There is a difference in the mechanism of conduction in metals from that in non-metals. Silver conducts heat about a thousand times better than card-board. Such a great difference in the conductivities of two solids points to a radically different mode of operation. A characteristic of metals is that they have one or two outer orbiting electrons which are so loosely linked with the nucleus, that they become separated and move in the inter-atomic space, only linking by chance momentarily with atoms. The motion of these 'free' electrons resembles the motion of gaseous atoms or molecules. These electrons absorb the energy at the locality where the metal is heated and convey it readily through the metal.

Free electrons are not found in non-metals and so the handing-on of increased vibration is not aided by the action of the free electrons.

As would be expected from the above there is a connection between thermal and electrical conduction in metals.

Convection

When a kettle of water is heated, all the water starts to boil at the same time, although the heat is supplied only through the bottom. The water is mixed by convection currents. The heated water at the bottom expands, becomes less dense, and is pushed up by the denser surrounding water. Its place is taken by cooler, and so, denser water, which, in its turn, behaves in the same manner. In this way, convection currents are created.

Convection in a solid is impossible.

Convection currents in water can be shown by dropping a few crystals of potassium permanganate to the bottom of a beaker of cold water.* If a small bunsen flame is applied to the beaker immediately under the crystals, the convection current is clearly indicated by the coloration produced as the water streams past the crystals (Figs. 19.4 and 19.5).

A simple experiment to show convection in a gas is illustrative of the old method of mine ventilation (Fig. 19.6). Two glass chimneys are fitted into the top of a wooden box provided with a glass front. In the box, just below one chimney, is a lighted candle. The heat from this creates a convection current, which can be demonstrated by holding a

---

* This is best done by dropping the crystals down a glass tube and then removing the tube and coloured water in it by holding the finger over the upper end.

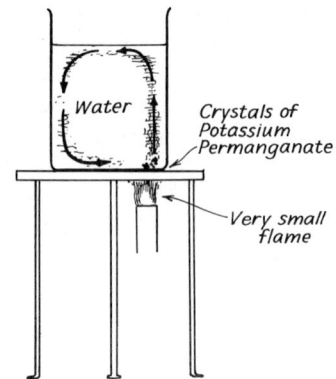

Fig. 19.4 Convection currents in water

Currents shown by coloration produced by potassium permanganate crystals

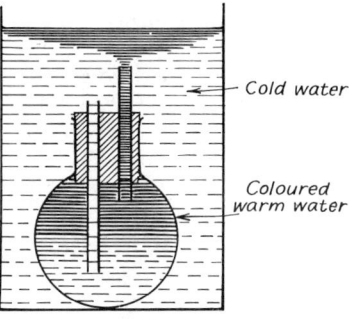

Fig. 19.5 Convection current

Flask of hot, coloured water placed in beaker of cold water. Convection current produced by cold, dense water pushing hot water out of flask

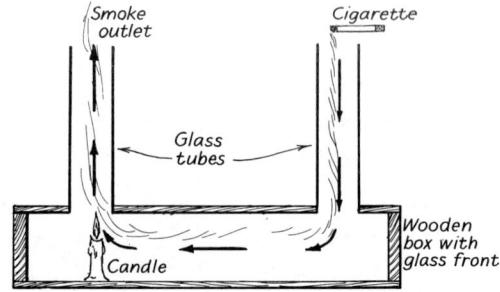

Fig. 19.6 Primitive mine ventilation

Burning candle sets up convection currents shown by the path of the smoke

piece of smouldering paper over the top of the other chimney.

When a glass tube, of diameter about 4 cm and length 30 cm, is placed over a lighted candle, it soon goes out. This does not happen if the tube is provided with a central partition, for this facilitates the production of a convection current which keeps the candle supplied with oxygen (Fig. 19.7).

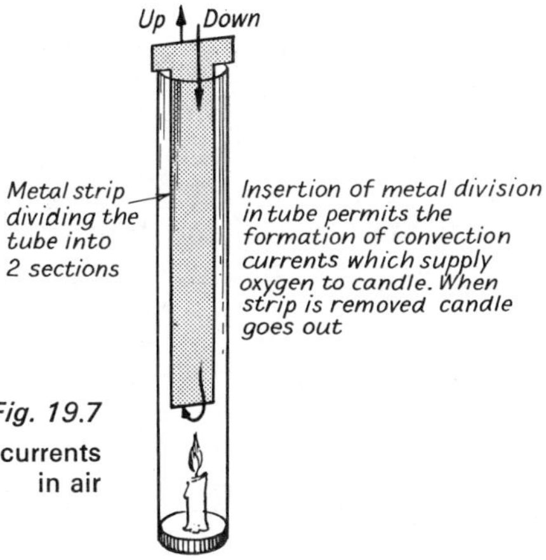

Up ↑ ↓ Down

Metal strip dividing the tube into 2 sections

Insertion of metal division in tube permits the formation of convection currents which supply oxygen to candle. When strip is removed candle goes out

Fig. 19.7
Convection currents in air

Convection currents produce ventilation, which is the process by which polluted air is replaced by fresh air. A coal fire creates a convection current in the chimney, cold air entering the room through the spaces around the doors and windows.

The air near the ceiling is always warmer than that near the floor. Two simple experiments show this.

Hold a burning candle near the gap between the door and floor. The flame will indicate air coming into the room. Now hold it at the top of the door, and it will show warm air leaving.*

Tie a box of matches to a balloon filled with coal gas so that it just 'floats' at a height of about 1.5 m in a heated room. (This can easily be done by adjusting the number of matches in the box.) Release the balloon near a wall and it will indicate the circulating air currents.

The main wind system of the earth is a particularly good example of convection. Where the sun is directly overhead, the ground below becomes the warmest place on earth (Fig. 19.8). The hot ground, in turn, warms the air in contact with it,

_____
* Provided the chimney draught is not too strong.

thereby setting up a large convection current. At the time of year when the sun is directly over the Equator, the cold air comes in from the north and south. But the rotation of the Earth causes a deflexion of the currents, which consequently become the famous north-east and south-east Trade winds. As the sun moves between the Tropics, the wind system also moves.

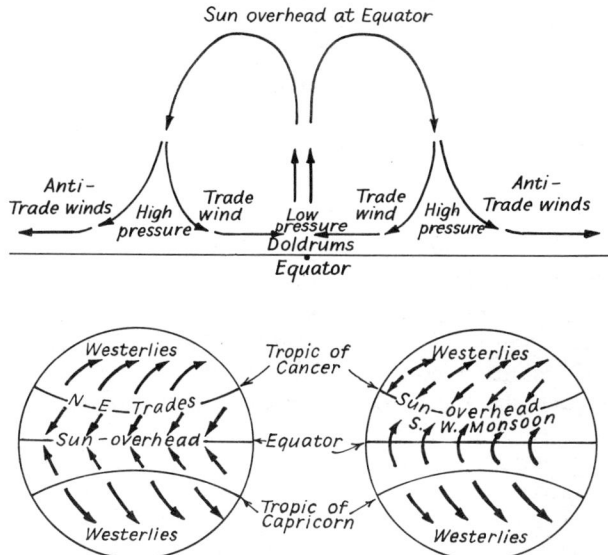

Fig. 19.8 The main wind system of the earth

Where the upward convection current occurs, there is no wind across the earth's surface and so there is a belt of calm produced known as the Doldrums (Fig. 19.8). Mariners in the old sailing ships experienced difficulty in crossing this belt.*

When the upward stream of warm air becomes cooled in the upper regions, it descends to earth again in two parts, producing high-pressure regions. From these areas flow the winds of the temperate latitudes of both hemispheres. These are known as the Prevailing Westerlies, for they suffer a deflexion of nearly 90°.

The winds are greatly affected by the land

_____
* Day after day, day after day,
  We stuck, nor breath, nor motion;
  As idle as a painted ship
  Upon a painted ocean.
          From *The Rime of the Ancient Mariner*
                    by Samuel Taylor Coleridge.

masses, and so the resulting system is far more complicated than the brief outline above would indicate.

Ocean currents

Ocean currents are very large convection currents in the sea. The warm currents travel along the surface and the cold ones below the surface. The Gulf Stream—a warm stream—reaches the shores of the British Isles and exerts a moderating influence on our climate. It is interesting to note that the Gulf Stream is higher in the middle than at its edges which are marked by flotsam weeds that have drifted 'downhill'.

Radiation*

We have mentioned previously on page 166 that radiation is a form of energy. An important type of radiation consists of electromagnetic waves† which include most of the radiations that affect our everyday lives and with which, therefore, we are familiar. E.M. waves within a particular range of wavelengths are visible.

E.M. waves are generated by accelerating electric charges. Vibrating atoms create radiations and so all material bodies are emitting, and also absorbing, E.M. radiation. The objects in a room maintain their temperature constant by absorbing and emitting at equal rates. The rate of emission increases rapidly with temperature; also, shorter waves are emitted as the temperature rises, until, when a body is very hot, it emits waves short enough to affect the human eye thus making the body visible.

When radiation falls upon a material substance, it may be absorbed. The absorption of radiation, within a particular range of wavelengths causes thermal vibration of the molecules, and so the substance is heated. The radiant energy is converted into kinetic energy of the molecules. The radiation itself is not hot, the heating being the result of the absorption. The terms 'thermal radiation' and 'heat rays' do not imply that the waves are hot, but that they are of a wavelength‡ lying within

the range of E.M. waves emitted by a hot body, which may be absorbed by a body, and produce a rise in temperature. The sun's radiation does not warm the space through which it travels to the earth, for there are very few molecules in space to absorb the radiation.

For absorption to occur, the frequency of the radiation should 'match' that of the atoms or molecules. It is in some respects like pushing a swing or a pendulum; to maintain the oscillations, the impulses must be given at the correct intervals.*

Little of the sun's radiant energy is absorbed by the air. (The air is heated by conduction from the warmed ground.) Some substances are more transparent to thermal radiation than others, e.g. rock salt is much more transparent than glass.

The range of E.M. waves which produce heating effects extends from the visible far out into the infrared, i.e. the region beyond the red end of the visible spectrum. The most effective of these rays are those in the infrared region.

Detection of heat rays

The human body will detect heat rays, but it is not sensitive enough for experimental purposes. Photographic plates which are affected by the long heat waves (infrared) are now made (Figs. 19.9 and 19.10). An ordinary mercury thermometer with a blackened bulb will detect radiation in the infrared region.

Though there are many other more sensitive and complicated instruments for detection, the radiometer is simple and easy to use for elementary demonstrations (Fig. 19.11). It consists of a glass bulb in which four light aluminium vanes are pivoted so that they may revolve with little friction. One corresponding side of each vane is blackened. Nearly all the air is exhausted from the bulb. When radiant energy falls upon the vane system, the blackened sides become slightly hotter than the shiny sides (see below). The air molecules hitting the vanes on the blackened sides rebound with increased speeds. The reactions to the molecular rebounds on the two sides, being unequal, cause the rotation of the vanes with the shiny surfaces leading.

The best known instrument for measuring thermal radiation quantitatively is the thermopile.

---

* Some of the detail concerning radiation included in this section can best be understood after studying Chapter 20.

† See page 253.

‡ See Chapter 20.

---

* This is allied with 'resonance' which is also mentioned in Chapter 30.

Fig. 19.9 Photography by infrared
The picture was taken in a dark room to demonstrate the efficiency of an EMI camera. Instead of relying on light reflected from objects in focus the camera creates pictures from the heat given off by them. In this case, the man's glasses, being relatively cold, do not show up as well as his body and clothing. But because the electric blanket he is holding is switched on, the wires in it can be clearly traced

(*a*) Normal photograph of bust

(*b*) Photograph taken in darkness with bust illuminated by two electric irons as infrared sources (Exposure 1 hour, *f*/4.5)

Fig. 19.10 Infrared photographs

This consists of a large number of thermocouples (*see* page 130) arranged in series, with all the corresponding junctions massed into two small areas. The radiation is allowed to fall onto one of these areas, which is blackened so as to absorb the radiation better. The resulting small rise in temperature causes an e.m.f. of sufficient magnitude to be measured by a low-resistance galvanometer (or micro-voltmeter).

A modern piece of equipment which is now rapidly replacing the thermopile for demonstrating the properties of heat radiation is the *thermistor*.

Fig. 19.11 A radiometer

This is a solid-state device which generates a small potential difference when E.M. radiation, within a limited band of wavelengths, falls upon it.

Properties of heat radiation

The properties* of infrared radiation differ little from those of visible light.†

Heat rays travel with the same speed as visible light.

This is proved by a simple observation. During a total eclipse of the sun, the light and heat from the sun are cut off at the same instant on the earth's surface.

Heat rays travel in straight lines.

Again, a simple observation proves the statement. When we move into the shade of a building, our bodies notice the stopping of the heat radiation as well as of the light.

Heat rays obey the inverse square law.

The intensity of the heat rays falling normally on a surface varies inversely as the square of the dis-

---

* These properties can be made more convincing if the visible rays from the source are cut off by using an infrared filter.

† Many excellent experiments are given in Nuffield *G. to E.*, II, Nos. 89–95.

tance from a small heat source. This can be proved experimentally using a constant source of radiation such as a small bowl-fire element in conjunction with a thermopile. Readings are taken with the thermopile at different distances from the source. The inverse square law is verified by showing that the product of the voltmeter reading and the square of the distance is a constant. A similar experiment for visible light is described on page 197.

Heat rays can be reflected.

The reflection of heat rays obeys the laws of reflection (*see* page 200). This can be shown by the

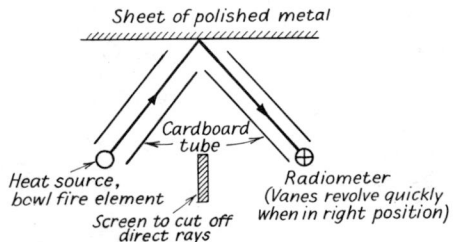

Fig. 19.12 Reflection of 'heat' rays at a plane surface

simple arrangement in Fig. 19.12. More spectacular experiments can be performed using

Fig. 19.13 The largest solar furnace in the world

This concave mirror which is in the Pyrenees, has a diameter of 11 m and is composed of 3500 small plane mirrors

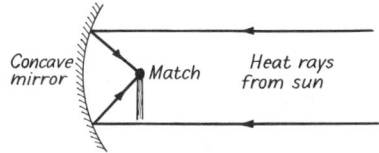 is part of Fig 19.15 diagram; is Fig 19.16 diagram.

Fig. 19.14 Solar furnace constructed by US Army, which can attain a temperature of nearly 3000 K

Sun's rays are reflected by plane mirror *C* on to concave mirror *E* which brings them to a focus at *F*

Inset: iron girder *B* held at focus of rays

concave or parabolic mirrors, for the heat rays can then be focused. If the sun's rays are allowed to fall upon a concave mirror, and a match is held where the light is brought to a focus, the match is ignited (Fig. 19.15). This shows that the heat rays

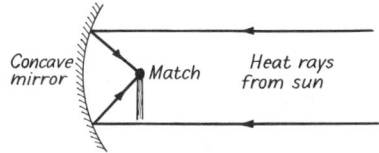

Fig. 19.15 Reflection of 'heat' rays

The match is ignited by absorbing radiant energy focused by the concave mirror

are focused at the same point as the visible light rays.

A similar experiment (Fig. 19.16) can be carried out using two parabolic reflectors. (These can be made by bending two sheets of 'tin' about 50 cm by 20 cm.) A bowl-fire element at the focus of one produces a parallel beam of heat rays which can be brought to a focus by the other placed several metres away. A radiometer can be used to show this. If a good *point* source, such as an arc, is used in place of the radiator element, a match can be ignited at the focus of the second reflector.

Some electric fires are constructed with curved

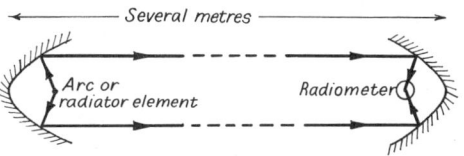

Fig. 19.16 Reflection of 'heat' rays by parabolic mirrors

metal reflectors behind the elements in order to distribute the heat about a room.

The earth loses much of its heat by radiation. On a clear night, radiation goes on unhindered, with a consequent drop in temperature. On a cloudy night, however, some of the radiated heat energy is absorbed and re-radiated back by the clouds, so that cooling is not so pronounced.

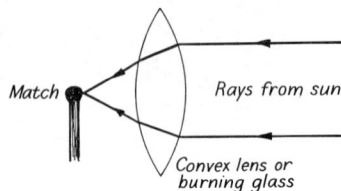

Fig. 19.17 Refraction of 'heat' rays
Convex lens focuses the radiation

Heat rays can be refracted.

The burning-glass illustrates this property (Fig. 19.17). This is a simple convex lens which focuses the heat rays by refraction.

Black surfaces are good absorbers: shiny surfaces are poor absorbers.

Obtain two similar-sized rectangular tins (Fig. 19.18): blacken the outer surface of one and leave the other bright. Fill both of them with equal

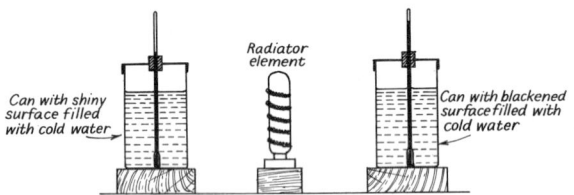

Fig. 19.18 Absorption of 'heat' rays
Can with blackened surface becomes warmer much quicker than can with shiny surface

quantities of cold water and insert thermometers through holes in the lids. Arrange the tins one on each side of a radiator element at equal distances. Note the temperature changes. The reading of the thermometer in the blackened tin increases at a much greater rate than the other. If this experiment is done in summer time, the two tins can be put outside in the sun (stand them on wooden blocks) and the effect will be shown in a few minutes.

Another simple way of showing this property is by holding a piece of heat-sensitive paper in the sun or near a fire. If half the paper is painted with dull black paint, the reverse side of this soon shows green while the other half is little affected.

The hand can be used as a detector for this effect. Two neighbouring patches of the palm are painted, one with aluminium paint and the other with dull black paint. The hand is then held in front of an electric radiator and a marked difference is felt in the heating of the two patches.

Employing the same two cans as in the above experiment, fill each with hot water, equal quantities, equal temperatures. Leave the cans to cool under identical conditions and regularly take the temperatures. That of the blackened can falls more quickly. This shows that a black surface is a good radiating surface. (The cans lose heat by conduction and convection at the same rate.)

This experiment is performed quantitatively by using a thermopile to measure the radiation from a source known as a Leslie cube. This consists of a cubical tin of about 15 cm side, which can be filled with boiling water. The four vertical faces of the cube have surfaces of different types, e.g. dull black, shiny black, dull white and brightly polished. The thermopile is arranged at an equal distance from each surface in turn, and the steady readings of the meter connected to it are taken. An estimate of the relative radiating powers of the different surfaces is obtained by comparing the meter readings.

Black surfaces are good radiating surfaces.
Shiny surfaces are bad radiating surfaces.
Good radiators are good absorbers and bad reflectors.
Bad radiators are bad absorbers and good reflectors.

Silver teapots are used, not only because they can be made to look attractive, but also because they do not cool very quickly. In hot countries, butter is usually kept in a silver container along with ice. The shiny surface does not absorb the heat rays readily. The roofs of single-storey factories are sometimes whitewashed in summer to keep the inside cooler.

Fig. 19.19
Aluminium foil being laid between the over joists for heat insulation

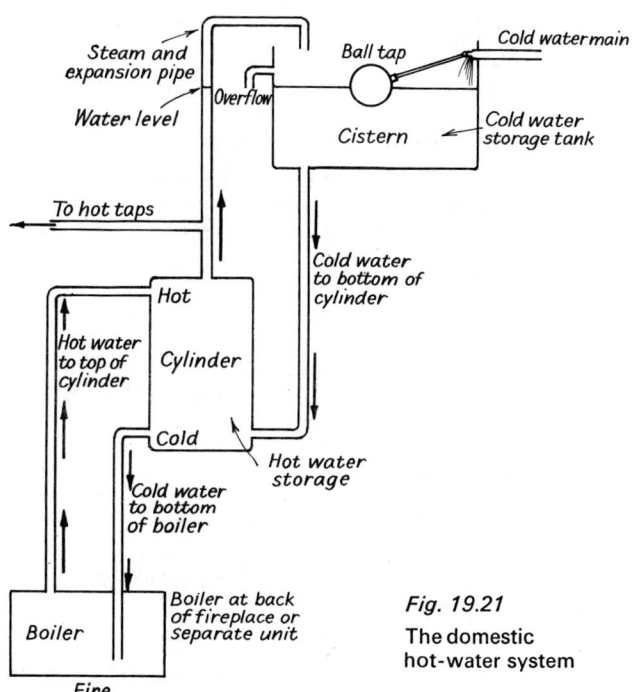

Distribution of thermal energy in the spectrum*

A spectrum is produced on a screen using a powerful small source such as an arc or a quartz-iodine lamp. If the bulb of a blackened thermometer is transversed across the spectrum, it will register a rise in temperature. This rise will not be constant: it will start at zero in the green and blue regions and attain a small value at the red end of the visible spectrum. Beyond this end, that is in the infrared region, it will show a greater response. Eventually, it attains a maximum value and then drops back to its original reading, i.e. room temperature.

This demonstration can be more convincingly performed using a phototransistor. Such a device has a small heat capacity and so is quick-acting and functions well with a suitable quick-acting galvanometer. A thermopile can also be used.

Transparent and opaque substances

The transparency of a substance to thermal radiation may be investigated by placing a sheet of the substance between the source and a detector, e.g. a thermopile or a radiometer. Glass is found to be only slightly transparent. Some substances, e.g. rock-salt and calcite are very transparent.

The greenhouse

If glass absorbs heat rays, how do we explain the heat developed in a greenhouse? Glass is transparent to heat rays from a very high temperature

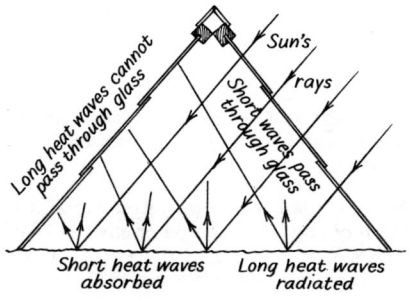

Fig. 19.20 Glass of greenhouse traps energy from sun

source, but opaque to those from a cooler source. The heat rays from the sun (the short waves only)

are transmitted by the glass and absorbed by the plants, etc. inside. These, in turn, re-radiate the absorbed energy in the form of rays of long wavelength, to which glass is opaque, hence the heat is trapped (Fig. 19.20).

Hot-water system

The hot-water system in a house illustrates the modes of heat transmission. Sufficient water has to be heated and stored to supply the domestic needs, e.g. for a bath. The diagram (Fig. 19.21)

Fig. 19.21
The domestic hot-water system

shows how this is accomplished. Cold water is stored in the cistern, the inflow being controlled by a ball tap. From here, it flows to the boiler by way of the cylinder. The boiler is either contained in a stove or behind a fireplace. The heated water sets up a convection current and travels to the top of the cylinder, where it is stored. If no water is drawn off, the circuit from cylinder to boiler and from boiler to cylinder is maintained. A pipe has to be provided to allow for the expansion of the enclosed water, and for the escape of dissolved air and steam should the water boil.

---

* See Nuffield *G. to E.*, II, No. 97.

The pressure at the taps is obtained from the head of water produced by the high level of the cold supply in the cistern.

In a building heated by low-pressure hot water, the system is similar in principle to the domestic system, the radiators and pipes replacing the cylinder. The water is not consumed so there is no trouble from furring.

Conduction and radiation both play a part in the boiler. Convection is responsible for the circulation of the water through the pipes. Conduction through the walls of the pipes and radiators heats the air and, once more, convection 'spreads' the heat by creating air currents. This is the chief way in which the air is warmed, although radiation plays some part in heating the room.

The rate of flow of water in the gravity-controlled heating system depends upon the slope of the pipes and their diameters. Modern central-heating systems now use small-diameter pipes (for ease in installing and for appearance) and drive the hot water round the circuit by means of a pump.

The vacuum flask

A vacuum flask is used to keep liquids hot or cold. To prevent a liquid cooling, we must minimize heat losses by conduction, convection and radia-

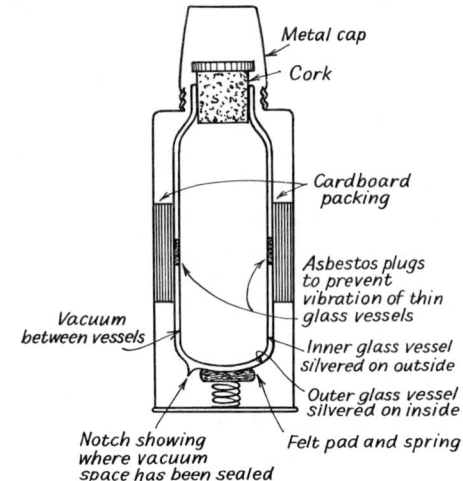

Fig. 19.22 The vacuum flask

tion. Conduction and convection are eliminated by containing the liquid in a double-walled vessel, the air in the space between the walls being evacuated. Radiation is reduced by silvering the inner (facing) sides of the double walls. Loss of heat by evaporation is minimized by corking the vessel (*see* Fig. 19.22).

QUESTIONS

1. How would you show that copper conducts heat better than aluminium.

2. Explain the following:
   (*a*) Steam pipes are often lagged with asbestos cement.
   (*b*) Pieces of broken glass bottles left on the ground sometimes cause fires during sunny weather.
   (*c*) The roofs of mills are sometimes painted white in summer.

3. Give experiments to illustrate the statement: 'Good absorbers are good radiators; bad absorbers are bad radiators.'

4. What are the chief features that distinguish conduction from other means of heat transference?
   Describe an experiment to show that water is a bad conductor of heat.
   Give a labelled diagram of a Davy safety lamp, and explain how it works. (O)

5. What are the chief characteristics of radiation as a method of heat transference? Describe one piece of apparatus for the detection of heat radiation, and explain briefly how it is used.

How would you demonstrate experimentally that a blackened surface is a good radiator of heat, and also a good absorber? (O)

6. Give a short account of the methods by which a vessel containing a hot liquid loses its heat.
   Draw a diagram of a vacuum flask and explain how the rate of loss of heat by a hot liquid placed in the flask is reduced to a minimum. (L)

7. A red-hot brick is placed on an iron tripod which stands on a large slab of copper. Describe briefly the ways in which the brick loses heat, and state the effect on them of (i) replacing the copper slab by a stone slab, (ii) placing on top of the brick a thin sheet of polished metal of the same area. (C)

8. Describe experiments to determine which is the greater in each of the following:
   (i) the thermal conductivity of wood or of copper,
   (ii) the radiating power of a brightly polished metal surface or of a dull black surface.
   State the result you would expect in each experiment, and mention a practical application of each result. (L)

9. Distinguish fully between *conduction, convection* and *radiation* of heat.

Give a diagram of a 'Thermos' flask and describe its special features. Explain its action in keeping liquids hot.

A 'Thermos' flask is found to be inefficient. Suggest a possible cause for this.   (S)

10. How would you show:

(*a*) that the sun's spectrum contains invisible infrared radiation;

(*b*) that a blackened surface is a better absorber of heat radiation than a polished one;

(*c*) that a blackened surface is a better radiator than a polished one?   (O)

11. State and explain the methods by which a hot body may lose heat to its surroundings.

Describe the part played by convection in (*a*) some form of domestic heating, (*b*) the freezing of a pond.   (O & C)

12. Describe and explain an experiment to show that heat can be transmitted through mercury by conduction.

Mercury can dissolve certain other metals to give liquid alloys. How would you find by experiment whether such an alloy was a better or worse conductor of heat than mercury?   (O & C)

13. Describe the ways in which a heated metal block cools when it is supported by a thin wire in a vessel containing air.

Draw a sectional diagram of a Dewar ('Thermos') flask and explain the function of each important feature. (O & C)

14. Explain the following observations:

(*a*) Water in a test-tube can be heated at the top to boiling-point without melting a lump of ice at the bottom.

(*b*) If a thick copper rod and thin copper wire, each about a foot long, are heated separately in a Bunsen burner the heated end of the wire becomes red hot while that of the rod does not; the cool end of the wire is, however, cooler than the cool end of the rod.   (O & C)

15. In certain circumstances the cooling of a body in air occurs at a rate which is similar to the rate at which radio-active decay occurs. The following observations show the result of testing this in the particular case of a body cooling in a steady current of air; the average air temperature was observed, and the values of θ are the temperatures of the body *above* this.

$\theta/°C$	80.0	62.5	49.0	38.1	29.5	18.5	11.2	7.0	2.5
Time/min	0	$\frac{1}{2}$	1	$1\frac{1}{2}$	2	3	4	5	7

Plot the graph of θ against time. Use the graph to find out whether θ falls to half its value in equal intervals of time; do this for a series of values for θ, e.g. 80 and 40 degrees, 70 and 35 degrees, etc. Give a summary of your results and draw a conclusion from them.

Show, by means of a labelled sketch, the apparatus you would use to conduct such a cooling experiment.   (C)

20 Wave motion

The nature of wave motion

We have already mentioned that energy can be transmitted by means of radiation (Chapter 19). We must now consider this form of transmission of energy in greater detail, because it is a dominating feature of everyday life.

A wave motion consists of oscillations generated by a rhythmic disturbance. We shall see that a vibrating rod, string or column of air creates a sound wave and that the disturbance causing a light wave lies within the atom.

Two types of progressive waves

1. *Transverse.* With this type of wave motion the oscillations are at right angles to the direction of motion of the wave. The oscillations may be, for example, those of a varying electric field or of water particles on the surface of a pond: the former produce light waves and the latter ripples over the surface.

2. *Longitudinal.* Here, the oscillations of the medium through which the wave passes are along the direction of propagation of the wave. Sound waves are longitudinal waves.

Progressive waves

The displacements in both types of wave motion follow the same pattern. The medium itself does not move in the direction of propagation. It is the *waveform* which progresses.

This progression means that energy is handed on and so a **progressive wave radiates energy.** This contrasts with a *standing* wave (*see* page 189), which does not radiate energy.

A clear idea of wave motion can be obtained by studying ripples on water. Observation of a cork floating on a pond on which ripples are produced, shows that the cork simply bobs up and down while the waveform passes over the surface. Each particle of water at the surface is moving up and down. The motion of the water particles resembles that of a mass on the end of an oscillating spiral spring, and is *simple harmonic* (*see* Chapter 11). If an oscillating body displaces an elastic medium, the local displacement of the medium will be communicated to the surrounding medium. When a finger is thrust through the still surface of a bowl of water, the local displacement of the water affects that near to it and the displacement is spread over the surface. A continuous succession of such pulses can be produced by using a dipper vibrating with constant frequency. The water at the dipper disturbs that next to it and so a continuous succession of ripples is generated. The phase of the disturbed water is behind that of the disturbing water. Thus it reaches full amplitude a little later. As the lagging behind is carried on through the water the waveform progresses but the

[handwritten annotations in top margin: "= up + down movement of particles.", "These aren't the same !!", "= ripples in the medium"]

water is moving only up and down. If A, B, C, etc. in Fig. 20.1 represent a number of neighbouring particles of water on the undisturbed surface, then if A is disturbed, it will begin to vibrate *[circled]*

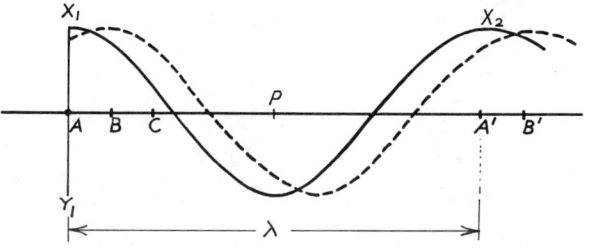

Fig. 20.1 Production of a transverse wave

between the limits $X_1 Y_1$. But A will disturb B which will describe the same motion as A, but *it will start a little later than A.* There will be a phase difference between A and B. The disturbance will be handed on from particle to particle, each one being a little behind its predecessor. When the displacement of A is AX_1, that of B is a little less, and so on. The full line in the figure shows the displacements at a particular instant. After the lapse of a short interval of time, the dotted line will show the displacements. Thus the waveform progresses from left to right while the displacements are up and down. The particles at A and A' have the same displacements, but A' is a complete vibration behind A. Such particles are in the same *phase* and are one wavelength λ apart. If it takes

time t for the waveform to travel from A to A', then,

[handwritten: eq .005]

$$\lambda = vt \text{ where } v \text{ is the velocity}$$

but $1/t$ = number of vibrations *[circled]* per second *[handwritten: $\frac{1}{.005} = 200$]*

$$= \text{frequency, } f$$

$$\therefore \qquad v = f\lambda$$

[handwritten: Velocity = no. vibrations per sec × wave length]

Sound waves are longitudinal, and when they pass through a solid, a liquid, or a gas they cause the molecules to vibrate along the direction of the waves. If an iron bar is struck with a hammer at one end, the molecules at that end are pushed by the hammer. They, in turn, push those next to them, and so on to the other end of the bar. Sound passing through a medium does the same kind of thing, but delivers a rapid succession of blows to the medium. The prongs of a sounding tuning-fork are vibrating and cause the molecules of air to oscillate in pendulum fashion. The handing-on from molecule to molecule, however, takes some time and consequently each layer of molecules is a little late in the state of its oscillation compared with its predecessor.

Both forms of wave motion can be illustrated with the model wave apparatus shown in Fig. 20.2. To show transverse waves (Fig. 20.3), the white balls (which are attached to strings and slide up and down rods) are made to execute oscillations above the centre line. There is a phase difference between successive balls—the total phase difference between the first and the last is 360°.

Fig. 20.4 shows a longitudinal wave motion. The balls execute S.H.M. on the ends of the rods.

A longitudinal wave can also be demonstrated

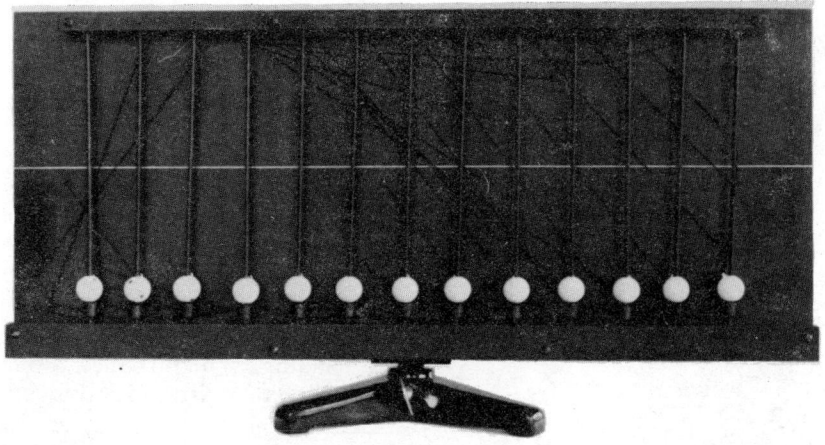

Fig. 20.2 Apparatus to demonstrate transverse and longitudinal waves (*see* Figs 20.3 and 20.4)

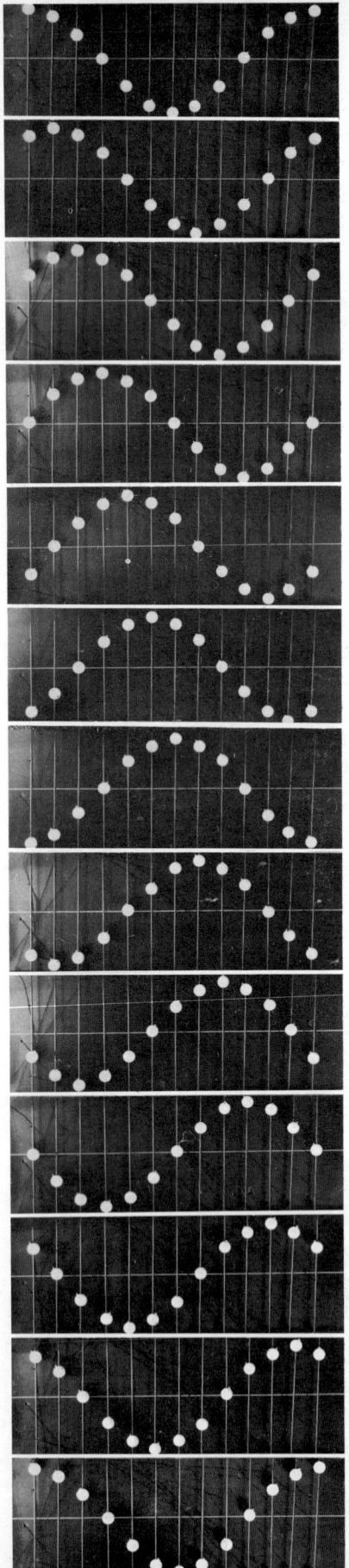

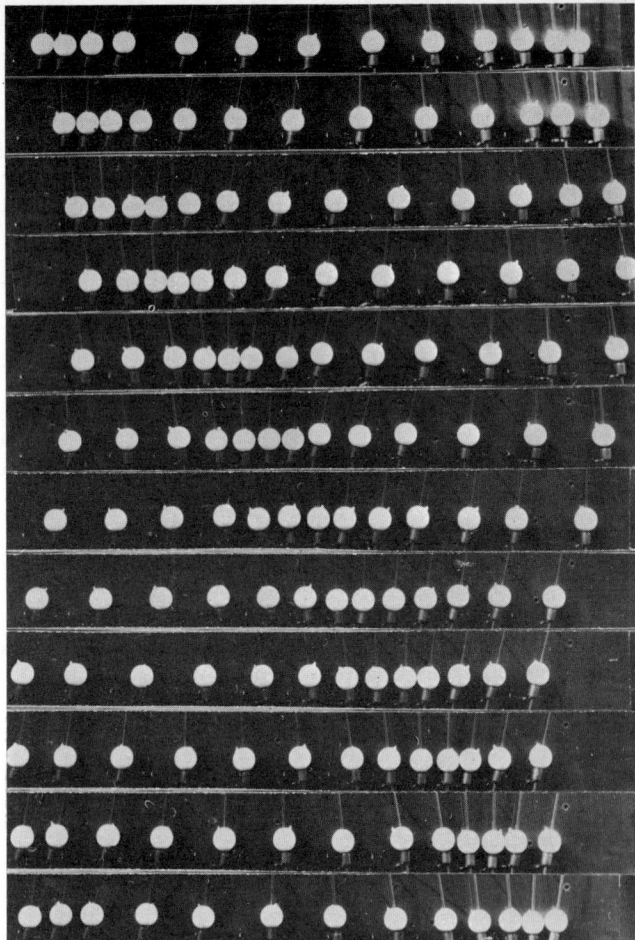

Fig. 20.4 Longitudinal waves

Successive photographs taken on the machine illustrated. in Fig. 20.2. The compression can be seen crossing from left to right. The simple harmonic motion of each sphere can also be seen.

by using a very long spiral* of suitable gauge wire supported horizontally, either by numerous long lengths of thread or simply resting on a table with a very smooth surface. (The reflection of the wave can be seen better when the spring is held vertically with the upper end as high as possible and the lower end resting on the floor.) An impulse given at one end and directed along the axis of the spiral passes along the whole length of the spiral. The handing-on of the vibration can be clearly seen as the progression of compressions and rarefactions. The change in velocity when a longitudinal wave passes from one medium to another can be shown by sending a disturbance along two spirals of different gauge wires joined together end-to-end.

---

* These spirals can now be purchased and are known as 'slinky' coils.

Fig. 20.3 Transverse waves

Successive photographs taken with the wave machine illustrated above. Note the simple harmonic motion of each sphere as the waveform progresses from left to right

General characteristics of a progressive wave

Methods of demonstrating waves

Ripple Tanks

The general characteristics of wave motion can be demonstrated experimentally using a ripple tank.* In its simplest form it consists of a shallow water tank with a glass bottom and shelving wooden sides to eliminate reflection. The waves can be generated by using a finger or a ruler, but a mechanical method is advantageous. A small electric motor loaded eccentrically is commonly used to produce vibrations.

It is mounted on a wooden bar to which various dippers may be attached as shown in Fig. 20.5. A shadow of the ripples can be cast on the ceiling by placing a lamp underneath the tank. If a large plane mirror is available, it can be adjusted to project the light upon a wall screen. The shadow picture shows the ripples moving, but they can be apparently slowed down or stopped, and so made easier to examine, by viewing them through a stroboscopic wheel. This is a simple disc about 20 cm in diameter provided with equally spaced radial slots and mounted so that it can be spun round by hand.

More elaborate tanks (Fig. 20.6) are available. These make the waves appear stationary by incorporating a stroboscope which is rotated by a mains motor at the correct speed to match the vibrations of the dipper, which is also actuated by the a.c. mains. Such tanks are quite small and the ripple patterns are projected on to a screen.

Short radio waves

A ripple tank illustrates the properties of water waves, but similar properties are displayed by other wave motions. These may be shown using visible light, but because the wavelength of visible light is so small, the effects are so minute that, in many cases, instruments are required to see them (*see* later in this chapter). We can, however, demonstrate many wave effects by using radio waves of about 3 cm wavelength.† There is no need to understand the construction of the transmitter and of the receiver (Fig. 20.7). The transmitter radiates radio waves of wavelength 3 cm.

---

* *See* Nuffield *G. to E.*, III for full details of the manipulation of a simple ripple tank.

† *See S.S.R.*, No. 155.

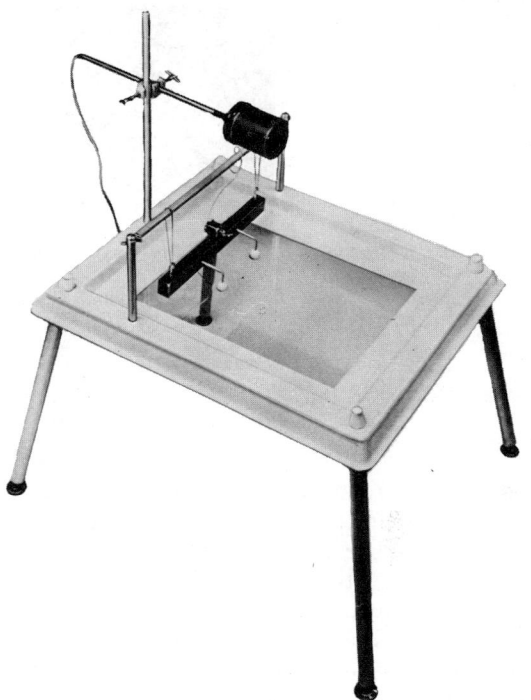

Fig. 20.5 Simple ripple tank

Fig. 20.6 Ripple tank

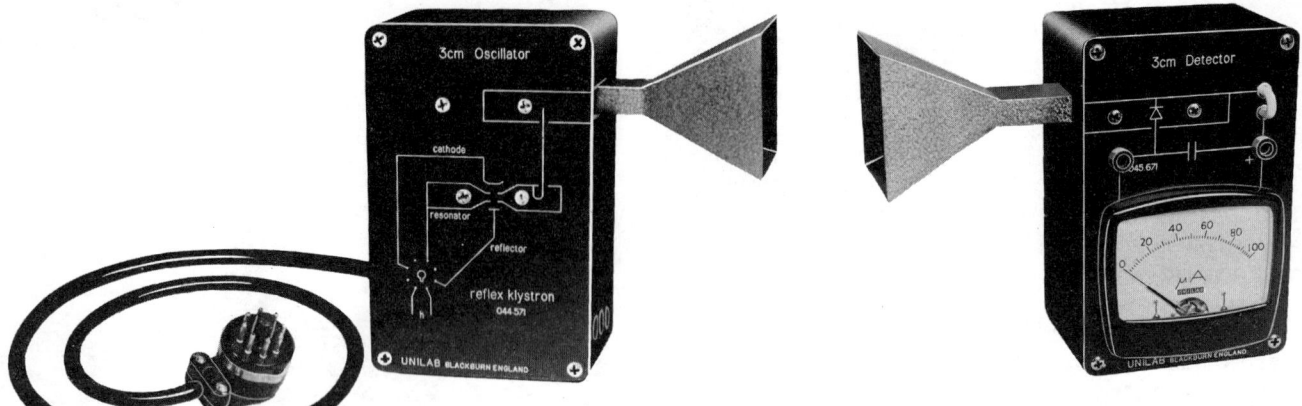

Fig. 20.7 3-cm EM wave transmitter and receiver

These have a frequency of 10^{10} Hz and so cannot be heard when they are received and passed through a loudspeaker. By what is called modulating the waves with waves of frequency 1000 Hz, we can, in effect, make the waves audible in a loudspeaker. The waves are received using a small dipole aerial. The signal after amplification is passed through a loudspeaker and also through a microammeter.

Wave properties

Independence of wave trains

Two trains of waves can pass through each other without affecting each other. This is borne out by all the radio waves of different wavelengths passing through each other without producing any effect upon the reception from one particular station. Two trains of ripples produced in the ripple tank pass through each other without affecting each other in any way. Similarly, two visible light beams pass through each other without any inter-effect.

Reflection of waves

When a wavefront AB (Fig. 20.8), is incident upon

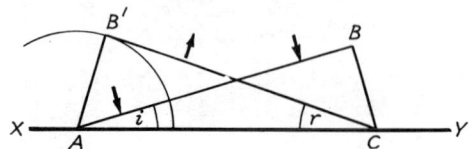

Fig. 20.8 Reflection of a plane wave (at a plane surface)

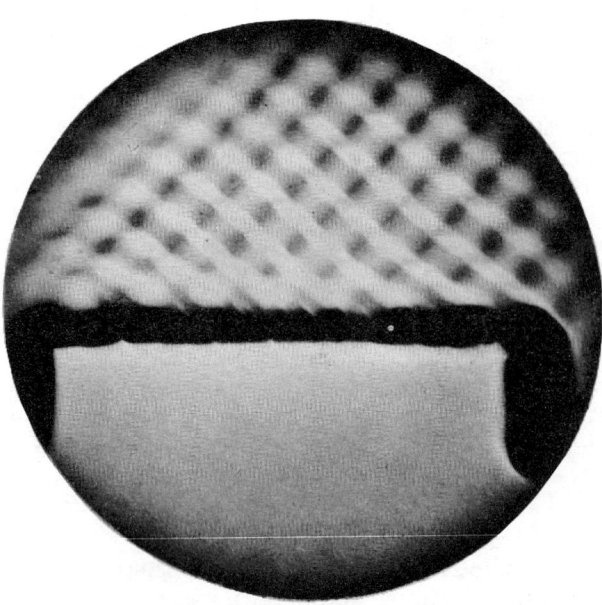

Fig. 20.9 Reflection of ripples at a fixed plane surface

a reflecting surface XY, the disturbance created at A will have a radius AB' ($=BC$) when that at C is just beginning. The reflected wavefront will be $B'C$. From the geometry of the figure, the incident and reflected beams are seen to make equal angles with the surface XY.

This can be demonstrated in the ripple tank by generating a plane wave with a line dipper and reflecting it at a vertical surface set up in the tank (Fig. 20.9).

A point dipper can be used to illustrate spherical waves emitted by a point source. After reflection at a plane surface, the ripples radiate as from a point as far behind the reflecting surface as the dipper is in front.

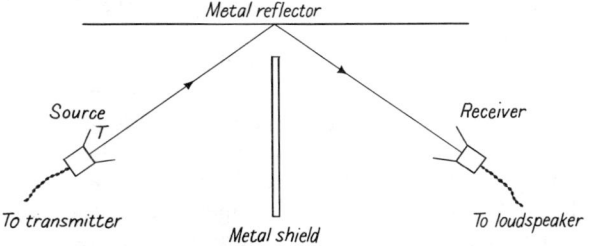

Fig. 20.10 Reflection of 3-cm radio waves

Reflection of radio waves is shown by using the arrangement in Fig. 20.10. The source is arranged so that the radiation is directed on to a flat metal plate, e.g. hardboard covered with aluminium foil. The waves are reflected by the metal plate and are received as shown. The path of the reflected waves is found by trial by adjusting the position of the receiver so that the sound in the speaker or the reading of the meter is a maximum.

Refraction of waves

When a wave passes from one medium into another in which its velocity is different, a change in direction results from the change in velocity. This effect is illustrated by a section of troops marching in line on hard ground and crossing obliquely on to grass. Because of the change in speed on the softer ground the direction of motion is changed.

This can be understood from Fig. 20.11. AB is a

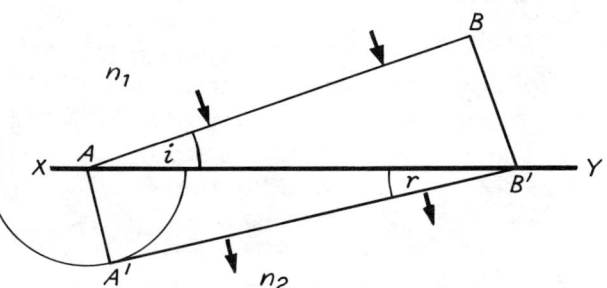

Fig. 20.11 Refraction of a plane wave (at a plane boundary)

wavefront of a train of waves approaching the boundary XY between two media. If the velocity in the second is less than that in the first, the disturbance arising at A will have travelled only as far as A' when B has reached B'.

The ratio $\dfrac{BB'}{AA'} = \dfrac{\text{velocity of wave in medium 1}}{\text{velocity of wave in medium 2}}$

By dividing numerator and denominator by AB',

we have, $\dfrac{BB'/AB'}{AA'/AB'} = \dfrac{\sin BAB'}{\sin AB'A'} = \dfrac{\sin i}{\sin r}$

Thus we see that the inclinations of the incident and refracted wavefronts to the boundary depend on the ratio of the velocities in the media. This ratio for visible light is called the refractive index.

Refraction is demonstrated in the ripple tank by changing the velocity of the ripples by changing the depth of the water. A glass plate is put in the tank so that the velocity of the ripples is reduced in the shallow water. The bending of the wavefronts as the ripples pass from the deeper to the shallower water can be observed (Fig. 20.12).

Fig. 20.12 Refraction of plane waves produced in a ripple tank

The change in direction of the plane waves is produced by changing the depth of the liquid, which changes their velocity

The refraction of 3 cm waves is shown by passing them through a medium such as paraffin wax in which their velocity is less than it is in air. To show refraction through a prism the transmitter is arranged as in Fig. 20.13 so as to direct the waves

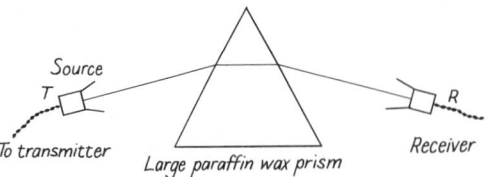

Source
T
To transmitter   Large paraffin wax prism   Receiver   R

Fig. 20.13 Refraction of 3-cm radio waves through a prism of paraffin wax

on to one face of the large paraffin wax prism. The receiver has to be placed as shown to pick up the waves which have been refracted on entering and leaving the prism.

Diffraction of waves

Diffraction is the bending of waves round a corner. We have noted that light travels in straight lines and we have observed that sound bends round corners.

These two statements require some explanation; but let us examine the behaviour of waves on water. Fig. 20.14 shows plane ripples passing through a small opening and Fig. 20.15 shows the effect of using a wider opening. In the second case the wavefront is only slightly curved at the edges and straight in the middle, but in the first case the wavefront is curved and extended. This shows that bending at the edges is greater for the narrower slit. The bending shown with the ripple tank can be increased either by using a narrower slit or by using ripples of longer wavelength. At a straight edge there is, therefore, just a little bending which is smaller the smaller the wavelength of the waves.

This result is important for it tells us that the diffraction of light waves with their extremely short wavelengths will be extremely small. On the other hand, sound waves of long wavelengths should be diffracted appreciably.

Diffraction is easily demonstrated with 3 cm waves for the effect is appreciable in waves of this

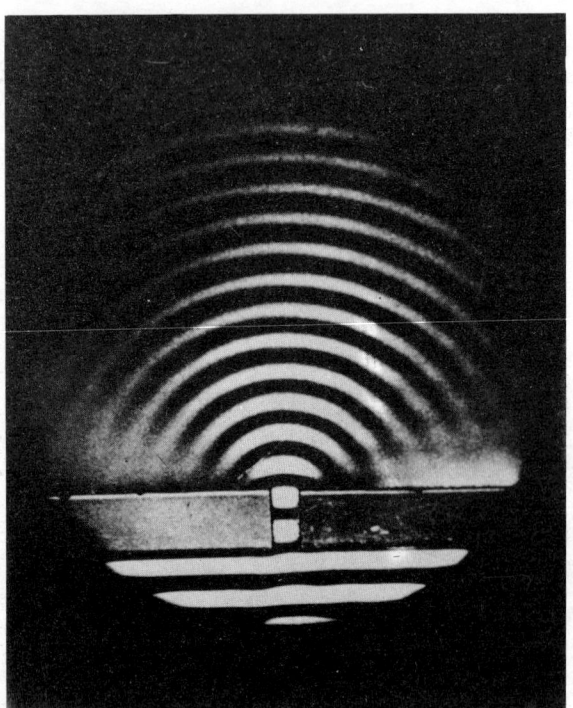

Fig. 20.14 Diffraction of plane waves in passing through a small aperture

Note the large amount of bending (cf. Fig. 20.15)

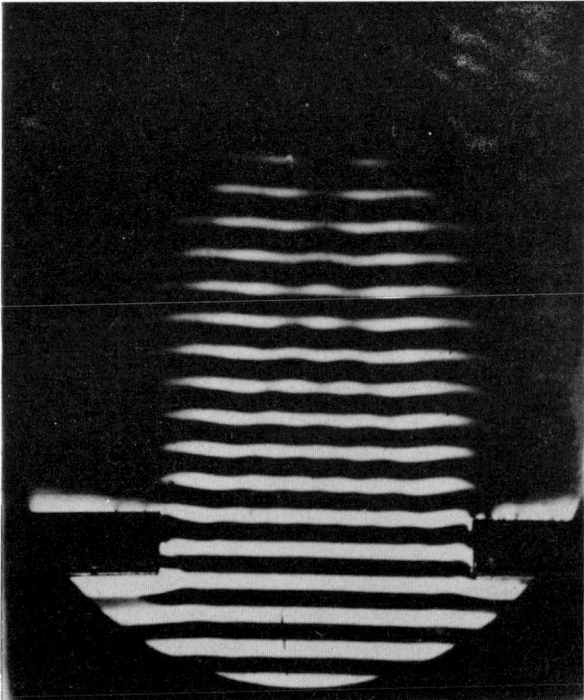

Fig. 20.15 Diffraction of plane waves in passing through a wide aperture

Note only a slight bending of the light (cf. Fig. 20.14)

Fig. 20.16 Diffraction of water waves round harbour wall of model
Note the greater diffraction of the shorter wavelength ripples

wavelength. The transmitter T (Fig. 20.17) is arranged some distance from an aperture formed by straight edges X and Y of metal screens. The small dipole receiver R is placed in the geometrical

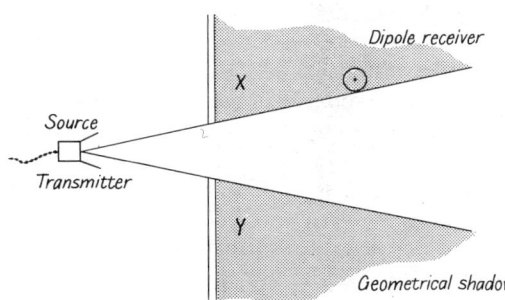

Fig. 20.17 Diffraction effect produced with 3-cm radio waves

shadow. When the width XY of the aperture is reduced by bringing Y nearer to X, there is a marked increase in the diffraction round X indicated by the strength of the signal received by R. (Compare with Figs. 20.14 and 15.)

An important conclusion obtained from the study of diffraction is that waves will pass round a very small object without being affected by it. For example, a thin vertical stick in a pond does not influence ripples passing round it. We can then conclude that in order to see a tiny object, the light waves must be affected. When the object is smaller than the wavelength it does not affect the waves and hence cannot be seen. No matter how effectively we design an optical instrument we shall never be able to see with it an object smaller than the wavelength of the light we are using.

Interference of waves

Consider two wave sources S_1 and S_2 (Fig. 20.18), emitting waves of the same frequency and in the same phase. In the diagram wave crests are shown by full lines and troughs by dotted lines. Where two full lines or two dotted lines intersect, the resultant maximum displacement (obtained by adding the two displacements) will be double that produced by one wave (Fig. 20.19a), but the resultant displacement at the intersection of the

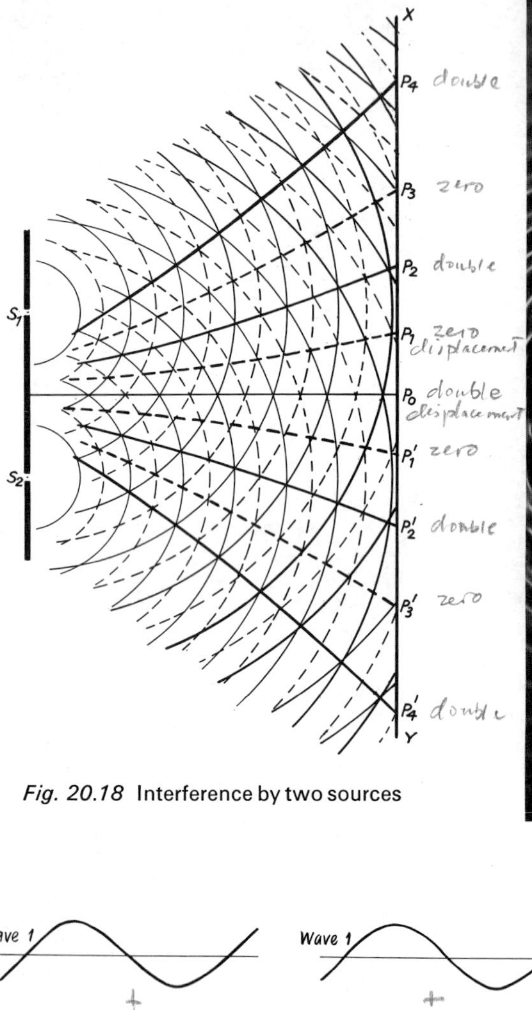

Fig. 20.18 Interference by two sources

Fig. 20.20 Interference of ripples produced by a Double Dipper

full and dotted lines will always be zero (Fig. 20.19b). Thus there will be regions of double and zero displacements which are shown in Fig. 20.18 by the slightly curved lines which cut the line XY at P_0, P_1, P_2, P_3, etc. At these places are bands alternately of double and zero displacement. The effect is called *interference* and the bands, *interference fringes*.

This effect can be shown using a two-prong dipper in the ripple tank. Fig. 20.20 shows a photograph of these ripples and the lines along which constructive interference occurs are clearly visible.

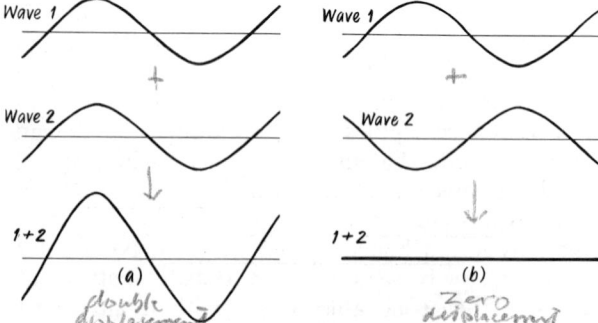

Fig. 20.19 Effect of phase on addition of displacements

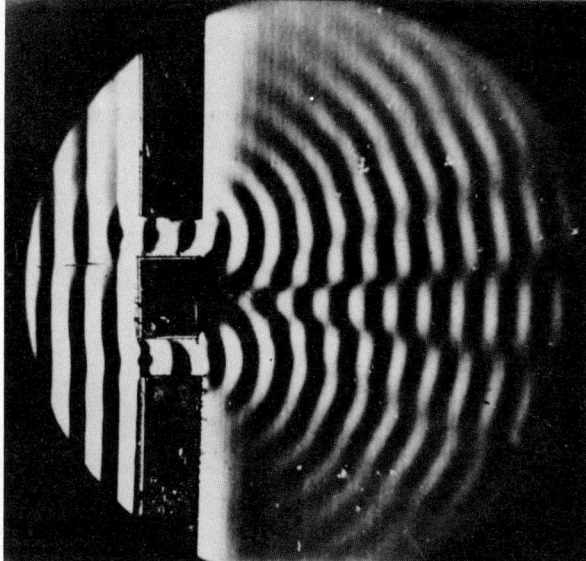

Fig. 20.21 Young's interference fringes produced in ripple tank

Plane waves are incident at two slits. The diffracted waves from the narrow slits interfere

A similar demonstration is illustrated in Fig. 20.21 which shows plane waves passing through two apertures close together. A similar pattern of interference fringes is produced.

For the demonstration of interference with 3 cm E.M. waves, the arrangement shown in Fig. 20.22 is used. Waves from the transmitter T pass through

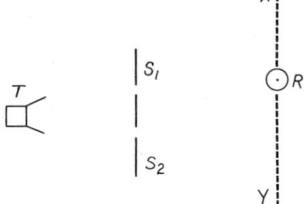

Fig. 20.22 Interference—Young's slits using 3-cm waves

the slits S_1 and S_2 formed between vertical screens. (The screens are metal or hardboard covered with aluminium foil.) These act as near sources and the fringe system is located by moving the receiver R along the line XY which can be at any distance from the slits (50 cm is a convenient value).

Many of the examples of interference can be demonstrated with 3 cm waves using simple apparatus. For example, interference is produced by reflection from two surfaces. The waves from T (Fig. 20.23) are incident upon the hardboard

surface S_1. This reflects some of the radiation but allows most to pass through. The transmitted radiation is reflected at the metal screen S_2. The

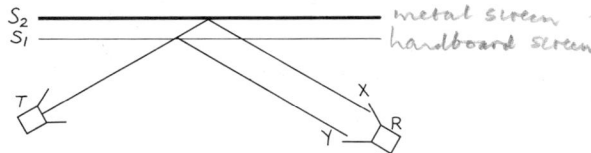

Fig. 20.23 Experiment illustrating interference at thin films using 3-cm waves

two emergent beams X and Y are received by the receiver R. If the distance between S_1 and S_2 is such that X and Y arrive at R in phase, a strong signal will be registered. On the other hand, if X and Y are out of phase, the signal will be weak. This experiment explains the production of colours in thin films, e.g. soap bubbles, oil on water. (*See* Chapter 27.)

The diffraction grating

If plane waves in a ripple tank are incident at a barrier with several apertures, all similar and equally spaced, the diffracted wavelets produce wavefronts at particular angles to the incident wavefront (Fig. 20.24).

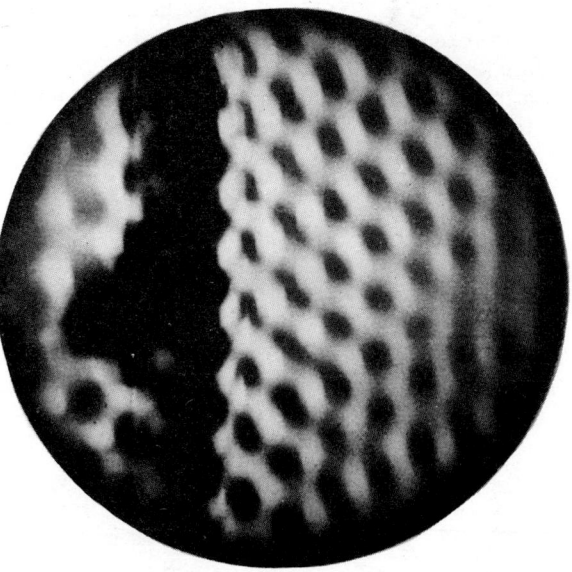

Fig. 20.24 Diffraction grating. Interference between a number of wavelets

This arrangement of many equally spaced parallel slits is called a diffraction grating. The same effect can be shown with 3 cm waves but the grating space must be about 10 cm. The grating should be about a metre square and made with strips of metal foil stuck on a transparent film. Further particulars and the theory of the diffraction grating are given in Chapter 27.

Polarization

The displacements of the transverse water wave in a ripple tank are always perpendicular to the water surface. In the case of a transverse wave in space such as an E.M. wave, the displacements may be in any direction in the plane normal to the direction of the wave. In a ray of visible light, the displacements are in all directions normal to the direction of the light. The waves are said to be unpolarized. It is possible to polarize E.M. waves, i.e. to arrange that the displacements are all in a particular direction (*see* Chapter 27). The 3 cm E.M. waves generated by the transmitter are polarized in a direction determined by the aerial. Polarized waves will pass through a polarizing device only if it is correctly orientated. If the incident waves are polarized in a direction at right angles to the polarizing direction of the device, the waves will be cut off. A grill made of equally spaced straight wires about 1 cm apart will polarize 3 cm waves or will cut them off if their direction of polarization is parallel to the wires. Using the arrangement shown in Fig. 20.25, the transmitter generates waves polarized in the vertical direction. These waves pass through the wire grill arranged as shown. When the grill is

ments are along the rods. Hence, in the diagram, the vertically polarized waves are not affected by the horizontal rods and the waves are received at the receiver R.

Longitudinal waves cannot be polarized. *(Sound)*

Change in phase

Lay a stretched slinky coil along a smooth bench and fasten one end. Holding the other end in the hand, give it a quick sideways flick. A pulse or kink will travel along the coil which will be reflected at the fixed end. Note that the displacement of the reflected pulse will be opposite to that of the incident pulse, i.e. a crest will be reflected as a trough. If, instead of fastening the end of the slinky coil directly, a length of string is interposed so that the end is free, the reflected pulse will be in the same direction as the incident pulse, i.e. a crest will be reflected as a crest.

In the terms used for simple harmonic motion, in one case there is a phase change of 180° and in the other, no phase change.

This can be demonstrated using a long length of rubber tubing (preferably filled with sand). In the first case one end is tied and the other held in the hand so that the tube is in tension. Pulses produced by vibrating the hand are reflected at the fixed end and suffer a phase change of 180°. If a long length of tube is allowed to hang from the hand, vibrations are now reflected at the free end without any phase change.

The change in phase also takes place with longitudinal waves and is discussed again in connection with the vibration of strings and air columns (Chapter 30).

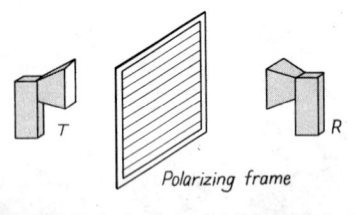

Fig. 20.25 Polarization with 3-cm radio waves

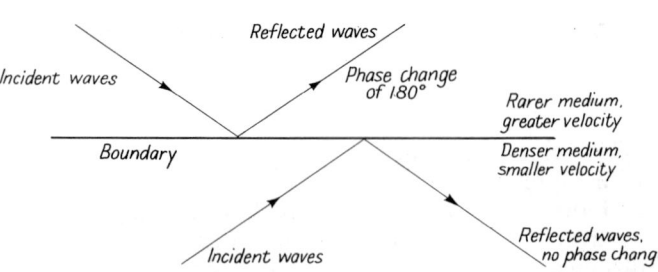

Fig. 20.26 Phase change on reflection of wave

turned through 90°, the waves are completely cut off and the loudspeaker sound is brought to zero.

The conducting rods behave like aerials and 'receive' the waves when their electrical displace-

The change in phase when an E.M. wave is reflected is determined by whether at the boundary the wave is passing from a rarer to a denser medium or vice versa, i.e. whether the velocity of the wave will decrease or increase in passing through the boundary. In the first case there is a phase change of 180°, but in the second case there is no phase change. This is referred to again in connection with visible light (Chapter 27).

Stationary waves

If a plane wave dipper D_1D_2 is set up parallel to a straight barrier B_1B_2 in the ripple tank, as in Fig. 20.27, a series of ripples will be produced which

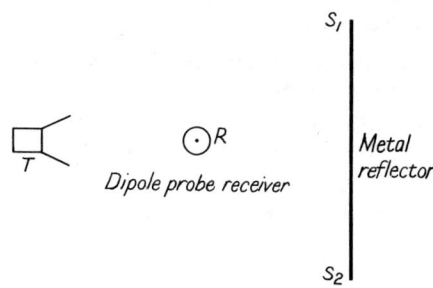

Fig. 20.28 Stationary waves demonstrated with 3-cm waves

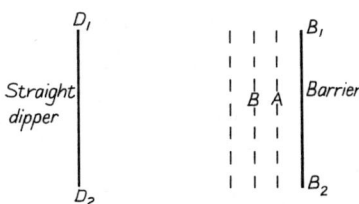

Fig. 20.27 Stationary waves demonstrated with ripple tank

will be reflected. If the distance between B_1B_2 and D_1D_2 has a certain value, a series of waves will be formed which are not moving. There will be lines through points such as A and B along which the water will be still. These lines are separated by a distance equal to half a wavelength. Points midway between the lines suffer maximum displacement.

The waves are called stationary or standing waves. They are the result of interference of the reflected wave with the incident wave.

They can be effectively demonstrated with 3 cm E.M. waves. The transmitter T and receiver R are placed as shown in Fig. 20.28. The waves from T are reflected at the metal plate S_1S_2 and interference between the incident and reflected waves takes place. By moving the receiver along a line perpendicular to the reflector, the signal received, indicated by the loudspeaker, goes through maxima and minima. The distance the receiver is moved between successive maxima is half a wavelength. With this arrangement it is easy to determine the wavelength of the E.M. waves.

Stationary waves are important in the theory of musical instruments and are discussed in Chapter 30.

A stationary wave does not emit energy, e.g. a vibrating wire on a stringed instrument would vibrate indefinitely if it did not lose energy to the supports and to the air.

This fact is of importance in atomic theory, for it is supposed that the orbiting electrons are in some such state of vibration when they are not emitting radiation.

The Doppler effect

When either the source of radiation or the receiver is in motion, the apparent frequency of the waves received is changed. When the source moves towards a stationary receiver, the wavelength of the waves is reduced and so that the frequency received is greater than that of the source. If the receiver is moving towards a stationary source, the frequency received is again greater. This effect is known as the 'Doppler' effect.

The change in wavelength produced by a moving source can be demonstrated with the ripple tank. A point dipper producing circular ripples is moved over the surface. The wave pattern in front of the moving dipper is crowded together while that behind is widened out.

The effect cannot be demonstrated with 3 cm radio waves, as their velocity is too great to show the influence of a comparatively insignificant change in velocity of the source. The effect is shown in the spectra of fast receding stars, page 245. Doppler effects produced with sound waves are mentioned in Chapter 29.

QUESTIONS

1. Explain with diagrams, the difference between transverse and longitudinal waves.

2. Explain, with diagrams, the refraction of a wave at a boundary of two media resulting from a change in velocity of the waveform.

3. Derive a relation between wavelength, velocity and frequency.

4. Explain the limitations of the statements, (i) light travels in straight lines, (ii) sound waves can bend round corners.

5. What effect has the length of a wave on the diffraction it suffers?

   Explain why small boats sheltering behind a harbour wall are safe from the small choppy waves outside, although they are still affected by the swell.

6. Explain any differences in the properties of transverse and longitudinal waves.

7. Explain the meaning of the terms *phase difference*, *wavefront*, and *frequency*.

8. What is the difference between a polarized and an unpolarized wave disturbance? Give examples to illustrate your answer.

9. Describe how you could demonstrate the polarization of a 3-cm EM wave.

10. Explain the difference between a transverse and a longitudinal wave. Give an example of each.

   Plane waves are incident on a narrow gap. Draw the appearance of the waves or draw the wavefronts before and after they have passed through the gap, (*a*) when the wavelength of the waves is large, (*b*) when the wavelength is very small. What name is given to this effect?   (L)

11. Draw three concentric circles of radius 1.2 cm, 2.4 cm, and 3.6 cm to represent the crests of ripples on a water surface. Draw two parallel lines *SS′* and *RR′*, each 3.1 cm from the centre of the circles, *SS′* representing the boundary of shallower water across which the ripples travel with only one-half their initial speed, and *RR′* representing a barrier which reflects the ripples. Show on the diagram (i) the ripplecrest which has entered the shallower water, (ii) the ripplecrest which has been reflected from *RR′*. Assuming that the initial speed of the ripples is 18.0 cm/s, find the frequency of the source of the ripples.   (C)

12. Describe any *one* experiment which shows the refraction of water ripples.

   Construct a diagram to show the refraction of plane wavefronts incident at 45° to a straight boundary, beyond which their speed of motion is half the initial speed. Given that the incident waves have a speed of 16 cm/s and a wavelength of 2.0 cm, find (*a*) the frequency of the wave motion, (*b*) the wavelength after refraction, (*c*) the angle between the refracted waves and the boundary.   (C)

13. Explain what is meant by a *longitudinal wave motion*. How does it differ from a transverse wave motion?

   Describe the use of some form of mechanical wave model (e.g. a 'Slinky' spring) which demonstrates the propagation of longitudinal waves. Draw diagrams showing how the vibrations of the individual parts give rise to compressions and extensions travelling outwards.

   Give your reason for believing that sound travels as a progressive longitudinal wave motion through the air.   (O)

14. A ripple tank is set up to produce plane water waves. Describe the adjustments which must be made to the apparatus in order to produce waves of longer wavelength.

   These waves may be observed more clearly with the aid of a stroboscope than without it. Explain the use of the stroboscope in this experiment and show how to obtain the velocity of the waves.

   Two point sources a few centimetres apart are placed in the ripple tank. They are caused to vibrate with the same frequency, and in phase with each other, and they produce two sets of circular waves. Comment on the resulting pattern and show how it demonstrates the phenomenon of interference.

   Give details of the experimental arrangement you would use to demonstrate interference of *either* light waves *or* sound waves.   (C)

15. In what circumstances can the wavefronts of a wave-motion be refracted, and what is the effect on the wavelength?

   Describe any one experiment which shows refraction in the case of water ripples. Explain the refraction in this case.

   Plane wavefronts of a light beam, of wavelength 1 unit, are incident at 45° to a plane boundary, beyond which their speed of travel is 1.3 times the initial speed. Construct a diagram to show the resulting refraction. Find (*a*) the wavelength after refraction, (*b*) the angle between the refracted waves and the boundary.   (C)

16. A straight vibrator causes water ripples to travel across the surface of a shallow tank. The waves travel a distance of 38.0 cm in 1.5 s and the distance between successive wave crests is 4.0 cm. Calculate the frequency of the vibrator.

   How would you modify the tank to show the phenomenon of refraction of water waves? Draw the wave pattern you would expect to observe. State the measurements you would make to obtain the 'refractive index' for water waves in this particular case.   (JMB)

Part Seven LIGHT, VISUAL ENERGY, GEOMETRICAL AND PHYSICAL OPTICS

Solar corona taken in Mexico during the total eclipse of 7 March 1970

21 Light; photometry

Seeing

The human eye is a wonderful organ. Its optical system, working with the brain, enables us to perceive the beauty and wonder of the things around us.

Plato and other Greek philosophers believed that the eye sees an object when light is sent out from the eye to the object. Pythagoras and Aristotle, however, were the only two philosophers of the Greek school to support the opposite view, that light is sent out by luminous bodies only. Plato's idea held first place for several hundred years, in fact, until the time of Alhazen, an Arabian philosopher of the twelfth century.

Now we believe that light is emitted by self-luminous bodies, and we see an object when light from that object enters the eye. A non-luminous object such as a book reflects some of the light which falls upon it, and when this reflected light enters the eye the book can be seen. An object always appears to be along the final direction from which the light enters the eye, e.g. an object appears to be behind a plane mirror because the light enters the eye after reflection from the mirror.

Theories of light

In Chapter 19 we mentioned the transfer of energy to a floating cork by direct impact and by means of wave motion. This example also serves to illustrate the theories of light. The early theory of light was that it consisted of a stream of tiny particles (corpuscles) sent out by the luminous body. This theory was replaced by the wave theory, proposed by Huyghens (1629–95).

The corpuscular theory requires certain assumptions:

1. The corpuscles behave as perfectly elastic spheres.
2. The corpuscles all travel with the same speed.
3. The corpuscles have a greater speed in a denser medium than in a rarer medium.

The phenomena of reflection and refraction were explained using this theory, but the discovery of interference effects presented a difficulty which only the wave theory could solve. Newton, however, did not support the wave theory of Huyghens; he was not convinced and so did not discard the corpuscular theory. The two theories were contradictory regarding the change in velocity in passing from one medium to another. The corpuscular theory demanded an increase in speed in passing into a denser medium, whereas the wave theory demanded a decrease. This deciding test could not be used, for the method of measuring the velocity in a medium was not known until the nineteenth century.

Geometrical and physical optics

Properties of light can be studied by considering the behaviour of pencils or fine rays of light. Their

Fig. 21.1 Searchlight beams travel in straight lines—beacons in New York

paths can be traced when they suffer reflection and refraction and, by geometry, the relationships between angles and distances can be worked out. This is what we call *geometrical* optics. Similar relationships can be obtained using the wave properties of light. This we call *physical* optics. Interference, diffraction and polarization are effects associated only with the wave properties of light.

Shadows

In the previous chapter we saw that the diffraction of a wave depends on its wavelength. With visible light covering a range of wavelengths from 4 to 7×10^{-7} m, the diffraction or bending is very small. Thus for most situations, we say

light travels in straight lines.

This is well known from experience. We cannot see round a corner. We have often also seen sunbeams 'like massive straight rods supporting the clouds' (Fig. 21.1).

That light travels in straight lines explains the formation of shadows. If a small source of light, e.g. a car head-lamp, is placed in front of an opaque sphere, a sharp circular shadow of the sphere is cast upon a screen set up behind it (Fig. 21.2*a*). When two lamps, placed close together, are used, their circular shadows do not quite coincide (Fig. 21.2*b*). Replacing the two lamps by a large or extended source, produces a shadow with a blurred edge. This results from the shadows cast by each point of the source not quite overlapping.

The shadow cast by a large source of light can be examined by using the arrangement in Fig. 21.3.

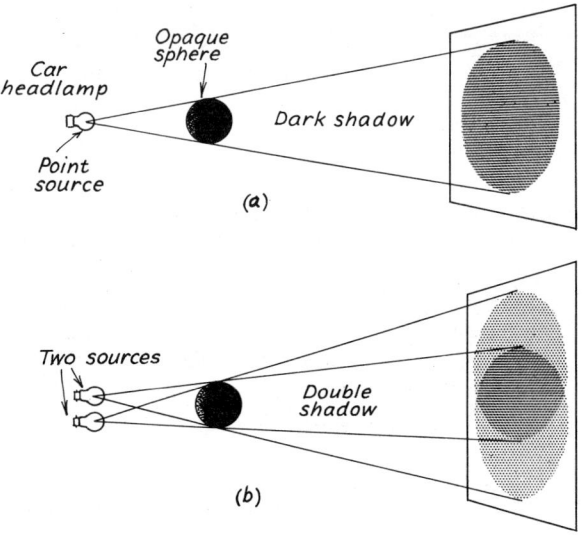

Fig. 21.2 Shadows

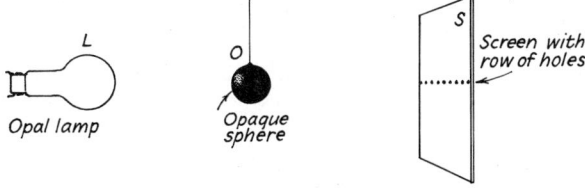

Fig. 21.3 Shadow cast by large source-eclipse experiment

L is a large opal lamp, O an opaque sphere and S a screen placed normal to the light. Each point of the surface of the lamp acts as a point source, producing its own circular shadow on S. The shadow cones produced by the limiting points

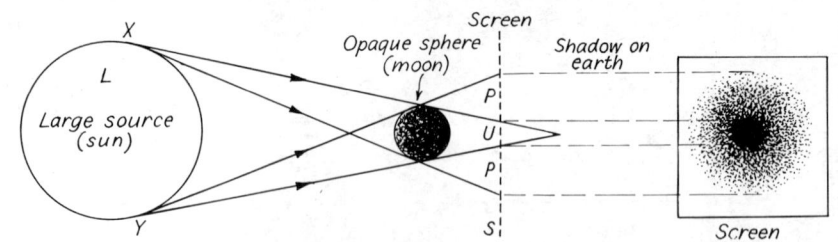

Fig. 21.4
Umbra and penumbra-eclipse

X and Y of the source are drawn (Fig. 21.4). The space U behind the spheres is a cone which is common to shadow cones from every point of the source. Thus, U is completely dark and it is impossible to see any part of L from within it. In the space P surrounding U, only part of the source L can be seen, hence, region P is semi-dark. Furthermore, the illumination here is not uniform, gradating from dark at the edge of U to maximum brightness at the outer edge of P. The whole of L can be seen from any point outside P.

The shadow U is called the umbra and P the penumbra.

Eclipses

This type of shadow is similar to that of the moon cast by light from the sun. If the moon's umbra reaches the earth's surface, a total eclipse of the sun is visible, since within the umbra the source is completely obscured. The penumbra provides a partial eclipse. Because of the rotation of the earth, the umbra and the penumbra pass across the earth's surface producing belts from which total and partial eclipses may be seen.

The successive stages of an eclipse as they would appear in reality can be seen by looking in turn through a series of small holes, about 1 cm apart, traversing the screen S (Fig. 21.3).

When the umbra of the earth falls upon the moon, a lunar eclipse is produced.

The pinhole camera

The pinhole camera (Fig. 21.5) depends upon the rectilinear propagation of light to produce the image. A useful form of pinhole camera for demonstration purposes consists of an adjustable wooden box, one end of which is covered with ground glass and the other provided with a large circular hole. This hole is covered with tinfoil and a tiny hole made in it with a needle. By directing

the camera towards the windows, an inverted image can be seen upon the screen. A more useful object for demonstration is a clear lamp with horseshow filament, housed in a box brought close to the camera. It is observed:

(a) the image is inverted,
(b) the image is always in focus,
(c) the size of the image is increased by lengthening the box.

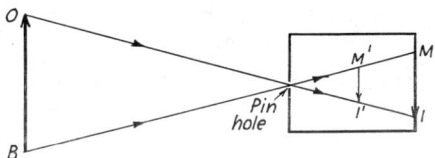

Fig. 21.5 Pinhole camera
OB is an object: *IM* and *I'M'* are images produced at different distances

Excellent pictures can be taken with a pinhole camera, though long exposures are necessary because of the small amount of light passing through the tiny hole.

If another hole is made in the tinfoil near to the first one, another image is produced nearly, but not quite, covering the first. By making many holes, images are produced which, because they do not quite coincide, give the effect of one blurred image. A large hole in the foil produces a patch of light which bears no resemblance to the lamp filament. This patch of light is, in effect, a large number of images of the filament, for the large hole can be regarded as a number of small pinholes.

The effect, then, of increasing the size of the pinhole is to produce a much brighter, but blurred image, which may be so blurred as to be unrecognizable if the hole is very large.

In order to take a photograph using a short exposure, a large hole must be used, to let through sufficient light. A lens is then necessary to produce a focused image. The lens brings all the images together so that they coincide and produce one brilliant image. The way in which the lens bends the light will be better understood after reading Chapter 25. If O is one point on the object OB (Fig. 21.6), light from O passing through the

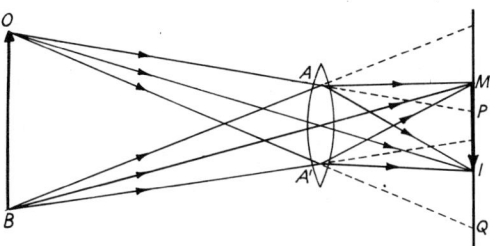

Fig. 21.6 Function of the lens in a camera

large hole AA' will cover the screen or plate from P to Q. For a focused image all this light should go to one point on the screen. The lens accomplishes this by bending the rays AP and $A'Q$ so that they, and all other rays from O, go to the point I. Similarly, all rays from B are brought to the point M by the lens, and IM becomes the image of OB. Note that the image IM will be focused only if the screen is placed in this one position. If it is either nearer or farther away from the lens, the rays from a point in OB will not go to one point on the screen.

Photometry

Photometry is the measurement of the strengths of light sources and the illumination they produce. Until recently, the importance of these measurements was not fully appreciated. Poor illumination gives rise to eye strain, and adequate lighting is one factor which contributes towards ideal working conditions.

Illuminating power or luminous intensity

Normally we speak of a source of light as being weak, dim or poor, but for scientific purposes we must measure the power of the source numerically. The illuminating power or luminous intensity of a source of light is measured by comparison with a standard source.

The old standard source was the *standard candle*—a wax candle of fixed size which, when burned under specified conditions, was said to have a luminous intensity of one candle power. This was later replaced by the pentane lamp, a 10 candle power source. The present standard, adopted in 1948, is called the *candela* and has a luminous intensity approximately equal to that of a standard candle.

The candela is one-sixtieth of the luminance (luminous intensity per cm^2) of platinum at 1773 °C.[*]

An ordinary electric lamp cannot be used because its power declines during use, though special lamps are used as substandards.

The difficulty of obtaining a good and simple unit is due to the fact that all the energy of a source is not emitted as visible light: some may be invisible ultraviolet and infrared light. Further, the ordinary source is a solid or gas heated to a high temperature, and nearly all the energy used is transformed into heat. Only a small percentage goes into visible energy.

Sources of light

A naked flame in some form or other was the only light source available until the end of the last century. These sources depended for their luminosity upon the production of high temperatures. A white source was obtained by the presence of carbon particles in a hot flame. A great step forward was the discovery of the gas mantle, consisting of fabric impregnated with the oxides of thorium and cerium. Such a mantle glows brilliantly when heated in a coal-gas flame.

Towards the end of the nineteenth century the electric filament lamp, consisting of a wire heated to incandescence by an electric current, was invented. In the first lamps, a wire or filament of carbon was contained in an evacuated glass bulb. The vacuum was necessary because the carbon filament disintegrated in air. Later the carbon was replaced by a metal filament; tungsten was found to be the most suitable. Developments have come very slowly, and today the electric lamp is still much the same as it was when first

---

[*] 1773 °C is the temperature of solidification of platinum.

invented. The filament now consists of a very fine coil of tungsten wire which is again coiled to produce what is termed a *coiled-coil* lamp, in which the coils are close enough to keep each other very hot. The glass bulb is filled with an inert gas. This permits the filament to be used at a higher temperature without the metal evaporating.

The common electric lamp, though very superior to the old light sources, is very inefficient. Over 80 per cent of the energy supplied is transformed into heat, and with high-powered sources the dissipation of this heat presents a special problem, particularly in optical projection instruments.

The study of light sources and lighting is now in the hands of the illuminating engineer, and his dream is the production of a cold light, a source emitting all its energy in the visible spectrum: he strives to attain the efficiency of the glow-worm!

The latest and most promising developments are in the field of discharge lamps. These consist of glass or quartz tubes, containing a gas at low pressure, through which an electric current is passed between two wires fused through the ends of the tube. The current causes the gas to glow. A disadvantage is that each gas gives its own particular coloured light, e.g. mercury vapour—blue-green; sodium vapour—orange. Such lamps have been adapted for road lighting, as their efficiency is more than double that of the filament lamp. For general use, the colour difficulty has been overcome by coating the tube with a fluorescent material which also increases the efficiency. This material absorbs the radiation lying outside the visible region and re-emits it with a wavelength within the visible spectrum.*

Extra-high-pressure mercury vapour lamps are very efficient. The current is passed through mercury vapour contained in a small quartz tube. For commercial purposes the tube is surrounded by a protective glass bulb. Coating this outer bulb with a fluorescent material increases the efficiency still further, and the colour is made to approximate more closely to sunlight.

Illumination

The illumination produced on a surface by a source of light depends upon the power of the source and its distance away. The illumination on a screen placed normal to the light is doubled when the power of the source is doubled, for there is double the amount of light energy falling on the surface in the same time.

Assuming the source to be a point, and emitting light equally in all directions, the same amount of light energy will be spread over an area four times as big when the distance away is doubled. This is simple to understand from Fig. 21.7, for the

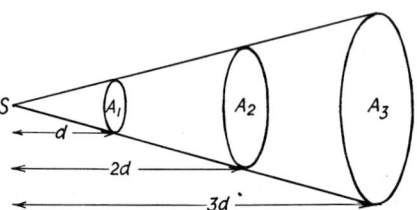

Fig. 21.7 Illumination ∝ 1/distance²

light energy in the cone illuminates area A_1 at distance d and area A_2 at distance $2d$. But, by geometry, $A_2 = 4A_1$. Similarly, at distance $3d$ an area A_3 ($= 9A_1$) is illuminated. So we conclude:

illumination on $A_2 = 1/4$ illumination on A_1
illumination on $A_3 = 1/9$ illumination on A_1

This should be recognized as an inverse square relationship.

To sum up:

Illumination E on a normal screen:
(1) varies directly as the luminous intensity I of the source,
(2) varies inversely as the square of the distance d from the source.

i.e.
$$E \propto \frac{I}{d^2}$$

We arrive at the unit for the illumination by making E equal to 1 when I and d are 1.

The meter-candle or lux: the lux is the illumination on a normal screen* one metre away from a source of one candela.

Candle power is the light-radiating capacity of a source compared with that of a source of one candela.

Luminous flux

It is now usual to speak of the flux, or rate of flow, of light energy. Although a point source radiates

---

* The spectrum is described in Chapter 28.

---

* i.e. a screen placed perpendicular to the direction of the light.

in all directions the luminous intensity of the source is measured by the luminous flux emitted per unit solid angle. A solid angle is measured by the area it cuts on a sphere of unit radius constructed with its centre at the apex of the angle. If we imagine a cone, its solid angle is the area it cuts on a unit sphere (Fig. 21.8). A unit solid

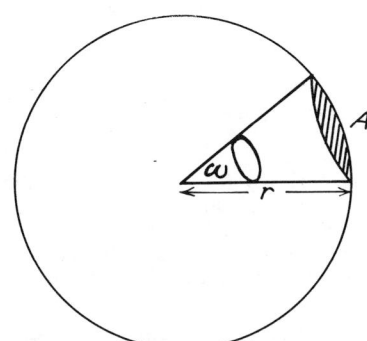

Fig. 21.8 Measurement of solid angle, $\omega = \dfrac{A}{r^2}$

angle (called a *steradian*) will cut an area of 1 m² on a sphere of radius 1 m. The total solid angle about a point is 4π steradians (area of the surface of a sphere is $4\pi r^2$).

The lumen is the luminous flux through one steradian from a point source of one candela.
The total flux from one candela is 4π lumens.

When flux equal to 1 lumen falls on an area of 1 m², the illumination is the same as that produced by one candela at a distance of 1 m. Hence,

$$1 \text{ lux} = 1 \text{ lumen/m}^2$$

We can now write

$$E = \frac{I}{d^2}$$

and E will be measured in m candles or lumens/m².

Note that the inverse square law can be applied only when the light is radiated equally in all directions from a point source. Where lenses or mirrors are employed a different treatment is required. For example, if a parallel beam is produced by a convex lens, the illumination on a normal screen placed anywhere in the beam remains constant. In practice, however, the illumination falls off, due to dust particles scattering the light: a beam will not penetrate far through a thick fog.

Measurement of illumination

Until recently, this determination presented much difficulty, for its accuracy depended upon visual judgment.

The instrument now used to measure illumination is a simple meter reading directly in lux. It incorporates a photoelectric cell connected to a microammeter. The photoelectric cell (of which there are now many types) converts light energy into electrical energy. The potential difference created when light falls on the cell is proportional to the illumination, hence, if a sensitive current meter is connected to it, the reading will be proportional to the illumination. The instrument is standardized and the meter calibrated to read directly in lux.

An important point has been passed over which should be explained. The photoelectric cell replaces the eye and, therefore, it should possess the same essential properties. The eye is not equally sensitive to all colours of the spectrum. It responds best to yellow-green light, in the middle of the spectrum, and is unaffected by ultraviolet and infrared light. For readings taken with a photoelectric cell meter to agree with those taken by visual adjustment, the photoelectric cell must have the same response curve as the normal eye. Such cells are now available: they are extremely useful and easy to operate.

This type of meter can be used as an exposure-meter for photography. A special scale on the meter enables the required exposure to be read off directly. The latest cameras have the exposure-meter incorporated within the camera.

Verification of the inverse square law

This experiment is quickly and easily carried out using a photoelectric cell illumination meter.

Arrange a clear lamp horizontally as shown in Fig. 21.9, and take readings of the meter at various distances from it. (It is usually advisable

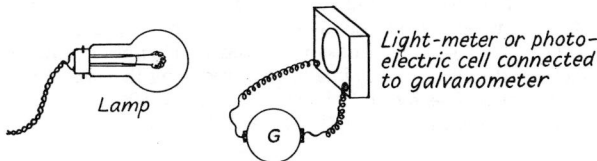

Fig. 21.9 Verification of the inverse square law

to lay black cloth along the bench to prevent reflected light falling on the cell.) The experiment is best performed in a darkened room, but by deducting the initial reading of the meter from subsequent readings, allowance for residual lighting can be made.

Tabulate the readings as shown in the next column.

The inverse square law gives $I = Ed^2$. In this experiment the luminous intensity I of the lamp

E/lux	d/m	Ed^2
110	0.50	27.5
48	0.75	27
27	1.00	27
17	1.25	26.5
12	1.50	27

is constant, hence, if Ed^2 is constant, the law is verified.

QUESTIONS

1. Describe and explain the effect produced upon the image in a pinhole camera by (a) increasing the size of the pinhole, and (b) increasing the distance of the screen from the pinhole.

2. When direct sunlight passes through the interspaces of a tree's foliage, *circular* spots of light are always seen on the ground. Explain this.

3. Explain, with the aid of diagrams, the function of a lens in a camera.

4. Describe and explain one laboratory demonstration of the rectilinear propagation of light.

A pinhole camera in which the hole is 1 mm in diameter and the screen 1 m from the pinhole, is pointed towards the sun and the diameter of the image is found to be 10 mm. Draw a ray diagram (not to scale) showing how the image is formed.

What would be the diameter of the sun's image if the size of the pinhole were negligible? Justify your answer. (O & C)

5. Explain how an eclipse of the sun occurs and draw diagrams to show how a total eclipse at some places may be observed as a partial eclipse at others.

Explain also why (a) an eclipse lasts only for a short time at one place, (b) there is not a partial eclipse of the sun every 28 days, (c) total eclipses are very rare. (O & C)

6. A spherical object of diameter 2 cm is held between a screen and a spherical source of light of diameter 3 cm. Draw diagrams showing the nature of the shadow cast on the screen (a) when the screen is very close to the object, (b) when it is at a considerable distance from the object. Explain how the shadows are formed.

Explain how eclipses of the sun occur, distinguishing between total and partial eclipses. (O)

7. Describe briefly a pinhole camera. State what it demonstrates and why it is rarely used for taking photographs.

The image of an object is produced in a pinhole camera. Explain the effect on the image of (i) reducing the size of the hole, (ii) increasing the object distance by 50%, and (iii) increasing the camera length by 50%. (L)

8. Describe and explain the action of a pinhole camera.

What changes would you expect to occur if the single pin-

hole were replaced by (a) a number of pinholes all close together, (b) a hole 1 cm in diameter?

The image of a building 50 m from the pinhole of a pinhole camera appears on the screen to be 5.0 cm high. If the pinhole is 12.5 cm from the screen, what is the height of the building? (D)

9. Describe an experiment you would carry out in the laboratory to show that light may be considered to travel in straight lines. If you were asked to double the accuracy of your experiment how would you attempt to do so?

Explain with a diagram how an eclipse of the moon occurs, and show how it is an illustration of rectilinear propagation.

Is it possible to observe an eclipse of the new moon? Give a reason for your answer. (O & C)

10. State and explain the properties which a photo-electric cell must possess to be suitable for illumination measurements.

11. The illumination on one side of a screen is balanced by that of a lamp placed 2.5 m from it on the other. When a sheet of transparent material is interposed on the first side, the lamp has to be moved up to 2.2 m in order to effect a balance. What fraction of the incident light is absorbed by the sheet?

12. Define *luminous intensity* (*illuminating power*) and *illumination* (*intensity of illumination*).

A certain lamp X, placed 1.0 m away from the screen of a photometer, gives the same illumination as a 60 cd lamp Y, placed 1.50 m away. What is the luminous intensity of X?

If a piece of glass which transmits 60% of the light falling on it is placed between X and the photometer, how far from the screen must Y now be placed in order to obtain an illumination match? (O)

13. A photographic print requires 6 seconds exposure at 1.0 m from a 100 W lamp. What exposure would be required with a 60 W lamp (a) at the same distance, (b) if the distance were 0.50 m? (JMB)

14. Two lamps of luminous intensities 90 cd and 40 cd are placed at distances of 3.0 m and 2.0 m respectively from the same side of a screen. Find the luminous intensity of the

lamp which will produce an equal illumination on the other side of the screen when placed 1.50 m from it. (JMB)

15. State the inverse square law used in photometry.

Describe how the luminous intensities of two lamps can be compared using a photometer.

A small source of light is 5.0 m from a wall. A thin transparent sheet of material which cuts off 36% of the light is placed between the lamp and the wall. Where must the lamp now be placed so as to give the same illumination as before on the part of the wall closest to the lamp? (L)

16. A pinhole camera is used to photograph an object several metres away from it. Explain with a ray diagram how the image is formed on the film.

State, with reasons, two effects on the above image of making the pinhole larger.

Draw ray diagrams to show how (i) the image is formed if the camera is improved by the addition of a suitable lens, (ii) the same lens could be used as a magnifying glass. State the nature of the image in each case. (JMB)

22 Reflection

In Chapter 21 we have stated that only self-luminous bodies send out light, and that we see objects by light which falls upon them and is reflected to our eyes.

We are apt to consider reflection from a good reflector, such as à mirror, as different from that which occurs at other surfaces. But this is not so; the same laws hold in both cases. The only difference is that a polished surface gives regular reflection, while other surfaces reflect light irregularly.

Allow a ray of light* to fall upon a plane mirror

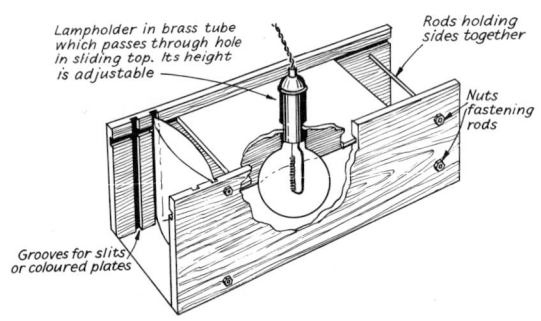

Fig. 22.1 The ray box : cut-away diagram

arranged on a sheet of white paper and observe the reflected ray. Mark the rays and measure the angles of incidence and reflection. These are shown in Fig. 22.2 and are the angles between the incident and reflected rays and the normal.

---

* The ray box (Fig. 22.1) designed by J. P. Stephenson Esq., is a convenient instrument for producing a ray or a wide beam of light required for elementary experiments.

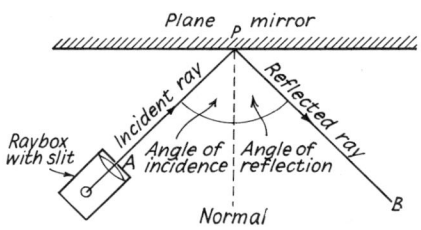

Fig. 22.2 Reflection at a plane mirror

The reflected light obeys the following laws which are, of course, the laws governing the reflection of all types of wave motion (Chapter 20).

Laws of reflection

(1) The incident ray, the reflected ray, and the normal at the point of incidence are all in the same plane.

(2) The angle of incidence is equal to the angle of reflection.

The periscope (Fig. 22.3) illustrates the laws of reflection.

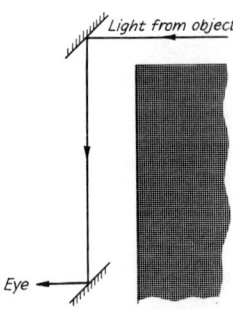

Fig. 22.3 Periscope

Observer can see over wall by arranging mirrors as shown

These laws state that light reflected at a mirror behaves very like a ball which is bounced on the ground.

Reversibility of a ray of light

A ray of light such as *AP* (Fig. 22.2) is reflected along the direction *PB*. Similarly, a ray along *BP* is reflected along *PA*. This reversibility holds for any beam and is a useful property.

Scattering

A beam of light falling on a rough surface is scattered. Each bit of surface is inclined differently and, although the laws of reflection are obeyed, the light is reflected in a haphazard manner. Fig. 22.4 shows this clearly.

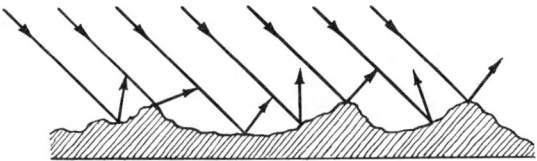

Fig. 22.4 Scattering at an unpolished surface

Not all the light incident upon any type of surface is reflected—some is always absorbed by the surface. The percentage absorbed may vary over wide limits. For a well-polished mirror, it is very small, whereas a rough black surface reflects very little.

When light falls upon the polished surface of a transparent substance such as a sheet of glass, some is reflected, but the greater part passes through. An ordinary mirror has a layer of silver deposited on the glass. As the outer surface of this layer soon tarnishes, we use the surface in contact with the glass for reflection, and paint the back surface with shellac varnish to preserve it. The light passes through the glass before and after reflection.

In some cases, e.g. episcopes, we use the silvered surface directly. A surface-silvered mirror should be used with care, and never touched by the hand.

Stainless steel is now used for mirrors for special purposes. It can be worked with precision and is much superior to the surface-silvered mirror, though very much more expensive.

A new method of surface coating has been devised, termed *sputtering*. Thin layers of certain metals can be put upon nearly any type of surface. An 'aluminized' surface does not readily tarnish and is of great use in optics.

Rotation of a plane mirror

Arrange a mirror in the path of a beam of light from a ray box (Fig. 22.5). Set the mirror normal to the incident light by making the reflected beam

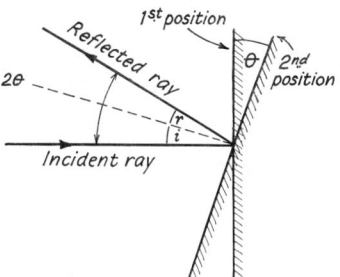

Fig. 22.5 Rotation of a plane mirror

coincide with the incident beam. Rotate the mirror through a known angle and measure the deflection of the reflected beam.

Note that the reflected beam is deflected through twice the angle of rotation of the mirror.

Tracing rays of light by the pin method

Although the modern method of using a ray of light produced by a ray box is much more realistic and convincing, the old method of using pins to define rays of light is more convenient in one or two instances.

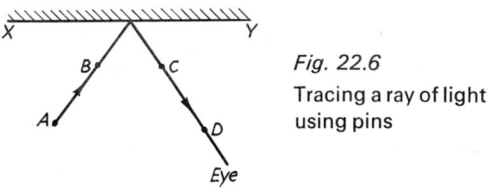

Fig. 22.6
Tracing a ray of light
using pins

If two pins *A* and *B* are erected on a soft-wood board (Fig. 22.6), two more pins *C* and *D* can be placed in line with *A* and *B* by sighting with the eye to give the direction of the ray *AB* later in its history. For example, if *AB* defines a ray incident upon a plane mirror *XY*, *CD* will give the reflected ray if pins C and D are placed in line with A and B seen through the mirror.

The image in a plane mirror

A plane mirror produces an image which differs from the image seen at the cinema, for it cannot be cast upon a screen. It is said to be virtual (not real). A real image cannot be formed behind a mirror, for no light can pass through a mirror. A real image is produced when light actually goes to the image and in such cases it can be formed on a screen.

Looking into a mirror, the eye receives light which has been reflected at the mirror so that the light appears to be coming from behind. The eye always interprets an object to be along the direction in which the light from it enters the eye.

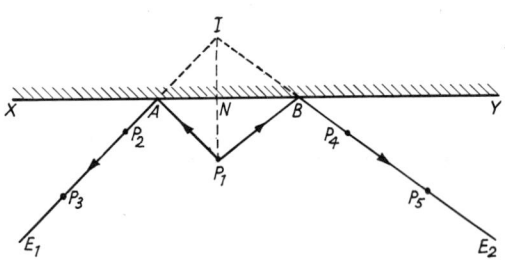

Fig. 22.7 Position of the image in a plane mirror

1. *Position of the image.* To locate an image in a plane mirror, the pin method (Fig. 22.7) is convenient.

Set up a pin P_1 in front of a plane mirror XY as shown. With the eye in some such position as E_1, set up a second pin P_2 and then a third pin P_3 which appears to be in line with P_2 and the image of P_1 seen in the mirror. Repeat with the eye in another position E_2, setting up two more pins P_4 and P_5. The image of P_1 must be along directions P_3P_2 and P_5P_4, hence it must be at I. Join P_1I and note that it is perpendicular to XY, and $IN = P_1N$. As the laws of reflection are obeyed at A and B, simple geometry gives triangles AIB and AP_1B congruent, hence, I and P_1 are at equal distances from XY.

The image in a plane mirror is as far behind the mirror as the object is in front of it.

Parallax. Set up two sticks A and B vertically and about 30 cm apart. View the sticks from a distance along the line joining them. If the eye is now moved from side to side, the sticks apparently move in opposite directions. This is called *parallax*. As A and B are brought nearer together, this movement gets less until, finally, when they are coincident, they appear to move together: there is then no parallax between them.

The method of parallax is often used for locating an image. It can be used for both real and virtual images.

Position of the image in a plane mirror by the method of parallax. Set up a pin P_1 as object in front of a strip of plane mirror. Behind the mirror set up a second pin P_2 and vary its position until there is no parallax between P_2 seen over the mirror, and P_1 seen through the mirror. The position of P_2 then gives that of the virtual image I of P_1.

2. *Size of the image.* Consider the image of AB in Fig. 22.8. Each point on AB will produce an image point as far behind the mirror XY as the object point is in front. Thus, the image of AB will be $A'B'$ and $AB = A'B'$.

The image in a plane mirror is the same size as the object.

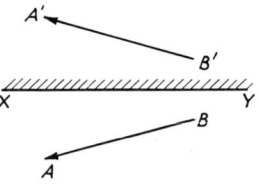

Fig. 22.8 Size of image in a plane mirror

3. *Inversion of image.* The image in a plane mirror is not identical with the object for it is 'turned round' and appears as in Fig. 22.9. This effect is commonly called 'lateral inversion', but the ordinary meaning of this term does not fully describe the effect. A right hand viewed through a mirror appears as a left hand and is not inverted. This effect occurs whenever an image is produced by reflection, although a second reflection, such as one gets with two mirrors at an angle (Fig. 22.10), produces a normal image.

Fig. 22.9 Inversion by reflection at a plane mirror

Mirrors at an angle

Many dressing-tables are fitted with hinged side-mirrors in addition to the large mirror in the centre. These can be adjusted for a person to see

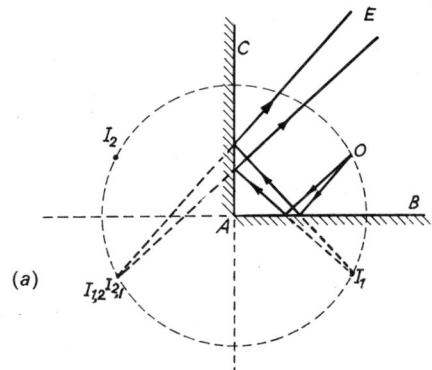

(a)

(b)

Fig. 22.10(a) and (b) Images produced by mirrors at right angles

the side of his head. This view is obtained by light which is reflected twice, first at the centre and then at a side-mirror.

To consider the formation of images in two mirrors, let us take a simple case first (Fig. 22.10). The mirrors AB and AC are at right angles. The object O will produce images I_1 in AB and I_2 in AC, in each case as far behind the mirror as the object is in front. But the image I_1 is in front of mirror AC and will, therefore, produce an image $I_{1,2}$ as shown. Similarly, I_2 will produce an image $I_{2,1}$ which, by symmetry, will be coincident with $I_{1,2}$. Both these images cannot be seen at the same time. $I_{2,1}$ is seen on looking into mirror AB, and $I_{1,2}$ is seen on looking into AC. The light reaches the eye in the position E by travelling along the path shown. The reflections obey the laws of reflection.

It must be understood that when we say an object or image is in front or behind a mirror, we mean it is so situated with respect to the plane of the mirror. For example, we say O (Fig. 22.11) is

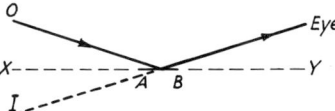

Fig. 22.11 Only a small portion AB of mirror XY is used to see image I of object O

in front of mirror AB. It will be noted that an image I of O can be seen although AB is quite small.

Referring again to Fig. 22.10, all the images and the object lie on a circle with its centre at the

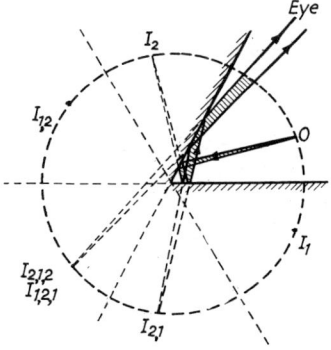

Fig. 22.12 Images produced by mirrors inclined at 60°

(a)

(b)

Fig. 22.13(a) and *(b)* Multiple images produced by two mirrors at an angle of 60°. The principle of the kaleidoscope

intersection of the mirrors. This simple condition must hold for mirrors arranged at any angle.

Fig. 22.12 shows the images produced when an object is placed between mirrors at 60°. The lettering of the images is similar to that used in the previous example. The symmetry of the image positions should be noted. The path of the light producing the image $I_{2,1,2}$ is shown.

If the mirrors are parallel, an infinite number of images is formed in a series extending indefinitely behind the mirrors.

The kaleidoscope is a simple toy using the symmetrical images produced by inclined mirrors. It consists of two mirrors arranged at 45° or 60° in a tube, with a few pieces of coloured glass of irregular shapes between them. One end of the tube is closed by ground glass, and the other has a small viewing hole in it. On looking down the tube, when it is held to the light, the images of the coloured glass form a pleasing symmetrical pattern which can be changed by shaking the tube (Fig. 22.13).

QUESTIONS

1. What is the size of the smallest plane mirror in which a man can see a full view of himself? Draw a diagram.

2. Two mirrors are arranged at an angle of 45°. How many images of an object placed between them are produced? Mark in these images on a diagram and show the path the light takes to the eye in forming the last image.

3. State the laws of regular reflection of light, and describe how you would test them experimentally.

An ornamental mirror consists of a plane sheet of thick plate glass which is silvered on the back. Show, by a ray diagram, how the chief image of a bright point-object is seen by someone looking into the mirror, and explain why several fainter ones can usually be seen as well. (O)

4. State the laws of reflection of light.

Prove that the image of a small object formed by a plane mirror is as far behind the mirror as the object is in front. How would you verify this experimentally?

What is meant by the statement that this image is virtual? (O)

5. State the laws of reflection of light. How would you show, by experiment, that a ray of light reflected from a plane mirror suffers a deflection of 2θ when the mirror is turned through an angle θ? A point source of light P lies between the reflecting surfaces of two plane mirrors AB and BC at right angles. P is 30 cm from BC and 40 cm from AB. Draw a scale diagram to show the positions of the images of P. On the diagram show three rays of light which leave P and are reflected to pass through a point Q which is 10 cm from AB and 60 cm from BC. (C)

6. Distinguish between regular and diffuse reflection of light. State the laws of reflection.

Draw ray diagrams showing how an image of a point object is seen by means of two reflections *(a)* between two mirrors at right angles to each other, and *(b)* between two parallel mirrors. (W)

7. State the laws of reflection of light and explain how you would verify them.

A converging beam of light is intercepted by a plane mirror and after reflection comes to a focus in a spot 1 m from the mirror. If the mirror is turned through 2 degrees, what is the deflection of the spot? (O & C)

23 Curved mirrors

We get much fun looking at the distorted images of ourselves in the curved mirrors in amusement fairs. Apart from entertaining us, curved mirrors have many other and more important uses. We use different types as shaving mirrors, dentists' mirrors, telescope mirrors, driving mirrors, electric fire reflectors, etc. They produce images more suitable for their purpose than those produced by plane mirrors. We shall consider spherical, cylindrical and paraboloidal mirrors.*

cross-sections, and the centre of this section is called the *centre of curvature*. The central point on the mirror is called the *pole*, and the line joining this point and the centre of curvature, the *principal axis*.

Allow a parallel beam of light from a ray box to fall upon a concave cylindrical mirror. (It is more convenient to use a cylindrical mirror than a spherical mirror, for the effect upon the light in one plane can be clearly studied. A spherical

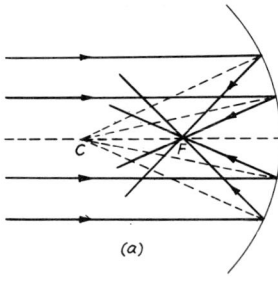 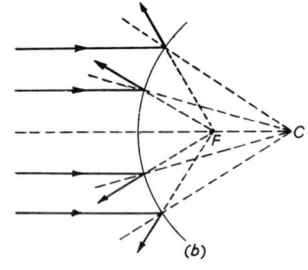

Fig. 23.1

(a) Concave mirror-real focal point

(b) Convex mirror-virtual focal point

Mirrors silvered to reflect from the surface which 'caves in' are called *concave* and those which are curved the other way, *convex*. Both spherical and cylindrical mirrors have circular

mirror produces a similar effect upon the light in all planes through its centre.) A multiple slit can be used. It will be noted that, after reflection, all the rays pass through one point on the principal axis of the mirror (Fig. 23.1a and Fig. 23.2).

The laws of reflection are always obeyed, and we can investigate this result geometrically. Consider the ray *OM* (Fig. 23.4), incident at *M*

* A paraboloid is the surface described by a parabola when rotated round its principal axis. Spherical and cylindrical mirrors are sections of the surfaces of a sphere and cylinder respectively.

206

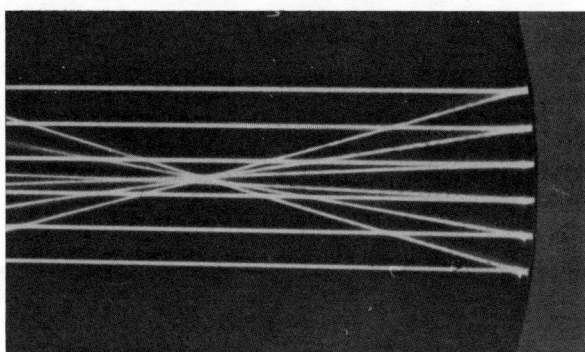

Fig. 23.2 Reflection at a concave mirror

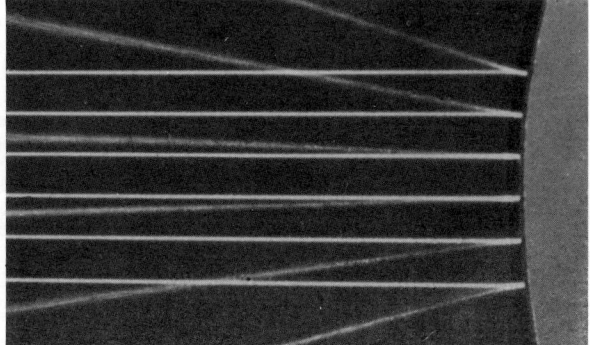

Fig. 23.3 Reflection at a convex mirror

Now if *M* is near to *P*, *MF* becomes very nearly equal to *PF* and *F* will then be midway between *P* and *C*.

As this result is independent of the position of *M* so long as it is not too far from *P*, it follows that any ray such as *OM* will, after reflection, pass through *F*.

This shows geometrically what we have observed to be true practically. The point *F* is called the *principal focus* or *focal point*.

If the concave mirror is now replaced by a convex one, the reflected light does not converge to a point focus, but diverges, apparently coming from a point *F* behind the mirror (*see* Fig. 23.1*b* and Fig. 23.3). This is a virtual focal point.

It will be seen, in a subsequent chapter, that lenses produce similar effects, and so we can define the focal point to cover all cases.

Principal focus or focal point

> This is the point from which a paraxial* parallel beam of light parallel to the principal axis diverges, or appears to diverge, after reflection or refraction.

The focal length is the distance from the focal point to the pole of the mirror or to the centre of the lens.

For a mirror:

focal length $=\frac{1}{2}$ (radius of curvature).

Cylindrical and spherical mirrors of large aperture

The geometrical deduction showing that the focal length of a mirror is half the radius of curvature is true only for rays falling on the mirror near to its pole, or, as we say, when using only a small aperture. If a mirror of large aperture is used, the rays near the edge or periphery are reflected so that they cut the principal axis at a point nearer the mirror than the focal point. The reflected pencils of the wide beam all touch a curve of the shape shown in Fig. 23.5, known as a *caustic*. The intersections of neighbouring reflected pencils lie on the caustic curve, which is shown brightly illuminated. The point or cusp of the curve is at the principal focus.

A caustic can be produced with a ray box, but a

upon the mirror *PM*. If *C* is the centre of curvature, the angle *OMC* is the angle of incidence and *CMF* the angle of reflection.

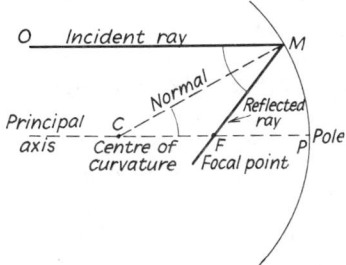

Fig. 23.4 Concave mirror

Thus,	$O\hat{M}C = C\hat{M}F$ (law of reflection)
but also	$O\hat{M}C = M\hat{C}F$ (alternate angles)
hence	$M\hat{C}F = C\hat{M}F$
and	$MF = FC$

---

* Near to the principal axis.

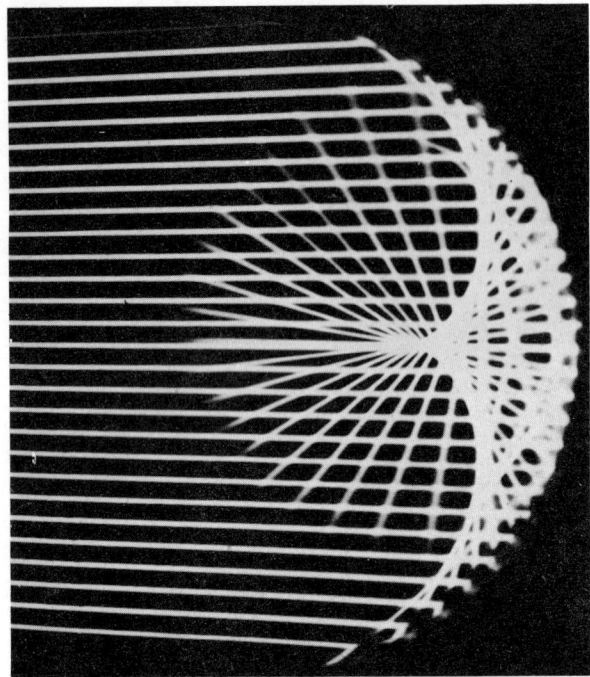

Fig. 23.5 Photograph showing the formation of a caustic curve—spherical aberration

very good specimen is to be seen on the surface of tea in a cup. The inside surface of the cup acts as the mirror.

This defect of spherical mirrors, i.e. the inability to bring a wide parallel beam to a point focus, is called *spherical aberration*.

As a beam of light is reversible, it follows that a point source of light placed at the principal focus of a concave mirror of large aperture will not produce a parallel beam of light (Fig. 23.6). Where a

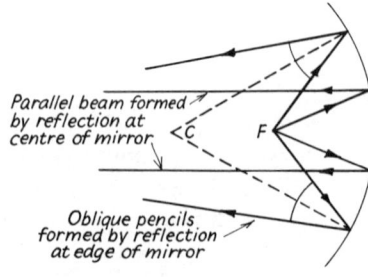

Fig. 23.6 Spherical aberration

Point source placed at focal point *F* of spherical mirror does not produce a parallel beam

parallel beam is required, as in a searchlight or a motor-car headlamp, a paraboloidal mirror is used (Fig. 23.7). Every pencil of light from a point

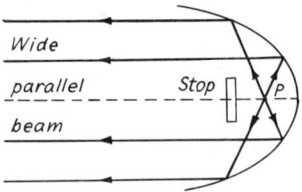

Fig. 23.7 Paraboloidal mirror

Point source at *P* produces parallel beam (What is the function of the stop?)

source placed at the focus of such a mirror is reflected parallel to the principal axis. A paraboloidal mirror is used in a reflecting telescope. In this case, a mirror of large aperture is very necessary, and all the light must be brought accurately to a point focus.

Images produced by spherical mirrors

A parallel beam of light can be regarded as coming from a very distant point source. Thus, the principal focus is the point image of a very distant object (Fig. 23.8a). Let us now consider the effect of bringing the object nearer to the mirror. The angle of incidence $O\hat{M}C$ and the angle of reflection $C\hat{M}I$ are both reduced (Fig. 23.8b). Hence, the image is formed farther away from the mirror than *F*. The image moves from *F* to *C* as the object moves from infinity to *C*. With the object at *C*, the light strikes the mirror normally and returns back along the same path, forming the image at *C* (Fig. 23.8c). Again, as the path of a beam of light is reversible, a point object between *C* and *F* forms an image beyond *C* (Fig. 23.8d), and as the object moves towards *F*, the image moves to infinity. An object at *F* produces its image at infinity (Fig. 23.8e).

An important difference occurs when the object is nearer to the mirror than *F*. The angle of incidence is then greater than *FMC* and so the reflected light is divergent, appearing to come from a point at the back of the mirror: in this case the image is virtual (Fig. 23.8f).

With a convex mirror, the image is always behind the mirror. From the geometry of Fig. 23.8g, you should be able to deduce that the image must

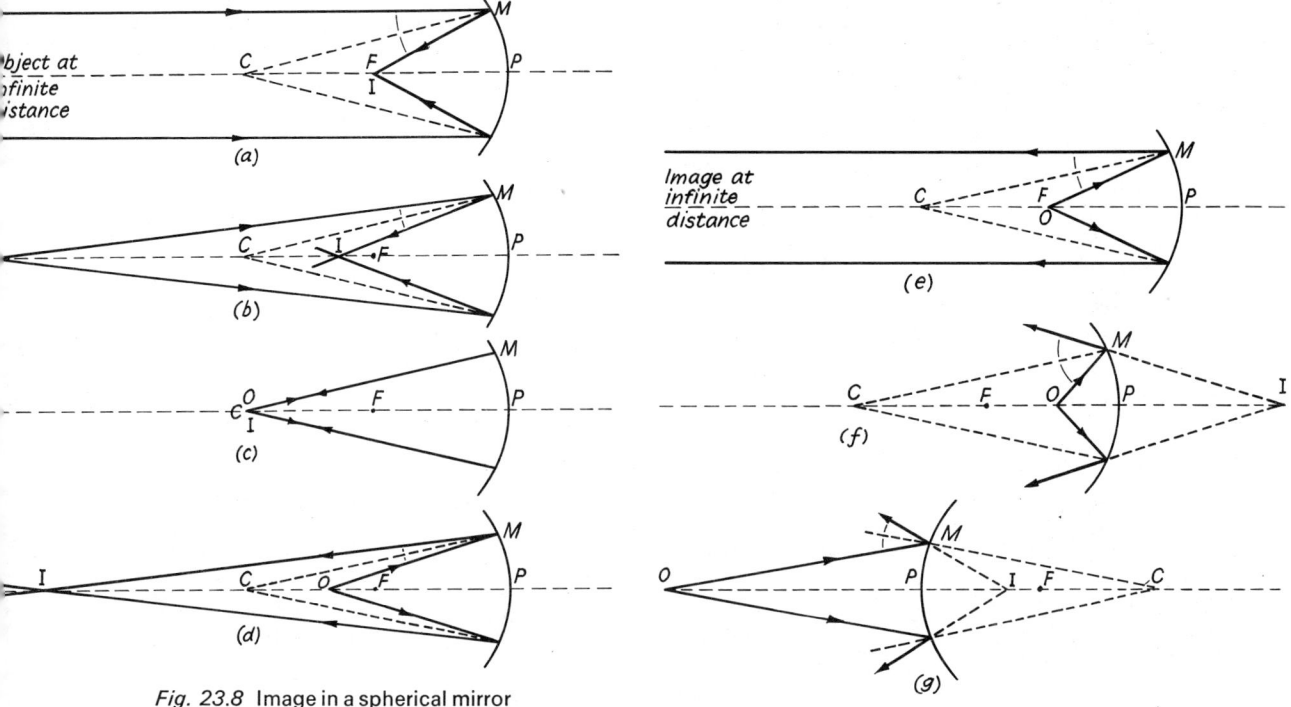

Fig. 23.8 Image in a spherical mirror

(a) to (f) show the change in image position due to a change in object position for a concave mirror. A convex mirror is shown at (g)

always lie between *F* and *P*. It is always erect and diminished.

The production of images in a concave mirror can easily be carried out practically, using a clear electric lamp as the object. The experiment yields interesting detail, for if a horseshoe filament lamp is used as object, the real images can be produced upon a screen, revealing that a real image is always inverted. The size of the image too can be observed. When between *C* and *F* it is diminished, but when beyond *C* it is enlarged. The virtual image, which of course cannot be produced upon the screen, is always erect and enlarged (Fig. 23.9*a*).

The above results can be summarized thus:

SUMMARY OF RESULTS

Position of object	Position of image	Character of image
Concave Mirror		
At ∞	At *F*	Real, inverted, diminished
Between ∞ and *C*	Between *F* and *C*	Real, inverted, diminished
At *C*	At *C*	Real, inverted, same size
Between *C* and *F*	Between *C* and ∞	Real, inverted, enlarged
At *F*	At ∞	Real, inverted, infinitely large
Between *F* and *P*	Between ∞ at back of mirror and *P*	Virtual, erect, enlarged
Convex Mirror		
At ∞	At *F* behind mirror	Virtual, erect, diminished
Between ∞ and *P*	Between *F* and *P*	Virtual, erect, diminished

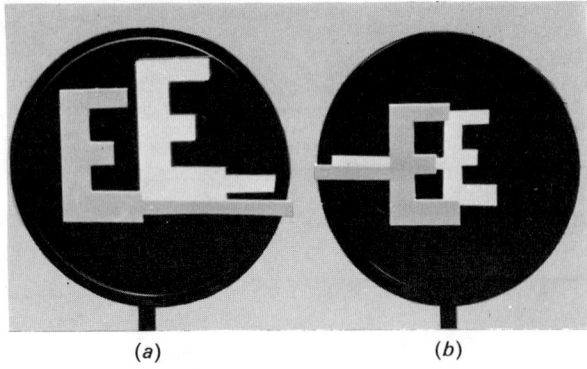

(a)   *(b)*

Fig. 23.9 Photograph of virtual image produced in
(*a*) a concave mirror, (*b*) a convex mirror

Fun-fair mirrors

It should now be clear how the amusing images of
ourselves are produced in fun-fair mirrors. These
are usually large mirror strips bent into irregular
concave and convex cylindrical shapes. They produce an image the same size as the object, in a
horizontal direction, but enlarged or diminished
in the vertical direction, according to whether they
are concave or convex. The irregular curvature
causes a vertically distorted image.

Graphical construction for images

Full detail of an image can be obtained by a comparatively simple graphical method. Consider an
object *OB* (Fig. 23.10) set up in front of a concave
mirror as shown. Each point of the object will be
giving out light in all directions. If the point *B* of
the object is on the principal axis, its image point
M will also be on the principal axis. Furthermore,
the image *IM* will be perpendicular to the axis, if
OB is perpendicular to it. We usually locate the
image point *I* of the object point *O* and then obtain
the image *IM* by drawing the perpendicular from
I on to the axis. Every ray of light sent out from *O*
must pass through *I* after reflection, if *I* is its image.
Of the infinite number of rays sent out from *O*, we
select a few special ones which are reflected along
definite directions, viz.

1. The ray from *O* parallel to the axis passes,
after reflection, through the principal focus *F*.

2. The ray through *F* becomes, after reflection,
parallel to the axis.

3. The ray through *C* strikes the mirror normally
and is reflected back along the same path.

4. The ray at *P* is reflected symmetrically with
respect to the principal axis.

As the image *I* must lie on all these reflected rays,
any two of the four rays are sufficient to give its
position. The same construction can be used for the
object in any position, and for convex as well as
concave mirrors.

This graphical construction reveals the complete
character of the image and can be used to solve
problems relating to images. It should be noted
that in adopting a scale for the diagram there is no

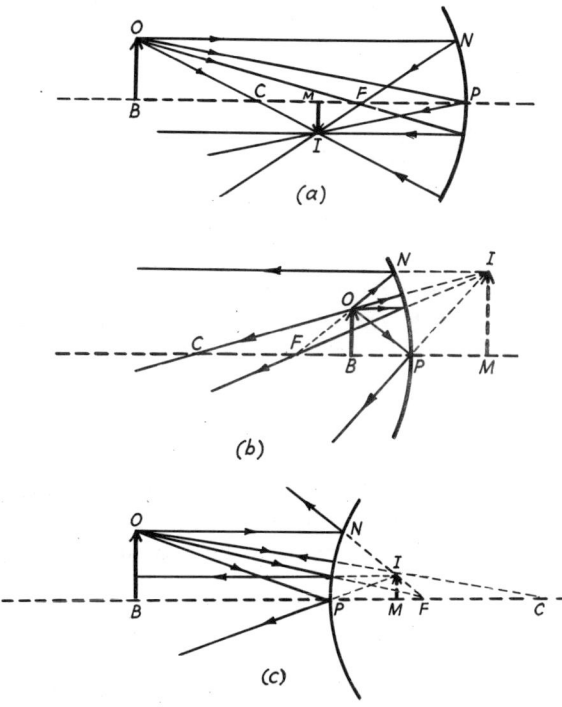

(a)

(b)

(c)

Fig. 23.10 Graphical construction for images

need to use the same scale for horizontal as for
vertical distances. It is usually advantageous to
use a much larger vertical scale.

Careful drawing of the special rays will reveal
that, after reflection, they are not concurrent at *I*
as they should be. This is due to the exaggeration
of the vertical scale. We have seen that all the
reflected rays pass through a point only when the
mirror aperture is small. This defect can be overcome by drawing the incident rays up to a line

through *P* perpendicular to the axis. The diagram (Fig. 23.11) shows this slight modification.

From the similarity of the triangles *OPB* and

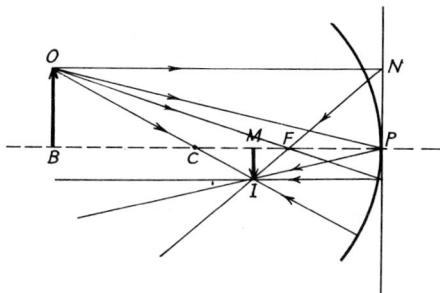

Fig. 23.11 Modified graphical construction

IPM, we can arrive at an expression for the magnification.

$$\text{Magnification} = \frac{\text{size of image}}{\text{size of object}} = \frac{\text{image distance}}{\text{object distance}}$$

Experimental determination of focal length

Concave mirror
The focal length of a concave mirror can be found directly by allowing a parallel beam of light to fall on it. Sunlight or a parallel beam produced by a ray box can conveniently be used. A screen is placed to receive a clear image, and the distance from the mirror to the screen is the focal length.

Another simple method consists of finding the radius of curvature and halving this to get the focal length. An object at the centre of curvature produces an image coinciding with the object. The centre of curvature can be located by a parallax method or by using an illuminated object. The ray box can be used if the lens is replaced by a piece of card with a small circular hole. Adjustment is made easier if the hole has two pieces of fine wire across it. This arrangement produces an image

which is easy to focus. The apparatus is set up as shown in Fig. 23.12. The position of the mirror is altered until a focused image is produced alongside

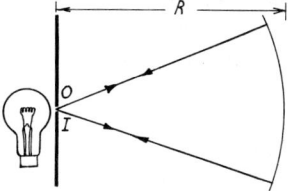

Fig. 23.12 Measurement of the radius of curvature and focal length of a concave mirror

the object. The light must then be falling normally on the mirror, and thus the distance from the object to the mirror is the radius of curvature.

Convex mirror
A good method of finding the radius of curvature, and so the focal length of a convex mirror, is illustrated in Fig. 23.13. *O* is an illuminated object

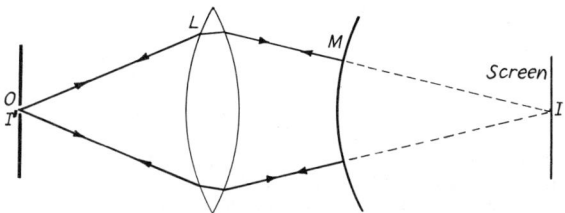

Fig. 23.13 Radius of curvature of a convex mirror

and *L* an auxiliary lens producing a real image *I* of the object on a screen. The convex mirror *M* is then interposed and its position altered until an image *I′* is produced alongside the object. For this to take place, the light must strike the mirror normally and be returned back along the same path. Hence, the radius of curvature is the distance of the mirror from the screen.

QUESTIONS

1. An object is placed (*a*) 18 cm, (*b*) 6 cm from a concave mirror of focal length 12 cm. Find the positions of the images.

2. An image is produced 3 cm behind a convex mirror of focal length 6 cm. Find the position of the object.

3. Define *magnification*. An image three times the size of an object is produced by a concave mirror of focal length 12 cm. What are the two possible positions of the object?

4. Define the *focal length* of a concave mirror.
    Describe an optical method for determining the focal length of a given concave mirror.
    When a small object is placed on the axis of a concave mirror at a distance of 12 cm from its pole, the image formed is real and the linear magnification is 3. What is the focal length of the mirror? (O)

5. Draw diagrams showing how a concave mirror can be used to form (*a*) a real enlarged image, (*b*) a virtual enlarged image, of a small object placed near its axis.
    Describe how you would determine experimentally the focal length of a concave mirror. (O)

6. Distinguish between *real* and *virtual* images. Illustrate your answer by diagrams showing the formation of images by pencils of rays reflected from (*a*) a concave mirror, (*b*) a plane mirror.
    A concave mirror, radius of curvature 20 cm, forms a *virtual* image of a small object placed at right angles to its axis, the image being four times as high as the object. What is the distance of the object from the mirror? How far, and in what direction, must the object be moved in order that a *real* image, having a magnification of four, may be produced? (L)

7. Define the terms *principal focus* and *radius of curvature* as applied to a concave mirror. Illustrate your answer by a ray diagram.
    Draw ray diagrams to show how such a mirror can form (*a*) an enlarged real image and (*b*) a virtual image of an object.
    An object 1 cm high is placed on and at right angles to the principal axis of a concave mirror 4 cm from the pole of the

mirror. A real image 2 cm high is formed. Find graphically or by calculation the focal length and radius of curvature of the mirror. (For a calculated solution state the sign convention used.) (JMB)

8. What is meant by (*a*) the radius of curvature, (*b*) the focal length, of a concave mirror? What is the relation between these lengths?
    A concave mirror of focal length 6 cm forms a real image 1.5 cm high of an object 0.5 cm high. Find by calculation or by drawing the position of the object and of the image.
    How would you verify your result by experiment? (JMB)

9. Distinguish between real and virtual images.
    Illustrate your answer by ray diagrams showing how a concave mirror can form (*a*) a real image and (*b*) a virtual image.
    Find, graphically or by calculation, the position of the image of an object of height 1 cm placed at right angles to the principal axis of a convex mirror of radius of curvature 9 cm and 3 cm away from the mirror.
    State the nature of the image and find its size. (JMB)

10. Starting from the laws of reflection, derive the formula connecting object distance and image distance from a concave mirror with its radius of curvature.
    An object, 0.6 cm high, is placed with its mid-point on the axis of a concave mirror, and 30 cm in front of it. A real image is formed 60 cm from the mirror. What are (*a*) the focal length of the mirror, and (*b*) the size of the image?
    If the eye is placed at 100 cm from the mirror, calculate the minimum size for the mirror in order that the image of the whole of the object may be seen. (C)

11. Explain the term *virtual image* as used in connection with a concave mirror, and illustrate your answer with a diagram.
    Make a diagram showing how an image of the sun can be formed by a concave mirror.
    What is the diameter of the image of the sun formed by a mirror of radius of curvature 4.0 m, if the sun itself is just obscured by a disc 1.0 cm in diameter when held at a distance of 1.0 m from the eye? (O & C)

24 Refraction

The following are four examples of the effect we call refraction:

(*a*) The apparent bending of a straight stick when held through the surface of water (Fig. 24.1*a*).

Refraction is the bending of light as it passes across the boundary of one medium to another. For light passing from an optically rarer to an optically denser medium, the refraction is toward

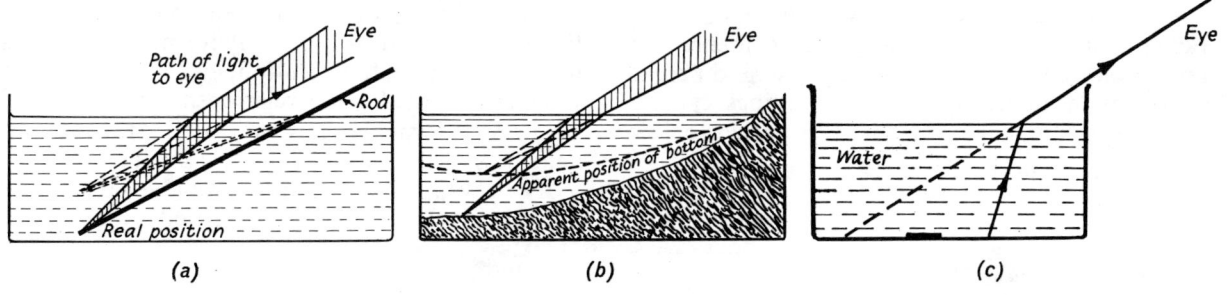

Fig. 24.1 Examples of refraction

(*a*) The apparent position of the rod in the water is indicated by a broken line
(*b*) Apparent reduction in the depth of a pond due to refraction
(*c*) Penny and basin experiment: with the eye in position shown, coin cannot be seen until water is poured into vessel

(*b*) The quivering of objects viewed through the hot air rising from a brazier.

(*c*) A pond or swimming bath appearing less deep than it really is (Fig. 24.1*b*).

(*d*) The apparent displacement of stars low down in the sky.

the normal, and vice versa. This is due, as was stated in Chapter 20, to the change in velocity of the waves as they pass through the boundary.

Using the ray box, pass a ray of light into a rectangular glass block with an angle of incidence about 60° (Fig. 24.2). (The glass should have the

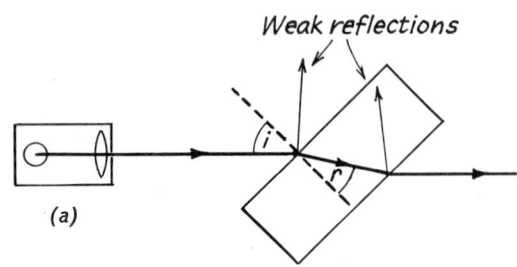

Weak reflections

(a)

Fig. 24.2 Passage of a ray through a glass block

(b)

lower face ground so that the path of the light in it is clearly visible.) It will be noted that the ray is refracted towards the normal on entering, making the angle of refraction (the angle between the refracted ray and the normal) less than the angle of incidence. On emerging, the ray is refracted away from the normal and is parallel to the direction in which it entered the block.

Partial reflections will be seen from the front and back surfaces. Reflection always occurs at a polished surface and, in lenses, may account for a serious reduction in efficiency.

If the position of the block and the path of the ray of light are marked, the angles of incidence and refraction can be measured. By using different angles of incidence, a series of readings can be obtained. Tabulate the readings as shown below and calculate the ratio $\frac{\sin i}{\sin r}$.

Angle of incidence i	Angle of refraction r	$\sin i$	$\sin r$	$n = \dfrac{\sin i}{\sin r}$
10°	6.5°	0.1736	0.1132	1.53
17°	11°	0.2924	0.1908	1.52
23°	14.5°	0.3907	0.2504	1.54
31°	20°	0.5150	0.3420	1.52
40°	25°	0.6428	0.4226	1.51
49°	30°	0.7547	0.5000	1.52
57°	33.5°	0.8387	0.5519	1.51

This ratio should be a constant which measures the optical power or optical density of the glass. It

is known as the *refractive index* and is usually denoted by n. (μ is also still widely used.)

Laws of refraction

(1) The incident ray, the refracted ray, and the normal at the point of incidence all lie in the same plane.

(2) The ratio of the sine of the angle of incidence to the sine of the angle of refraction is constant (Snell's law).

Although the phenomenon of refraction was known to the ancients, the simple relationship between the angles baffled investigators for many centuries until Snell gave the solution in 1621. The wave theory accounts for this relationship. On page 183 we have shown that this ratio is the ratio of the velocities of the waves in the two media.

If a ray apparatus is not available, the above experiment (Fig. 24.2) can be carried out by means

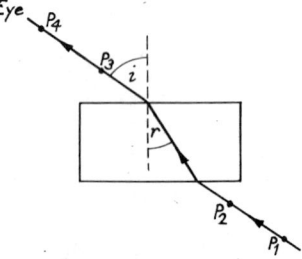

Fig. 24.3 Tracing the passage of a ray through a glass block, using pins. For accuracy, the pins P_1, P_2, and P_3, P_4 should be kept as far apart as possible

of pins. Two pins P_1 and P_2 are set up as shown in Fig. 24.3 on one side of a glass block. Two more pins P_3 and P_4 are then aligned with the images of the first two pins produced by refraction through the block. Knowing the directions along which the light enters and leaves the block, the ray through the block can be drawn and the experiment completed as above.

Measurement of the refractive index

If $ABCD$ is the path of a ray of light through a parallel glass block, traced by either the ray or pin method, the refractive index may be obtained more conveniently and accurately by finding the ratio of the distances BC and BF (Fig. 24.4).

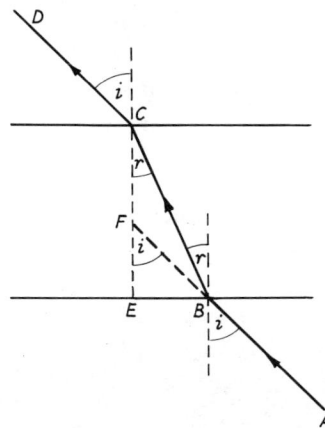

Fig. 24.4 Determination of refractive index

From the figure,

angle of incidence $i = B\hat{F}E$

refraction $r = B\hat{C}E$

Hence $$n = \frac{\sin i}{\sin r} = \frac{BE/BF}{BE/BC} = \frac{BC}{BF}$$

To obtain a good mean value for n, trace several rays through the block and tabulate the results.

BC cm	BF cm	$n = BC/BF$
6.5	4.35	1.49
6.8	4.5	1.51
7.1	4.65	1.53
7.5	4.9	1.53
7.8	5.15	1.52

Apparent thickness of a block

We have seen that a pond appears to be less deep than it really is, due to refraction. This effect can be made use of in determining the refractive index of a solid or of a liquid.

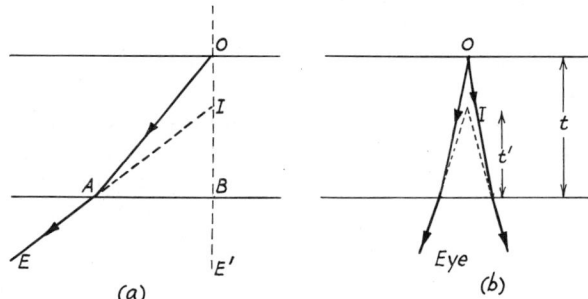

Fig. 24.5 Apparent thickness of a glass block

If a pin O (Fig. 24.5a) is placed in contact with one face of a parallel block of glass, the eye at E sees an image of O in the direction AI. From above $n = AO/AI$. If the eye is moved to E' and the object viewed normally through the block, a narrow cone of light is received by the eye which apparently comes from I (Fig. 24.5b). Thus, the glass block appears to be thinner than it really is.

For the normal viewing, the point A moves very near to B, hence:

$$n = \frac{AO}{AI} = \frac{BO}{BI} = \frac{\text{real thickness}}{\text{apparent thickness}} = \frac{t}{t'}$$

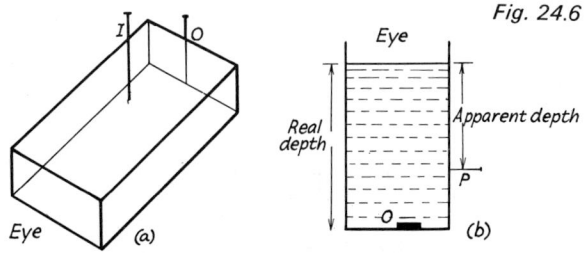

Fig. 24.6

(a) O is a pin stuck in paper at the end of the glass block, and I is a second pin mounted with its point on the top of the block.

(b) O is an object at the bottom of a vessel of liquid, and P is a pin held at the side of the vessel

To find n by this method (Fig. 24.6a) I has to be located, and this may be done by parallax. The pin at O is viewed through the block. A second pin is mounted vertically on the top surface of the block and adjusted so that there is no parallax between it and the image I of O.

For a liquid, the object must be placed on the bottom of the vessel (Fig. 24.6b) and viewed through the liquid. A pin at the side of the vessel is adjusted for no parallax with the image.

Determination of the refractive index of a liquid

Place a concave mirror, of radius of curvature about 50 cm, on a horizontal surface and locate the centre of curvature by a parallax method (Fig. 24.7a). The object pin at C is adjusted until it is co-

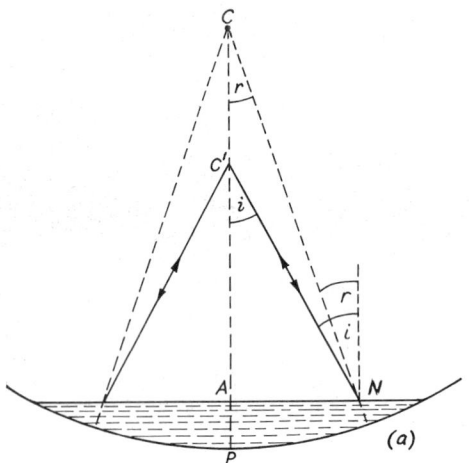

REFRACTIVE INDICES			
Glass (crown)	1.52	Turpentine	1.47
Glass (flint)	1.66	Glycerine	1.47
Diamond	2.42	Oil—cedar-wood	1.52
Ice	1.31	Carbon disulphide	1.63
Water	1.33		

is refracted so that it strikes the mirror normally and is reflected and refracted back along the same path.

From the figure

$$n = \frac{\sin i}{\sin r} = \frac{AN/C'N}{AN/CN} = \frac{CN}{C'N}$$

As N is near to P, we can write:

$$n = \frac{CP}{C'P}$$

Passage of light through a prism

The path of a ray of light through a prism can easily be traced using a ray box or by the use of pins (Fig. 24.8). The light on emerging does not travel in the same direction as on entering: it is deviated. The deviation D is different for different angles of incidence i and, for one special angle of

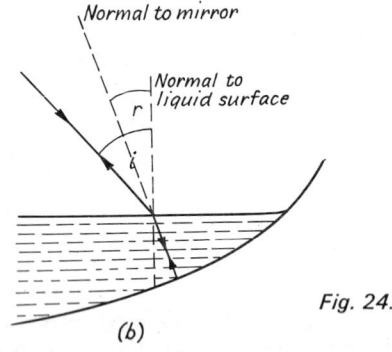

Fig. 24.7

(b)

incident with its image. The light is then returning along the same path and must be striking the mirror normally. Some liquid is now poured on to the mirror, and it is found that the pin has to be brought to position C' in order to be coincident with its image. This position is again located by parallax. Fig. 24.7b shows how the light from C'

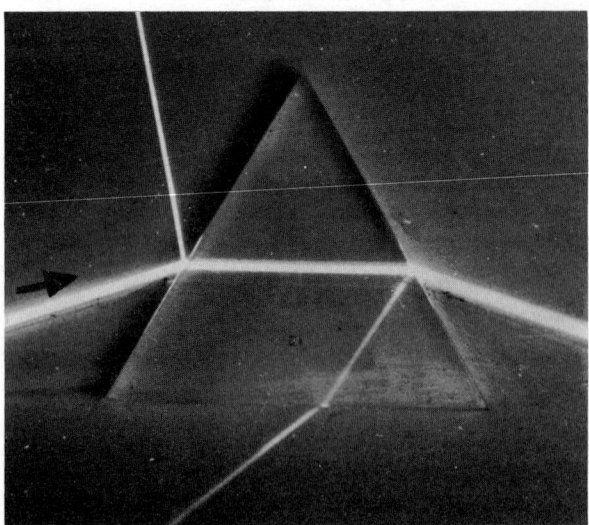

Fig. 24.8
Passage of light through prism

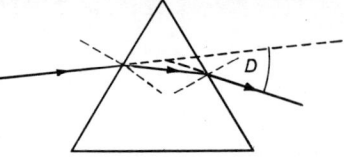

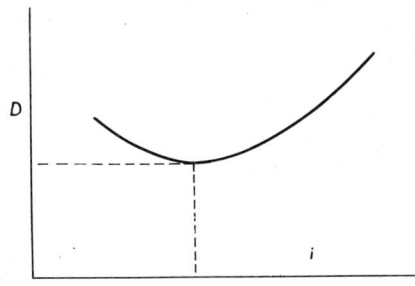

Fig. 24.9 Graph of deviation/angle of incidence
Deviation is a minimum for a particular
value of the angle of incidence

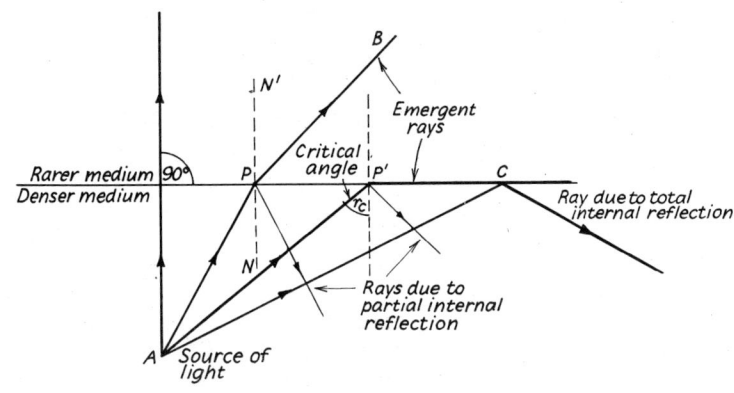

Fig. 24.11 Refraction at a rarer medium

incidence, is a minimum. This is shown on the graph of deviation against angle of incidence (Fig. 24.9). The minimum deviation occurs when the

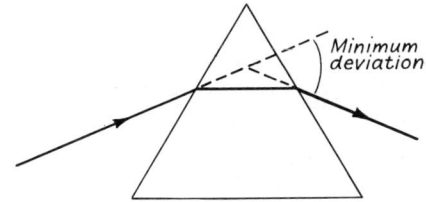

Fig. 24.10 Symmetrical ray suffers
minimum deviation

light passes symmetrically through the prism, as shown in Fig. 24.10.

The emergent light is coloured when white light is used, but this effect is dealt with in Chapter 28.

Total internal reflection

We have seen that, on refraction at a denser medium, a beam of light is bent towards the normal and, vice versa, on refraction at a rarer medium the bending is away from the normal. In Fig. 24.11 the ray *AP* is refracted away from the normal, making *BPN'* greater than *APN*. For any refraction at a rarer medium, the angle of refraction is greater than the angle of incidence. By increasing the angle of incidence, the angle of refraction will eventually become 90°, as is the case for the ray *AP'C*. A further increase in the angle of incidence should give rise to an angle of refraction greater than 90°, but this is impossible, and the ray is reflected at the boundary, remaining within the denser medium: it suffers *total internal reflection*. The reflection is perfect, none of the light passing through the

boundary. The effect in water can be easily and effectively demonstrated, using the arrangements shown in Fig. 24.12 and Fig. 24.13.

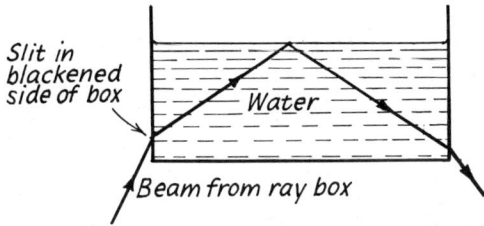

Fig. 24.12 Total internal reflection in water

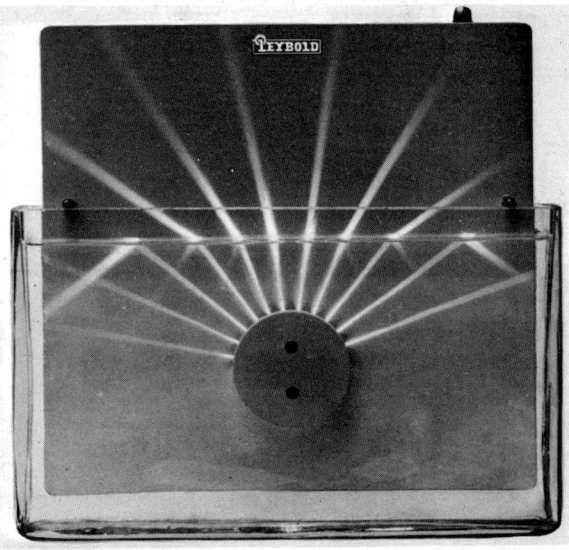

Fig. 24.13 Photograph showing partial and total reflection at a water surface

Conditions for total internal reflection

1. The light must be passing through a denser and towards a rarer medium.

2. The angle of incidence must exceed a certain critical value.

The critical angle

Consider the ray $AP'C$ in Fig. 24.11. As the ray is reversible we can consider the light passing in the direction $CP'A$. The angle of incidence is $90°$ (grazing incidence) and the angle of refraction is the critical angle r_c. Then we have:

$$\frac{\sin i}{\sin r} = n$$

$$\frac{\sin 90°}{\sin r_c} = n$$

but $$\sin 90° = 1$$

$$\therefore \qquad \sin r_c = \frac{1}{n}$$

For glass, $\sin r_c = \frac{2}{3}$ $\therefore r_c = 42°$
For water, $\sin r_c = \frac{3}{4}$ $\therefore r_c = 49°$

Experimental determination of the critical angle

The critical angle can be determined, using the ray box (Fig. 24.14). Direct a ray of light through a semicircular glass block towards the centre C. This

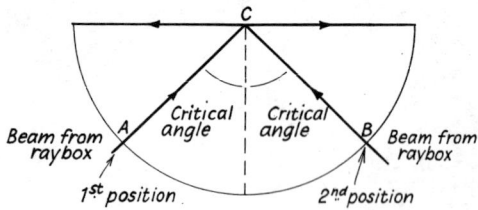

Fig. 24.14 Measurement of critical angle

light will always be normal to the curved surface and will not be deviated. By carefully rotating the block about C, and so increasing the angle of incidence of the pencil at C, a position AC will be found where all the light is reflected instead of passing out of the block. This direction will give that of the critical ray and the angle of incidence of the ray AC will be the critical angle. To measure

this it is easier to find (in the same way) the corresponding critical ray BC, on the other side of the normal. Then the critical angle is $\frac{1}{2}A\hat{C}B$.

Totally reflecting prisms

As the critical angle for glass is about $42°$, light passed normally into a $45°$, $90°$, $45°$ prism strikes the inclined face at $45°$, thereby fulfilling both the conditions for total internal reflection. There

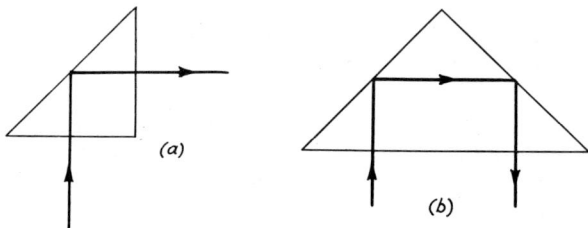

Fig. 24.15 Totally reflecting prisms

are two cases, the first being shown in Figs. 24.15*a* and 24.16 and the second in Fig. 24.15*b*. In the first case, the prism deviates the ray through $90°$ acting like a plane mirror at $45°$. In good optical instruments prisms are used instead of plane mirrors. With an ordinary mirror, silvered on the

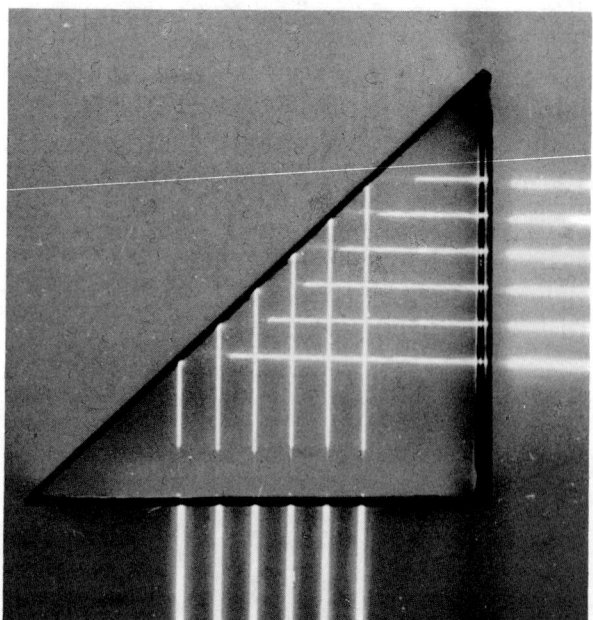

Fig. 24.16 Total internal reflection produced by a glass prism

back surface, there is always some light reflected from the front surface and, further, the silvering deteriorates. The disadvantage of the mirror is not due to this loss of light (for some light is lost by reflection at the surfaces of the prism), but is due to the reflected light proceeding in the same direction as the main beam (Fig. 24.17), thereby producing a faint image not quite coincident

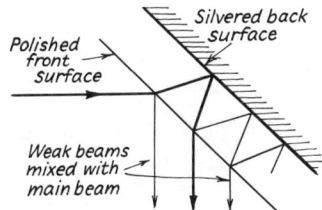

Fig. 24.17 Multiple reflections at a plane mirror

with the main image. Such 'ghost' images can ruin the definition obtained in an otherwise good optical instrument.

It can now be understood why surface-silvered mirrors are used in episcopes. In this case use of a prism is not practicable because of the large size required.

The refractive index of diamond is 2.4; thus its critical angle is very small. It is this and other optical properties of diamond which give it its brilliance. The cutting of a diamond is worked out mathematically to give the best results.

Further demonstrations and applications of total internal reflection

Light from a car headlamp is converged by a lens so that it enters the apparatus as shown in Fig. 24.19. This consists of a tapered tube *AB* with a T-junction at *C*. The end *B* is sealed by a glass window, and water from the mains is passed into the tube at *C* and streams out in the form of a

Fig. 24.18 Optical system of lighthouse lamp

As a large lens would be impracticable, glass rings which refract and totally reflect the light are employed

smooth jet at *A*. This emergent stream falls on a white tile at *D*. The light sent into the apparatus at *B* is trapped in the beam by total internal reflection and passes down the jet as indicated, finally illuminating the tile.

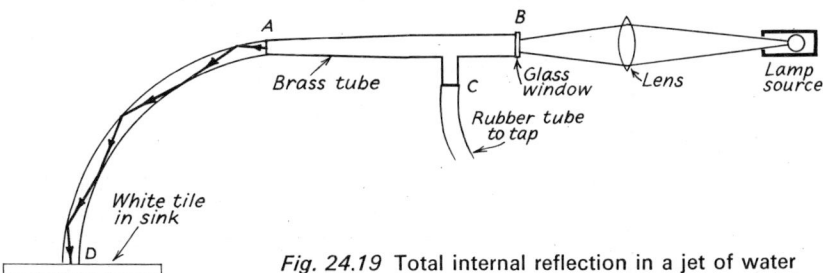

Fig. 24.19 Total internal reflection in a jet of water

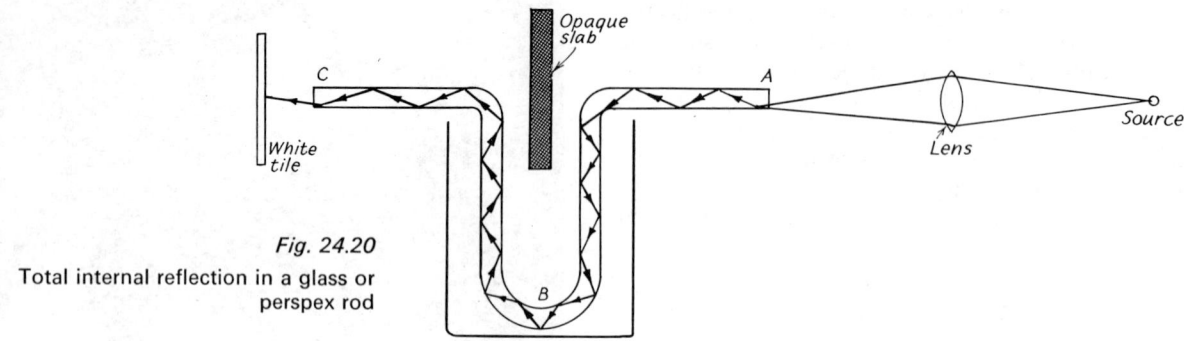

Fig. 24.20

Total internal reflection in a glass or
perspex rod

A similar demonstration can be done with a glass rod bent as shown in Fig. 24.20. Light is passed into the glass rod at *A*. Some of the light remains in the rod and passes round the bend, being totally internally reflected. It finally emerges at the end *C* where it falls on a white surface. If the loop is now immersed in a beaker of water, the quantity of light emerging is considerably reduced. The reduction is even greater if a liquid such as paraffin oil is used. The smaller the critical angle the greater will be the amount of light totally reflected and passing through the tube to *C*. The critical angle for a paraffin-glass boundary is quite large as the difference between the two refractive indices is small. This device is much used in medicine for illuminating internal organs during examination. A plastic rod is bent in the way required, the illumination being provided by a small lamp-house and condensing lens attached to one end. A further advance has been made by using bundles of very fine fibres of plastic material which together form, in effect, a flexible rod.

The fish-eye view

No light passing through the surface of a pond from above can have an angle of refraction greater than the critical angle for water, namely 49°. Thus, all light from objects above the surface must be refracted within the cone *XEY* (Fig. 24.21). A fish with its eye at *E* sees the entire scene above the surface in a cone of angle 98°. The image is distorted, for the displacement of objects near the horizon is greater than that of objects vertically above the surface. The fish sees the rising sun at *A'*!

Mirages

Mirages are produced by total internal reflection. The refractive index for air is very nearly unity, but it obviously depends upon the density of the air. Thus, cool air has a slightly higher refractive index than warm air. On a very hot summer's day, the air in contact with the warm ground is heated (Fig. 24.22). This sets up convection currents, but the hottest air at any instant is that in contact with the ground. An observer at *O* sees a distant object *T*, such as a tree, by light reaching his eye along *TO*. Other rays from *T* may be totally reflected at the layers of warm air near the ground and enter the eye as shown. The object then appears to be at *I* and is inverted. The observer seeing the object *T*, and its inverted image, concludes that there is water between him and the tree.

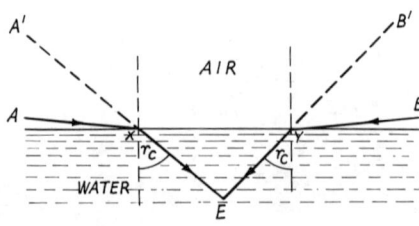

Fig. 24.21 Fish-eye view

Fish sees horizon *A* and *B* at *A'* and *B'*

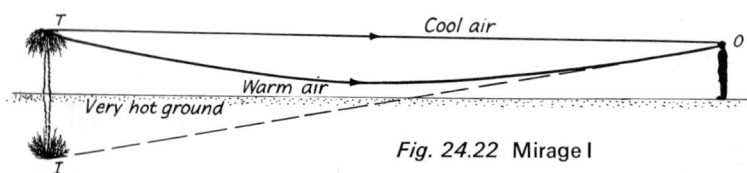

Fig. 24.22 Mirage I

The conditions necessary for a mirage to be formed are often produced in the desert, but mirages can sometimes be seen when driving along the road on a hot summer's day. These appear as

Mirage

Fig. 24.23
Mirage on road produced by total internal reflection

pools of water ahead, but they disappear on approach (Fig. 24.23).

Fig. 24.24 explains the formation of a mirage by light totally reflected in a horizontal plane at a region of very hot air.

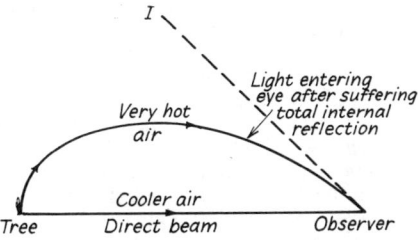

Fig. 24.24 Mirage II

A different type of mirage is often seen at sea, in polar regions, in the presence of icebergs, and is

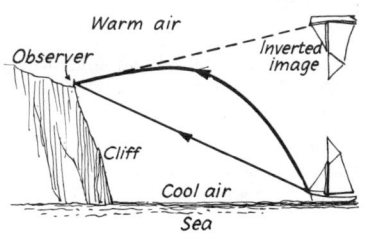

Fig. 24.25 Mirage III

occasionally reported from the north-east coast of England. If there are layers of warm air above the

cool air in contact with the cold sea, an observer may see a ship by light reaching him directly, also by light totally reflected at the bank of warm air. An inverted image of the ship appears in the sky (Fig. 24.25). Refraction in the air may cause

Fig. 24.26 Effect caused by atmospheric refraction

deviation and so produce phenomena similar to the above (Fig. 24.26).

QUESTIONS

1. Why does a pond appear to be less deep than it really is? Draw a diagram showing the path of light from a stone on the bottom to the eye.

2. A ray of monochromatic light is incident at an angle of 45° upon a 60° glass prism of refractive index 1.50. Draw an accurate diagram showing the path of the light through the prism and measure the deviation it suffers.

3. Define refractive index.
Sunlight making an angle of 60° with the horizontal enters a pool which is 50 cm deep. Find, by calculation, or by scale diagram, the distance travelled by the sunlight in the water.
[Refractive index of water=1.33.]

4. State the laws of refraction of light at the boundary between two transparent media.
A small circular disc lies on the bottom of a beaker containing water to a depth of 8.0 cm. Compare the angle it subtends at an observer's eye situated 22 cm vertically above it with that which it would subtend in the absence of the water. Illustrate your answer by a diagram.
[Refractive index of water=1.33.]

5. Define *refractive index* and explain briefly a method of measuring its value for water.
A small flat object, e.g. a penny, lies on the bottom of a tank containing water to a depth of 16 cm. Find by drawing or by trigonometry the apparent position of the object when viewed from directly above.
Does the object appear to be the same size as it does when the tank is empty and it is viewed from the same point? If not, does it seem larger or smaller? Give a reason for your answer.
[Refractive index of water, 4/3.]   (O & C)

6. How would you find the refractive index of glass?
Multiple images of a candle flame can be seen in a thick plate glass mirror. Explain this and illustrate your answer with a diagram showing how the first three images are formed.   (JMB)

7. Explain the terms *refractive index* and *critical angle*.
Describe a method of measuring the refractive index of a liquid.
A ray of light enters one face of a 60° glass prism, and is just totally internally reflected at the next face. Draw a diagram to illustrate this, and calculate the angle of incidence at the first face if the refractive index of glass is 1.53.   (O)

8. Sodium light falls normally upon one of the faces forming the refracting angle of a glass prism. Find the angle through which a ray is deviated in passing through the prism, given that the refractive index of the glass for sodium light is 1.5, and that the refracting angle of the prism is 30°.
With the help of diagrams describe and explain what would happen at the second face if (*a*) white light were used instead

of sodium light, (*b*) the refracting angle of the prism were 45° instead of 30°.   (JMB)

9. What is meant by the *critical angle* of a medium?
How would you find the refractive index and the critical angle of a given specimen of glass?
Show by a ray diagram how a right-angled glass prism may be used (i) to turn a ray through 90°, (ii) to turn a ray through 180°, (iii) to invert a beam of light.   (JMB)

10. State the laws of refraction of light, defining the angles of incidence and refraction and showing them on a diagram.
A slab of glass 1 cm thick is silvered underneath and rests on a table. A speck of dust is on the top face. Find by a drawing to scale or by calculation the position of the image of the speck by reflection at the silvered surface and refraction at the top.
[Refractive index of glass=1.5.]   (O & C)

11. Explain the terms *deviation* and *dispersion* and define *refractive index*.
Describe an experiment to show that the glass of a given prism has different values of refractive index for the red and blue components of white light.
A parallel beam of red light falls, at normal incidence, on one face of a prism of refracting angle 30°. Calculate the deviation of the beam when it emerges from the prism if the refractive index for red light is 1.50.   (O & C)

12. Explain what is meant by *refractive index, partial reflection, total reflection, critical angle*.
Draw a diagram to show how you would use an isosceles 90° glass prism to reflect a ray of light through an angle of 90°.
Explain carefully what would happen to the ray of light which you have drawn if the prism were immersed (*a*) in water, (*b*) in cedar wood oil.
[The refractive indices of water, cedar wood oil, and glass may be taken to be 4/3, 3/2, 3/2, respectively.]
What purpose, other than changing the direction of the light, is served by the two prisms in a field glass?   (C)

13. State the laws of reflection and refraction of light.
Explain under what circumstances reflection occurs without refraction at a water-air surface. How would you attempt to demonstrate that your answer is correct?
A small electric light bulb is supported 2.0 m below the surface of a lake. Calculate the radius of the circle on the surface through which light can emerge into the air.
[Refractive index of water=4/3.]   (O & C)

14. Explain the statement, 'the critical angle for glass is 42°'.
Why is it impossible to see a coin placed under a glass cube by looking through any side?

15. An insect hovers above the still water of a pond. Draw a freehand diagram, of rays or waves, to show approximately where it appears to be to a fish in the water vertically below it.
(O & C)

25 Lenses

We are all familiar with lenses, for they are used in many optical instruments. The commonest type is the bi-convex. This is a thin plate of glass which is ground so that its surfaces are spherical. Other types of spherical lenses are shown in Fig. 25.1.

The function of a lens is to produce focusing by refraction, in a similar way to that produced by spherical mirrors by reflection.

Consider rays of light *AP*, *BQ*, *CR* of a wide parallel beam incident on a bi-convex lens (Fig. 25.2). The ray *AP* is not deviated, for it passes through the centre of the lens where the two sur-

Bi-convex   Plano-convex   Concavo-convex   Bi-concave   Plano-concave   Convexo-concave

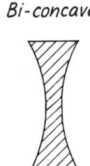

Fig. 25.1 Types of lenses

faces are parallel. The farther the rays are from *AP* the more they are deviated, because the lens surfaces become more inclined. All the rays of the beam are brought to a focus at *F*, the principal focus or focal point. It is real for a convex lens (Fig. 25.3) and virtual for a concave lens (Fig. 25.4).

> The principal focus or focal point is the point from which a paraxial parallel beam of light parallel to the principal axis diverges or appears to diverge after refraction.
>
> The focal length is the distance from the focal point to the optical centre of the lens.

Note that for a lens, the focal length depends on the refractive index of the material and the curvature

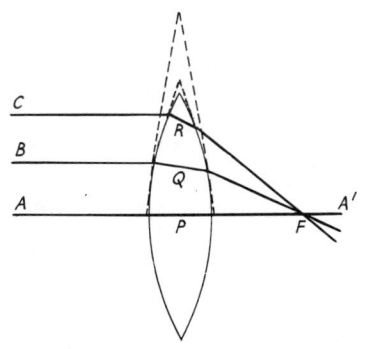

Fig. 25.2 Deviation through a lens

Fig. 25.3 Refraction through a convex lens

Fig. 25.4 Refraction through a concave lens

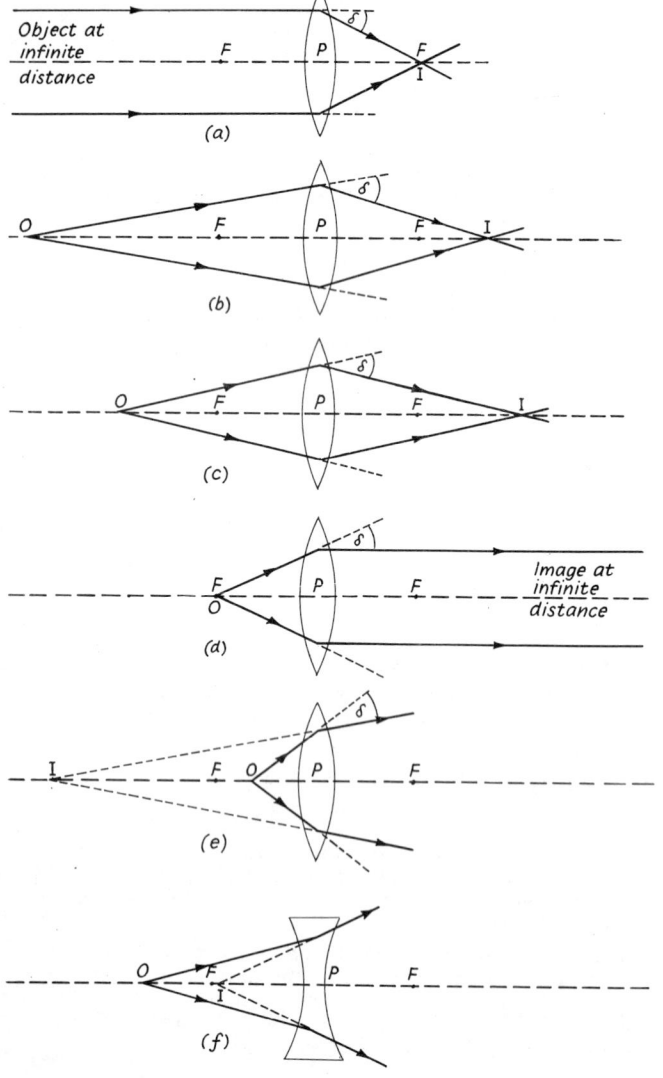

of the faces. The formula connecting these quantities is beyond the standard of this book.

A convex lens is sometimes termed a *converging* lens, since it converges light in passing through it. Similarly, a concave lens is termed a *diverging* lens.

Position of the image produced by a lens

Consider light from a distant object incident upon a convex lens. The light will be parallel and a real image will be produced at *F* (Fig. 25.5). The ray will be deviated through a small angle δ which will be the same for all directions of the incident light (so long as the angle of incidence is small). Let the distant object approach the lens. The image will be formed farther away. This will continue until the object reaches *F*, when the image will be at infinity. When the object is nearer to the lens than the principal focus, the emerging light is divergent, hence, a virtual image is produced on the same side of the lens as the object.

In a concave or diverging lens, a virtual image of a real object is always produced. The light from the object, which is already divergent, is made more so by the lens, and the image is produced on the same side, but nearer the lens than the object.

Two principal foci of a lens

In the above diagrams it will be noticed that a focal point is marked on each side of the lens. A

Fig. 25.5

Image produced by a lens

(*a*) to (*e*) show the position of the image produced by a convex lens; a concave lens is shown at (*f*)

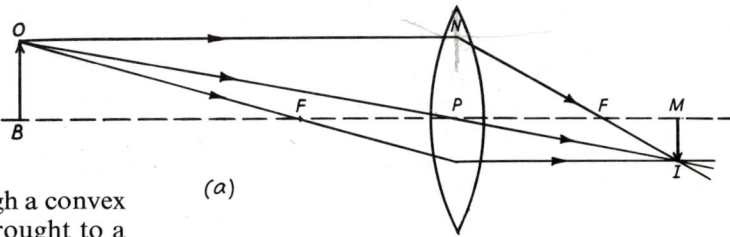

Fig. 25.6

Graphical construction
for images

parallel beam of light can be sent through a convex lens in either direction and it will be brought to a focus the same distance from the lens in each case.

By carrying out the above investigations practically—using an illuminated object as for mirrors—the results obtained can be summarized as shown at the foot of this page.

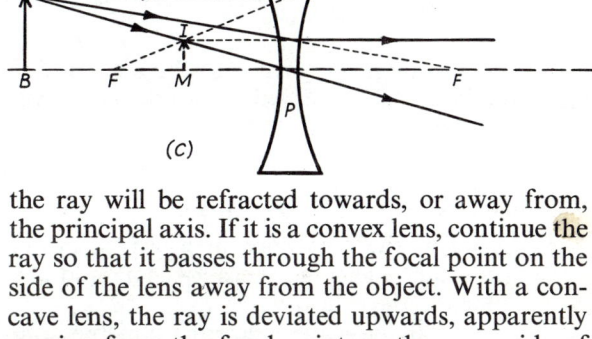

Graphical construction for images

We can again find the position and character of the image by construction. We consider an object OB, on the principal axis, and pick out three special rays of light from O. We trace these through the lens to locate I the image of O.

These three rays are:

1. The ray through the centre of the lens which continues in the same direction suffering no deviation.
2. The ray parallel to the axis which, after refraction, passes through one focal point.
3. The ray through the other focal point which, after refraction, becomes parallel to the axis.

The first ray is easy to construct, but with the other two rays, which of the two focal points to use may not be obvious. It becomes quite easy if the following procedure is adopted. First put in ray 1 and then ray 2. In constructing the latter, draw the parallel ray and then stop to consider whether

the ray will be refracted towards, or away from, the principal axis. If it is a convex lens, continue the ray so that it passes through the focal point on the side of the lens away from the object. With a concave lens, the ray is deviated upwards, apparently coming from the focal point on the same side of the lens as the object.

Only two rays are necessary to fix I, but it is useful if all three rays are known. Ray 3 is a little more difficult for beginners than the other two (Fig. 25.6).

SUMMARY OF RESULTS

Position of object	Position of image	Character of image
Convex Lens		
At infinity	At F	Real, inverted, diminished
Between ∞ and $2f$	Between F and $2f$	Real, inverted, diminished
At $2f$	At $2f$	Real, inverted, same size
Between $2f$ and F	Between $2f$ and ∞	Real, inverted, enlarged
At F	At infinity	Real, inverted, infinitely large
Between F and P	Between ∞ and P on same side as object	Virtual, erect, enlarged
Concave Lens		
At infinity	At F on same side as object	Virtual, erect, diminished
Between ∞ and P	Between F and P	Virtual, erect, diminished

f=focal length

Oblique parallel beam

It must be clearly understood that only a parallel beam directed along the axis is brought to a focus at the principal focus. An oblique parallel beam is brought to a focus in the focal plane, i.e. the plane through the focal point and normal to the principal axis (Fig. 25.7). This point is located by continuing

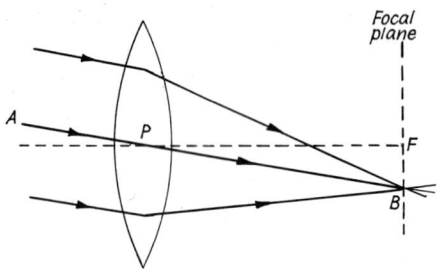

Fig. 25.7 Focusing of an oblique parallel beam

the ray AP through the centre of the lens until it meets the focal plane at B. All rays of the beam will then pass through B.

The passage of light waves through a lens

The change in the curvature of a wavefront in passing through a lens produces focusing and can be explained in the following way.

Consider a plane wave incident upon the converging lens, shown in Fig. 25.8. If XAY is the

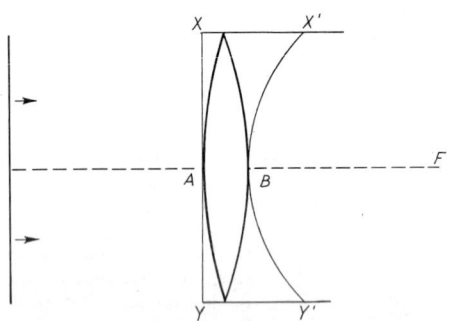

Fig. 25.8 Plane wave front made concave in passage through a converging lens

incident wavefront, the wave at A, in passing through the lens, will travel at a reduced speed and will only have reached B by the time the edges X and Y of the beam will have reached X' and Y',

($XX' = n \times AB$). The emergent wavefront will thus have the form $X'BY'$ which will be concave* with its centre at F, i.e. the emergent light will converge to the point F.

The lens formula

The position and character of the image, which has been determined above using graphical methods, can also be determined by calculation, employing a formula relating the object distance and the image distance with the focal length. But any formula must differentiate between an object or an image on one side of the lens from one on the other. We do this by adopting a sign convention.

Sign Convention—'Real-is-Positive'.

1. All distances are measured from the centre of the lens.
2. Distances measured to a real object or to a real image are positive: distances measured to a virtual object or to a virtual image are negative.
3. A lens having a real focal point has a positive focal length: a lens having a virtual focal point has a negative focal length. A convex lens has a positive focal length: a concave lens has a negative focal length.†

Now consider Fig. 25.6a.
Using the property of similar triangles, we have:

$$\frac{IM}{OB} = \frac{PM}{PB}$$

and
$$\frac{IM}{OB} = \frac{IM}{NP} = \frac{FM}{PF} = \frac{PM - PF}{PF}$$

Hence,
$$\frac{PM}{PB} = \frac{PM - PF}{PF}$$

All these distances are measured from the centre of the lens and, in Fig. 25,6a, all are positive. (We commonly represent the object distance, the image distance and the focal length by u, v and f respectively.)

---

* Approximately (due to spherical aberration).

† These statements apply only when the lens is in a rarer medium. In a denser medium, a convex lens has a virtual focal point and it diverges the light, e.g. a convex glass lens in carbon disulphide.

Writing $PB=u$: $PM=v$: $PF=f$, we have:

$$\frac{v}{u}=\frac{v-f}{f}$$

Dividing by v we get,

$$\frac{1}{u}+\frac{1}{v}=\frac{1}{f}$$

This formula may be deduced from the change in the curvature of the wavefront, as described in the previous section.

For the virtual image diagrams (Fig. 25.6b and c), similar geometry gives:

(i) $\dfrac{PM}{PB}=\dfrac{PM+PF}{PF}$ (ii) $\dfrac{PM}{PB}=\dfrac{PF-PM}{PF}$

But, on substituting $PB=u$: $PM=-v$: and $PF=f$ in (i), and $PB=u$: $PM=-v$: and $PF=-f$ in (ii), we get the same formula.

Size of the image: magnification. In the diagrams shown (Fig. 25.6) the magnification is given by:

$$\frac{\text{size of image}}{\text{size of object}}=\frac{IM}{OB}=\frac{PM}{PB}=\frac{\text{image distance}}{\text{object distance}}$$

Application of the formula

The lens formula requires the substitution of the known quantities with their appropriate signs (Examples 1 and 2). The sign of the unknown then locates the image.

Using the lens formula we can deduce some

Example 1

An object 0.5 cm high is placed 8 cm from a convex lens of 10 cm focal length. Find the position and size of the image.

In the formula

$$\frac{1}{u}+\frac{1}{v}=\frac{1}{f}$$

We have, $u=+8$ cm, $f=+10$ cm (u positive for real object; f positive for convex lens).
Substituting:

$$\frac{1}{8}+\frac{1}{v}=\frac{1}{10}$$

$$v=-40 \text{ cm}$$

$$\text{Magnification}=\frac{v}{u}=\frac{-40}{8}=-5$$

Size of image $=5\times0.5=\underline{2.5 \text{ cm}}$

The image distance being negative shows that the image is virtual, therefore, it must be on the same side of the lens as the object.

The image is virtual and 40 cm from the lens on the same side as the object: it is 2.5 cm high.

Example 2

Find the position and size of the image produced by two lenses, one a convex of focal length 12 cm and the other a concave of focal length 30 cm fixed 6 cm from the convex lens. The object is 2 cm high and is placed 20 cm from the convex lens.

For refraction through the convex lens, we have,

$$\frac{1}{u}+\frac{1}{v}=\frac{1}{f}$$

in which $u=20$ cm, $f=12$ cm.
Substituting:

$$\frac{1}{20}+\frac{1}{v}=\frac{1}{12}$$

$$v=\underline{30 \text{ cm}}$$

The first image becomes the object for the concave lens. In this case, because $v=30$ cm, the image would be formed 24 cm behind the concave lens. This real image

produced by the convex lens becomes a virtual object for the concave lens.

We have, $u=-24$ cm, $f=-30$ cm (u negative for virtual object; f negative for concave lens).
Substituting:

$$-\frac{1}{24}+\frac{1}{v}=-\frac{1}{30}$$

$$\frac{1}{v}=-\frac{1}{30}+\frac{1}{24}=\frac{1}{120}$$

$$v=120 \text{ cm}$$

Magnification $=$ product of separate magnifications

$$=\frac{30}{20}\times\frac{120}{24}=7\tfrac{1}{2}$$

The final image is 120 cm beyond the concave lens: it is real and 15 cm high.

important relationships concerning object and image.

When we put $u=v$ in the formula, we obtain $u=v=2f$. Hence, an object placed at a distance of twice the focal length from a convex lens forms a real image at the same distance on the other side, and the image is the same size as the object (unit magnification). Furthermore, in these positions, the real object and the real image are at their minimum distance apart.

Conjugate foci

An object at a position such as O, Fig. 25.9, will give a real image at I. In such cases O and I are called conjugate foci. As a ray of light is reversible,

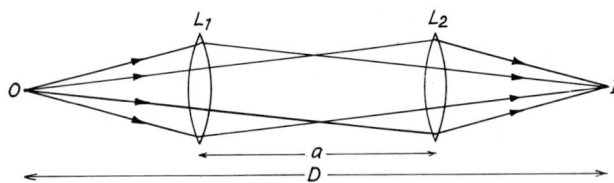

Fig. 25.9 Two positions for lens

an object placed at I will produce an image at O. This is equivalent to saying that an image can be produced on the screen at I with the lens at L_1 or at L_2.

From the symmetry of the lens positions it follows,
$$u+v=D \quad \text{and} \quad v-u=a$$
Hence,
$$u=\frac{D-a}{2} \quad \text{and} \quad v=\frac{D+a}{2}$$

Substituting, we get,
$$\frac{2}{D-a}+\frac{2}{D+a}=\frac{1}{f}$$
$$4fD=D^2-a^2 \quad \text{or} \quad f=\frac{D^2-a^2}{4D}$$
$$D=2f\pm\sqrt{(4f^2+a^2)}$$

D will be a minimum when $a=0$, i.e. when $D=4f$.

This result should be remembered when finding f experimentally.

Power of a lens

The power of a lens is measured by the reciprocal of its focal length. An optician usually quotes the power of a lens and not its focal length.

$$\text{Power in dioptres}=\frac{1}{\text{focal length in metres}}$$
$$=\frac{100}{\text{focal length in cm}}$$

A lens of focal length 100 cm has a power of one dioptre $(+1D)$. Convex lenses have a positive power and concave lenses a negative power.

If two thin lenses are placed in contact, the resulting combination has a power equal to the sum of the powers of the two lenses. Due regard must be paid to the signs of the powers ($+$ for a converging lens, $-$ for a diverging one).

Defects of lenses

Spherical aberration

Although in the simple theory above we have not mentioned the lens aperture, a similar defect is found in lenses as in mirrors. The focusing discussed is not true for rays which do not pass through the lens near to its centre, i.e. as for mirrors, we have always to assume that only the central portion of a lens is used. Rays parallel to the axis near the periphery, after refraction, cross the principal axis nearer to the lens than the focal point. More will be said about this in the section dealing with the camera. Very oblique rays do not obey the simple rules in passing through a lens.

Chromatic aberration

This defect arises due to white light being a mixture of light of different colours. Further reference to it is made in Chapter 28.

Experimental determination of focal length

Convex lens

The focal length of a convex lens can be determined directly by passing a parallel beam of light through it.

Another quite accurate and common method

involves the use of a plane mirror and an illuminated object. Referring to the apparatus shown in Fig. 25.10, the light, after passing through the

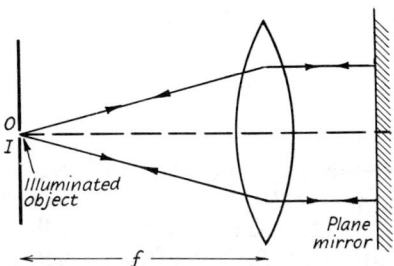

Fig. 25.10 Focal length of a convex lens

lens, is reflected at the plane mirror and returns through the lens. The position of the lens is altered until the image is focused alongside the object. For this to occur, the light must be returned back along the same path and must, therefore, strike the mirror normally. The object is then at the focal point of the lens, and the distance from the object to the lens gives the focal length. The adjustment is independent of the position of the plane mirror.

A parallax method may be used for this experiment. The position of the pin is adjusted until there is no parallax between it and its image produced by refraction through the lens and reflection at the plane mirror.

Method of conjugate foci
The most common method of finding the focal length of a convex lens is by locating the position of the image of an illuminated object. A convenient object is a cross-wire illuminated from behind by an electric lamp. The position of the lens is altered so as to form a focused image I on the screen (Fig. 25.11). The object and image distances are then measured. By alteration of the screen position, a series of corresponding values

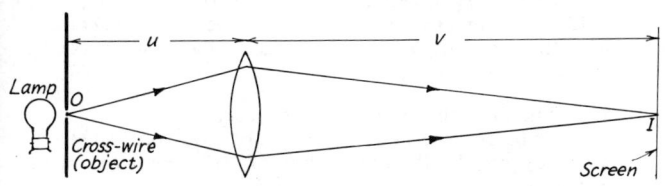

Fig. 25.11 Conjugate foci

for u and v may be obtained. (As shown theoretically above, a real image cannot be produced upon the screen if the distance between the cross-wire and the screen is less than $4f$.)

The position of the image can be located by a parallax method. In this case, a pin is used as object and a second is positioned so that there is no parallax between it and the real image of the object pin. By this method it is possible to locate the position of a virtual image. The object pin P_1 is arranged so that it is nearer to the lens than the principal focus (Fig. 25.12). With the eye in the

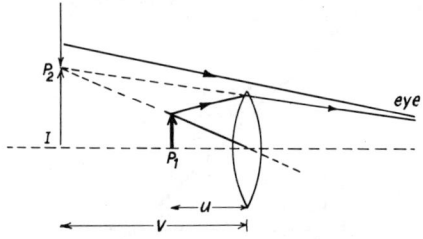

Fig. 25.12 Location of virtual image

position shown, a second pin P_2 is arranged so that there is no parallax between it, seen directly, and the image I produced by viewing P_1 through the lens. The image distance v in this case is negative.

The value of the focal length is deduced from

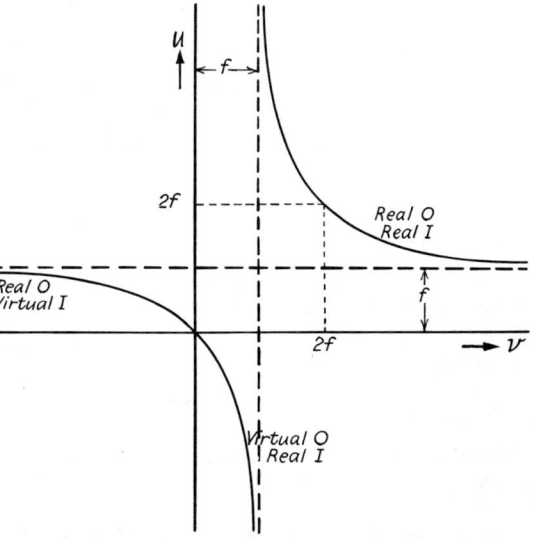

Fig. 25.13 Graphs of u against v for convex lens

the readings, either by substitution in the lens formula or, better, by graphing the readings. The graph of u against v (Fig. 25.13) is a hyperbola with asymptotes $u=f$ and $v=f$, or, in other words, $u=f$ when $v=\infty$, and $v=f$ when $u=\infty$. A better way of finding f from this graph is to read off the value when $u=v$. We have seen above that this occurs when $u=2f$.

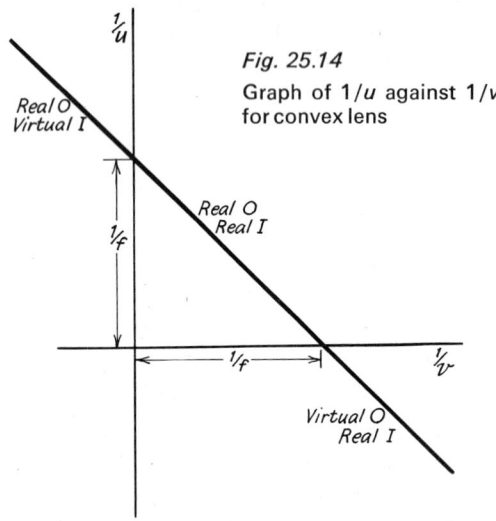

Fig. 25.14

Graph of $1/u$ against $1/v$ for convex lens

A more accurate graph is obtained by plotting $1/u$ against $1/v$ (Fig. 25.14). This is a straight line and the intercepts on the axes give $1/f$.

Concave lens

To determine the focal length of a concave lens, an auxiliary convex lens is employed (Fig. 25.15). This lens, L', is used to produce a real image I' of

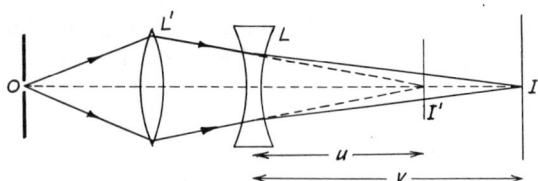

Fig. 25.15 Focal length of a concave lens

the illuminated object O. The concave lens L is then introduced and the screen moved back until a focused image I is produced upon it. So far as the lens L is concerned, the incident light upon it is proceeding to I', but after refraction, passes through I. Thus I' is a *virtual* object and I a *real* image. The distances u and v are measured, u is negative and v is positive. Substitution in the lens formula gives f, which should be negative. A series of values of u and v can be taken, and the value of f found from the graph of $1/u$ against $1/v$.

A slight modification of the above methods gives a direct determination of f. A plane mirror is placed as shown in Fig. 25.16 and the position of

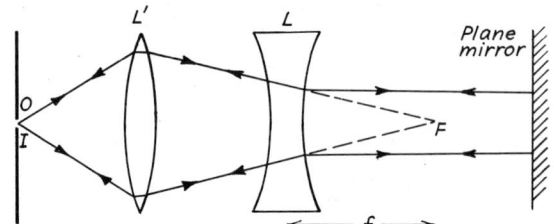

Fig. 25.16 Focal length of a concave lens

lens L is adjusted until a real image is formed alongside the object. The position of F is located as above, and so f can be found.

QUESTIONS

1. Show by means of a diagram how an oblique parallel beam of light is focused by a lens.

2. Draw ray diagrams showing how a virtual image is produced by (*a*) a convex lens, and (*b*) a concave lens.

3. How can a convex and a concave lens be distinguished by looking through them? Explain your answers.

4. Find the focal length of the concave lens which produces an image 1 cm high of an object 2 cm high placed 12 cm from the lens.

5. Find the position and size of the image of an object 1 cm high placed 18 cm from a concave lens of 12 cm focal length.

6. Where must an object be placed in front of a convex lens of focal length 12 cm to give an image with magnification 3?

7. An object is placed 25 cm in front of a lens of power $+10D$. A second lens of power $-2D$ is then placed in contact with the first lens. Find the change in the position of the image.

8. Define the terms *principal focus, focal length*.
   Describe an experiment for measuring the focal length of a convex lens, by the observation of object and real-image distances.
   When a small lamp is placed on the axis of a convex lens

and 15 cm from it, a parallel beam of light emerges from the lens. When the lamp is placed 30 cm from the lens, a real image of the filament 3 mm long is formed on a suitably placed screen. What is the actual length of the filament? (O)

9. Define the terms 'principal focus' and 'focal length' as applied to a converging lens. Illustrate your definitions by ray diagrams. Draw ray diagrams to show the two ways in which a convex lens could form an image about three times the size of the object. A convex lens forms an image of the moon 2.0 mm in diameter. What is its focal length? Take the diameter of the moon as 3200 km and its distance from the earth as 384 000 km. (JMB)

10. Define *principal foci* of a lens. How would you locate them for a converging lens?

A converging lens is used to produce a virtual image of an object 2 mm in height placed on, and at right angles to, the principal axis at a distance of 8 cm from the lens. If the image is formed 24 cm from the lens, calculate the focal length of the lens and the height of the image. Check your result by means of a scale diagram. (L)

11. Describe how you would determine the focal length of a convex lens.

A convex lens of focal length 6 cm is placed with its principal axis horizontal, at a distance of 18 cm from a vertical screen. When a point source of light is placed on the axis and 24 cm from the screen, a circular patch of light, of the same diameter as the lens, appears on the screen. Explain this. Find the position to which the source must be moved in order that an image of the source may appear on the screen. (C)

12. Define the terms *principal axis*, *principal focus* and *focal length* of a converging lens.

A disc is perpendicular to the principal axis of a converging lens with its centre on the axis and 10 cm from the centre of the lens. The focal length of the lens is 15 cm. Find graphically the position and diameter of the image.

Draw at least *two* pencils of rays to show how the object is seen by an eye on the axis of the lens at a point 20 cm away on the side remote from the disc.

What is the smallest diameter of lens which will allow the eye to see the whole disc without moving? Neglect the diameter of the pupil of the eye.

[Diameter of disc 1.0 cm.] (O & C)

13. Define *principal focus*, *focal length* of a converging lens and explain with illustrations the terms *real* and *virtual image*.

A converging lens forms an image of the sun. Draw a ray diagram, not to scale, to show (a) where the image is formed, (b) why it is not a point but has a definite size.

Given that a disc of 2 cm diameter held 2 m from the eye just covers the sun's disc, calculate the diameter of an image of the sun formed by a lens of focal length 50 cm. (O & C)

14. Define the terms *real image*, *virtual image*, *magnification*.

A small object is placed on the axis of a thin converging lens of 10 cm focal length at a distance of 9 cm from the optical centre. Find the position and magnification of the image. At what distance should the object be placed so that the lens may form a real image of the same size as the first, and what is the new image distance? (O & C)

15. Account for the action of a diverging lens on a parallel beam of light. Define the *focal length* of a diverging lens.

Describe how you would determine the focal length, accurate to the nearest one-fifth of a centimetre, of a converging lens whose focal length is about a fifth of a metre.

A parallel beam of light falls perpendicularly on a screen. A converging lens, near to the screen, intercepts *some* of the light and the illumination on the screen appears as a bright circular patch surrounded by a dark ring. Explain this with the aid of a diagram. Deduce the appearance on the screen when the distance between the lens and the screen is (a) equal to the focal length of the lens, (b) twice the focal length of the lens. Illustrate your answer with ray diagrams. (C)

26 Optical instruments

Vision: the eye

The mechanism of seeing can be divided into three processes:

1. The light enters the eye and falls upon receptors connected to nerve endings.
2. The stimulus produced by the light is transmitted to the brain.
3. The brain collects all the nervous impulses and interprets them.

Here, we are mainly concerned with the excitation of the receptors.

The human eye is a complex and wonderful optical instrument which has gradually evolved from very primitive beginnings.

The optical arrangement of the human eye is shown in Fig. 26.1. Light enters through the cornea, which is a small hump on the otherwise spherical eyeball. Within the eye, and dividing it into two unequal sections, is the crystalline lens. The refractive index of the material of the lens is about 1.5, and that of the two liquids filling the two compartments on each side of the lens about 1.33. The lens is pliable and is supported by the suspensory ligaments. Its power can be changed by muscles acting around it. A real inverted image of an object is produced on the back surface of the eye called the *retina*. Here are receptors of two kinds, rods and cones. About the centre of the retina is a very dense concentration of cones, each with a separate nerve attached. When the receptors are close together, a very

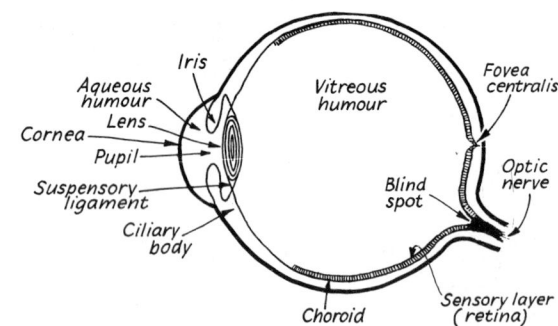

Fig. 26.1 Section through the human eye

sharp image is produced. From the centre of the retina (called the *fovea centralis*), the cones get fewer in number and some rods appear. In these parts, several receptors may be coupled to the same nerve-ending. All the nerve-fibres leave the eyeball at one spot in a single bundle. This is the blind spot for just here there are no receptors.

The result of this is that, if we wish to see an

object clearly, we look straight at it so as to form the image on the central portion of the retina.

Rods require less light than cones to stimulate them, and so a very faint star can be seen more easily by forming its image away from the fovea, i.e. by not looking straight at it.

Although colour vision is not fully understood, it seems probable that the cones are responsible and not the rods. If an image is formed well away from the fovea, where there are only rods, it is impossible to define the colour of the object.

In front of the lens is the iris, which is adjustable and controls the size of the opening or pupil. Too much light energy falling on the receptors is harmful, and so nature provides this automatic aperture.

In order to produce a focused image on the

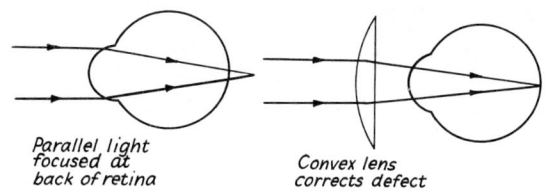

Parallel light focused at back of retina   *Convex lens corrects defect*

Fig. 26.2a Hypermetropia long sight (eyeball shortened)

Short-sightedness: myopia

In this case the power of the eye is too great and the image of a distant object is formed in front of the retina. A concave lens is required to compensate for it (Fig. 26.2b).

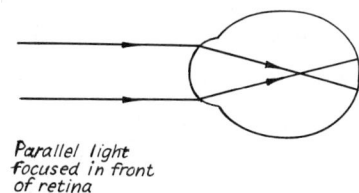

Parallel light focused in front of retina

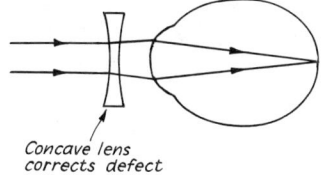

Concave lens corrects defect

Fig. 26.2b
Myopia short sight (eyeball elongated)

retina of an object at different distances, the power of the lens is changed by the ciliary muscles.* This process is called *accommodation* and has to be learnt by experience in childhood. This faculty is often lost or weakened in middle age. When in a restful condition and not accommodated, the eye is focused on a distant object. Accommodation enables the eye to focus on objects down to a distance of about 25 cm from it. This is called the minimum distance of distinct vision. An object brought nearer than this produces a slightly larger image, but it is not in focus.

Defects of vision

Long sightedness: hypermetropia

This is due to the power of the eye being too small. When the eye is not accommodated, it cannot produce an image of a distant object as near to the lens as the retina. The defect is remedied by the use of a convex spectacle lens (Fig. 26.2a).

Astigmatism

This is a more complicated defect and may be caused by the cornea being distorted, and so possessing different powers in different planes. Thus a person suffering from this defect may see vertical lines perfectly, but horizontal lines may be blurred. The optician tests for this defect with a fan chart,* consisting of a number of lines radiating from a point. The defect is corrected by using a cylindrical lens, which can bring up the power in the required plane and so make all the lines of the fan chart appear equally distinct.

Presbyopia

This is a defect suffered usually by older people and is the loss of the power of accommodation. It can be helped only by using two pairs of spectacles, one for distant and one for near objects. Sometimes bifocal lenses are used, which have a different power at the top from at the bottom.

---

* Note, however, the total power of the eye is the combined power of the cornea and of the lens.

* He now uses more sophisticated instruments which perform the same function.

The camera

The photographic camera, as mentioned on page 195, consists simply of a convex lens mounted in front of a photographic plate or film. A real image of the object is produced upon the film.

The variation of the brightness of the different parts of the image produces a chemical action of varying intensity upon the sensitive layer of the plate. In order to produce a focused image upon the plate, the lens position must be adjustable.

If the camera is so simple, why are some very costly? To answer this, we must again refer to that error of lenses known as spherical aberration. A sharp image can be produced with an ordinary lens only when a small aperture is used. But sometimes we wish to use a short exposure and allow light to pass into the camera for a small fraction of a second. If the aperture is small, very little light can enter in this short time and so, on a dull day, a larger aperture must be used. The best cameras allow this to be done and still produce a sharp image. They use an expensive lens which is a combination of several lenses, made in such a way as to minimize spherical aberration.

Depth of focus

Fig. 26.3 illustrates the formation of the image *IM*, but for clarity it is drawn much out of scale. Actually, the length of the camera is only a few centimetres, whereas the object is usually a few metres away. The position of the film has to be altered to focus objects at different distances from the lens. It is sometimes an advantage to produce a picture showing images of objects distributed over a limited range of distances all in focus. This

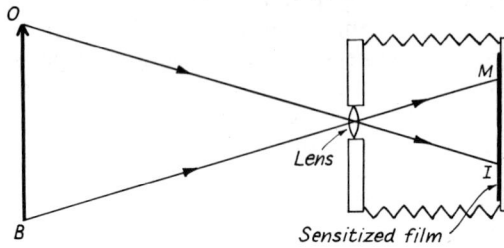

Fig. 26.3 The camera

is best achieved by using a small aperture: it is referred to as *depth of focus*.

Looking at Fig. 26.4a, the object O produces an image I, O_1 an image I_1, and O_2 an image I_2. Thus, the rays from O_1 will not produce a perfectly sharp image on the film in the position shown, i.e. a point object O_1 will produce a point image at I_1, but a small circular image on the film (Fig. 26.4b). The size of this circle determines how much the image of I_1 is out of focus on the film: it is governed by the size of the aperture (Fig. 26.4c). Hence, by using a small aperture, as in Fig. 26.4b, objects at O_1, O and O_2 will all produce reasonable focused images on the film, and the photograph will have a good depth of focus.

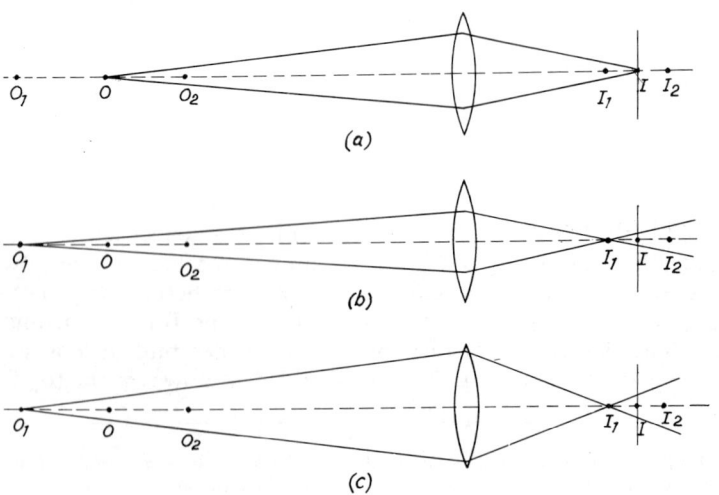

Fig. 26.4
Effect of aperture on depth of focus

Optical instruments

Optical instruments, which are held to the eye in use, produce a virtual image. The simplest of these is the magnifying glass.

The magnifying glass

This is a convex lens used as shown in Fig. 26.5. The glass must be held so that the object is nearer to it than the focal point. The virtual image is produced at the minimum distance of distinct vision, 25 cm.

With this type of instrument, the *actual* size of the virtual image does not matter. The important factor is the angle it subtends at the eye, for this determines the actual size of the image on the retina. The advantage gained by using an optical instrument is measured by,

$$\frac{\text{angle subtended by image}}{\text{angle subtended by object at the unaided eye}}$$

Most optical instruments first produce a real image of an object and then view it through a magnifying glass. In order to reduce the errors of simple lenses, lens systems are used. For example, an eyepiece, consisting of at least two lenses, replaces the simple magnifying glass.

The compound microscope

The microscope is used for viewing very small objects. A real image of the object is produced by a convex lens with a very short focal length. This image is then viewed through an eyepiece with a short focal length acting as a simple magnifying glass.

The first lens or objective has to collect as much light as possible from the object, hence it must have a relatively large aperture. Instead of a single lens, a complicated system of lenses is used. Fig. 26.6 shows the passage of light through a microscope: for simplicity, only single lenses are shown for the objective and eye-lens.

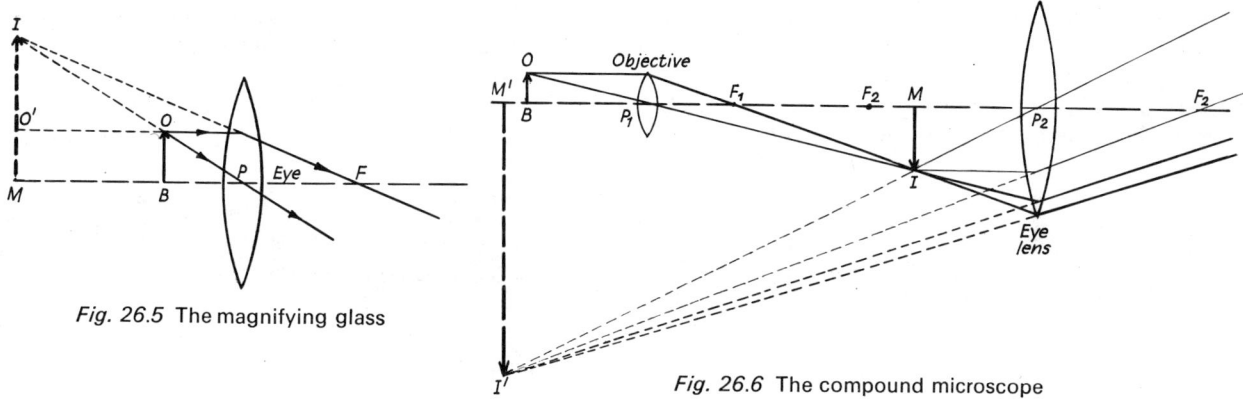

Fig. 26.5 The magnifying glass

Fig. 26.6 The compound microscope

This ratio is called the *magnifying power* or *angular magnification*. On examining Fig. 26.5, it appears that the two angles are the same, hence the magnifying power is unity. But the denominator states *at the unaided eye*. IM is produced at the minimum distance of distinct vision and so, when the lens is taken away, the object will have to be moved back to the position $O'M$ for the unaided eye to see it clearly. The magnifying power is then $\dfrac{I\hat{P}M}{O'\hat{P}M}$ and, as these angles are small, this therefore becomes $\dfrac{IM}{OB}$ or $\dfrac{v}{u}$.

To trace the light through the system, first find the position of the real image IM, using the ray through the centre of the lens and the ray parallel to the principle axis. Next regard IM as the object for the second lens and find its image $I'M'$, again tracing similar rays. Note IM is nearer to the eye-lens than the focal point. Having located $I'M'$, we know that all the light from I, after refraction, appears to come from I', hence the two rays traced through the objective will appear to come from I' and so can now be completed. Any other ray from O can be traced through the instrument, for it must pass through I after refraction through the objective and appear to come from I' after refraction through the eye-lens. The angular magnification is the angle subtended by IM, i.e. $I'\hat{P_2}M'$ divided by the angle OB subtends at a distance of 25 cm.

The position of the final image in an optical instrument is governed by the position of the eyepiece relative to the real image produced by the objective. The eye, when applied to a microscope, is usually accommodated for viewing a very small object at the minimum distance of distinct vision, and so the final image in a microscope is produced at a distance of about 25 cm from the eye. If the eye, however, is fully accommodated for long periods, it suffers strain, and so microscopists get accustomed to adjusting their instruments so that the final image is produced much father away, and the eye is relaxed.

The telescope

The objective of the telescope first forms a real image of the object, which is viewed through an eye-lens. Thus, in principle, the telescope resembles the microscope but, whereas the microscope is used for examining very small, near objects, the telescope is used to view distant objects. The light which enters the telescope is very nearly parallel to the principal axis, therefore spherical aberration is quite small. The objective,

however, has to be corrected for chromatic aberration (*see* page 249).

Consider light coming from a distant object such as the moon (*see* Fig. 26.7). The telescope will be directed with its principal axis pointing towards the centre of the moon's disc. Light from this point will be brought to a focus at the focal point. Light from a point on the edge of the moon's disc will form an oblique parallel beam and will be focused at a point in the focal plane. This point I is where the ray through the centre cuts the focal plane. Thus, a real image of the moon's disc is produced with semi-diameter IM. This real image is viewed through an eyepiece and the enlarged virtual image $I'M'$ produced. Note that the image is inverted.

(*Note.* Do not look at the sun through a telescope as the high intensity of the image will damage the retina.)

It is customary to produce the final image in a telescope at infinity. The eye when viewing it is relaxed and is in the same state as it was when viewing the object unaided. A telescope so adjusted is said to be in *normal adjustment* (Fig. 26.8). The focal planes of the objective and of the eyepiece are coincident.

The angular magnification is a/β. When the telescope is in normal adjustment this is equal to

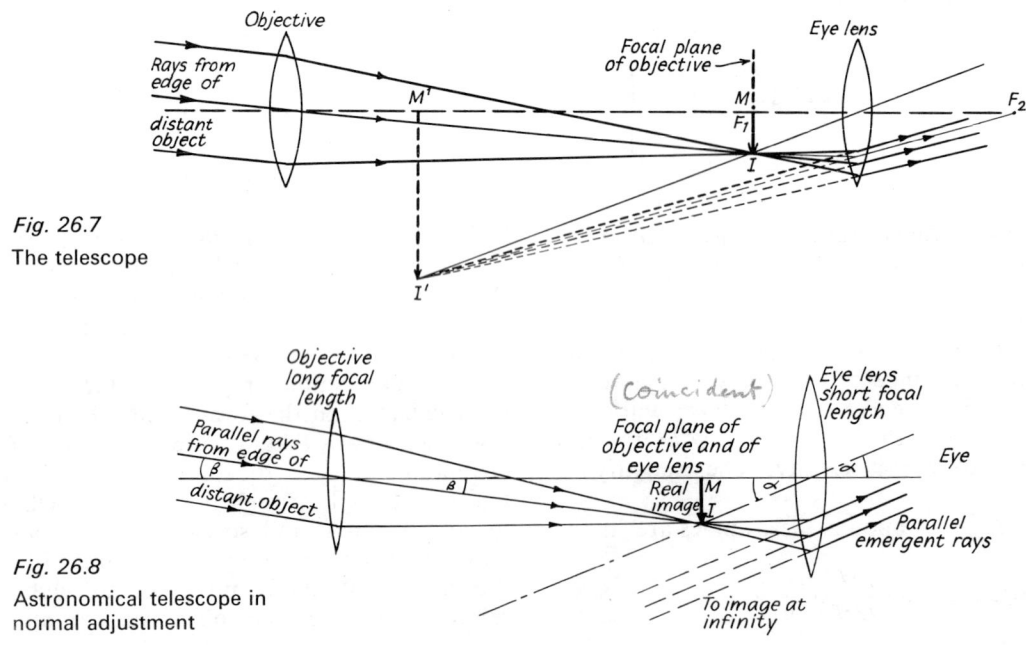

Fig. 26.7
The telescope

Fig. 26.8
Astronomical telescope in
normal adjustment

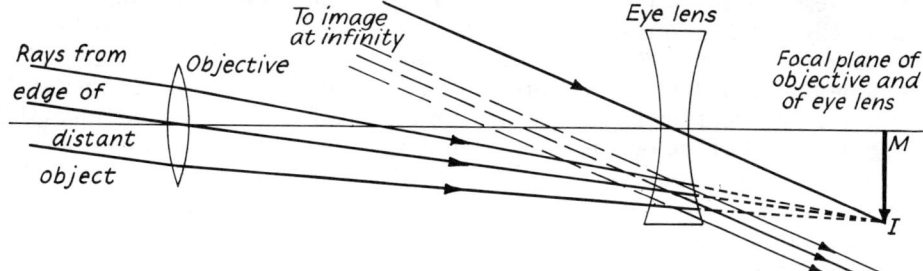

Fig. 26.9
The opera glass

Rays from edge of distant object

Objective

To image at infinity

Eye lens

Focal plane of objective and of eye lens

M

I

$$\frac{\text{focal length of objective}}{\text{focal length of eye-lens}}$$

Hence, the telescope has a long focal length objective and a short focal length eye-lens.

The opera glass

In this case, an erect image is produced by using a concave lens as eye-lens. This must be inserted before the real image *IM* is produced. The ray diagram (Fig. 26.9) is difficult, for the image of a virtual object *IM* has to be found. In addition to the erect image, it should be noted that a shorter tube is required than for the astronomical telescope.

Disadvantages arise due to the use of a concave lens. Spherical aberration cannot be effectively reduced and the field of view is small. These disadvantages are overcome in prism binoculars.

Prism binocular

The optical system of a binocular is shown in Figs. 26.10, 26.11 and 26.12.

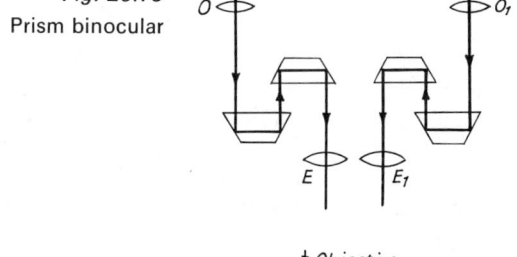

Fig. 26.10
Prism binocular

O O₁
E E₁

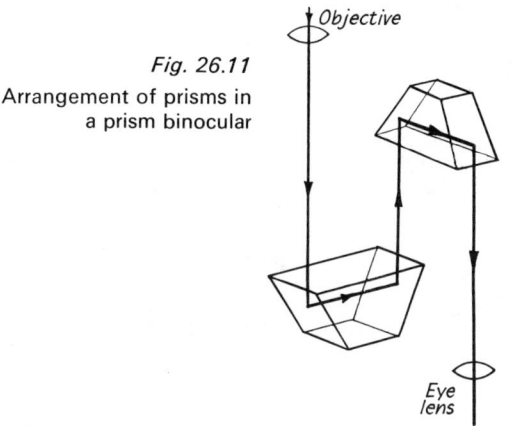

Fig. 26.11
Arrangement of prisms in a prism binocular

Objective
Eye lens

Fig. 26.12
Prism binocular

Light entering the objective is reflected by a double totally reflecting prism, so that its direction is turned through 180°. It is then turned through another 180° by a second prism at right angles to the first, and finally passes through the eyepiece. The effect of the prisms is to produce an erect image, i.e. the eyepiece is a positive-lens system. The reflection of the light up and down the tube enables a longer focal length lens to be used with a consequent increase in angular magnification. The instrument is, in effect, an astronomical telescope with prisms added.

Rangefinders

We judge distance by estimating the angle an object makes with our two eyes. The accuracy of this estimate can be improved by increasing the effective distance between our eyes. In the rangefinder (Fig. 26.13), four plane mirrors, M_1, M_2,

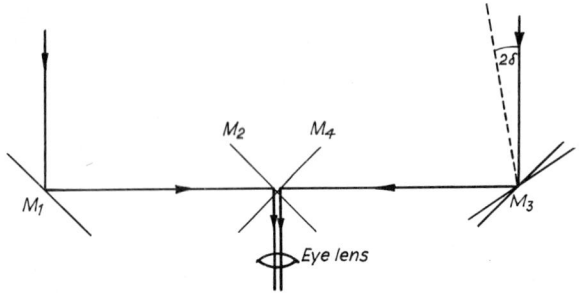

Fig. 26.13 A simple rangefinder

M_3 and M_4 are arranged as shown. M_2 is above M_4 so that light reflected by mirrors M_1 and M_2 illuminates the upper half of the field of view of

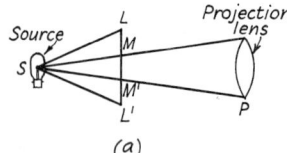

(a)

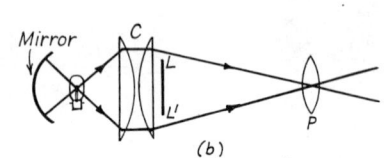

(b)

the eye-lens and that reflected by M_3 and M_4, the lower half. The two images of a distant vertical object are adjusted to be in the same vertical line by the rotation of mirror M_3. Thus the position of this mirror determines the distance of an object and the instrument is calibrated to read this distance directly on a scale attached to M_3. The accuracy depends upon the distance between M_1 and M_3.

The lantern or diascope

In the optical lantern, light is passed through the transparent slide, an image of which is focused on a screen by a converging projection lens.

The lens system used for illuminating the slide requires explanation. It is called a *condenser*. Consider the arrangement shown in Fig. 26.14a. Light from the source S in the cone LSL' will illuminate the slide LL', but of this, only MSM' will pass through the projection lens. Hence the image on the screen will have a bright central patch corresponding to MM', which is surrounded by a faint region due to a little scattered light from LL'. The condensing system C (Fig. 26.14b and c), collects a much larger cone of light from S and passes it through the whole of LL' and P. Further improvement in the illumination is obtained by placing a concave mirror behind the lamp at a distance equal to the radius of curvature.

Ciné projectors and film strip projectors use optical systems similar to that of the ordinary lantern.

The episcope

The episcope is an instrument for projecting opaque pictures. Very often the same instrument, with slight modifications, can be used to project lantern slides also. It is then called an epidiascope.

In the episcope, only scattered light from the picture passes through the projection lens. There

Fig. 26.14

The projection system. The lantern or diascope

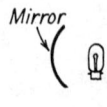

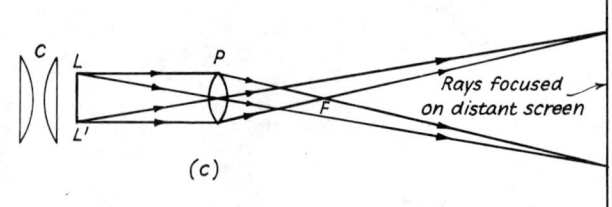

(c)

is a complicated system of mirrors, which is used to send as much light as possible from the lamp or lamps on to the picture. To get a reasonably bright image, the illumination of the object has to

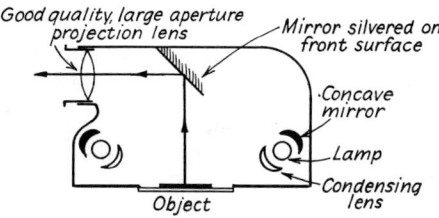

Fig. 26.15 The episcope

be high, also a good quality lens of wide aperture has to be used. This accounts for the high cost of these instruments. The image is laterally inverted

and so the plane mirror, shown in Fig. 26.15, is needed to correct this. The mirror must be surface-silvered to prevent the production of multiple images.

Heating produced in optical instruments

Powerful light sources are used in optical instruments in order to obtain brilliant pictures. Nowadays special electric lamps are nearly always used, for they are more convenient than arc lamps. All these sources give out a large amount of heat (*see* page 195). Sometimes, in addition to good ventilation, special precautions are taken to keep instruments cool. A special glass plate may be inserted between the source and the condenser, or a fan may be fitted to extract the hot air.

QUESTIONS

1. Explain *long* and *short* sight and, with the aid of diagrams, show how they can be corrected.

2. Give an account of the eye as an optical instrument, and explain the process of accommodation.
   Why do many elderly people wear bifocal spectacles? Explain how these are constructed, and draw diagrams to illustrate the correction provided by each section.   (O)

3. Make a labelled sectional drawing of the human eye and explain how the eye (*a*) accommodates itself to changes in the distance of an object, (*b*) reacts to changes in the light intensity.
   A man can see clearly objects at a distance of 40 cm or more. What type of spectacle lens must he wear and what must be its focal length to reduce his nearest distance of distinct vision to 25 cm?   (O & C)

4. Describe the optical system of the eye, explaining the function of each part. Illustrate your answer by a labelled diagram.
   A man cannot see clearly objects nearer than 75 cm. What type of lens should he use in order to see objects up to 25 cm away? Give a ray drawing of the action of the lens. Assuming that it is placed close to the eye, find its focal length. (O & C)

5. Explain how a normal human eye can focus objects at varying distances away. Give an account and explanation of the defects of vision known as 'long sight' and 'short sight'.
   Draw *one* ray diagram to show how a convex lens can, form a virtual image further from the eye than the object, and *one* ray diagram to show how a concave lens can form a virtual image nearer to the eye than the object. Explain how your diagrams illustrate the use of such lenses in spectacles.   (C)

6. What is meant by the terms *principal focus of a converging lens, real image, virtual image*?
   Explain, with the help of a diagram, how a converging lens may be used as a magnifying glass.
   A man whose nearest distance of distinct vision is 25 cm holds a converging lens of 10 cm focal length very close to his eye. Find, by drawing or calculation, how far from the lens a small object must be placed in order that the image may be formed at his nearest point of distinct vision. What is then the magnification?   (O & C)

7. Make a careful drawing of two rays from a point *P* on a distant object through a simple astronomical refracting telescope when *P* is not on the axis. Show clearly the type of lens used in each part of the instrument and the nature and position of the final image.
   If the object glass and eyepiece of this telescope have focal length 50 cm and 5 cm respectively, find (i) the diameter of the image of the moon formed by the object glass, (ii) the angular magnification when the telescope is adjusted for the unaccommodated eye.
   [Diameter of the moon: 3.5×10^3 km. Distance of the moon from the earth: 380×10^3 km.]   (O & C)

8. Draw a ray diagram, half-scale, to show the action of a compound microscope consisting of two lenses. Use the following information.
   The object, of length 1.0 cm and perpendicular to the axis, is situated 4.0 cm from the object glass which forms a real image of length 3.0 cm. This real image is 4.0 cm from the eye-lens which forms a virtual image 12.0 cm in length.
   Deduce, *either* from the diagram *or* by calculation, the focal length of each lens.   (C)

9. Two biconvex lenses of focal length 5 cm and 20 cm are to be used to form an astronomical telescope. Draw a ray

diagram showing how the final image is formed when the telescope is in normal adjustment.

Draw the path of a pencil of rays through both lenses.

What is the distance apart of the lenses if (a) object and final image are both at infinity, (b) a final virtual image is formed 25 cm from the eye-lens?

10. Describe and explain the optical arrangement of a compound microscope. Illustrate your answer by a ray diagram.

The focal length of the objective of a compound microscope is 1 cm and that of the eyepiece 10 cm. The object distance is 1.2 cm. Find by drawing or calculation how far apart the lenses must be placed for the final image to be at infinity.   (O & C)

11. Draw a ray diagram for a compound microscope showing how an image of a point not on the axis is formed at about 30 cm from the eyepiece, and where the crosswires should be placed if the microscope were used for making measurements.

The focal length of the objective of a compound microscope is 0.80 cm and that of the eyepiece is 5.0 cm. If the observer relaxes his eye so that the final image is at infinity, and the lenses are 25.0 cm apart, find the distance between the object and the objective lens. What magnification does the objective lens produce?   (O & C)

12. Define the term *focal length* as applied to lenses. Illustrate your answer with ray diagrams for (a) a converging lens, (b) a diverging lens.

In a projection lantern, an object on the slide is 2.0 cm in length; the slide is 12.0 cm from a lens which forms a real image magnified 4 times. Draw a scale diagram to show the arrangement. Use a scale which is one-sixth full scale along the principal axis and which is full scale in a direction at right angles to the principal axis. Draw rays to show how the image is formed; deduce the focal length of the lens.

Draw a labelled diagram to show how you could illuminate the slide evenly and brightly.   (C)

27 Diffraction, interference, polarization

The effects with visible light of diffraction, interference and polarization cannot receive ray treatment as has been used in the previous chapters for reflection and refraction. We have to resort to the ideas contained in Chapter 20 dealing with the properties of wave motion.

Diffraction

It was pointed out that the extent of the diffraction or bending of waves round a corner depends on the wavelength. Thus, in the case of visible light with such small wavelengths, the diffraction effect is very small and we have to use an eyepiece or other optical instrument to observe the bending at the edge of a shadow. Not only does the light bend slightly into the shadow, but light and dark bands are produced. Figs. 27.1 and 27.2 show examples of these bands or fringes which form pretty patterns round suitably shaped obstacles.

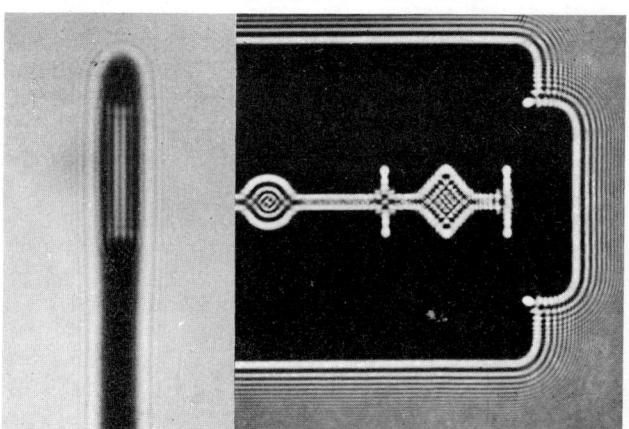

Fig. 27.2 Diffraction patterns in shadow of a needle and a razor-blade

Further treatment of diffraction is beyond the scope of this text.

Interference

Early in the nineteenth century, Thomas Young was able to show interference effects with visible light. The arrangement he used was similar to that shown in Fig. 27.3. A monochromatic source

Fig. 27.1 Diffraction pattern at a straight edge
The photograph shows the slight bending of the light into the shadow, causing a blurred edge and the system of diffraction fringes

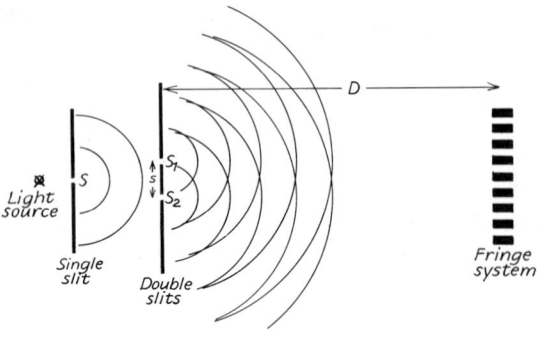

Fig. 27.3 Young's fringes

illuminates the slit S which, in turn, illuminates the two slits S_1 and S_2. These slits are extremely close together and the light from them is diffracted. Interference effects are produced in the region where the diffracted beams overlap. An eyepiece is usually necessary to view the fringe system, for the fringe width (the distance between consecutive fringes) is small: it depends on the distance from the slits.

Conditions for interference. A source of sound produced electronically can be maintained constant for a long period of time. The waves sent out will be continuous. This is not so with an ordinary source of light, for visible light waves are not continuous. The emission is irregular, with pulses of radiation from different atoms randomly selected, so that visible interference of light from two different sources is impossible. Young's fringes are produced with light from the one source. Other methods of producing similar fringes reflect or refract the light from one source, so

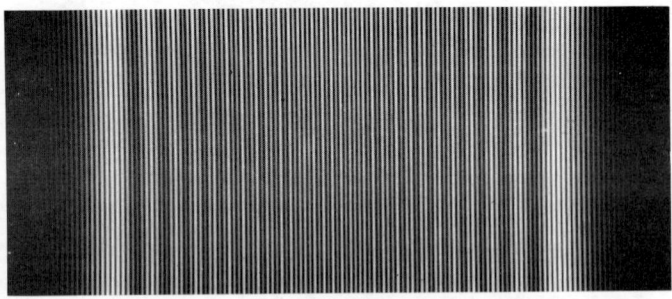

Fig. 27.4 Interference fringes

These fringes are produced by passing the light from a single slit through a biprism (two prisms of small angle with their bases together). On the equally spaced interference fringes are superimposed diffraction fringe systems

that it appears to come from two identical sources very near together.

We can summarize the conditions required for visible interference:

1. The two sources must be emitting waves of the same wavelength.

2. There must be a constant phase difference between corresponding points on the sources, i.e. they must be coherent.

3. The distance between the sources must be very small if the wavelength of the waves is very small.

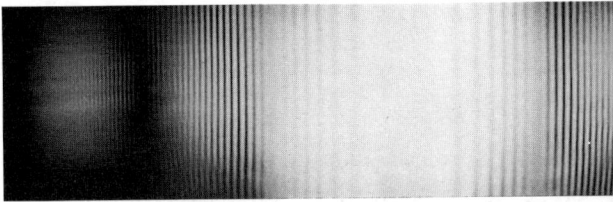

Fig. 27.5 Diffraction pattern produced by a thin wire

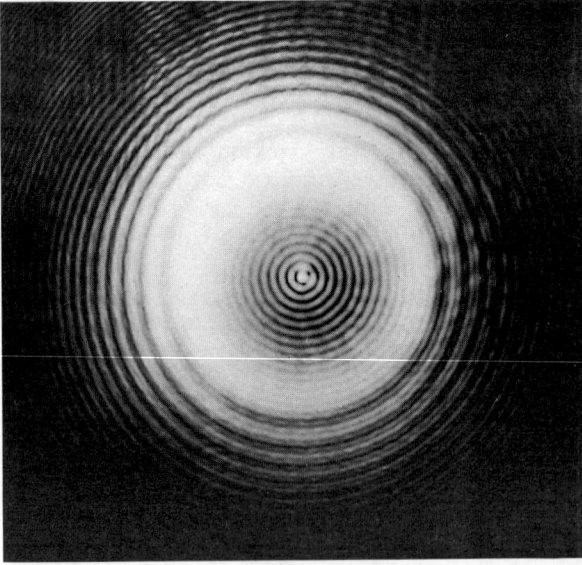

Fig. 27.6 Diffraction pattern produced by a ball-bearing

It is interesting to note that recently a method of obtaining a constantly emitting or coherent source has been discovered. Such a source is called a *laser*.* Very striking interference and

---

* Light Amplification by Stimulated Emission of Radiation.

diffraction effects can be produced with this apparatus (Figs. 27.5 and 27.6).

White-light fringes. If a source of white light is used in place of a monochromatic one, there is produced a separate fringe system for each wavelength present in the source. As the fringe width depends on the wavelength, only the central fringe will be similarly placed for all wavelengths. Thus a white-light fringe system has a central white fringe with coloured fringes on each side. Not many coloured fringes can be seen as the effect produced by their overlapping is white light.

Interference at thin films. The colours which we observe in a soap film, in a thin layer of oil on water, in the feathers of some birds and in the fissures of crystals such as Iceland spar and mica are all produced by the interference of light.

Let us consider light incident upon a thin film

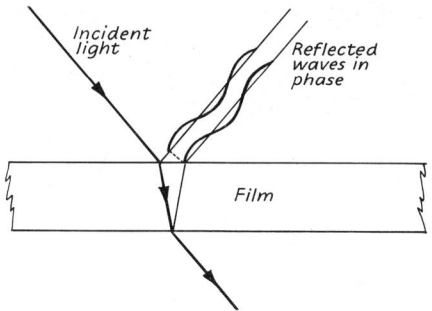

Fig. 27.7 Interference at a thin film

as shown in Fig. 27.7. Most of the incident light will pass through the film, but small percentages will be reflected at both the front and the back surfaces, as shown. There will be a phase difference between these two reflected beams which will depend upon (i) the angle of incidence, (ii) the thickness of the film, and (iii) the refractive index of the material of the film. Under suitable conditions, then, the two emergent rays will be in phase and so constructive interference will occur. By varying (i) or (ii), conditions may produce alternately constructive and destructive interference and so it is possible to obtain a pattern of interference fringes.

A very simple case occurs when interference is produced in a wedge-shaped air film. In Fig. 27.8 two plane glass plates are separated down one edge by a strip of very thin material so as to produce a wedge-shaped air film. If a monochromatic source illuminates the plate, a system

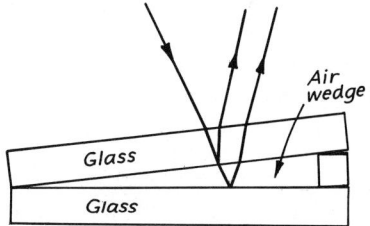

Fig. 27.8 Interference at an air wedge

of straight light and dark fringes is observed. These are produced by the combination of light reflected at the two surfaces of the air film. If the thickness of the air film is too great, so causing a large path difference between the rays, no fringes are obtained. This is why the light reflected from the upper surface of the top plate does not interfere with the other reflected rays.

The fringes are straight and evenly spaced only when the surfaces of the plates are perfectly plane. Any deviation amounting to only a fraction of a wavelength can be detected, for it distorts the regularity of the fringe system. In fact, the fringe pattern provides a perfect contour map. In this way a plane surface can be tested for accuracy against a standard plane surface.

If one of the plates forming the boundary of an air film is mounted in such a way that it can be moved away from the other plate, the system of fringes moves across the field of view. The apparatus is set up as in Fig. 27.9 in order to observe

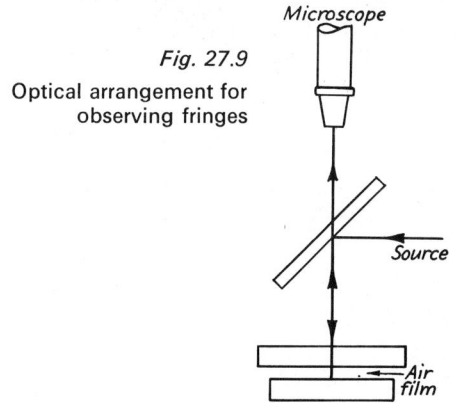

Fig. 27.9

Optical arrangement for observing fringes

this. A movement of $\lambda/2$ of one plate causes a shift of one fringe in the microscope. In this way very small distances can be measured in terms of the wavelength of light.

It is by using more advanced interference

methods that the number of wavelengths of a particular coloured light in one metre has been found (*see* Chapter 1).

If we now consider interference at a soap film, we must first note that the thickness of the film will vary as it drains down. Conditions for constructive interference for many different colours (wavelengths) will be satisfied if the film is more than a few wavelengths thick. Thus if a soap film is formed on a frame, and a beam of white light reflected from it on to a screen, no colours will be seen if the film is thick but, as it drains, horizontal bands of colour will be observed. These colours will get purer as fewer wavelengths satisfy the conditions simultaneously. Eventually, the film will appear black, when it is very thin. This means that when the path difference is nearly zero, destructive interference occurs, when we would expect constructive interference. This arises due to a change in phase of 180° on reflection at a denser medium (there is no phase change on reflection at a rarer medium).

This phase change occurs in other cases. A disturbance along a stretched string suffers a phase change at a fixed end, but not at a free end. A crest travelling along a rope is reflected at a fixed end as a trough. If a rope is held with a free end hanging down, a crest sent downwards is reflected as a crest from the free end, i.e. there is no phase change. (*See* Chapter 20.)

Diffraction grating

We have seen that diffraction takes place at a narrow slit. Consider now many equally spaced slits, such as A, B, C, D, etc. in Fig. 27.10. Wavelets from a monochromatic beam of light will radiate as shown from these slits. We obtain the resulting wavefront by enveloping the surfaces of wavelets

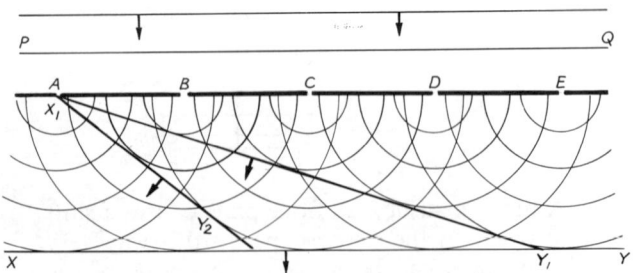

Fig. 27.10 First two orders of spectra formed by diffraction grating

in the same phase. Thus we get a wavefront XY in the same direction as the incident wavefront PQ. But we can also join wavelets in the same phase (or with phase difference of 360° which is equivalent to zero phase difference) by the line X_1Y_1: and again by the line X_1Y_2. Thus strong beams of light should be seen in these directions.

We can carry out this experiment on a grand scale, for we can buy *diffraction gratings* which consist of many thousands of slits or lines accurately spaced, very often with 6000 to the centimetre. The original gratings are ruled on metal, but replicas can be cast from the originals and mounted on glass.

The angles which X_1Y_1 and X_1Y_2 make with XY depend on the wavelength. Thus, when white light is used, the red light will be diffracted more than the blue light, and so spectra will be seen along directions normal to the wavefronts X_1Y_1 and X_1Y_2. These are called the first and second order spectra and can be seen on each side of the central white image.

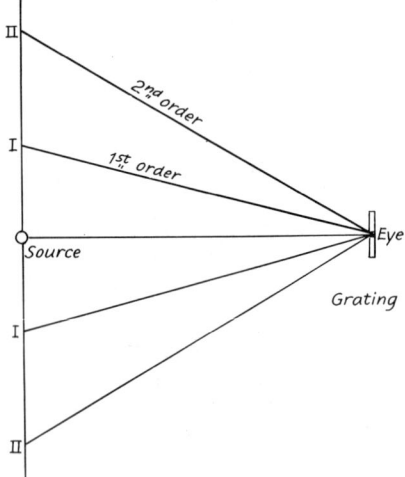

Fig. 27.11 Arrangement for viewing spectra by diffraction grating

The strip source is normal to the plane of the diagram

Measurement of wavelength. If a monochromatic line source (a mercury strip light does well, but has two strong lines in its visible spectrum) be viewed through a grating held close to the eye, the spectra are clearly seen (Fig. 27.11). The angles which these subtend at the eye can easily be found by marking the positions and measuring the distances. In Fig. 27.12, the first order spectrum will be produced by wavefront AB' at

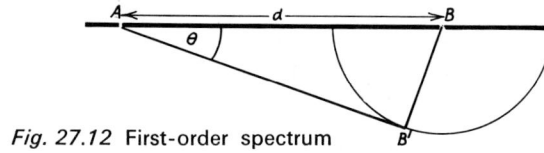

Fig. 27.12 First-order spectrum

angle θ, where $\dfrac{BB'}{AB} = \sin\theta$

but $\qquad BB' = \lambda$ for first order
$\qquad\qquad\quad = 2\lambda$ for second order
$\qquad\qquad\quad = n\lambda$ for nth order.

Hence $\quad d\sin\theta = n\lambda$

where d is the grating space and is always supplied with the grating. Substitution in the formula gives the wavelength of the light.

This simple experiment can be carried out with increased accuracy using an instrument called a *spectrometer*. Fig. 27.13 gives the general idea of the

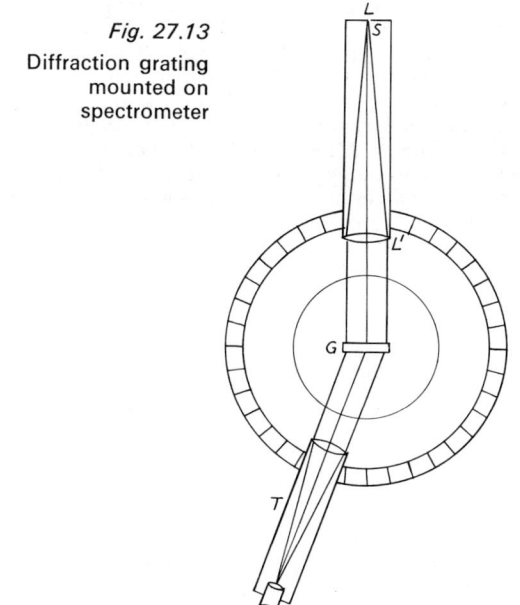

Fig. 27.13

Diffraction grating mounted on spectrometer

instrument (*see also* Fig. 28.5). The slit S, which is illuminated with monochromatic light L, is at the focal point of a lens L'. This provides a parallel beam of light which falls on the grating G. The telescope T, which can be rotated about a circular scale, receives the diffracted light. By viewing the direct light and the light giving the first order spectrum, the angle θ in the above formula is found accurately.

Polarization

The waves emitted from an ordinary source of light are unpolarized, i.e. the displacements are in all directions in the plane normal to the direction of the light. These displacements can be polarized by passing the light through certain crystals. The difficulty of producing large natural crystals has been overcome to some extent by using certain needle-like crystals which can be aligned and so made equivalent to a large crystal. 'Polaroid' is the name given to such a prepared 'crystal'. The small crystals are aligned mechanically by using a ribbed roller and sealed in a transparent sheet. Passing an ordinary beam of light through a polaroid disc produces light with vibrations in one direction only. If this polarized light is passed through a second disc, a complete cut-off can be obtained by rotating the second disc so that its axis is at right angles to that of the first disc.

It is interesting to observe the effects which some substances have upon polarized light—many transparent solids and solutions possess the property of rotating the plane of polarization. For example a sugar solution displays this property and it is used in differentiating between the types of sugars (Faraday discovered that a magnetic field can produce this effect). If a piece of cellophane is folded to produce different thicknesses and placed between crossed polaroids, the cut-off is affected and the colourless cellophane appears multi-coloured. This effect is due to the plane of polarization being rotated by the cellophane by different amounts for different colours. When glass or perspex is strained, the strain lines are revealed by placing between crossed polaroids. This has a useful application in engineering for the strain lines in structures can be investigated by experimenting with perspex models of the structures.

The Doppler effect

The Doppler effect can be observed with visible light. The lines in the spectra of some stars are displaced from their normal positions towards the red end of the spectrum, indicating an increase in wavelength. This is thought to be due to the star moving away from the solar system. Measurement of the shift enables the speed of separation to be calculated.

QUESTIONS

1. Describe, with experimental details, how you would demonstrate an *interference* effect with (*a*) water ripples, (*b*) light. State how, in each case, the effect is related to the wavelength of the waves. Hence show how, in the optical effect you describe, it is possible to deduce which of two colours has the greater wavelength. (C)

2. Describe how Young demonstrated the interference of monochromatic light.

3. Describe and explain the appearance of Young's fringes produced by white light.

4. Describe how you could set up an optical arrangement to measure an extremely small movement, such as the expansion of a short metal rod on heating.

5. Explain why it is impossible to observe interference effects produced by the light from two small electric lamps.

6. Explain the formation of coloured fringes produced by a soap film.

7. Calculate the angle between the first and second order spectra of sodium light produced using a diffraction grating with 6000 lines/cm, if the wavelength of the light is 589 nm.

8. How many orders of spectra can be obtained with light of wavelength 650 nm, using a grating of 2000 lines/cm?

9. Light from a sodium vapour lamp falls perpendicularly on a diffraction grating. Beyond the grating a telescope is swung round and when it makes an angle θ with the line from the lamp to the grating a yellow line is seen. Explain with a diagram why this is so, and why a yellow line may also be seen at a larger angle than θ. (O & C)

10. A small sodium lamp is placed on a wall and viewed by an eye 200 cm away placed directly opposite to the lamp. On inserting a grating having 3000 lines per cm just in front of the eye, several images of the sodium lamp appear on the wall on each side of the lamp. If the nearest ones are x cm from the lamp, calculate x, given that the wavelength of sodium light is 5.90×10^{-5} cm. (O & C)

11. Why are the elements of a diffraction grating regularly spaced? How may a grating be used to study the spectrum from an electric lamp with a straight filament?

What is meant by different *orders* of spectrum?

A narrow detector of infrared radiation, insensitive to visible light is moved through the spectrum. It is found that a strong response occurs in the second-order spectrum for green light of wavelength 4.9×10^{-5} cm, yet there is no response in the first-order spectrum for this wavelength. Suggest an explanation for this and deduce the wavelength of the radiation responsible for the strong response of the detector. How could you check your explanation if you were given a piece of material which absorbed green light but was transparent to infrared? (O & C)

12. Explain the action of a diffraction grating in producing a spectrum of white light, and show how the wavelength of light of a given colour can be determined.

When illuminated normally with green light of wavelength 6×10^{-5} cm, a diffraction grating with a grating spacing of 2×10^{-4} cm gives a bright line on either side of the normal in a direction making an angle of 17.5° with the normal.

(*a*) When the grating is illuminated normally with white light would you expect to find (i) the corresponding yellow light, (ii) the corresponding blue light at angles greater than or less than 17.5° with the normal to the grating?

(*b*) In a demonstration to show that similar effects can be obtained with microwaves, a grating is made from strips of aluminium foil. A maximum is observed on the detector at an angle of 17.5° to the normal for microwaves of wavelength 3 cm. What is the grating spacing? (O)

28 The spectrum; colour

In tracing the path of a ray of light through a prism, using a ray box, did you notice that the emerging ray was coloured? Perhaps you used pins instead of a ray of light. If so, did you observe that the pins, when seen through the prism, were fringed with colour?

We shall now investigate the production of these colours. Although the effect must have been

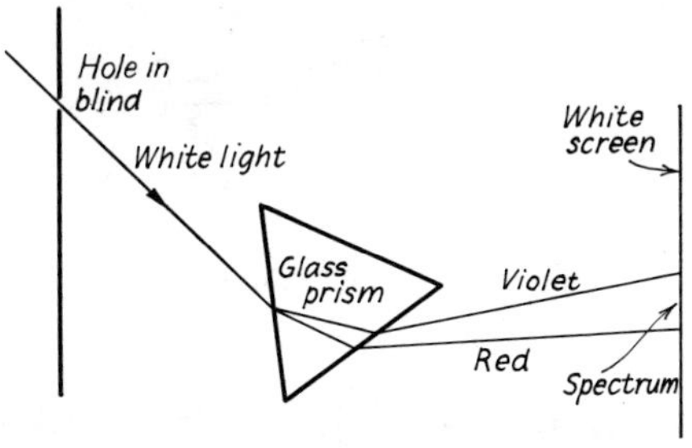

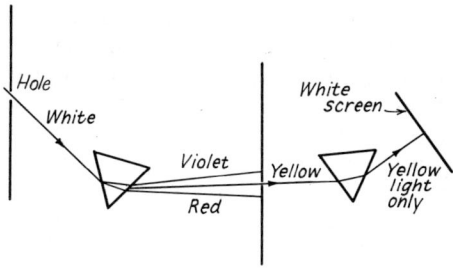

Fig. 28.1 The production of a spectrum—Newton's experiment (1)

observed in glass and crystal ornaments by many people long before Newton's time, it was he who first investigated it. He allowed a beam of sunlight to enter a darkened room through a hole cut in a blind and to pass through a prism as shown in Fig. 28.1. A white screen was fixed to receive the light. Instead of a circular white patch appearing on the screen, there was a broad multi-coloured band, which we call a *spectrum*. Newton stated that it contained seven colours—violet, indigo, blue,

green, yellow, orange and red. Most people, seeing a spectrum for the first time, say there are four distinct colours, but as these colours merge together there is really an infinite number.

Newton's next task was to explain the origin of the colours. Did they originate in the prism or in the light? To solve this problem, he performed the following three experiments.

Fig. 28.2 Newton's experiment (2)

(*a*) A slit was cut in the screen to allow light of one colour in the spectrum to pass through (Fig. 28.2). This light was refracted again by a second prism, but only a broadening was observed without further production of colour.

(*b*) The light proceeding to the screen was passed through a second prism, inverted with respect to the first, so that the light was refracted in the opposite direction. The emergent light was white (Fig. 28.3).

247

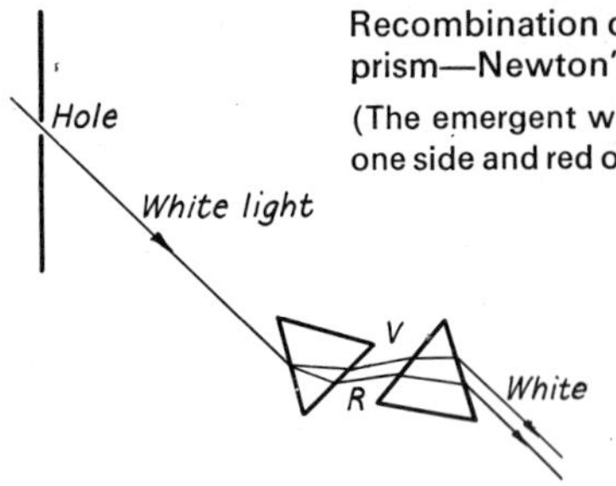

Fig. 28.3

Recombination of colours by inverted prism—Newton's experiment (3)

(The emergent white ray is tinged blue on one side and red on the other)

Hole

White light

V

R

White

(*c*) The spectrum colours were painted in sectors on a circular disc. When illuminated and rapidly rotated the disc appeared white.

These experiments indicate that the origin of the colours is not in the glass (experiment (*a*)) but in the white light (experiments (*b*) and (*c*)), for white light can be produced by the combination of the colours of the spectrum.

White light is not simple, but composite, constituting an infinite number of colours ranging from red to violet.

Production of a spectrum

White light is split up or dispersed in passing through a glass prism because glass has a different refractive index for each colour or wavelength. Newton's arrangement of passing a parallel beam through a prism does not produce a good spectrum because, if the beam is broad enough to give reasonable illumination, the colours overlap in the spectrum. Such is called an 'impure' spectrum. A pure spectrum is one in which there is no overlapping of the colours. To produce a pure spectrum, a parallel beam of white light must be passed through a prism and then focused by a convex lens to bring each colour to a focus without overlapping. The arrangement shown in Fig. 28.4 is the best to use. The narrow slit is illuminated by white light, preferably from a carbon arc. The white light from the slit is

rendered parallel by a convex lens, passed through the prism, and then brought to a focus by a second convex lens. The lenses should be achromatic (*see* below). This optical arrangement is like that incorporated in the spectrometer shown in Fig. 28.5 (also refer back to previous chapter). For

Fig. 28.5 A spectrometer

ordinary demonstration or qualitative experiments one convex lens produces a reasonably pure spectrum (*see* Fig. 28.6). A lantern is convenient to obtain a large spectrum. The slit is put in the

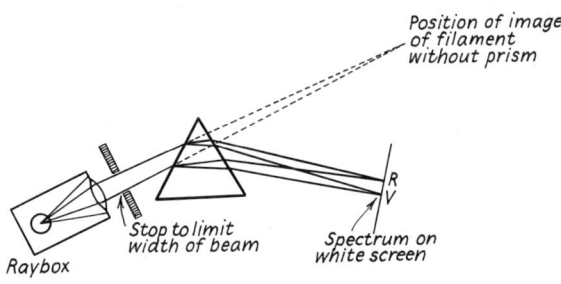

Fig. 28.6 Spectrum (slightly impure) produced with a ray box

slide carrier and the projection lens used to focus the spectrum. A ray box arranged as shown, gives a splendid spectrum for simple experiments.

It should be noted that a spectrum really consists

Fig. 28.4

Method of producing a pure spectrum

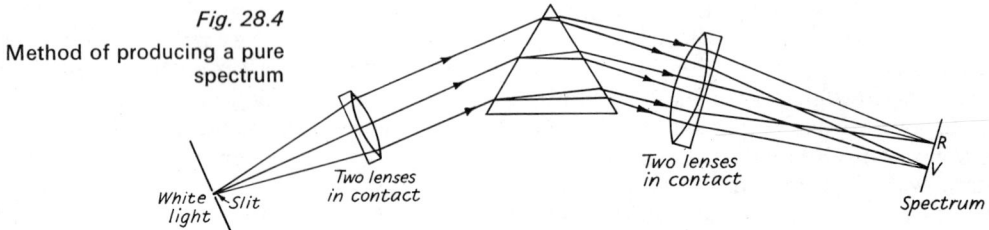

White light *Slit* *Two lenses in contact* *Two lenses in contact* R V *Spectrum*

of a large number of images of the slit placed side by side in perfect order of wavelength.

Achromatic prisms and lenses

We sometimes wish to deviate a beam of white light without producing dispersion. In Fig. 28.3 the dispersion is corrected, but the emerging beam is parallel to the incident beam. We can accomplish the desired effect, however, by using two prisms of different glasses with particular angles. The prisms are usually made of crown and flint glass.* Such a pair of prisms, which produces deviation without dispersion, is said to be *achromatic*.

As a lens is, in effect, a prism of varying angle, its focal length for blue light is less than that for red light. This leads to the coloration often seen on the edges of images, the defect being known as *chromatic aberration*. By combining a crown glass converging lens with a suitably chosen flint glass diverging lens, a lens system, having approximately the same focal length for all colours, is produced.

Colour

Why does a particular book appear red and another blue? Red light is a constituent of white light and so, when white light falls upon a red object only the red constituent must be reflected, the remainder must be absorbed. For this absorption to take place, the white light must penetrate a little into the colouring or pigment of the book cover. If the surface has a shiny finish, some light will be reflected at the surface. There will be no absorption of this light and it will be white. In performing these colour experiments, objects with rough or matt surfaces give the best results since they give no surface reflection.

If the red book will reflect red light only, illuminating it with any other coloured light except red should make it appear black, for it should absorb all the coloured light. It does appear black in all but the red light. Try this experiment with other coloured specimens. You will find that it does not work very well in every case. This is because the colours of dyes used are not pure spectral colours, that is they are not monochromatic but a mixture of many colours.

---

* The ratio $\dfrac{\text{dispersion}}{\text{deviation}}$ is greater for flint glass than for crown glass, hence when the dispersions are equal, the deviations are not equal.

Analysis of light and the examination of the purity of colours

The purity of a transparent coloured plate* can easily be examined by passing a white light through it and then through a prism. The resulting spectrum shows the pure colours transmitted. This experiment can easily be done with a ray box.

It will be found that gelatines, which appear to be one definite colour, transmit several colours, and the experiment shows conclusively the impurity of common pigments.

Pure colours can be produced fairly easily in the laboratory. A pure colour is obtained by putting a small quantity of any sodium compound in a bunsen flame, which turns to a golden yellow. If this flame is analysed, it will be found to give one narrow line in the spectrum. Other substances give characteristic colours, but usually their spectra consist of several lines. The positions of these spectral lines can be used to identify a substance (*see* page 251). Each line in a spectrum is a pure colour and can be separated out by using an appropriate filter.

Colour mixing

We have seen above that the colours of common objects produced by dyes are really very impure from a scientific point of view. Use is made of this in colour mixing.

There are two ways in which colours can be mixed, and these must not be confused.

Subtractive combination

(*a*) Pass a beam of white light through a peacock-blue and a yellow filter placed together, or simply hold them up to a bright light. Note that a good green light is transmitted. Try also magenta and yellow, and magenta and peacock-blue. Reference to the results for the experiment in which the purity of these gelatine filters was examined provides the explanation of these results. A peacock-blue filter, as the name implies, filters all colours out of the incident light except blue and green. The yellow filter permits yellow, red and green light to be passed, hence the only colour which is transmitted by both is green.

---

* For these colour experiments, a set of coloured plates or filters is required. The six colours blue, green, red, peacock-blue (or cyan), magenta and yellow are the most useful.

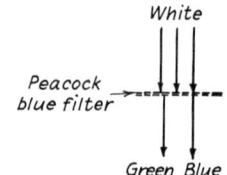

Fig. 28.7
Subtractive combination
of colours

This is called *subtractive* combination of colours (Fig. 28.7) as each filter subtracts some colour from the incident light. Obviously, the ultimate result will be the absorption of all the light, as is the case when white light is passed through peacock-blue, yellow and magenta filters.

(*b*) Mix together two pigments from an ordinary paintbox, or two solutions, e.g. copper sulphate and potassium dichromate. Similar results are obtained as in experiment (*a*), for the solution mixture will transmit only the components of the incident light which *both* solutions transmit. If the pigments in the ordinary paintbox were spectral pure colours, what would be the result of mixing any two?

(*c*) Using a filter in front of a ray box, illuminate specimens of coloured paper or wool with coloured light, or examine coloured specimens, in daylight, through coloured gelatines. In this case, the pigment in the paper will reflect only certain colours contained in the incident light. For example, peacock-blue light falling on a yellow paper will make it appear green, for the paper will reflect yellow and green light, but the yellow is absent in the peacock-blue, hence green only is reflected.

This experiment suggests many amusing trick experiments in which the appearance of multi-coloured objects can be completely changed by a change in the colour of the incident light.

Results: subtractive colour mixing

Peacock-blue (cyan) and **yellow** combined subtractively give **green.**

Yellow and **magenta** combined subtractively give **red.**

Magenta* and **peacock-blue** combined subtractively give **blue.**

Peacock-blue,* **yellow** and **magenta** combined subtractively give **black.**

---

* Magenta contains blue and red: peacock-blue (cyan) contains blue and green.

Additive combination

(*a*) Using two ray boxes with coloured gelatines, shine two coloured beams of light upon a white screen so that they partly overlap. The white screen reflects all the light which falls upon it and, therefore, if the blue and red filters are used, all the constituents of the red and blue light will be received by the eye. The colour of the screen where the coloured beams overlap will be magenta. The illuminations of the beams may be varied to give the best effect by adjusting the lenses. Repeat with the other filters.

A final impressive experiment may be carried out using three ray boxes with blue, red and green filters. If an object, e.g. a wooden pyramid, is

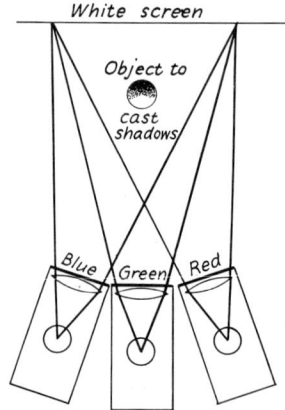

Fig. 28.8 Additive combination of colour using ray boxes

placed as shown in Fig. 28.8, a colour pattern is produced on the white screen revealing the results obtained with different combinations of the three colours taken two at a time. The general illumination on the screen is produced by all three colours, which is white.

Blue, green and red are *primary* colours. They cannot be produced by additive mixing, but they can be mixed additively to give any desired colour.

Colours produced by mixing two primary colours are called *secondary* colours.

(*b*) Newton's colour disc is an example of another method of colour addition. If a disc has sectors of two different colours, then provided it is rotated rapidly while being illuminated by white light, its colour is that of the additive combination of the two colours used.

Results: additive colour mixing
   Blue plus **green** give **peacock-blue.**
   Green plus **red** give **yellow.**
   Red plus **blue** give **magenta.**
   Red plus **green** plus **blue** give **white.**

Peacock-blue, yellow and magenta are secondary colours.

The colour effects produced with stage lighting are obtained by employing both additive and subtactive methods of combination.

Colour matching in day and artificial light should now be quite clear. Differences are due to the different spectra of daylight and artificial light. The latter is richer in yellow than the former.

Spectra

We have mentioned above that certain metallic salts put into a bunsen flame give characteristic colours which, when analysed, show spectra which enable us to identify the metallic constituents.

We can divide spectra into two main classes, *emission* and *absorption* spectra.

1. Emission spectra

For simple treatment the spectra in this class can be subdivided into three types.

(*a*) *Continuous spectra.* A continuous spectrum is produced by light from an incandescent source such as the sun, an arc, an electric-lamp filament or a very hot molten metal. Such sources contain all the colours of the visible spectrum, and the light produced results from the extremely rapid motions of the atoms due to the very high temperature.

(*b*) *Line spectra.* Line spectra are produced by what is termed the 'excitation' of gaseous atoms. This effect is achieved by heating the gas or by passing an electric current through it. Under such conditions, atoms of one kind emit light of certain wavelengths only.* This light, when analysed,

---
* *See* Chapter 43.

reveals a line spectrum which is characteristic of the particular substance. An element can easily be recognized by its spectrum. When an arc is struck between metal electrodes, the incandescent gas between the electrodes emits a line spectrum. If a metallic salt is put into a cavity drilled in a carbon electrode, the arc will show the line spectrum of the metal. A simple method of exciting the line spectrum of a gas is to pass a current through the gas contained at low pressure in a sealed tube. Incandescent vapours of sodium, mercury, cadmium, krypton, etc. can be used as standard sources as they emit light of certain accurately known wavelengths.

(*c*) *Band spectra.* The inner structure of molecules permits of a complicated mechanism of excitation. The emitted light may have extended ranges of wavelengths instead of single values and instead of *lines* the spectrum contains *bands.*

2. Absorption spectra

When white light is passed through a coloured plate and then examined, it is found that certain colours are missing from the continuous spectrum. The dark lines or bands on the continuous background form what is termed an absorption spectrum. The continuous spectrum of the sun is found to be crossed by a number of black lines. These are called Fraunhöfer lines, after their discoverer, and occupy positions in the spectrum coincident with the lines of many known substances. Such spectral lines are produced when the light from a very hot source passes through a slightly cooler vapour. The vapour extracts from the light the particular wavelengths which it emits. Hence, from the Fraunhöfer lines we know what substances are present in the sun's atmosphere.

During a total eclipse the direct light from the sun is cut off. The light from the glowing gases of the sun's atmosphere, which is called the *corona,* then gives a line spectrum corresponding to the absorption lines of the Fraunhöfer spectrum.

The absorption spectrum of sodium may be produced by passing light from a carbon arc through a very intense sodium flame, or better, through hot sodium vapour produced by vaporizing a little sodium in a partially evacuated tube.

The absorption spectrum of chlorophyll is interesting. A solution of chlorophyll in alcohol contained in a glass cell is introduced in a powerful beam of white light which is then passed through a prism. The continuous spectrum now has a dark band across it. This is the absorption spectrum of chlorophyll. Light with the particular wavelengths of the dark band is absorbed by the chlorophyll. The absorbed light is energy from the sun which supports the life of the plant.

The complete spectrum

When examining the spectrum of white light, it will be noticed that the edges of the spectrum are not sharp. This indicates the possibility that there are radiations beyond the ends which the eye is unable to perceive. This is indeed so, for we can detect the presence of such radiations.

The radiation beyond the violet end of the visible spectrum is called *ultraviolet* radiation. It is present in the emission from very high temperature sources, such as the electric arc and the sun. This radiation can be detected photographically using ordinary plates and film. It causes *fluorescence* of many minerals and dyes. This effect is the absorption of radiation of a wavelength which is outside the visible range, and the re-emission of part of the radiation with a wavelength which lies within the visible range. This property is exploited for advertising and theatrical purposes. Spectacular effects are produced by the brilliantly coloured fluorescence of objects on a completely darkened stage. The

Fig. 28.9 Examination of eggs in ultraviolet light
An old egg glows purple, but a fresh egg glows scarlet

Fig. 28.10
Drying of paint on car body by means of infrared light

effect is accomplished by using ultraviolet lamps. These are mercury vapour lamps surrounded by 'black glass' envelopes which absorb practically all the visible light and transmit the 'near' u.v. light (i.e. the u.v. light just beyond the visible spectrum). It is the addition of fluorescent materials to washing powders which creates the 'brightness' of clothes washed with them. When a mercury vapour lamp is enclosed in a quartz envelope, the u.v. light much beyond the visible light is transmitted. This radiation is present in the sun's radiation and causes sunburn. Its health-giving properties have to be carefully employed for, in any but small doses, the radiation can be dangerous. The lamps sold for experiments on fluorescence are not dangerous as the more powerful u.v. rays are absorbed by the special black glass envelope.

It is necessary to employ prisms and lenses of quartz when experiments are performed with u.v. light, for glass and many other substances which are transparent to visible light are opaque to u.v. light.

greater than that of red light and when it is absorbed by material bodies it heats them, the radiant energy being converted into kinetic energy of the molecules. As explained earlier, infrared light can be detected by special photographic plates, special photo-electric cells, radiometers and by several electrical instruments, the best known of which is the thermopile.

Extending beyond the ultraviolet and infrared regions of the spectrum there are other rays which have the same character and yet display different properties and require different methods of generation. Fig. 28.11 shows the complete spectrum of electromagnetic waves. They all have the same character, i.e. they are electrical and magnetic in nature and they all possess the same velocity in a vacuum, viz. 3×10^8 m/s. Their properties depend upon their wavelength and we classify them by groups based on the methods used for their excitation and detection. The longest waves are wireless waves generated by transmitters and detected by receivers. Using special electrical cir-

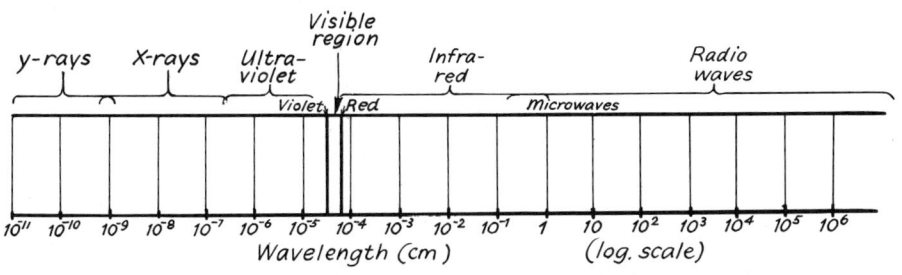

Fig. 28.11
Spectrum of electromagnetic waves

Photoelectric cells, which are mentioned in Chapter 41, can be made sensitive to u.v. light and so used for its detection and measurement.

Beyond the red end of the visible spectrum is the *infrared* region. Mention has already been made of this radiation in Chapter 19. Its wavelength is

cuits, waves of much shorter wavelength can be generated. These electrical waves join up with the infrared band of waves. Beyond the ultraviolet on the other side of the visible spectrum we have X-rays and γ-rays. These rays are described in Chapters 41 and 42.

QUESTIONS

1. How did Sir Isaac Newton produce a spectrum and what experiments did he carry out to show that white light is composite?

2. (i) Ordinary (impure) yellow light is passed through (a) a yellow, (b) a red, (c) a blue filter. What is the result?
(ii) A piece of blue paper is illuminated by (a) yellow, (b) red, (c) blue light. What colour does it appear?

3. Describe two methods of demonstrating (a) the subtractive combination and (b) the additive combination of colours.

4. Why is colour matching in artificial light different from that in daylight?

5. Explain how the Fraunhöfer lines in the sun's spectrum are produced.

6. Given a source of white light, how would you project a pure spectrum on to a screen?

How may colours of the spectrum be recombined to give white light again?   (O)

7. Draw a diagram to show how you could project a reasonably pure spectrum on a screen, using a filament lamp as the source of light. Compare the spectrum obtained in this case with the spectrum you would expect to see if the source were a metallic salt (such as common salt) heated in a colourless bunsen flame.

A beam of white light is passed first through blue glass, then through yellow glass, and finally projected on a white screen. State and explain the appearance of the screen. Explain also what you would see if two separate beams of white light were passed, one through blue glass, and one through yellow glass, and the two beams then projected together on the screen.   (C)

8. White light is said to be composite (or compound). Explain the meaning of this statement, and describe carefully with the aid of a diagram, how you would demonstrate its truth.

A scene includes a white house with a red roof, yellow flowers, and green trees. This is viewed first through red glass and then through green glass. Assuming that all the colours mentioned are pure, insert in the form of a table, under the headings of red glass and green glass, the appearance of each part of the scene.   (W)

9. Explain what happens when a beam of white light passes through a glass prism and illustrate your answer by a diagram.

'A piece of purple cloth reflects both blue light and red light, and absorbs light from all other parts of the spectrum.'

Describe how you would test this statement experimentally.   (O)

10. Describe how you would project a pure spectrum of white light on to a screen. Illustrate your answer by a good ray diagram.

A white envelope bears a red stamp cancelled by a black postmark. State, with explanations, how the whole appears when held in (a) the red, (b) the green part of the spectrum.   (O)

11. Given an apparatus which produces an intense parallel beam of ultraviolet and infrared rays, how would you detect the presence of each component of the beam and how would you show that the components of the beam can be separated by a prism if it is suitably placed? Illustrate your answer with diagrams.   (C)

12. A tube contains a small white-hot source fitted with a reflector which concentrates the energy in a parallel beam. It can be fitted with two covers, A and B; these transmit only infrared and ultraviolet radiations respectively. Describe how you would detect the infrared and the ultraviolet radiation emitted by the source. How, if at all, would the detectors be affected if the tube were used without either A or B in place?   (C)

13. Two pieces of yellow glass, labelled A and B, look alike; but one transmits only yellow light, the other a mixture of red and green. Describe briefly how you would use a piece of pure red glass to identify each of the yellow pieces of glass. Give an explanation of the method. Describe any *one* other experiment you could do to confirm the identity of the two pieces of glass.   (C)

14. Explain the terms *dispersion* and *pure spectrum* with reference to white light passing through a prism. Explain how you would arrange to produce on a white screen a pure spectrum of the light of an electric filament lamp. Draw a ray diagram showing the passage of red and blue rays from lamp to screen, and mark on your diagram whereabouts on the screen you would expect to find ultraviolet and infrared rays, assuming that they were present. Name ONE material which allows ultraviolet rays to pass through it.   (O & C)

15. The heater coil of an electric radiator and the filament of a projector lamp are each rated at 1000 W. State the difference between the radiations from these two sources and give the reason for the difference.   (C)

16. An electric filament lamp is put close to a thin sheet of red glass. The reflected light is *white*, the transmitted light is *red*. Account for this difference, and state what becomes of those parts of the lamp's radiation that are not transmitted by the sheet of glass.   (C)

17. Describe how you would attempt to find out whether the spectrum of white light from an intense source, e.g. a carbon arc, extends beyond the visible region. What would you expect to find?

Give a brief account of the electromagnetic spectrum.

(O & C)

Part Eight VIBRATIONS AND WAVES

Setting up equipment in the anechoic room (room without an echo) to measure the noise from a vacuum cleaner. The plastic foam wedges are designed to prevent internal echoes

29 Sound: production and propagation

The production of sounds

If we sit quietly for a few moments, we may hear many familiar sounds, for example, the crackling of the fire, a door being opened, a running tap, and an orchestra playing over the radio. Every one of these sounds is produced by something moving.

A sound is produced by a vibration. But sounds can roughly be divided into two classes, musical and non-musical. The latter we call noises.

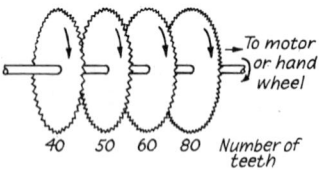

What, then, is the difference between a noise and a musical sound? If something is dropped to the floor, a noise results. The object and the floor vibrate for a short interval of time.

Hold one end of a ruler on the edge of a table and flick the free end. The ruler vibrates so slowly, perhaps, that we can see the vibrations but not hear them. Shorten the vibrating length of the ruler and again flick it. This time the vibrations will be

quicker and the ruler may emit a low note. Again shorten the length and a higher note is emitted. In this way musical notes are produced. The vibrations of the ruler are regular and sustained.

A noise is caused by irregular vibrations.

A musical note is caused by sustained regular vibrations.

The nature of these vibrations will be dealt with more fully later.

Fig. 29.1

Savart's toothed wheels

A musical note is obtained when a card is held against teeth and the wheels are rotated. The above wheels give doh, me, soh, doh'

A tuning fork provides a simple means of obtaining a musical sound. Its vibrations can easily be detected by allowing one of the prongs to touch a hard surface. The note emitted is governed by the number of vibrations per second, i.e. the frequency. This can also be shown with Savart's toothed wheels (Fig. 29.1). Four toothed wheels, with 40, 50, 60 and 80 teeth, are mounted on one shaft, which is rotated. The wheels are touched by a thin piece of card, and the scale, doh, me, soh, doh′, is heard. The more quickly the wheels are rotated the higher the notes,* but the relationship between the four notes remains the same: that is, they are still the four notes of a major chord (i.e. doh, me, soh, doh′).

The siren (Fig. 29.2) produces a note by a rapid succession of puffs of air. Air is blown through a

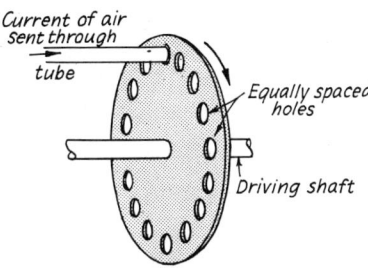

Fig. 29.2 Simple form of siren

tube, the end of which is very near a rotating disc with holes arranged equidistant round a circle. As each hole passes the tube, a puff of air escapes. If these puffs occur many times each second, a note is emitted. No separate device is needed to rotate the disc if a second disc is placed near the first, having holes drilled slant-wise. The jet of air passing through the holes will then cause rotation.

Characteristics of a musical note

When two instruments are in tune and play the same note, they generate sound waves of the same frequency or pitch. But the two sounds, although in tune, are different. This is due to a difference in quality.

There are three characteristics of a musical note: (1) pitch, (2) quality (or timbre), (3) loudness.

Pitch is the technical term defining whether a note is high or low. In the natural scale, doh, ray, etc., we say the pitch of each succeeding note is higher. With Savart's toothed wheels we have already seen that the vibrations are quicker for a higher note. Thus pitch is determined by the number of vibrations per second or the frequency. When the frequency of the vibrations is doubled,

the pitch of the notes rises an octave. Middle C* on the piano has a frequency of about 262 hertz (vibrations per second).

The quality of a sound is obtained from the presence of notes of higher frequencies along with the main frequency or fundamental (*see* page 267).

The amplitude of the vibrations determines the loudness. If a tuning fork is struck hard, the prongs vibrate through a greater distance (but not quicker) and give a louder sound.

The mechanism of hearing

When the minute compressions and rarefactions in the air which constitute a sound wave enter the ear (*see* Fig. 29.3), they fall upon a stretched membrane, the ear-drum, which is made to vibrate in accordance with the pressure changes in the wave. The drum separates the outer ear from the middle ear, where the vibrations are conveyed by three bones (the hammer, the anvil and the stirrup)

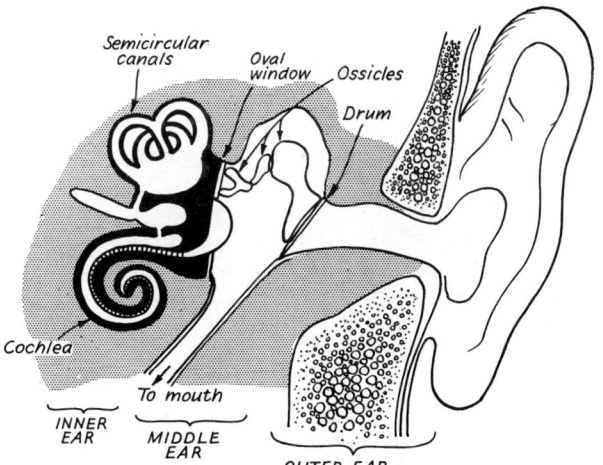

Fig. 29.3 Section through ear showing ossicles (hammer, anvil, and stirrup)

to the inner ear, the hearing organ proper. Here are situated nerve-endings, which are stimulated by the bone vibrations and convey the impulses to the brain. The energy required to stimulate the ear is approximately equal to that required to stimulate the eye.

---

* This may be shown very simply by holding a card so that it touches the spokes of a rotating cycle wheel.

* The internationally recommended standard musical pitch is based on a frequency of 440 Hz for A, which gives middle C an absolute frequency of 261.6 Hz.

Fig. 29.4 Cleaning with ultrasonic vibrations

Glassware being cleaned by immersion for a few seconds in water through which ultrasonic vibrations are sent from the quartz-crystal transducer in the base

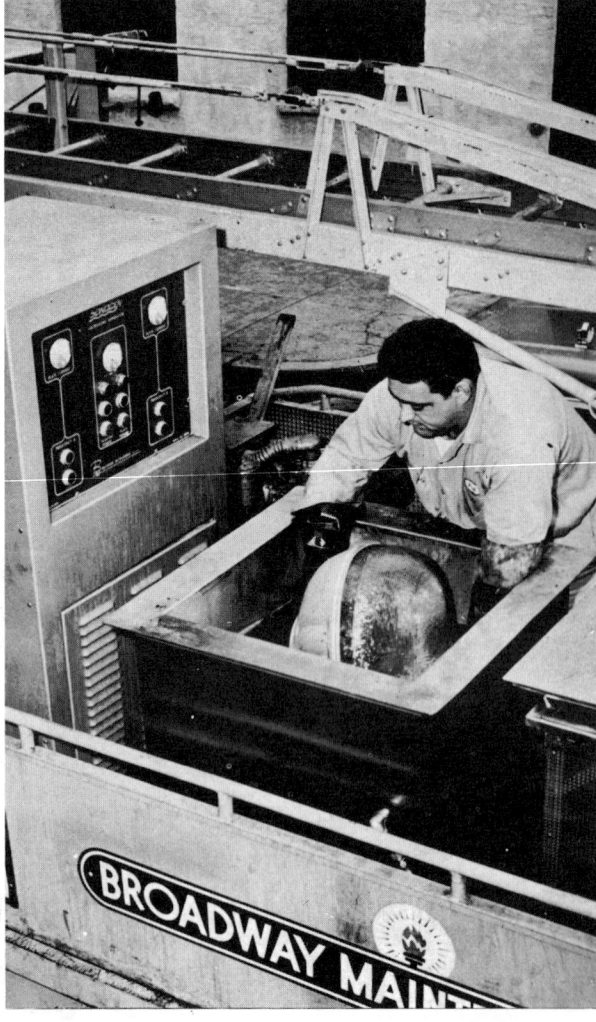

Fig. 29.5 Cleaning with ultrasonic vibrations

The shades of street lights being cleaned by mobile plant in Philadelphia

A normal ear is sensitive to frequencies of sound waves varying from about 20 to 20 000 Hz, and has a maximum sensitivity for about 2000 Hz. Beyond the upper limit the sound is inaudible, but causes acute pain in the ear.

Ultrasonics

Ultrasonics is the study of sounds beyond the range of the human ear. These very rapid compressions and rarefactions are generated by means of a transducer, which is a type of loudspeaker unit or a vibrating crystal energized by an oscillatory electric circuit. Ultrasonic waves can be transmitted through solids and through liquids, and have special properties, for example, they can be used for cleaning purposes and as flaw detectors and thickness gauges (Figs. 29.4 and 29.5).

The use of two ears: the binaural effect

With the help of two ears we judge the direction of a sound source. The head is turned in the correct direction by balancing the pressures of the sound waves on the two ears. Early sound direction indicators consisted of two long horns with the wide ends a few metres apart and the narrow ends fitting in the ears. This was simply an arrangement for artificially increasing the distance between the ears so as to amplify the pressure difference.

The velocity of sound

Direct measurement of the velocity of sound is today a fairly straightforward experiment. It involves a method of accurate timing and, with

distant explosion and hearing it. Sound travels at 330 m/s in air. Thus, if the sound of an explosion is timed over a distance of about 30 km, the interval is long enough to be measured with reasonable accuracy. As light travels at a speed of 3×10^8 m/s, the time it takes to travel a few kilometres can be neglected compared with that taken by sound.

This simple method is open to many errors, in particular that due to the wind. A system of reciprocal timing was introduced to minimize this. A gun was fired at one post, and when the explosion was heard at a second post some kilometres distance, a second gun was fired. When the sound from this was received at the first post, the time between the firing of the first gun and the reception of the sound from the second gun, gave the time for sound to travel over the double distance. The effect of a *steady wind blowing from one post to the other* was minimized.

The accuracy of such simple experiments as the above depends largely on the quickness of the observer to respond to what he sees or hears. This we call the 'personal equation' of the observer.

Modern methods detect the sound by using microphones which eliminate such errors.

A simple and quick laboratory determination of the velocity of sound in air uses a small speaker connected to an audio-frequency generator (an electronic instrument which provides a pure note of known frequency). The sound is received with a microphone connected to an amplifier and second speaker. This is arranged as shown in Fig. 29.6. By suitably arranging the source (first speaker) stationary waves can be set up and the positions of maximum and minimum amplitude

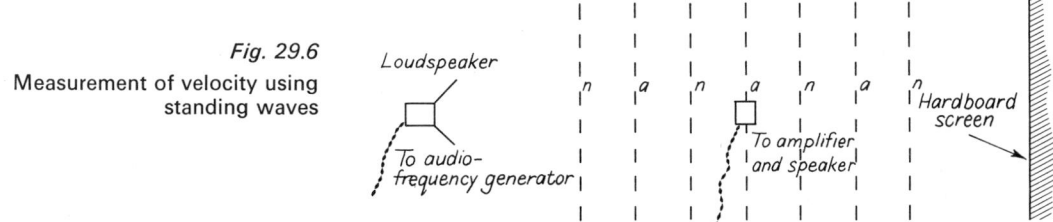

Fig. 29.6

Measurement of velocity using standing waves

modern electrical instruments, this is comparatively simple.

Before the advent of electrical timing instruments the velocity of sound was measured by determining the time interval between seeing a

(or the nodes and antinodes) can be located by moving the microphone along a direction perpendicular to the screen. By measuring the distance between the antinodes, the wavelength of the source is found and hence the velocity.

Fig. 29.7 Sound waves used to control ship's position while drilling

In order to keep ship in fixed position while deep-sea drilling, two beacons placed on sea-bed send signals which are received by hydrophones and fed into a computer which automatically monitors the ship's position

Wavelength $\lambda = 2 \times$ distance between antinodes
velocity $= f \times \lambda$.

The velocity of sound in water has been accurately determined, and is used in the *Asdic* device for determining the depth of the sea, shoals of fish, etc. A sound wave sent out from the ship's hull is reflected from the bed of the sea and received by a microphone, also under the hull. The time taken gives a measure of the depth when the velocity of sound in the water is known. Many ships have such appliances which can be made to record automatically the depth on a dial (*see* Fig. 29.7).

Effect of temperature on velocity

Newton deduced a mathematical expression for the velocity of sound in a gas. This formula shows that the velocity depends on the square root of $\dfrac{\text{pressure}}{\text{density}}$ for the medium. But this fraction varies as the absolute temperature, hence the velocity varies directly as the square root of the absolute temperature.

$$\frac{V_1}{V_2} = \sqrt{\frac{T_1}{T_2}}$$

where V_1 is the velocity of sound at T_1 K, and V_2 is the velocity of sound at T_2 K, i.e. the velocity increases with increase in temperature.

Effect of wind

Sounds can be heard better when the wind is blowing in the same direction as the sound is travelling. The wind velocity increases upwards from the ground, and the sound waves are bent

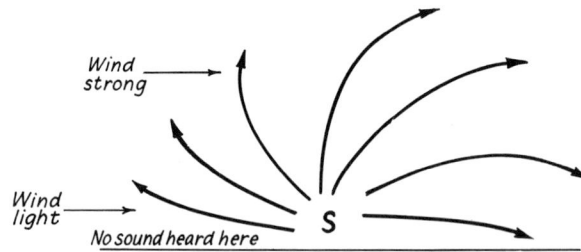

Fig. 29.8 Diagram to show the effect of wind on the direction of sound waves

downwards as shown in Fig. 29.8. With the wind against the sound, the waves are bent upwards.

Properties of sound waves

We have already mentioned (Chapter 20) that sound, like light, is transmitted by means of waves. But sound waves differ fundamentally from light waves. They are longitudinal waves and a material medium is necessary for their propagation. In other words, sound cannot pass through a vacuum. This can be shown by suspending an electric bell inside a bell jar on the plate of an air pump (Fig. 29.9). If the bell is ringing, the sound will become fainter and fainter as the bell jar is evacuated, until it is scarcely audible. A little sound is conducted through the connecting wires and the supporting cord. If the latter is made of rubber, it minimizes the sound conducted.

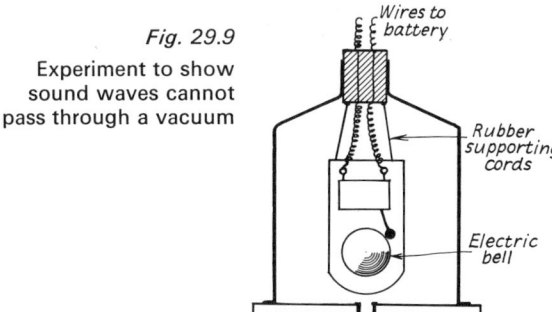

Fig. 29.9

Experiment to show sound waves cannot pass through a vacuum

Fig. 29.10

Reflection of sound waves round Whispering Galley of St Paul's Cathedral

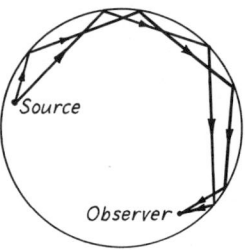

Reflection

Echoes are produced by the reflection of sound waves. Thunder is caused by many reflections in the clouds of sound resulting from the lightning spark. Reflection at a concave surface brings sound waves to a focus. Architects, when designing large halls or theatres, must avoid concave ceilings or alcoves where these may focus the sound in one spot. The acoustical properties of a hall can be investigated before it is built by examining reflections using a sectioned model in the ripple tank.

A whisper can be heard all round the Whispering Gallery in St Paul's Cathedral because of the reflections of the sound at the wall (Fig. 29.10).

Refraction

Refraction of a wave is produced when it passes from one medium to another in which its speed is different. Sound travels with very different speeds in different gases. The usual experiment to show refraction of sound employs a balloon filled with carbon dioxide which, acting like a convex lens of light, focuses the sound (Fig. 29.12).

Sound waves can pass through solids and liquids;

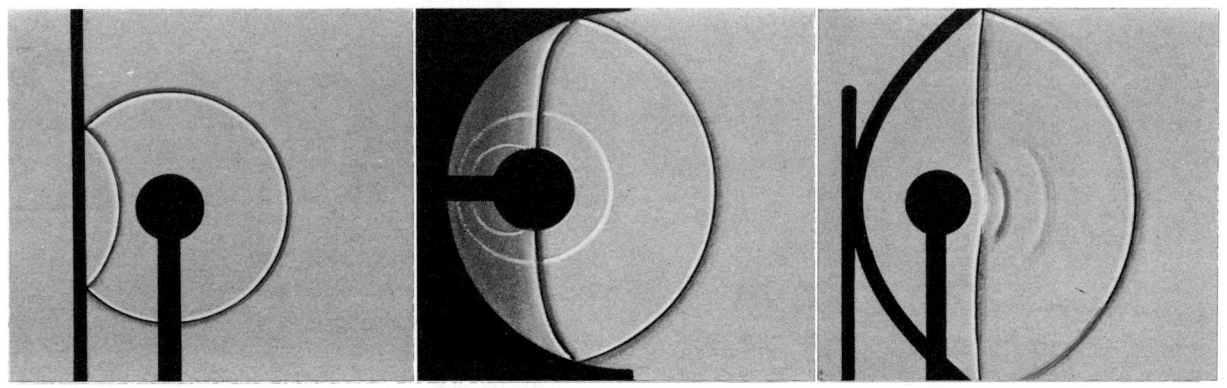

| (*a*) (after Foley) | (*b*) (after Davis) | (*c*) (after Foley) |
| Reflection at a plane surface | Reflection at a spherical surface | Reflection at a parabolic surface |

Fig. 29.11 Spark photographs of sound pulses

(*a*) The spherical wave from the source (black circle in photograph) is reflected by the plane surface as a spherical wave which appears to come from an image point behind the mirror

(*b*) The spherical wave from the source is reflected by the spherical mirror. The reflected wave is plane near the principal axis, but curved a little way from it

(*c*) The spherical wave, after reflection by the parabolic mirror, is plane. These reflections of sound waves can be compared with those of light waves (*see* Chapter 23)

in fact, the speed of sound in these media is greater than in a gas. If the ear is held against a very long iron pipe which is struck at a point some distance away, two distinct sounds may be heard: one which has travelled through the pipe, and a later one which has travelled through the air. A sound wave falling upon the boundary of a solid (less dense medium) may, therefore, suffer total internal reflection. The critical angle for most solids is small. The sound in a speaking tube easily passes round the bends by undergoing repeated total internal reflections.

Photography of sound waves

The spark produced by an electrical discharge gives rise to a momentary sound caused by the compression wave which radiates from the source. The compressed air has a slightly different refractive index and, with suitable illumination, can be photographed to give an instantaneous picture of the sound wave (Fig. 29.11).

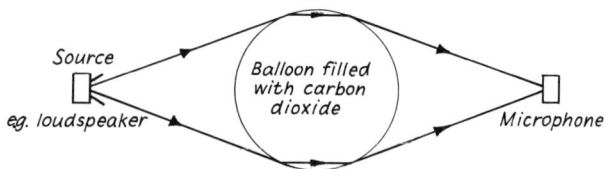

Fig. 29.12 Refraction of sound waves

Effect of motion on pitch

We notice that the pitch of the warning siren of a fast-moving vehicle rises as it approaches and then falls as it recedes. The waves emitted are crowded together by the approaching source so that more pass the observer per second, that is, the frequency is increased. Just the opposite occurs when the source is moving away from the observer. A similar effect is produced when the observer is moving towards or away from the source of sound. It is called the Doppler effect (*see* pages 189 and 245).

QUESTIONS

[Velocity of sound in free air = 332 m/s.]

1. (i) What distinguishes a noise from a musical sound? (ii) State the three characteristics of a musical note.

2. If the sound of thunder is first heard 20 seconds after the flash of lightning is seen, how far away is the storm? Explain your method of working.

3. Describe the motion of the particles of air between a loud-speaker, which is giving out a continuous steady note of constant pitch, and an observer.

How would this motion alter if (*a*) the pitch of the note were raised one octave; (*b*) the note were louder; (*c*) the temperature of the air increased? (JMB)

4. Describe an experiment to verify the relation between the length and the frequency of vibration of a stretched wire.

How would you make a wire vibrate so that it emits a note whose frequency is double that of its fundamental?

Describe the method of vibration of the wire and give diagrams to illustrate your answer. (JMB)

5. Explain why a card held against the spokes of a revolving bicycle wheel emits a musical note. How would you determine its frequency using a sonometer and tuning-fork of known frequency?

A card is fixed to the frame of a bicycle so that it touches the spokes as the wheel revolves. Find the speed in km/h at which the bicycle must be ridden to produce a note whose frequency is 96 Hz if the wheel has 48 spokes and its overall diameter is 70 cm. (L)

6. Explain how sound is transmitted through the air.

Describe an outdoor method for finding directly the velocity of sound in air, pointing out any precautions which are necessary.

What will be the effect on the velocity, of (i) a rise in temperature, and (ii) replacing the air by a denser gas? *A* and *B* are two observers in a line between a church bell and a wall. *A* hears the echo from the wall 1.00 s after hearing the bell while *B* hears it 0.20 s after the bell. Find the distance between *A* and *B*. (L)

7. Describe and explain *one* experiment to show that sound is a type of wave-motion and *one* to show that a material medium is necessary to transmit sound.

How would you measure the wavelength in air of the note of a tuning-fork? Give details and draw a diagram. (O & C)

8. State the laws of reflection of sound and explain how an echo may be produced.

Two men stand facing each other, 200 m apart, on one side of a high wall and at the same perpendicular distance from it. When one fires a pistol the other hears a report 0.60 s after the flash and a second report 0.25 s after the first. Explain this, and calculate (*a*) the velocity of sound in air, (*b*) the perpendicular distance of the men from the wall. Draw a diagram of the positions of the men and the wall. (O & C)

9. Explain the terms *frequency*, *wavelength* in connection with a pure musical note and find how they are related to the velocity of sound in a medium.

Describe how you would prove in the laboratory that two

sounds of different frequency travel in air with the same velocity of approximately 300 m/s.   (O & C)

10. How are echoes produced? Explain how you would use the formation of echoes to obtain a value for the velocity of sound in air.

A man standing some distance in front of a flight of stone stairs may, if it is quiet, hear a faint ringing note when he claps his hands. Give an explanation of this and make an estimate of a frequency he hears if the width of the step is 22 cm and the velocity of sound in air is 330 m/s.   (O & C)

11. Explain how sound waves travel through a gas and describe carefully an experiment which shows that a material medium is necessary for the transmission of sound.

A and B are two observers 1 km apart. There is a steady wind blowing. When a gun is fired at A the time interval between the flash and the report observed at B is 3.04 s. When a gun is fired at B the interval between flash and report observed at A is 2.96 s. Calculate the velocity of sound in air and the velocity component of the wind in the direction BA.   (O & C)

12. Derive by simple reasoning the relation between the velocity of a wave, the frequency, and the wavelength. Explain the ways in which the sound of one musical note can differ from the sound of another, and indicate the physical causes of the differences.

A man standing close to a long fence of narrow vertical posts 27 cm apart claps his hands, causing a single compression to spread outwards. A small portion of the compression wave is reflected back to him by each post in succession. If these reflections arrive at such a rate as to cause him to hear a note of frequency 600 Hz, what is the velocity of sound in air on this occasion?   (O & C)

13. State the characteristics of a musical note on which its *pitch*, *loudness* and *quality* depend.

Describe a simple method of measuring the frequency of a tuning-fork.

A lighthouse sends simultaneous signals to a ship by sound waves *in air* and *under water*. Both signals are heard on the ship and they arrive 4 s apart. Calculate the distance of the ship from the lighthouse, given that the velocities of sound in air and in water are 3.31×10^2 m/s and 14.00×10^2 m/s respectively.   (O & C)

14. Explain what is meant by *longitudinal waves*, *transverse waves*, and give one example of each.

Make a freehand drawing in pencil of a sine wave showing two complete waves. Taking this as a graph of pressure against time for a pure tone of frequency 1000 Hz, draw on the same diagram in ink (full line) the graph for a pure tone of about the same loudness but an octave higher in pitch: and also in ink (dotted line) the graph for a much louder note of frequency 1000 Hz. Indicate the time scale used, on the x axis.   (O & C)

15. Define the term *frequency* with reference to a tuning-fork and with reference to the sound wave it produces in the air. Prove the relation between the frequency, wavelength and speed of travel of a sound wave.

A small source of wave-motion sends out waves from the bottom of a tank of oil. Deduce the change in wavelength which occurs, when the waves emerge into the air, for the case of (a) sound waves, (b) light waves. Explain the fact that total internal reflection could occur for the light waves; but not, with air above the tank, for the sound waves.

[Refractive index of the oil = 1.5; speed of sound in the oil is 4 times the speed of sound in air.]   (C)

16. Two loud-speakers connected to the same radio receiver are emitting a note of fixed frequency. Explain why the sound picked up by a microphone may be less intense than it would be if one of the loud-speakers were completely enclosed in a sound-absorbing box.   (O & C)

17. Two small sound transmitters, A and B, send out identical notes to a point C which is not equidistant from them. What are the conditions that C will receive (a) no sound at all, (b) sound much louder than would be given by A or B separately?   (C)

18. Two small identical loud-speakers are set up on posts in a field so as to face one another. They are wired in series to an a.c. generator so as to emit the same pure tone. A microphone which is moved along the line joining them indicates loud and quiet points. Explain this and show how the wavelength of the emitted note can be determined.

Suggest a path along which the microphone could be moved so that the note picked up would not fluctuate in loudness. How, if at all, would the loudness vary?

Explain why the experiment could not be done in an ordinary laboratory; and state what other fact you would need to know, to determine the velocity of sound in air.   (O & C)

30 Vibrations of strings and air columns

The vibration of strings

We are all very familiar with the production of sound by a stringed instrument. The symphony orchestra has an array of 'strings', from the violins to the double-bass, and the dance band uses the guitar, etc. These sounds are produced by the creation of stationary waves* in the strings.

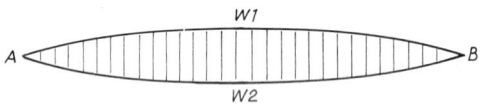

Fig. 30.1 Vibration of a stretched string

Stationary or standing waves in strings

If a wire is tightly stretched between two supports and then plucked, the crest extends the whole distance between the supports, i.e. this distance is half the wavelength of the transverse wave in the wire (Fig. 30.1). At each end, the waveform

---

* For stationary waves refer back to Chapter 20.

suffers a phase change. Crest W_1 by reflection at B becomes trough W_2 and trough W_2 by reflection at A becomes crest W_1.

This simple vibration of a string consists of a transverse wave passing along the string and being reflected at each end in turn. At B the displacements of the incident and reflected waves are always equal and opposite, and so B is stationary.

No energy is radiated from a standing wave, but the vibrating string loses some energy through the supports.

It is perhaps worthwhile to explain the formation of these stationary waves in detail. Let $ABCDE$ (Fig. 30.2b) be the displacement of a waveform travelling along a string attached to a hook at O. No displacement can occur at O, and so the displacement OE which would have occurred must be neutralized by a reaction at the hook. The form of this reaction must be such that it would neutralize the displacements $EFGH$, which

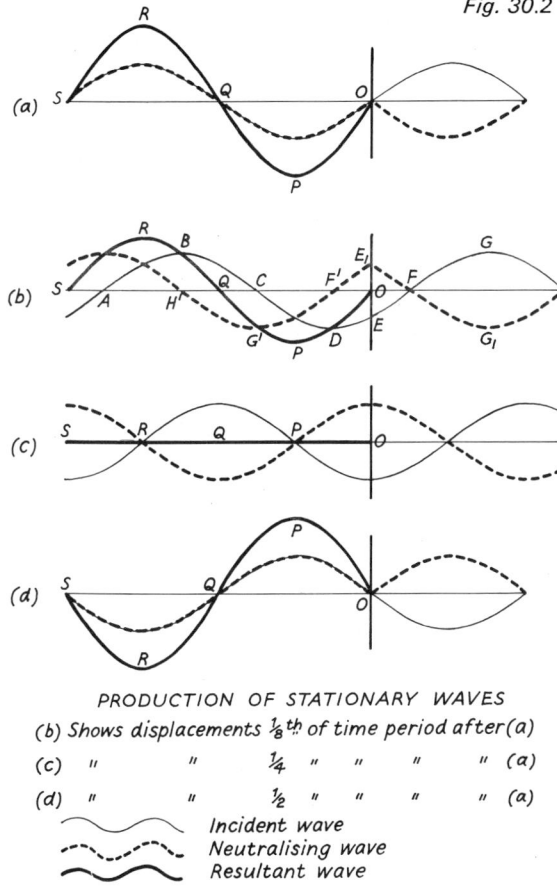

Fig. 30.2

PRODUCTION OF STATIONARY WAVES

(b) Shows displacements $\frac{1}{8}$th of time period after (a)

(c)   "   " $\frac{1}{4}$ "   "   "   " (a)

(d)   "   " $\frac{1}{2}$ "   "   "   " (a)

〜〜〜 Incident wave
------ Neutralising wave
〜〜〜 Resultant wave

the string would have had in the absence of the hook. E_1FG_1H is the necessary compensating disturbance. But the reaction at the hook must emit this disturbance, in both directions, along the string, viz. E_1FG_1H and $E_1F'G'H'$. This disturbance $E_1F'G'H'$ is, in effect, a reflected disturbance, and it combines with the incident disturbance to produce the resultant disturbance $OPQRS$. The incident, neutralizing, and resultant disturbances at different times are shown in Fig. 30.2a, c and d. In all cases it will be observed that the resultant wave has no displacement at O, Q and S, while maximum displacement occurs at P and R. Such a disturbance is said to form *stationary* or *standing* waves.

The points O, Q and S, where no displacements occur, are called *nodes*, while P and R, where maximum displacements occur, are called *antinodes*.

Melde's experiment

With the arrangement shown in Fig. 30.3, the string may be made to vibrate and to produce stationary waves of different wavelengths (*see*

Vibration showing nodes and antinodes

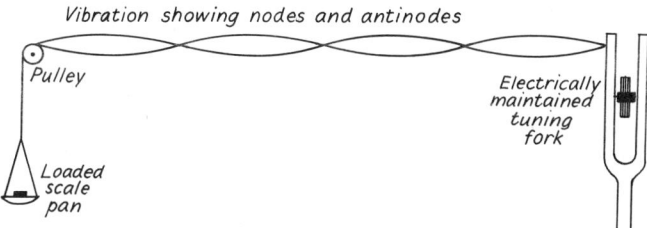

Fig. 30.3 Stationary waves in a string—Melde's experiment

Fig. 30.7). The electrically maintained tuning fork vibrates with a constant frequency. The number of loops in the string is changed by altering the length and tension of the string.

With the fork in the position shown in the figure, displacement at the antinodes is a maximum when the prong of the fork is displaced its maximum amplitude to the left. When the string is straight (undisplaced at all points), the prong of the fork is displaced its maximum to the right. Hence, while the fork is making half a complete oscillation, the string is making only one-quarter of a complete oscillation. The frequency of the stationary waves in the string is half that of the fork.

If the fork is now rotated so that its prongs are along the direction of the string (Fig. 30.4) a

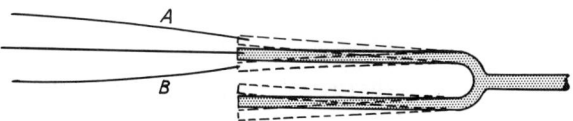

Fig. 30.4 Frequency change on rotation of fork

crest is produced when the prong is displaced to A, and a trough when it is displaced to B. In this case the frequency of the waves is the same as that of the fork. Turning the fork round while it is oscillating doubles the number of loops produced, i.e. doubling the frequency halves the wavelength.

Any type of vibrator can be used. One working on the a.c. mains gives a frequency of 50 Hz and is very suitable. A vibrator (transducer) energized by an electronic oscillator can be used to show how the wavelength of the wave in the string depends on the frequency.

The sonometer

If a thin wire, hanging vertically, is kept taut by a heavy load, it will vibrate when plucked, but the vibrations will not be audible. If it is mounted upon a board then its vibrations are transmitted to the board which, because of its large area, sets a much greater amount of air in vibration; hence, the sound emitted is louder.

For studying the vibration of strings, we employ a sonometer (Fig. 30.5), which consists of a board or box about 1 m long equipped with two wires which are stretched between pegs. The tensions

Tuning is best done by ear, but for those who have not the ability to do so, it can be achieved by placing a light, paper rider on the centre of the string. If the base of a sounding fork is now pressed against the sonometer board, the paper rider will vibrate freely and may be thrown off the string when the string and fork are in tune. Now take the next highest fork and decrease the distance between the bridges until the wire is in tune with this fork. Continue for all the forks, finding the length for each frequency, keeping the tension the same throughout.

Tabulate the results as shown and note that *the frequency* (stated on each fork) *is inversely proportional to the length of the wire.*

$$\text{Frequency} \propto \frac{1}{\text{length}}$$

frequency × length = constant

Fork	Frequency of fork	Length of wire	Freq. × length = constant

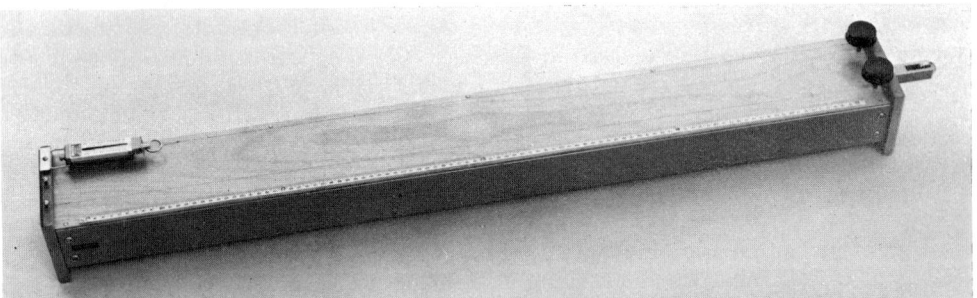

Fig. 30.5
A sonometer

in the wires are adjustable, and one wire has one of its ends attached to a spring balance* to enable the tension in it to be measured. The effective length of the wires can be altered by means of movable bridges.

Factors influencing pitch or frequency

1. *Length.* By altering the tension, tune a sonometer wire to the lowest of a series of tuning forks and measure the length between the bridges.†

---

* An alternative method is to pass the free end over a pulley and to load it with known weights.

† This length is the distance between two nodes and, hence, $\lambda = 2l$.

2. *Tension.* As in the previous experiment, tune the sonometer wire to the forks of known pitch, this time by altering the tension while keeping the length constant. Tabulate the results and note that *the frequency is proportional to the square root of the tension*, i.e. to double the pitch, the tension must be increased four times.

$$\text{Frequency} \propto \sqrt{(\text{Tension})}$$

Fork	Frequency of fork	Tension	√(Tension)	$\frac{\sqrt{(\text{Tension})}}{\text{Frequency}}$

3. *Type of Wire.* By replacing one of the sono-meter wires, first with one of the same gauge but of different material, and then with one of similar material but of different gauge, it is found that the frequency varies as the square root of the density × cross-section of the wire.

$$\text{Frequency} \propto \frac{1}{\sqrt{(\text{density} \times \text{cross-section})}}$$

$$\propto \frac{1}{\sqrt{(\text{mass per unit length})}}$$

We thus find experimentally,

$$\text{frequency} \propto \frac{1}{\text{length}} \sqrt{\left(\frac{\text{tension}}{\text{mass per unit length}}\right)}$$

It can be shown theoretically that the velocity of a disturbance along a stretched string is given by

$$v = \sqrt{\frac{T}{m}}$$

where T = tension measured in newtons, m = mass per unit length measured in kg/m

but,
$$v = f\lambda$$

hence,
$$f = \frac{1}{\lambda} \sqrt{\frac{T}{m}}$$

Further experiments with the sonometer: overtones

Pluck a sonometer wire and then gently touch it with a light object, e.g. a piece of string or a

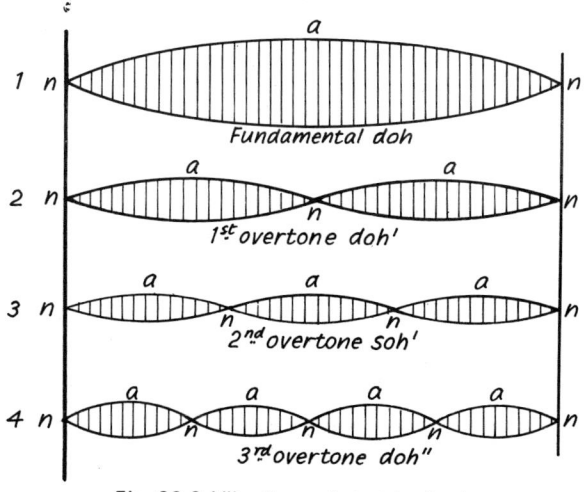

Fig. 30.6 Vibrations of stretched string

feather, exactly at its centre point. Listen care-fully and you will hear a note an octave higher than the original. The wire then vibrates in two sections (Figs. 30.6). Repeat the experiment, this time touching the wire one-third the way along. A higher note is now produced. Where the wire is touched, it ceases to vibrate, forming a node.

The higher notes are called the *overtones* of the lowest note, the *fundamental.*

Hence, for a string sounding its fundamental when $\lambda = 2l$, we have,

$$f = \frac{1}{2l} \sqrt{\frac{T}{m}}$$

For the first overtone,

$$f_1 = \frac{2}{2l} \sqrt{\frac{T}{,m}}$$

and for the second overtone,

$$f_2 = \frac{3}{2l} \sqrt{\frac{T}{m}}$$

Quality. We have already mentioned that the quality of a sound is due to the presence of addi-tional higher notes. These are overtones, and give to each instrument its characteristic quality. When a string vibrates it does so in a complicated

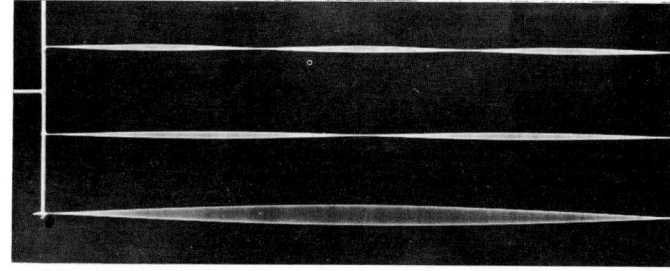

Fig. 30.7 Photograph of vibrating strings
Different tensions cause different modes of vibration

manner, and some of these quicker vibrations are added to its main vibration (Fig. 30.7). A good player of a stringed instrument is able to bow the strings of his instrument so that certain overtones are brought out to give richness to the fundamental note.

We shall see later that overtones are also present when air columns vibrate, as in organ pipes and wind instruments.

In Chapter 31 a simple method of analysing sounds is given, and the presence of overtones is shown in the form of the vibrations.

Vibration of air columns

Stationary waves in air columns

We have all produced a sound by blowing across the mouth of a test-tube and probably noticed, also, that a higher note is obtained with a shorter tube.

To understand the production of a note in a tube, perform the following simple experiment.

AB (Fig. 30.8) is a glass tube dipping into a jar of water, and *F* a vibrating tuning fork held over the upper end *A*. The effective length of *AB* is

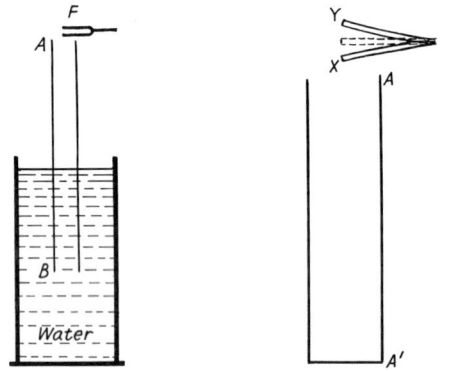

Fig. 30.8 Arrangement of apparatus for resonance experiment

Fig. 30.9 Stationary wave in a tube

altered by raising or lowering it. While keeping the sounding fork near the end *A*, some position can be found when the tube *responds* to the vibration of the fork and emits a loud note. This effect is called *resonance*.

When the prong of the fork moves from *Y* to *X* (Fig. 30.9) a compression wave is sent down the tube. This is reflected at the closed end *A'* and returns to *A*. As it emerges from the end, a rarefaction starts down the tube, as it does when a cork is pulled out of a bottle. If this occurs as the fork sends out a rarefaction (i.e. when the prong is moving from *X* to *Y*), the effect of the fork is amplified and the tube sings out. Thus, the vibrations of the air will be in sympathy with those of the fork if the sound wave takes the same time to travel down the tube and back as the fork does to execute half a complete vibration. A stationary wave is set up in the tube. The air in

contact with the closed end cannot vibrate freely, and so is a node. At the open end maximum vibration occurs, and at this end is an antinode. The vibration can be represented diagrammatically as in Fig. 30.10a. It will be seen that the length of

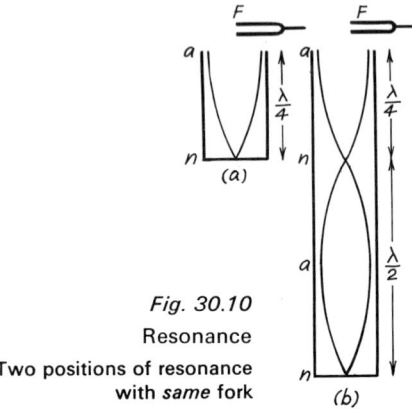

Fig. 30.10

Resonance

Two positions of resonance with *same* fork

the tube is one-quarter of a wavelength. If the tube is long, and the fork of high frequency, a second position of resonance can be found. As the closed end must be a node, the vibration is then represented by Fig. 30.10b. This experiment can be used to measure the velocity of sound, for $v = f\lambda$.

Overtones in closed pipes

The fundamental of a closed pipe has a wavelength four times the length of the pipe (Fig.

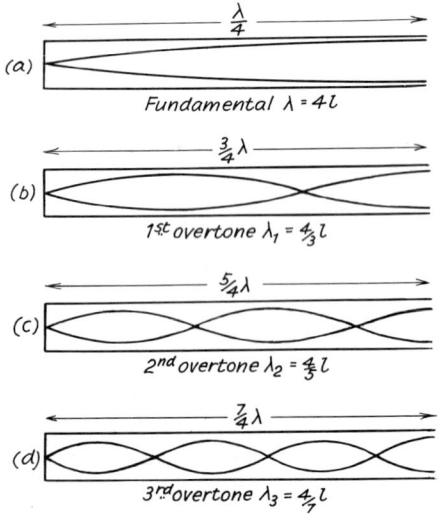

Fig. 30.11 Overtones produced in a closed pipe

30.11*a*). The pipe will also resound to higher frequencies for vibrations such as (*b*), (*c*) and (*d*), which are the first, second and third overtones respectively.

Overtones in open pipes

With an open pipe, both ends must be antinodes, and the fundamental has then a wavelength equal to twice the length of the tube. The first three overtones are shown diagrammatically in Fig. 30.12.

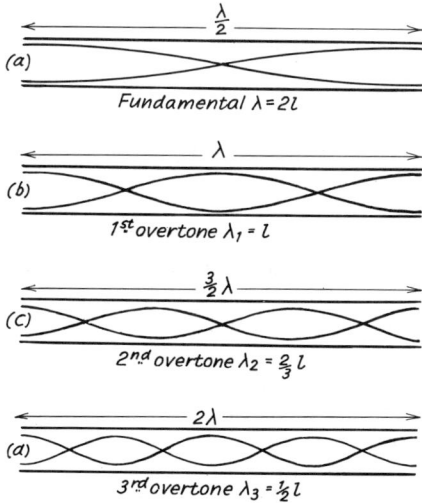

Fig. 30.12 Overtones produced in an open pipe

It will be noticed that more overtones are possible with open pipes than with closed pipes. Because of this, the quality of the sound emitted by closed pipes is not as rich as that emitted by open pipes.

Beats

When two wave disturbances are received simultaneously at the ear, the resulting displacement is the sum of the separate displacements. For two waves of frequencies 260 and 263 Hz, the disturbances will be in phase three times in each second. Midway between these states, the motions will be out of phase. Assuming the two amplitudes to be equal, the resultant displacement will be a maximum (double that of the individual displacements) three times each second, and it will be zero three times each second at intermediate times. This variation in the amplitude causes a waxing and waning of the intensity (loudness) of the sound—three times per second in the case quoted. These volume maxima are called *beats*. They are easily demonstrated by using two similar tuning forks. If these are struck and then held on a sounding board, their combined effect will be similar to, but louder than the individual effects. Repeating the experiment after loading one of the prongs of one fork with soft wax, which lowers the frequency of the fork, produces beats which are quite easily discerned.

A more elaborate method of producing beats is to employ two audio-frequency oscillators connected to two loudspeakers (or to the same speaker). By adjusting the difference in the frequencies of the oscillators to be only a few cycles per second, this difference is clearly heard as beats.

A musician, in tuning his instrument, listens for beats and he produces accurate tuning by eliminating them.

Fig. 30.13 shows the addition of the displacements of two waveforms to give a resultant form showing beats. Such a waveform can be produced on a cathode ray oscilloscope by feeding in the output from two oscillators as mentioned above.

Fig. 30.13
Production of beats

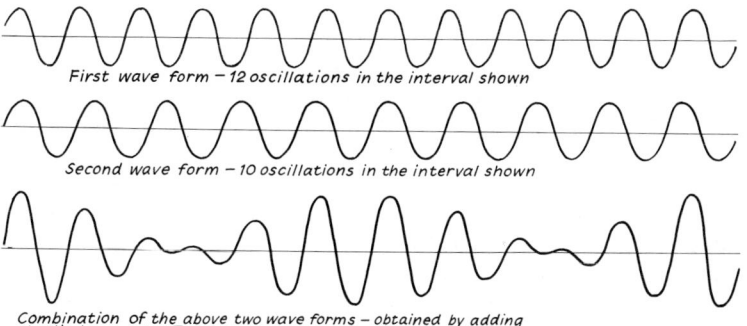

First wave form – 12 oscillations in the interval shown

Second wave form – 10 oscillations in the interval shown

Combination of the above two wave forms – obtained by adding the displacements. Two beats produced in the interval shown

Stringed instruments

All stringed instruments have their strings, made of catgut, nylon or metal, mounted on a sounding chamber, usually a wooden box of special design. The string is plucked or bowed and emits a note with a pitch dependent upon its length, tension and mass. To make the note audible, the vibrations are conveyed to the soundbox by way of a bridge over which the wires pass. The violin has a sound-box of characteristic design. The front or belly is of soft wood and the back of hard wood. The bridge rests upon the belly. The soundbox has a natural frequency and resounds strongly to a note of this frequency. This is quite undesirable, for equal resonance should be obtained for all frequencies. To produce uniform resonance, a small wooden peg, called the sound post, is wedged inside the soundbox between the front and back. This prevents excessive resonance for the natural frequency of the soundbox. To play different notes, the violinist shortens the effective length of the strings by pressing them down, with his fingers, on the finger-board.

The viola, 'cello and double-bass are large models of violins, and cover a lower range of notes than is possible on the violin.

The piano is the most common stringed instru-ment. For each note it has one or more wires which are tuned by adjusting the tension. All the wires are stretched across an iron frame which is fastened to a soundboard. Striking a key causes a padded hammer to strike the particular wire and so to vibrate it.

Wind instruments

The organ

Fig. 30.14 shows a simple open flue pipe. It has a fundamental of definite pitch, and to get it to resound or 'speak', vibrations of this particular frequency have to be supplied to it. A stream of air is sent into the pipe and impinges upon a vibrator which may be of wood or metal. The vibrations of this cover a range of frequencies, which includes the fundamental frequency of the pipe. Thus the pipe singles out its natural frequency from the mixture of frequencies supplied by the vibrator.

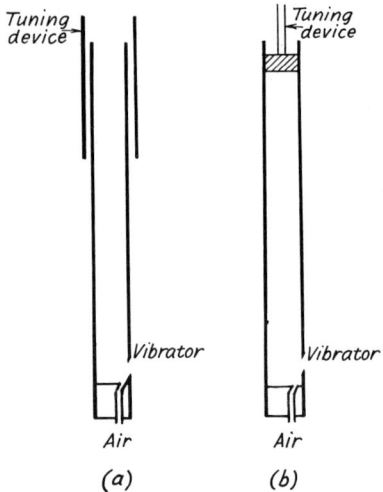

Fig. 30.14 Organ pipes (a) open, (b) closed

Woodwind

The woodwind section of the orchestra includes the flute, clarinet, oboe and bassoon. The flute is a cylindrical tube and is played by blowing across a small hole. A mixture of frequencies is supplied to the tube by the vibrating lips disturbing the air, and the flute responds to a particular frequency. The effective length of the tube is decreased by successively opening holes covered by padded keys. Over-blowing causes the flute to sound its first overtone and, being a pipe open at both ends, this is the octave of the fundamental.

The clarinet is a conical tube and is sounded by a reed which is vibrated by blowing.

The oboe and its bass counterpart, the bassoon, are sounded by blowing through double reeds. The piercing quality of the notes of the oboe is due to the strength of a large number of high over-tones. The flute, on the other hand, gives a much purer note with a few high overtones present.

Brass

Brass instruments like the trumpet, horn and tuba, are metal tubes coiled for convenience. The lips are vibrated in the mouthpiece and the tube resounds to a particular frequency. The length of the tube is varied by introducing extra lengths of tube by means of valves, usually three in number. The length of the trombone tube is altered by a slide—one tube sliding in another.

The brass instruments make great use of

overtones. The bugle is a tube of fixed length and the notes obtained with it are all overtones. The trumpet gives the additional notes by altering the length of the tube with the valves.

The musical scale

When two notes are sounded together the effect may be pleasant or unpleasant. We find that if the ratio of the frequencies of the two notes is a simple fraction, they harmonize when sounded together. The octave of a note has double the frequency of the note.

C and G give pleasant harmony when sounded together. The frequency of G is 3/2 that of C.

F (Fah) has a frequency 4/3 that of C (Doh)
E (Me) has a frequency 5/4 that of C
A (La) has a frequency 5/3 that of C

Note	Doh C	Me E	Fah F	Soh G	La A	Doh' C'
Frequency of note compared with that of Doh	1	$\frac{5}{4}$	$\frac{4}{3}$	$\frac{3}{2}$	$\frac{5}{3}$	2

In this way the musical scale has been evolved.

These notes were insufficient for the musicians, and so others were added in the gaps between C and E, and A and C' to give the eight-note scale shown below.

The interval between two notes is the ratio of their frequencies. The intervals shown in the musical scale fall roughly into two classes, one twice the value of the other. The former is called a *tone* and the latter a *semitone*. To get a similar succession of intervals when starting a scale on G, an extra note between F' and G' is required. This is called F sharp. For other scales all the tones

have to be interspaced with semitones. These extra notes are the black keys on the piano. There is none between E and F, and B and C', as these intervals are only semitones in the natural scale.

Scales built up assuming the 9/8 and 10/9 intervals to be equal are not quite the same as the natural scale with the succession of intervals shown above. To average out the discrepancies, a piano is tuned so that the intervals between successive notes are all equal. This is called the scale of 'equal temperament'. There are twelve intervals in an octave and therefore the common interval is $^{12}\sqrt{2}$ or 1.059.

$$\frac{C\#}{C} = \frac{D}{C\#} = \frac{D\#}{D} \text{ etc.} = 1.059$$

Analysis of sound vibrations

We have already seen that a pendulum-like vibration gives rise to a sound wave by making the air molecules vibrate similarly. The mode of

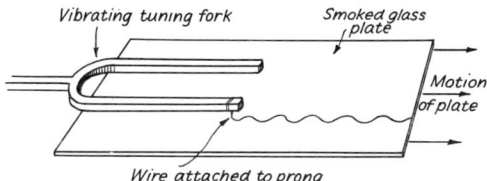

Fig. 30.15 Vibration of prong of tuning-fork revealed by trace on moving smoked glass

vibration of the prongs of a tuning fork can be shown by attaching a short length of wire to the end of one prong and letting it touch a smoked glass plate. If the plate is held still, the wire will scratch a short line on it. But when the plate is moved quickly across the wire, a wavy trace, as shown in Fig. 30.15, is obtained. A similar trace can

Note	Doh C	Ray D	Me E	Fah F	Soh G	La A	Tee B	Doh' C'
Frequency of standard pitch/Hz	264	297	330	352	396	440	495	528
Frequency of equal temperament/Hz	261.6	294	330	349	392	440	494	523.2
Interval compared with Doh	1	$\frac{9}{8}$	$\frac{5}{4}$	$\frac{4}{3}$	$\frac{3}{2}$	$\frac{5}{3}$	$\frac{15}{8}$	2
Interval between successive notes		$\frac{9}{8}$ tone	$\frac{10}{9}$ tone	$\frac{16}{15}$ semi-tone	$\frac{9}{8}$ tone	$\frac{10}{9}$ tone	$\frac{9}{8}$ tone	$\frac{16}{15}$ semi-tone

be obtained on a strip of paper by attaching a light, inked brush to a pendulum.

Such is the vibration of a pure note. But musical instruments obtain their quality by adding overtones. What, then, is the resulting type of vibration? Let us first see the effect of adding vibrations of an octave overtone to the fundamental. Curve (a), Fig. 30.16, shows the fundamental vibrations, curve (b) the first overtone vibrations, and curve

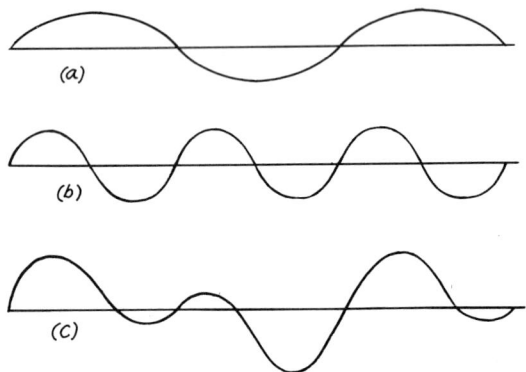

Fig. 30.16 Resultant vibration produced by two vibrations of different frequencies

Curve (c) is obtained by adding the displacements of curves (a) and (b)

(c) the combination of the two obtained by adding the displacements and paying due regard to their signs.

A loudspeaker can easily be modified to make a moving-coil oscillograph by attaching some device to magnify and make visible the oscillations of the coil. The usual method is to convey the oscillations of the coil to a small pivoted plane mirror so that it is rocked about a horizontal

axis. The vibrations are studied using a ray of light reflected from the mirror on to a screen. To observe the form of the vibrations, the spot of light must be moved horizontally across the screen. This is done by reflecting the light at a rotating mirror.

Sound wave analysis is most easily accomplished by using a cathode ray oscillograph. The electrical pulses from a microphone are fed into the oscillograph and the waveform is produced on the screen.

Fig. 30.17 Sound curve of the simple tone from a tuning-fork

The note is of frequency 256 hertz and the dots indicate intervals of 1/100 second

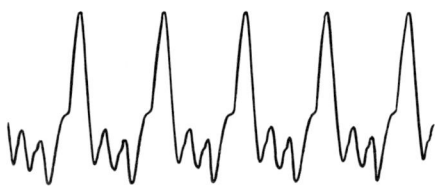

Fig. 30.18 The sound curve of a saxophone

The trace, which might be that of a tuning fork (Fig. 30.17) or of any musical instrument sounding a constant note (Fig. 30.18), can be made stationary on the screen. The trace of an orchestra playing a piece of music passes across the screen, but a photographic record of this can be made. Likewise the human voice can be analysed and so the quality of a singer's voice can be investigated scientifically.

QUESTIONS

1. Two strings of the same material and length, weighing 1.00 and 2.00 g respectively, are kept taut under the same tension. If the lighter string emits a note of frequency 282 Hz, what will be the pitch of the heavier string?

2. The fundamental frequency of a string is 144 Hz. What will be the frequency of its third overtone?

3. Describe a sonometer. How may it be used to compare the frequencies of two notes?

Two cog wheels, A and B, in a piece of machinery, revolve at 15 and 8 revolutions per second respectively. A has 28

teeth, and a card held against it emits a note of unison with 16.8 cm of the sonometer wire. When the card is transferred to B, the length of the wire has to be changed to 25.2 cm for unison. Calculate (a) the frequency of the card when touching A, (b) its frequency when touching B, (c) the number of teeth on B.   (L)

4. How does the frequency of vibration of a stretched string depend on (a) its length, and (b) the tension?

Describe an experiment to verify the first of these statements.

The frequency of a note given by a stretched string is 256 Hz. Calculate the new frequency if (a) the length is reduced by a third, and (*b*) the tension is then increased by a third. (Give the result in (*b*) to nearest integer.)　　(L)

5. Describe the experiment which you would carry out to determine the relation between the tension in a stretched string and the length of the string, when the frequency of the note emitted is constant. What result would you expect to get?

A sonometer wire has a frequency of 320 Hz when the distance between a fixed bridge, *A*, and a movable bridge, *B*, is 80 cm. Through what distance and in which direction must *B* be moved to tune the wire to a frequency of 256 Hz, if the tension in the wire remains unchanged?　　(JMB)

6. How would you tune a sonometer wire so that it gives the same note as a tuning-fork and how would you show that you have succeeded?

A wire of fixed length is correctly tuned to various forks by altering its tension, with these results:

Frequency of fork, n/Hz	256	320	384	512
Tension of wire, T/N	16	25	36	64

Plot n against $\sqrt{T}$ and from the graph find the tension when the frequency is 480 Hz　　(JMB)

7. Write a short account of sound echoes and describe *one* practical use of echo formation.

When a sonometer wire is adjusted to a length of 60 cm and plucked it produces the same note as a tuning-fork of frequency 256 Hz.

What is the frequency of a tuning-fork which would be in tune with the same wire adjusted to a length of 40 cm and at the same tension?　　(JMB)

8. Define *frequency*, *amplitude* and *wavelength*, as used in connection with a sound wave.

Describe an experiment to show how the frequency of transverse vibration of a stretched wire depends on its length. State the result you would expect. What other factors affect the frequency of transverse vibration of the wire?

A length of 28 cm of a wire is in transverse vibration under a tension of 40 N. If the tension is increased to 90 N to what value must the length of the vibrating segment be altered to keep its frequency the same as before?　　(L)

9. Describe an experiment to demonstrate that the prongs of a tuning-fork vibrate when it emits sound and explain the increased loudness of the sound when the stem of the vibrating fork is held in contact with a table.

A tuning-fork has the number 256 stamped on it. What does this signify? A sonometer wire is adjusted to be in unison with this fork. State exactly two different ways in which the wire could be readjusted so as to be in unison with a fork marked 384.　　(L)

10. Explain what is meant by *resonance* and give *three* examples from any branches of physics to illustrate your answer.

Given a sonometer, describe how you would 'tune' the wire to each of a number of tuning-forks of known frequencies by altering the length while keeping the tension constant. What relation would you expect to find? Show how you would set out and use your results to demonstrate this relation.

A sonometer wire in tune with a fork of frequency 320 Hz

is lengthened by 10 cm (tension constant) and is then in tune with a fork of frequency 256 Hz. Find the first length of the wire.　　(L)

11. What do you understand by *frequency*, *musical interval*? Describe an experiment to show how the frequency of the note obtained by plucking a stretched wire depends on its length.

Why does the note given by the wire differ in quality from that given by a tuning-fork of the same frequency?　　(S)

12. What is meant by the term *frequency* as applied to a tuning-fork? Explain the meaning of the term *wavelength* as applied to a note sounding in air.

An organ pipe, open at both ends, is sounding its fundamental note. Draw a diagram to show the position of *one* node and *one* antinode; state what you understand the terms *node* and *antinode* to mean. Calculate the frequency of the note emitted by the pipe, the length of the pipe being 50 cm.

[Neglect end-corrections to the pipe. Speed of sound in air=322 m/s.]　　(C)

13. What is meant by *resonance*? Illustrate your answer by four examples from different branches of physics or from everyday life.

A tuning-fork is struck and held over a tall jar full of water. The water is run out until the note given becomes very loud. Explain, with the aid of diagrams, exactly how the prongs of the fork and the air in the jar are vibrating.

How can this effect be used to determine the velocity of sound in air?　　(L)

14. Define *frequency*, and describe how a sonometer can be used to determine the frequency of a tuning-fork.

The wavelength in air of the sound from a tuning-fork is 170 cm when the velocity of sound in the air is 340 m/s. What is the frequency of the tuning-fork? If the fork is held over an adjustable resonance tube, closed at one end, containing air and the length of the air column in the tube is gradually increased from zero, calculate approximately its length when the tube is first in resonance with the fork. Explain this resonance and why the result is only approximate.　　(L)

15. An open pipe and a closed one emit fundamental notes of the same frequency. Explain the difference in the quality of the sounds made by these two pipes.

16. Explain how echoes arise (in sound), and describe *one* practical application of echo production.

A vibrating tuning-fork is held above the open end of a cylindrical tube the lower part of which dips into water. A loud response is obtained from the tube when the length from the open end to the water is 29.0 cm and again when it is 93.5 cm but not at any point in between. Explain this and calculate the frequency of the fork.

[Velocity of sound in air 330 m/s.]　　(O & C)

17. Explain the meaning of the term *resonance* illustrating your answer by one example taken from mechanics and one from sound. Describe, deriving any formula you employ, how you would make use of resonance to measure the velocity in air of the sound emitted by a vibrating tuning-fork of known frequency. When there is resonance, will the vibrations of the fork die away more or less slowly than when there is not? Give a reason for your answer.　　(O & C)

18. A steel wire of length 60 cm is under constant tension and fixed at both ends. When plucked, it vibrates and emits a fundamental note of frequency 480 Hz. What is the velocity of the waves travelling along the wire? Explain your answer.
(L)

19. (a) What is meant by resonance in sound?

Sketch a diagram of a resonance tube, closed at one end, and mark the position of the nodes and antinodes when it is resonating at its (a) fundamental frequency, (b) second overtone.

State how the frequency is related to the length in each case.

(b) A sonometer wire 100 cm long vibrates with a fundamental frequency of 256 Hz when under a tension of 120 N. Calculate how this frequency may be adjusted to 384 Hz (a) by changing only the length of the wire, (b) by adjusting only the tension.
(JMB)

20. What is meant by frequency, musical interval?

How does the frequency of a note obtained by plucking a stretched wire depend on (a) the length of the wire, (b) its tension?

Describe an experiment to verify the statement given in either (a) or (b).

Why is the intensity of the sound emitted by a plucked wire increased when it is mounted on a board?
(L)

21. Calculate the frequencies of three notes which will harmonize with middle C (264 Hz).

22. Describe the resonating chamber of a violin. Explain the function of the sound post.

23. Explain why the same note played on a flute and on a clarinet sounds different.

24. Describe and explain the character of the vibrations of a loud-speaker diaphragm or the form of the track of a gramophone record.

25. How is sound produced using (a) a violin, (b) an organ pipe, (c) a drum? State in each instance, how the pitch of the note can be raised. Why do the qualities of sounds of the same pitch differ when emitted by different instruments?

Describe, in a general way, how the sound is transmitted from an instrument to the listener.
(L)

Part Nine ELECTRICAL ENERGY

Powerful electromagnet specially designed for the
stacking of steel girders at Dorman Long Ltd

31 Magnetism

Introduction

Most people at an early age know the property possessed by a magnet, of attracting small pieces of iron. It is thought that the Chinese, as long ago as 2000 B.C., were acquainted with this property, which they found was possessed by a natural iron ore. They used the magnetic rock to make a simple form of compass for navigational purposes.

There are several legends connected with the derivation of the word 'magnetism'. Some say it comes from the name of a town in Asia Minor, called Magnesia. Others, from Magnus, the name of a shepherd who, in the legend, had his iron crook violently drawn from his hand as he passed a large boulder of magnetic iron ore.

This iron ore is an oxide of iron, Fe_3O_4. It is called *magnetite* or *lodestone*; this latter name means leading stone and was given to the ore because it was found that a piece, when suspended, always pointed in the same direction.

When a piece of lodestone is dipped into iron filings, many stick to it, but not uniformly over its surface. It can be seen that there are two points on the lodestone round which the majority of the filings cluster. These are called magnetic *poles*.

It was later found that when a piece of steel was stroked with lodestone, it too became magnetized with a pole at each end.

If a bar magnet is pivoted* or suspended by a piece of unspun silk, it always comes to rest pointing in a definite direction. Not only does the magnet always set along the same line, but it also always has the same end pointing the same way. This can be shown easily by rotating the magnet through 180°, when it swings back into its original direction.

This simple experiment shows that the two poles of a magnet are different. To distinguish between the two poles, we call them *north* and *south poles*. These terms are contractions of *north-seeking pole* and *south-seeking pole*, given because a suspended magnet sets approximately in the north–south direction with its north pole pointing towards the north.

A very simple experiment, which should now be carried out, is to test other common materials to see if they are affected by a magnet. Investigate whether there is any attraction between a magnet and such things as wood, glass, iron, brass, ebonite, etc. The result of this experiment is important: it will be found that, of the common materials available, only iron is affected. (Cobalt, nickel and gadolinium are also magnetic, but to a lesser extent.)

Iron, cobalt and nickel are called 'magnetic'

---

* A compass needle is a pivoted magnet.

substances and, apart from certain special alloys, are the only substances showing any appreciable magnetic effect.

Hold sheets of cardboard, glass, etc. between a piece of iron and a magnet. They do not influence the magnetic effect. An iron plate, however, considerably cuts down the attraction. (*See* magnetic shielding, page 281.)

The law of magnetism

Bring the N-pole of a magnet near to the N-pole of a suspended magnet or compass needle, as shown in Fig. 31.1, and observe the result. Repeat,

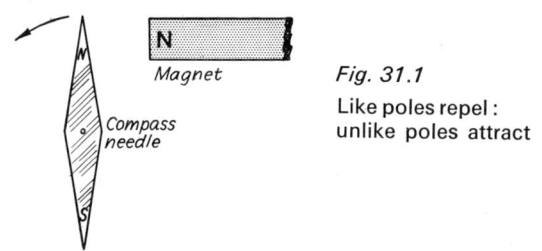

Magnet

Compass needle

Fig. 31.1

Like poles repel:
unlike poles attract

using the S-pole of the magnet. Then try both poles in turn on the S-pole of the compass needle. Your results should lead to the following conclusion:

like poles repel: unlike poles attract.

Making a magnet

We have already mentioned that a magnet can be made by stroking a piece of steel with lodestone. Instead of lodestone, a magnet may be used (Fig. 31.2). Try this experiment, and be careful to stroke

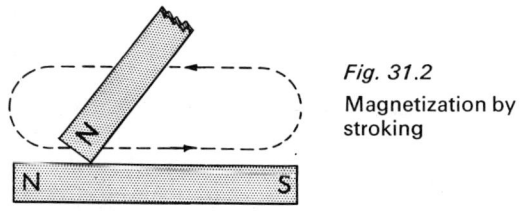

Fig. 31.2

Magnetization by stroking

the steel with one end of the magnet one way only. Afterwards test the polarity of the new magnet. The finishing end always has the opposite polarity from that of the end of the magnet used.

This method does not produce very powerful magnets.

It will be seen in Chapter 33 that an electric current produces a magnetic effect and to make a strong magnet an electrical method is employed. A

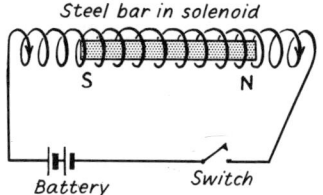

Steel bar in solenoid

Battery

Switch

Fig. 31.3 Magnetization by an electric current

long coil of wire is made by winding many turns of insulated copper wire on a cardboard tube. This long coil is called a *solenoid*. When a large current is passed through the wire, a steel rod placed inside the tube becomes strongly magnetized (Fig. 31.3).

Induced magnetism

If an iron nail is held near one pole of a magnet, it becomes magnetized and is capable of supporting another nail which, in turn, also becomes a magnet and can support more nails (Fig. 31.4). This effect

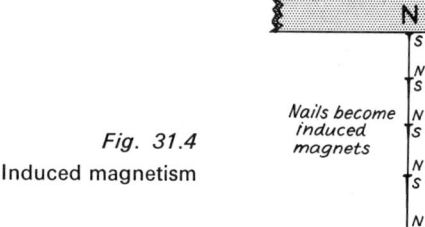

Fig. 31.4

Induced magnetism

Nails become induced magnets

is called *induction*, and each nail is an induced magnet. If an electromagnet (the magnetizing coil with a soft-iron core) is used in place of the magnet, 15-cm iron rods can be used instead of nails. If the nails are attached to the N-pole of the magnet, the extreme end of the chain will be found to be a N-pole also. It will repel the N-pole of another magnet brought near to it. If the nails are removed, they will no longer be able to support each other: they lose their magnetism.

There is no need for the first nail to be actually in contact with the magnet. The effect still takes place, though less strongly, when the nail or iron bar is held a little way from the pole of the magnet.

Induction explains the attraction of unmagnetized iron by a magnet. On bringing the magnet to the iron, it becomes an induced magnet with

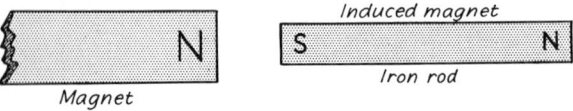

Fig. 31.5 Polarity of induced magnet

the near end an opposite pole (Fig. 31.5). Thus, attraction takes place. Hence we conclude,

Induction precedes attraction.

Attraction, then, does not indicate that the attracted specimen is a magnet, but only that it is magnetic.

Repulsion is the test for magnetization

If small pieces of steel instead of iron are used for the above experiment, a much weaker effect is shown. But another important difference is found: the steel does not lose all its induced magnetism when the magnet is removed.

This difference between the magnetic properties of iron and steel can be shown experimentally by winding two similar solenoids round similar sized bars of iron and steel. The solenoids are connected in series and the same current passed through both. The magnetism induced is estimated by counting the number of iron rivets (or nails, etc.) which will stick to one end of each bar when dipped into a box of the rivets. Counts are taken for different values of the current up to a maximum and then down to zero. An increase in the current beyond a certain value does not cause any increase in the magnetization. This state is called *saturation* (Fig. 31.6). It is

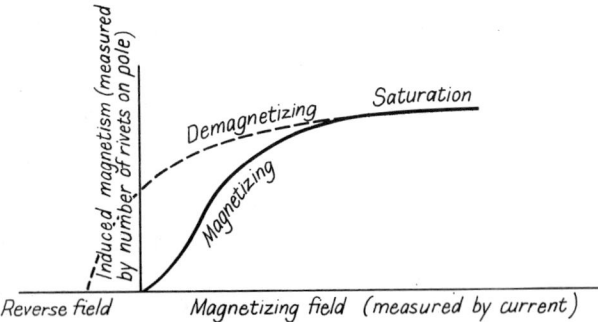

Fig. 31.6 Magnetization of iron and steel
The dotted curve shows the effect of decreasing the current to zero and then reversing it. The shape of this magnetizing–demagnetizing curve is different for different magnetic substances. From it the magnetic properties of a substance can be assessed

found that the iron bar holds very many more rivets than the steel bar at maximum current, but when the current is brought to zero, the rivets all fall off the iron, while an appreciable fraction remain on the steel bar. This shows that steel possesses the property of retaining a good fraction of the induced magnetism, whereas the soft iron loses all the induced magnetism when the magnetizing field is removed. Hence, steel is best for permanent magnets and iron for temporary magnets.

Modern research methods have produced alloys whose magnetic properties are much better than those of ordinary iron and steel.

Magnetic fields

We have seen that a magnet affects magnetic substances placed in its vicinity. This is like a mass being affected when near to another mass—another case of invisible forces. In Chapter 3 we saw that round a mass is what we call a *gravitational field*—the space in which another mass is acted upon by a force. A large mass, e.g. the earth, has a large gravitational field round it in which small masses are acted upon by relatively large forces. In the same way magnetic fields are created round magnets and round electric currents. A magnetic substance placed in a magnetic field is acted upon by a force. We shall see later that a current-carrying conductor placed in a magnetic field is also acted upon by a force. We shall describe more fully these forces and account for the effects, but we cannot *explain* why the forces exist, no more than we can *explain* gravitation.

Magnetic field is measured by the *magnetic flux density*, which is a vector quantity: i.e. it has magnitude and direction. The measurement of the magnetic flux density is beyond the scope of this textbook: we only concern ourselves with its direction. The form of a magnetic field can be represented by a pattern of the magnetic flux. In the case of a gravitational field, the field lines are the paths taken by a small free mass placed in the field. Likewise the direction of a magnetic field at any place is that which would be taken by a free north pole placed there. It is simply by mutual agreement or convention that we choose a N-pole and not a S-pole. A S-pole in a magnetic field would move along the same path as a N-pole, only in the opposite direction. (In gravitation this would correspond to some body having a negative mass!)

We define a magnetic field line as a line along which a free N-pole would move in a magnetic field.

Where the magnetic flux density is great we say the magnetic field is strong and the field lines are concentrated.

> Converging field lines indicate an increasing field; diverging field lines indicate a decreasing field; parallel field lines indicate a uniform or constant field.

Plotting magnetic flux patterns or magnetic fields

We cannot obtain a N-pole without a S-pole and so we cannot plot magnetic field lines as easily as we can plot gravitational field lines (i.e. by simply releasing a small mass).

A small piece of iron in the field of a magnet (Fig. 31.7) becomes an induced magnet with poles

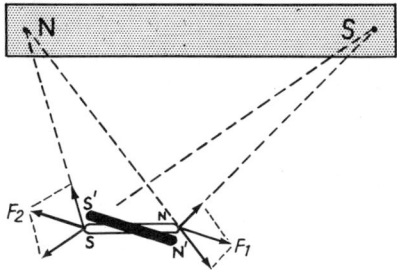

Fig. 31.7 Magnetic forces acting on an iron filing

as shown. The two poles of the magnet act on each pole of the iron. The diagram shows these forces, F_1 being the resultant force on the N end of the iron and F_2 that on the S end. Their effect is to tend to rotate the iron in the direction of F_1 and F_2. This shows the magnet's effect at the point where the iron is placed. It can, of course, be shown by placing a compass there, but we can scatter pieces of iron and show the field all round a magnet very easily.

Place a piece of stiff paper over a bar magnet and support the edges of the paper, if necessary, to keep it perfectly flat. Sprinkle iron filings thinly and evenly over the paper. (A tin with the lid perforated like a pepper-pot should be used.) Tap the paper gently. (This lifts the filings slightly and so reduces the friction preventing their rotation.) Note all the filings set in a definite pattern showing the field of the magnet. Repeat with various arrangements of magnets (*see* Fig. 31.10).

Magnetic field lines were first employed by Faraday to represent the effects between magnets. The experiment shows only the lines in the plane of the paper. You should remember there are field lines in all planes about a magnet and not just in the plane of the paper. Fig. 31.8 shows a small magnet enveloped in fine nickel powder. It illustrates clearly the disposition of the field lines about a magnet.

Iron filings will not set in weak fields, and the plotting of such fields has to be done using a small compass needle. To plot the field, the magnet is placed on a sheet of white paper. Starting near one pole of the magnet a mark is made on the paper. The compass needle is positioned so that one of its poles covers the mark and then a second mark is

(*a*)

(*b*)

Fig. 31.8 Magnetic field pattern produced by a horseshoe magnet

Nickel powder is used instead of iron filings enabling the pole patterns to be seen more clearly in 3 dimensions; (*b*) is the view of (*a*) seen from above

made at the other pole. The compass needle is then moved along to repeat the process from dot to dot until either the other pole of the bar magnet or the edge of the paper is reached. More field lines are likewise plotted to cover the whole paper.

That the magnetic force of a magnet is along such lines can be shown by releasing a movable N-pole near to the N-pole of the magnet and noting its track. We cannot get a N-pole without a S-pole, but we can overcome the difficulty in the following way. Float a cylindrical magnet about 10 cm long on water by passing it vertically through a cork (Fig. 31.9). Bring it near to the N-pole of a bar

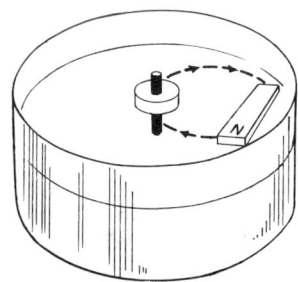

Fig. 31.9 Experiment to show that a field line is the path of a free N-pole

magnet held on the surface and then release it. It will not go directly to the S-pole but will move along a field line.*

A field line is the path a free N-pole would take in a magnetic field.

Points to remember about magnetic field lines are:

1. Field lines cannot cross. If they did, this would mean that a free N-pole at the point of intersection could move in two directions at once.

2. Field lines must begin on a N-pole and end on a S-pole. (In some cases one or both of the poles may be outside the diagram.)

3. Field lines behave as if they were in tension, that is, they always tend to shorten their length.

4. Field lines behave as if they repelled each other sideways.

5. Where the field is strong, the field lines crowd together: in a weak field they are farther apart. A uniform field is indicated by straight parallel field lines.

6. Field lines crowd into a magnetic substance. *See* Fig. 31.10.

---

* Its motion along a field line is interesting. Can you account for its variation in speed?

Neutral points

The field between two similar and equal magnets is shown in Fig. 31.11. The field lines surround a point midway between the poles. A N-pole placed

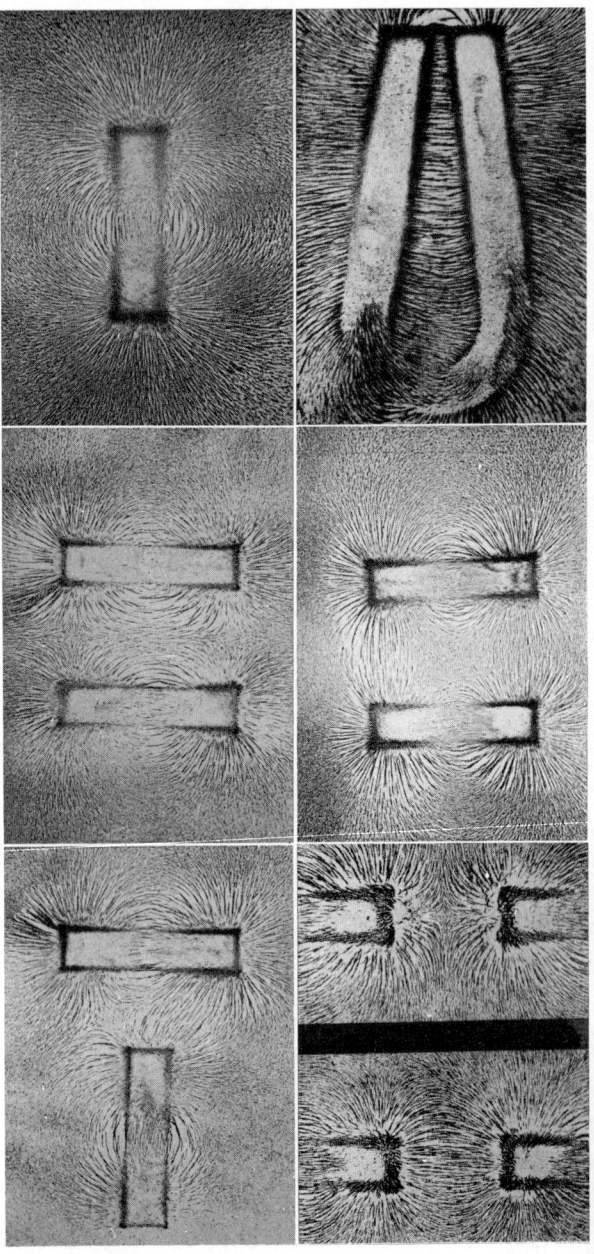

Fig. 31.10 Photographs of field lines for various arrangements of magnets

As an exercise, name the respective poles of the magnets

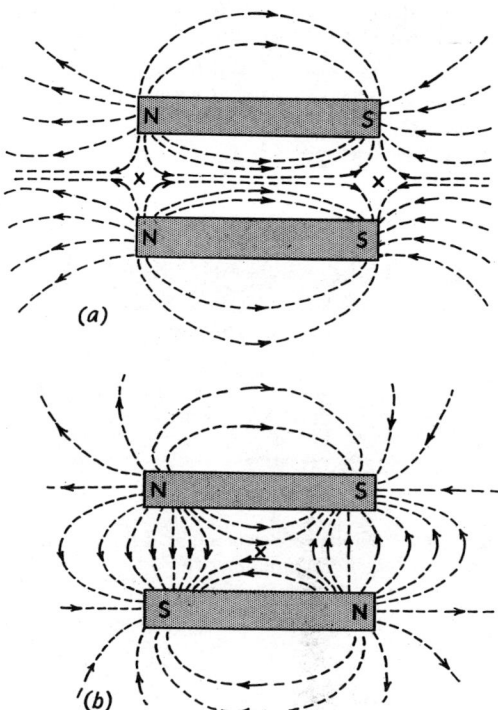

Fig. 31.11 Field line diagrams
X = neutral point

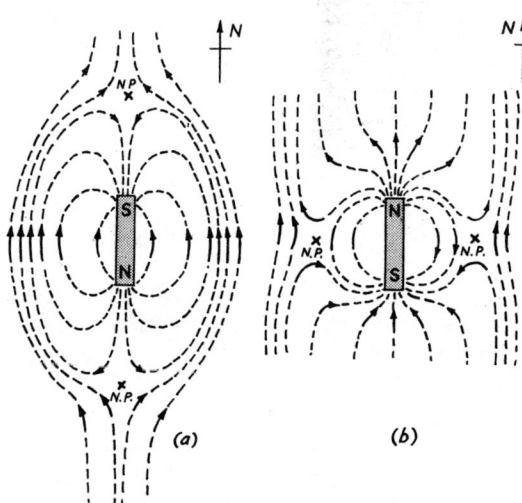

Fig. 31.12 Fields due to a bar magnet in the earth's field

(a) with its N-pole pointing south; (b) with its N-pole pointing north

there will be acted on by equal and opposite forces, because the two poles make the resultant force zero. Such a point we call a *neutral point*.

> A neutral point is a place where the magnetic flux density (magnetic field) is zero.

Combined fields of magnets and the earth

Figs. 31.11*a* and *b*, show the magnetic field lines very near to the magnets. The field as we move from a magnet is weaker and becomes comparable in strength to the earth's field. In fact, neutral points are formed where the earth's field is equal and opposite to the magnet's field. The plotting of these weak fields has to be done using a small compass needle. Figs. 31.12*a* and *b*, show two examples of such fields.

Magnetic shielding

When an iron bar is placed in a magnetic field, it becomes an induced magnet. The field lines are drawn into it and pass along its length creating poles at the ends. Fig. 31.13 illustrates this and shows that, in concentrating within iron, the field lines leave empty spaces. These regions, *x*, in which

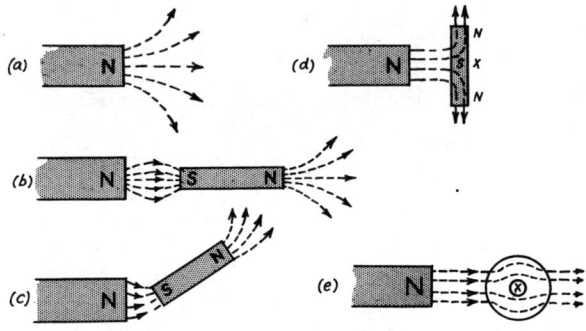

Fig. 31.13 Field lines in iron and magnetic shielding
X = shielded regions

the magnetic field is zero, are said to be screened (*see* Fig. 31.13*d*). Fig. 31.13*e* shows how an iron washer in a field deflects the lines field and leaves the inside area screened.

A magnetic substance always tends to move from where the field is weak to where it is strong. An iron bar is drawn into a solenoid when a current is passed.

Theories of magnetism

Early theories put forward to explain magnetism were fluid theories and they were very fashionable in science until the end of the eighteenth century. It was said that magnetization was produced by the addition to a piece of steel of two fluids, north and south polarity, in equal amounts. This theory quickly gave place to a simpler version of the same idea, a one-fluid theory. Magnetic substances were supposed to contain a definite amount of magnetic fluid. On magnetizing, the fluid became piled up at one end (north), leaving a deficiency at the other (south). The fluid was hypothetical: it could not be detected by any method.

The single-fluid theory explained quite well a few elementary facts such as the equality of the poles, but it just could not explain the effect of breaking a magnet. If a piece of magnetized clockspring is broken in two, each piece is itself a magnet (Fig. 31.14). The fluid theory demands that each half should have one polarity only.

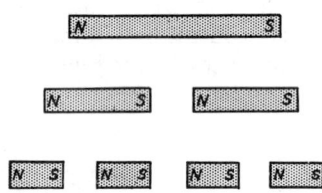

Fig. 31.14 Breaking a magnet makes small magnets

The experiment of breaking a magnet and so producing two magnets gives a clue to the latest theory. If it were possible to continue the breaking of a magnet indefinitely, the ultimate result of division would be a magnetized elementary particle. This leads to the theory of magnetism, first proposed in the nineteenth century, which assumed each molecule of a magnet to be itself a small magnet. In an unmagnetized specimen the molecular magnets were arranged haphazardly (Fig. 31.15). Under the influence of a magnetizing field the elementary magnets were aligned: the mutual interaction of the molecules caused them to show some opposition to perfect alignment in a weak field. The modern conception of magnetization is a modified form of the above theory. It is now established that the molecular magnets do not orientate individually, but in groups known as

domains (Fig. 31.16). In iron all the molecules in a domain have their magnetic axes pointing in the same direction. The domains are quite small—smaller than the individual grains of the iron—but their existence can be revealed under a high-powered microscope using a technique developed

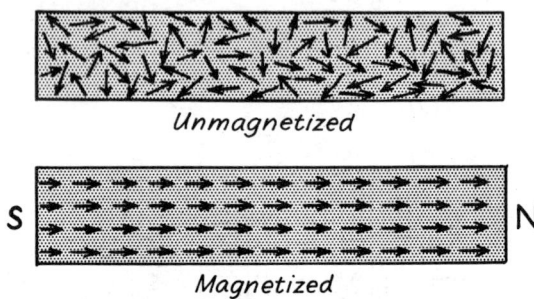

Fig. 31.15 To illustrate molecular arrangements in unmagnetized and magnetized iron

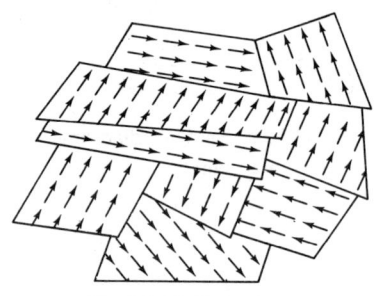

Fig. 31.16 Domains

by Bitter. The micro-photographs (Bitter patterns) reveal the domains by showing the field lines between them. They are made visible by covering the prepared surface of the specimen with a suspension of finely divided magnetite (Fig. 31.17).

We can take the molecular theory just a little further, for when we search within the molecule for the cause, we find moving charges, electrons (Chapter 41). A moving charge is a current and an electric current has a magnetic effect (Chapter 33).

The orbiting electrons produce magnetic fields similar to those which would be produced by a pair of equal and opposite magnetic poles close together. Such an arrangement we call a *dipole*. (A bar magnet is a dipole.) Electrons moving in similar orbits, but in opposite directions, produce fields which neutralize. The spinning of the electrons also contributes to the magnetic field. A

(a) specimen unmagnetized

(b) specimen magnetized

Fig. 31.17 Domains in a cobalt crystal

molecule containing many dipoles may produce a field of zero when the fields of the individual electrons are combined. On the other hand, the resultant field may be small or large. We thus have three categories of substance classified according to their magnetic properties.

1. *Diamagnetic substances*. These substances have no resultant magnetism. When they are placed in a very strong magnetic field the electron orbits are affected, with the result that they show a weak magnetic effect which is opposite in performance to that of iron. A short bar of a diamagnetic substance, such as bismuth, when suspended in a strong magnetic field sets across the field, i.e. with its long axis at right angles to the field lines.

2. *Paramagnetic substances*. These substances behave like iron but with a much smaller effect. A small bar of paramagnetic substance, e.g. platinum, becomes a weak dipole in a field and sets with its principal axis along the field lines of the magnetizing field.

3. *Ferromagnetic substances*. This is a special group and the substances behave as if they were strongly paramagnetic. In this group we have iron as the outstanding member together with cobalt, nickel and gadolinium, and compounds of these elements. As we have seen above, a bar of a ferromagnetic substance becomes a powerfully induced magnet when placed in a magnetic field. The molecular dipoles are already aligned in the domains with their axes parallel. In a magnetic field the domains are orientated into alignment with the field lines. Para- and ferro-

magnetic substances always move from weaker to stronger parts of a field, i.e. towards parts of greater magnetic flux density. Diamagnetic substances behave in an opposite manner and move to parts of weaker flux density.

Liquids and gases show magnetic effects, but their measurement requires delicate detection. Liquid oxygen shows a relatively large paramagnetic effect.

We arrive at the conclusion that magnetism is not a particular physical phenomenon, but it is due to the magnetic effects of the electric currents of the orbiting and spinning electrons.

Demagnetizing a specimen

The magnetization of a specimen is destroyed by breaking up the domain alignment. Severe hammering slightly reduces it, but an effective way is to heat the specimen and allow the thermal agitations to destroy the arrangement of the molecules. This is easily shown experimentally by strongly heating a piece of magnetized clockspring, which completely destroys its magnetization.

A convenient method of demagnetizing which is used in the laboratory is to put the specimen into a solenoid (such as is used for magnetizing) which is arranged in an E–W direction. An alternating current is passed which is slowly reduced to zero. Alternatively, the specimen can be withdrawn along the axis of the solenoid to a distance of about two metres while the current is kept steady.

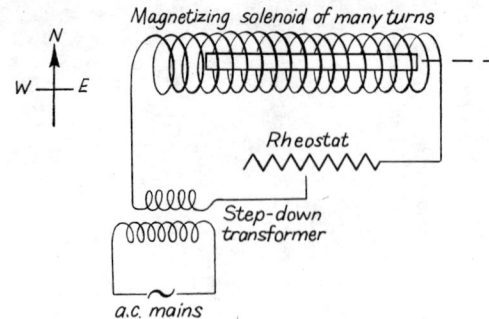

Fig. 31.18

Apparatus for demagnetization

To demagnetize specimen either withdraw it from the solenoid along the axis or reduce current to zero using rheostat while specimen remains in solenoid

In the magnetizing process a direct current is used which, when brought to zero, does not reduce the magnetization of the specimen to zero. (*See* experiment and Fig. 31.18.) The alternating current takes the specimen through cycles, magnetizing it first in one direction and then in the other. As the current is reduced to zero, the magnetization cycles get smaller and smaller eventually becoming zero.

The Curie point

The temperature to which a magnetic alloy must be heated so that its magnetic properties become very small, is called its *Curie point*. For special purposes, alloys have been developed with Curie points differing over a large range. The Curie point for iron alloys is in the region of 600 °C, while that for cobalt is about 1200 °C. Alloys, however, may have very low Curie points. One special alloy JAE* metal has a Curie point about 70 °C. Ferroxcube† is an iron alloy with special properties, and developed specially for the cores of inductors and transformers. It has a very high resistivity (*see* page 317), so that eddy current losses (*see* page 384) are low. The Curie points of different types of this alloy lie between 100 and 250 °C.

To demonstrate these low Curie points, a specimen can be held by a permanent steel magnet while immersed in cold water (Fig. 31.19).‡ Heating eventually results in the specimen leaving the magnet when the Curie point is reached and the magnetic properties of the specimen fall to a low value.

---

* JAE metal is an alloy of 70 per cent nickel and 30 per cent copper developed by International Nickel Ltd.

† A Mullard product.

‡ Oil is used if the Curie point is over 100 °C.

Magnetic substances and their properties

In recent years many new substances have been introduced with much stronger magnetic properties than those of ordinary iron and steel.

We have already seen that iron is more suitable as a temporary or induced magnet and steel as a permanent magnet. These properties depend upon the ability with which the domains may be aligned or their alignment destroyed.

The ability of a magnetic substance to become strongly induced in a weak field and to lose its induced magnetization readily when the field is removed, renders it very suitable for use as the core of an electromagnet (page 309) or of a transformer (page 369). Two alloys possessing such properties are known by their trade names Stalloy (96% Fe, 4% Si) and Mumetal (73% Ni, 22% Fe, 5% Cu). Powdered magnetic materials (ferrites) are being used for cores of small transformers. These have the added advantage of being electrical insulators.

Good permanent magnets are made of materials which retain much of their induced magnetization

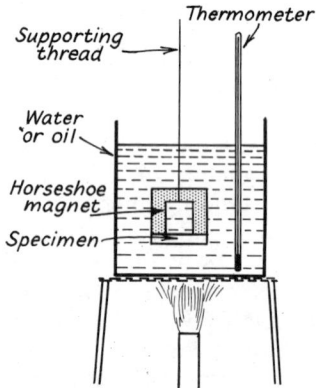

Fig. 31.19 Apparatus for demonstration of the Curie point

when the magnetizing field is removed and, more important, can retain it against harsh treatment. Good examples of these alloys are Alni (13% Al, 25% Ni, 4% Cu, 58% Fe) and Alnico (10% Al, 19% Ni, 12% Co, 6% Cu, 53% Fe), Alcomax, Ticonal, etc. In recent years ceramic materials have been made with powerful magnetic properties. Magnets of such materials are very brittle and break easily. They are very hard and are very difficult to drill. They are made by bonding a powdered mixture and are moulded under high pressure. Very recently, in America, permanent magnets have been cast from materials containing the rare earths samarium and cerium. A mixture of copper, iron, cobalt and the rare earths is melted in an electric furnace in an argon atmosphere. Tiny magnets made of this material possess great strength and also have a high coercive force (ability to resist demagnetization.)

Terrestrial magnetism

Variation or declination

It was not until the seventeenth century that a satisfactory explanation of the behaviour of a compass needle was forthcoming. Sir William Gilbert, in 1600, published his famous treatise on magnetism, and in it he described the earth as behaving like a huge spherical lodestone. Although the field produced over the earth's surface is weak, it is sufficient to cause a compass needle to set in the north–south direction. That the magnetic poles are not situated at the geographical poles was probably known to the Chinese about A.D. 1100.

Fig. 31.20 $\boxed{\text{S} \qquad \text{N} | \text{S} \qquad \text{N}}$

(a)

$P.$ $\boxed{\begin{array}{cc} \text{S} & \text{N} \\ \text{S} & \text{N} \end{array}}$ $P.$

(b)

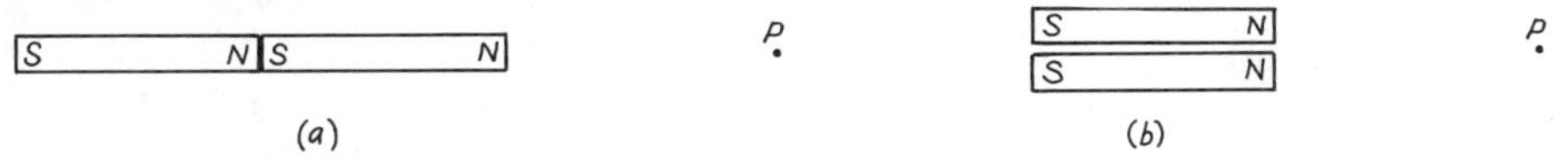

Work done in magnetizing a specimen

The magnetic field produced by the two equal magnets at the point P (Fig. 31.20) has the same strength when the magnets are placed in either of the positions (a) or (b).

The effect or 'power' of a magnet is not only dependent upon the strength of its poles, but also on their distance apart.* A magnet possesses potential energy resulting from the work done in magnetizing it. This depends on the work done in separating the poles, i.e. upon the distance between the poles. When a magnetic substance is in a rapidly changing field, such as that inside a solenoid through which an alternating current is passing, work is being done on the substance as the domains are repeatedly orientated. This work is shown by the specimen becoming hot. The heating is less in a substance like iron in which the domains can rotate freely, than in steel which has a high coercive force. It is therefore necessary to employ a substance like iron for the cores of transformers and motors in order to reduce this heating to a minimum.

At any point on the earth's surface the magnetic field lies in a vertical plane called the *magnetic meridian* (*see* Fig. 31.21).

The magnetic meridian is the vertical plane through the magnetic axis* of a compass needle.

The geographical meridian is the plane through the point and the two geographical poles of the earth.

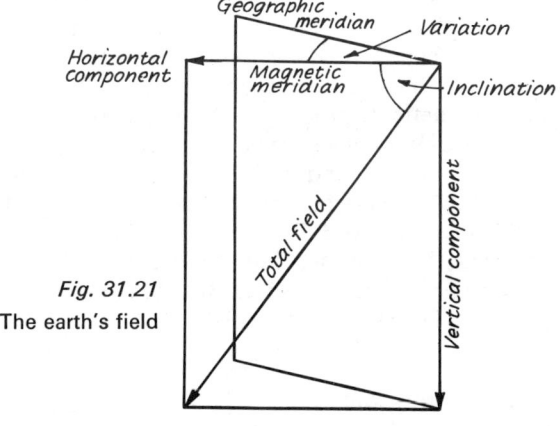

Fig. 31.21
The earth's field

---

* The product of the pole strength and the distance between the poles is known as the *magnetic moment* of a magnet.

---

* Magnetic axis is the line joining the two poles of a magnet.

These two planes (the magnetic and geographical meridians) are inclined at an angle—termed the *variation*—which, in England, is about 8° West (Fig. 31.22).

The magnetic variation or declination is the angle between the magnetic and geographical meridians.

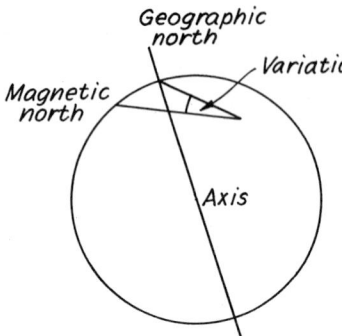

Fig. 31.22 Variation

The variation is measured by first aligning a telescope in the magnetic meridian. This is done by observing a compass needle through the telescope. For this setting the needle is provided with a simple optical system. A plane mirror is fixed to reflect the sun's rays into the telescope. (*See* warning note on page 236). The passage of the sun through the magnetic meridian is watched through the telescope, and the time noted. Reference to astronomical tables gives the position of the sun at the noted time and this gives the variation.

Columbus, in 1492, on his voyage to America, noted that this angle varies over the earth's surface. Charts are now constructed showing lines joining places having the same variation. These lines are called *isogonic lines*. They run roughly north–south from one magnetic pole to the other. The magnetic N-pole is in the north of Canada. There are two lines (*agonic lines*) along which the variation is zero, and one of these passes through N. and S. America, and the other through the U.S.S.R., Iraq, and W. Australia. In one region between these agonic lines the variation is west, and in the other, east. The British Isles is in the former, where the variation is west.

Inclination or dip

If a compass is accurately mounted about its centre of gravity so that it can rotate in a vertical instead of a horizontal plane, it sets with its N-pole dipping downwards in the northern hemisphere. When such a compass rotates in the magnetic meridian in England, it sets at an angle of 67° with the horizontal. This angle is called the *inclination or dip*.

A compass needle mounted so as to take this reading is called a *dip circle* (Fig. 31.23). The needle must always be in the magnetic meridian to allow the inclination to be read, for, in any other plane, the reading is greater and reaches 90° when at right angles to the magnetic meridian.

Fig. 31.23 A dip circle

As the compass needle sets along the field lines we see, therefore, that the earth's field lines are not horizontal but inclined. The ordinary compass is acted upon by the horizontal effect of the earth's total field.

We can explain inclination by considering the field lines produced by a bar magnet within the earth (Fig. 31.24). This shows the change in the inclination over the earth's surface. At the magnetic N-pole, a dip needle sets vertically with its N-pole downwards, and at the S-pole with its S-pole downwards. Roughly along the equator the inclination should be zero.

Figures 31.22 and 31.24 illustrate only the ideas of variation and inclination. Variation is not the angle made by the directions of the true N-pole and the earth's magnetic N-pole at the point. It is the angle between the direction of the true N-pole and that in which a compass points. This latter direction, because of the irregularity of the earth's magnetic field, may be very different from the direction of the magnetic N-pole.

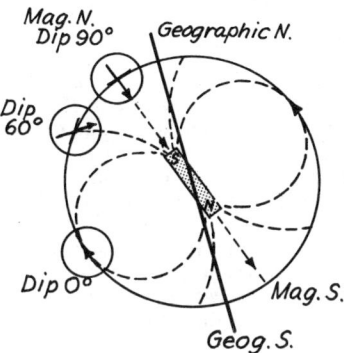

Fig. 31.24 Dip or inclination

Lines joining places having the same inclination are called *isoclinic lines*. The line of zero inclination is called the *magnetic equator*.

The presence of the vertical effect of the earth's field is shown by the magnetization of iron railings. These are found to be induced magnets with, in the northern hemisphere, N-poles at the bottom and with S-poles at the top. A soft-iron bar becomes an induced magnet when held in the earth's field. The maximum effect is produced when the bar is held along the direction of the earth's total field. Hammering the bar helps the induction.* It is for this reason that in demagnetizing a

---

* A Mumetal bar becomes induced readily and without hammering.

specimen the solenoid should be arranged with its axis E–W.

The variation and the inclination are not constant over the years, and so magnetic charts have to be re-drawn at intervals. The principal change in the magnetic elements is a slow change which goes through a cycle in about 1000 years, ranging from 24° East to 24° West in England. The change is not regular.

Theories of terrestrial magnetism

Modern ideas attribute the earth's magnetism to internal causes. It has long been established that the inside of the earth contains a molten mixture of nickel and iron. Owing to temperature variations and the earth's rotation, convection currents are set up which give rise to magnetic fields. The changes in the magnetic field on the earth's surface are caused by subsidiary currents which are superimposed upon the main circulatory motion of the central fluid mass.

In addition to the changes just described, there are smaller changes in the earth's field which are due to external and not internal causes. First, there is a small daily change which becomes greatest at noon, the magnitude of which also depends upon the season of the year. It is due to tidal movements in the ionosphere which cause magnetic fields.

At any time a magnetic storm causing severe and spasmodic fluctuations in the earth's field may occur. Such magnetic storms are accompanied by a 'black-out' of short-wave communication, intense cosmic ray showers and displays of aurora. All these effects appear to originate from solar flares which occur in the vicinity of sun spots.

QUESTIONS

1. Describe an experiment which shows that the two poles of a magnet are of equal strength.

Describe how you would magnetize a bar of steel, and how you would demagnetize it. How may these processes be explained in terms of what happens inside the material? (O)

2. Describe two methods, only one of which needs the use of a bar magnet, by which a piece of steel may be magnetized. Draw a diagram to illustrate each method and name the poles formed in the steel.

Two similar bar magnets are placed in line with opposite poles adjacent and a short distance apart. Sketch the magnetic field lines between these two poles when a piece of soft iron is placed between but not touching them.   (JMB)

3. What are the chief differences in the magnetic properties of soft iron and steel? How would you demonstrate them experimentally? For each substance, name an instrument or piece of apparatus in which it is used because of its magnetic properties. Explain why the load which an electromagnet can lift has a maximum value.   (JMB)

4. Describe, with the aid of diagrams, and explain how you would magnetize a steel knitting needle so as to obtain a south pole at a marked end of the needle (*a*) by using a permanent magnet, (*b*) by using an electromagnet.

State, giving a reason, which method of magnetization you would expect to produce the stronger magnet and describe an experiment to test your conclusion.

Describe how a magnetized knitting needle can be effectively demagnetized.   (O & C)

5. Give brief explanations of the following:

(*a*) A piece of soft iron is attracted by a magnet.

(*b*) A magnetized needle, floated on a water surface, sets in the magnetic meridian but does not move bodily in any direction.

(*c*) A steel ring can be magnetized so that it has no free magnetic poles.

In the last case explain how you would magnetize the ring and how the magnetization could be destroyed.   (O & C)

6. Define *magnetic field*, *magnetic field line*.

Describe how you would plot a magnetic field using a small compass.

Give diagrams of the fields which would be obtained near (*a*) a U-shaped magnet lying flat on a table, (*b*) an air cored solenoid carrying a direct current, in a plane containing its axis.

Explain the action of the keeper which is placed across the poles of the magnet in (*a*) above when it is not in use and state two ways in which the solenoid in (*b*) could be made into a stronger electromagnet.   (L)

7. Explain the following:

(*a*) Two steel needles hang from the lower end of a vertical bar magnet, but do not hang vertically.

(*b*) A horseshoe magnet is often supplied with a 'keeper'.

(*c*) When a bar magnet is placed on a table there are two points near it where there is no horizontal magnetic field.   (JMB)

8. Describe an electrical method of magnetizing a steel bar. Draw a diagram showing the direction of the current and the polarity of the magnet produced by it.

How would you check the polarity experimentally without the use of another magnet?

Describe how you would demagnetize a magnet by an electrical method.

Why is a steel bar not suitable for use as the core of an electromagnet?   (JMB)

9. In connection with the earth's magnetism, explain the terms *horizontal component*, *vertical component*, *angle of variation* (*declination*), *angle of inclination* (*dip*).

What is understood by the term *neutral point*? Draw a diagram showing some of the magnetic field lines due to a horizontal bar magnet with its north pole pointing to magnetic north and indicate the positions of the neutral points.

State the direction in which you would expect a small suspended magnetic needle to set when it is free to pivot in any direction about one of these neutral points. Give a reason.   (O & C)

10. Explain the terms *magnetic meridian*, *angle of inclination* (*dip*), and *magnetic variation* (*declination*).

How would you attempt to determine the magnetic variation at a given place?

Of what practical importance is a knowledge of the change in the variation with latitude and longitude?   (O & C)

11. Define *magnetic meridian* and *magnetic inclination* (*dip*).

Describe the construction of the dip-circle and explain how it is used (*a*) to locate the magnetic meridian, (*b*) to measure the angle of dip. Illustrate your answer with diagrams.   (O & C)

12. Explain the following observations:

(*a*) A bar magnet supported in a large vessel of water by a cork raft rotates until it points north and south but does not move bodily in any direction.

(*b*) A dip needle with its axis horizontal and in the magnetic meridian points with its north pole vertically downwards when in the northern hemisphere.

(*c*) If a magnet is broken in the middle the two pieces are also magnets.

(*d*) If the pole of a strong bar magnet is in contact with the end *A* of a soft-iron bar *AB* iron filings can be picked up by *B*.   (O & C)

13. What is meant by the total intensity of the earth's field?

State the relation between the angle of dip (inclination) and the horizontal and vertical intensities.

Describe an experiment to find the angle of variation (declination) in your town.

Describe how the angle of variation varies over the surface of the earth and how it changes with time at any one position.   (JMB)

32 Electrostatics

It is difficult for young people to appreciate that electricity, as encountered in our homes, is a modern convenience. Very few homes were so equipped before 1900. Yet electricity was known to the ancients and, in fact, the name 'electricity'* dates from its early discovery. It was found that amber, when rubbed with fur, possessed the property of attracting small light objects. This is the same experiment as you perform when you rub your fountain pen and pick up small pieces of paper with it. We now know that this rubbing process produces electrical charge. Gilbert, in 1600, pursued the subject further and showed that friction may produce two kinds of electrical charge. This may readily be shown in the following manner. Rub a polythene (Alkathene) rod with a woollen cloth and suspend it in a stirrup having a silk or nylon suspension. Rub and likewise suspend a cellulose acetate rod. Now charge a second Alkathene rod and bring it up to the two suspended rods in turn.† (It is more convenient to use pivoted strips of Alkathene and cellulose acetate which are now available. In this form the strips resemble large compass needles but their difference must be kept in mind. The strips are charged by rubbing and carry one type of electricity only which is not necessarily uniformly distributed. The two ends are definitely not oppositely charged.) You will find that the Alkathene rods repel but the suspended acetate rod attracts the charged Alkathene rod. Repeat the experiment by bringing a charged acetate rod up to the two suspended rods. Also bring the charged rods near to small light objects. These simple experiments show that the changes on the Alkathene and acetate rods are different and that

like charges repel and unlike charges attract.

Uncharged bodies are attracted by both types of charged rod.

Further experiments with other charged rods lead to the conclusion that there are **two and only two types of electricity**. These we call positive and negative.*

The charging of bodies is very common. Toy balloons suspended by nylon thread can be charged by rubbing with the hands or by flicking with a cloth. They then repel each other. A charged acetate strip will attract a charged balloon: a charged Alkathene strip will repel a charged balloon. On a much smaller scale, tiny

---

* ηλεκτρον (electron) meaning amber.
† This experiment used to be carried out using an ebonite rod rubbed with fur (negative), and a glass rod rubbed with silk (positive). The new materials are very superior.

* Originally termed *vitreous* and *resinous*.

289

spheres of pith or foamed plastic suspended by nylon or silk thread may be charged, and if a pair of such spheres is mounted so that they touch when hanging normally, when given a charge they

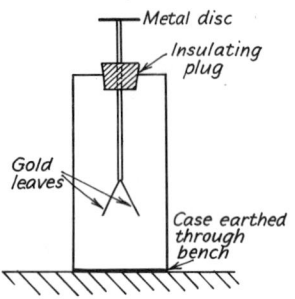

Fig. 32.1 The electroscope

repel each other and take up an equilibrium position. The spheres are usually sprayed with aluminium paint, or covered with Aquadag (colloidal graphite), to make them conductors. Such an arrangement can be used to detect, and roughly to measure, charge. An instrument which does this is called an *electroscope*, and one such satisfactory instrument—a gold-leaf electroscope*—is shown in Fig. 32.1.

A simple form of electroscope can be made by sticking one end of a strip of Melinex to the upper end of a brass strip which must be supported to hang vertically from an insulating rod. The strips can be made any size but 2×20 cm is suitable. The Melinex strip is a very thin metallized plastic.†

The gold-leaf electroscope

This instrument consists of an earthed‡ case of metal or wood, with opposite sides of glass. Through the top of the box, and well insulated from it, passes the leaf system. This consists of a metal rod with a metal disc mounted on the outside end, as shown. The inside end of the rod carries two light leaves of gold or aluminium foil

stuck together at their upper ends. (Very often one leaf is replaced by a plate.) When the inside system is given a charge, e.g. by drawing a charged Alkathene rod across the disc,* the charge spreads

over the surface of the conductor, and the forces between the like charges on the leaves cause them to open or diverge. The magnitude of the divergence is increased by increasing the charge. A divergence may likewise be produced by using a charged acetate rod. If a positively charged acetate rod is brought *near* to the disc when the electroscope is charged negatively, the divergence will decrease.† Similarly, bringing a negatively charged rod near to the disc when the electroscope is negatively charged causes an increase in the divergence.

Conductors and insulators

If we touch the disc of a charged electroscope with a metal rod, the leaves will collapse, but if we touch it with an uncharged glass rod, nothing

---

* Most of the experiments described in this chapter can be demonstrated using a d.c. amplifier. Full details are given in *The Teaching of Modern Physics* (Association for Science Education).

† Nuffield *G. to E.*, III, No. 99b.

‡ Joined to the ground by a conductor. Earthing is fully explained under the heading 'Potential'.

---

* Some of the charge is scraped off the charged rod by contact.

† An earthed plate also decreases the divergence when held near to the cap. *See* later, under 'Capacitance'.

happens. In this way we can divide materials into two classes, *conductors* and *insulators*.

Charge can move freely over the surface of a conductor but remains stationary on the surface of an insulator.

You should test a number of common substances yourself and find out which are insulators and which are conductors, e.g. try a piece of dry thread and then wet it and repeat the test. You will find that wood will conduct the charge away from a charged electroscope. This may surprise you if you are familiar with electric toys and gadgets, for dry wood is often used as an insulator. An electric current is simply a flow of electric charge. When charge passes from one object to another, the moving charge constitutes a current. In electrostatics we deal with very small charges, whereas in current electricity we deal with the movement of relatively large charges. There is always a little, but usually insignificant, leak of charge through wood: this is not noticed in current electricity simply because it is so minute.

When we charge a body by rubbing we give it one type of charge and leave the material used for rubbing with an equal and opposite charge. We shall describe later an experiment which shows this.

When a conductor is earthed by touching, a spark may pass between the finger and the charged conductor. This spark may, in certain cases, be dangerous. For example, when a plane is refuelling, the fuel pipe is always connected to the plane by a metal chain. Any generation of charge by the flowing liquid will not be allowed to accumulate. In this way the passage of a spark from the fuel pipe to the plane is eliminated.

The use of rubber sheets in hospitals presents some risk as they become charged by friction, and a spark may result. In operating theatres, where flammable vapours may accumulate, this risk is dangerous. This, and similar dangers have been overcome by the manufacture of 'conducting' rubber for use as hospital sheets, floor coverings, and aeroplane and trolley-bus tyres.

We have mentioned above that a spark may pass between a charged conductor and the finger when held close to it, The spark signified the passage of charge, in this case from the conductor to earth. Before a gas will permit the passage of a spark, it must be made conducting. The molecules of the gas can split up into two parts which are positively and negatively charged. This ionization of a gas can be produced by heating. A flame contains ions. The flame of a candle placed between two insulated metal plates is divided into two parts when a potential difference of a few thousand volts is put across the plates. This shows the presence of both positive and negative ions. An electroscope across the plates will show the discharge when the supply is disconnected.*

Most liquids contain numerous ions and so readily conduct away the small charges encountered in static electricity.

We can use the conduction of a flame to advantage. In performing many of the demonstrations outlined in this chapter, it is often necessary to start with both conductors and insulators uncharged. It is easy to discharge a metal sphere mounted on an insulated rod simply by touching it, but the supporting rod may be charged. The modern materials, such as Alkathene, readily acquire a charge simply by handling. An effective way of discharging the supporting insulating rods is by passing them quickly through a bunsen flame. Another method is by dipping them in water, but this obviously is not a suitable method to use.

The presence of ions in water causes much inconvenience in demonstrating electrostatics. When the humidity (dampness) of the air is high, the water vapour present causes the charge to leak away from charged bodies.

Electrostatic induction

We have seen that soft iron in the vicinity of a magnet becomes an induced magnet. This we call *induction* and we observe a similar effect in electrostatics. In many respects, both qualitatively and quantitatively, electrostatic effects resemble magnetic effects. There is, however, one outstanding difference. We can isolate positive and negative charges, but we cannot isolate north and south magnetic poles.

We can temporarily charge a conductor, just as we can temporarily magnetize a piece of iron. (*See* Fig. 32.2a.)

Field lines or lines of electric flux show the fields in the two cases. We use field lines in

---

* Nuffield *G. to E.*, V, Nos. 120 and 121.

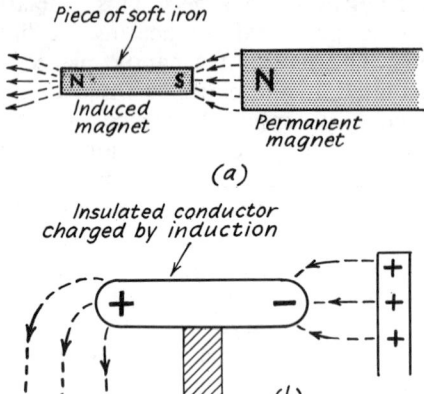

Fig. 32.2 Induction in (*a*) magnetism, (*b*) electrostatics

Fig. 32.3 Electric field lines

(Produced using reeds suspended in an insulating liquid)

(*a*) Field between two rods positive and negative
(*b*) Field between parallel plates
(*c*) Field produced by positive rod at centre of negative ring
(*d*) Positive ring surrounded by negative ring, showing no field inside the hollow charged conductor

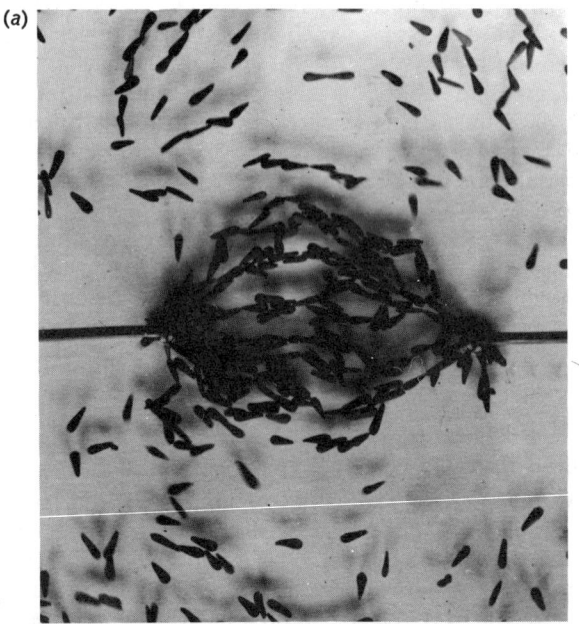

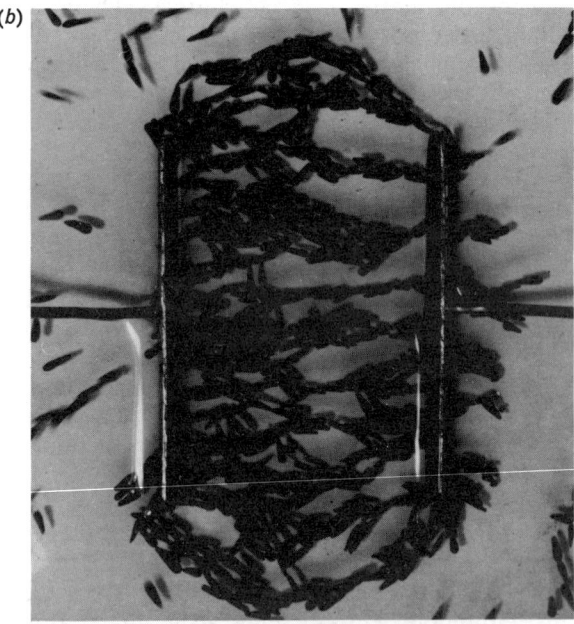

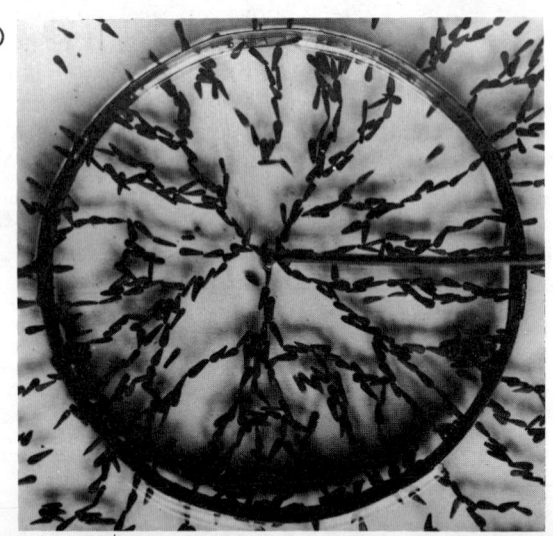

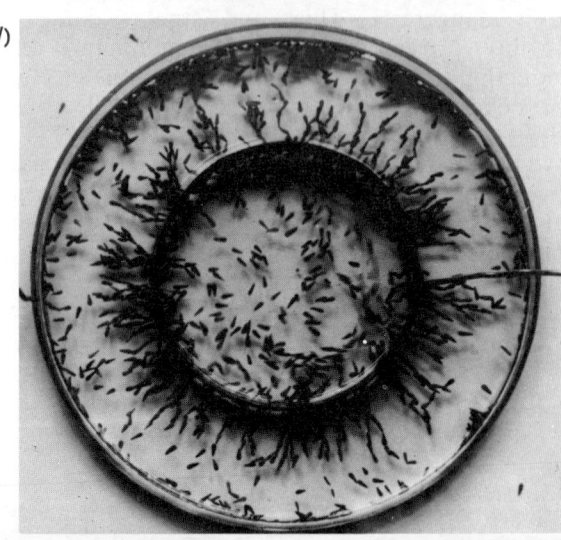

electrostatics just as we did in magnetism (Fig. 32.3).

A field line is the path a positive charge would take if it were free to move.

From a charged body the field lines must pass to an oppositely charged body or to the earth. In Fig. 32.2*b*, some of the field lines from the charged rod pass through the conductor on their way to earth. In this experiment we can show the induced charges by separating them. In place of the long conductor we use two spheres which are touching. While the charged rod is held near (Fig. 32.4) we separate the spheres and test their charges using an electroscope. These are found to be positive and negative as shown.

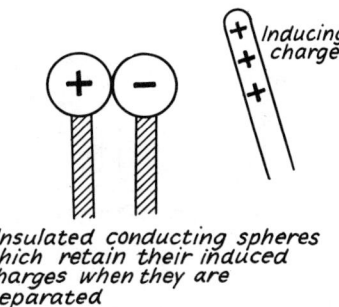

Insulated conducting spheres which retain their induced charges when they are separated

Fig. 32.4 Charging by induction

Potential

Work must be done against the force of repulsion when a charge q is brought up to a similarly charged body. The energy expended is possessed by the charge q for it will move away from the charged body if it is released. It therefore possesses potential energy by virtue of its position relative to the charged body. We can liken the arrangement to the pushing of a mass up a hill against gravity. It will acquire potential energy as it is pushed up, which will be equal to the work done in pushing it up.

The force between two charges varies inversely as the square of the distance between them. The work done on the charge q will not be great until its distance from the charged body is small. This is like gravitation, for the pull of gravity on a body is not appreciable until the body is relatively near the earth.

So we have in electrostatics, forces which, in some ways, resemble gravitational forces. The space round a charged body in which it exerts an influence on other charges, we call an *electric field* which corresponds to the gravitational field surrounding a mass. At any point in this field there is an electric potential which is a measure of the work done in bringing a charge to the point.

The potential due to a charged body decreases as the distance from it increases: it becomes zero at an infinite distance.

A charge put upon an isolated conductor raises its potential. Its potential may be changed by bringing another charged body near to it. Thus a positively charged conductor can have a negative potential if the positive potential due to its own charge is overwhelmed by a negative potential resulting from a negatively charged body in its vicinity. This can be demonstrated with an electroscope.

The potential of a charged conductor can be influenced by the presence of a charged body in its vicinity.

The charge on a conductor can be changed only by contact* with another conductor.

Electric potential is a scalar quantity and resembles *temperature* and *height*. For practical purposes we employ arbitrary zero standards for these quantities, namely the freezing point of pure water and sea-level respectively. Likewise, in the case of potential we adopt that of the earth as the zero of potential.

Potential determines the direction in which charge will flow from one conductor to another,

just as temperature determines the direction of heat flow. When positive charge flows *from* the earth to a conductor connected to it, the conductor is said to have a negative potential: similarly, when positive charge flows *to* the earth from a conductor it is said to have a positive potential.

The electroscope really measures potential difference and not charge. A divergence can be caused by holding a charged rod near to the cap. No charge passes and the divergence again becomes zero when the charged rod is removed. If a can is stood on the cap and a charge is given to the can

---

* 'Contact' is taken to include 'close proximity', when the passage of the charge is indicated by a spark.

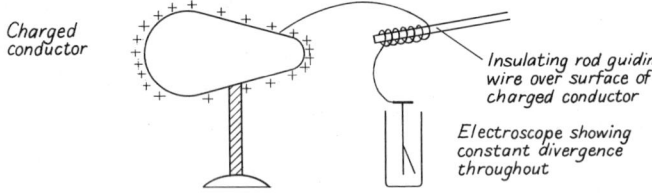

Charged conductor

Insulating rod guiding wire over surface of charged conductor

Electroscope showing constant divergence throughout

Fig. 32.5

Potential of conductor same at all points

by touching its inside surface with a charged sphere (*see* Faraday's ice-pail experiments below), the leaves diverge. If the experiment is repeated, using a larger can and an equal charge, the divergence is smaller. This shows the divergence does not depend only on the charge. If the electroscope is used without modification* its divergence will be proportional to the charge put on it.

By connecting one end of a wire to the electroscope plate and traversing the other end of the wire over the surface of any charged conductor, the divergence of the leaves remains the same (Fig. 32.5). (To do this, the wire must be held by an insulating rod.) This shows that

the potential is the same at all points on the surface of a conductor.

Charge does not distribute itself uniformly over the surface of a conductor. The surface charge density, i.e. the charge on unit area, varies with the curvature—the greater the curvature the greater the surface density. Hence, at a sharp end of a charged conductor the surface density is very high.†

The distribution of charge over the surface of a conductor can be investigated using a *proof plane*.‡ This is a small metal disc about 2 cm diameter mounted on the end of an insulating rod. When the disc is brought into flat contact with the

surface it picks up the charge it covers and this can be conveyed to a measuring instrument.

In carrying out the experiment, the pear-shaped conductor (Fig. 32.6) is charged to a certain

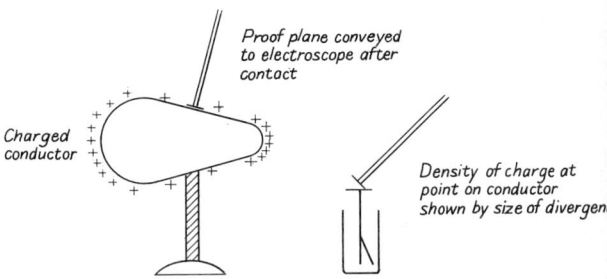

Proof plane conveyed to electroscope after contact

Charged conductor

Density of charge at point on conductor shown by size of divergence

Fig. 32.6 Surface density of charge on conductor depends on curvature

potential by connecting it momentarily to a high-tension source. The proof plane is then brought into contact with the 'sides' of the conductor and afterwards with an electroscope. In this way a sample of the charge covering an area equal to that of the proof plane is conveyed to the electroscope. Although little charge will have been removed from the conductor, it should be once again recharged to the same potential as previously. The electroscope and the proof plane should also be discharged. The procedure is repeated by touching the conductor at different points with the proof plane and conveying the charge to the electroscope. In this way the surface charge density on the more pointed end of the conductor is shown to be the greatest.

When the surface charge density exceeds a certain value the charge streams off. On a charged conductor, the mutual repulsions between the like charges causes this ever-present tendency for the charge to leave the surface. It causes an outward pressure, and a soap bubble, when charged, becomes measurably larger. The charge streams off sharp points, and so conductors which are raised to a high potential must have smooth well-

---

* i.e. its capacitance (*see* page 298) is maintained constant.

† The effective capacitance per unit area (*see* page 298) of the surface is great where the curvature is great, hence, in order to keep the potential the same at all points, the charge per unit area must be great.

‡ The working of a proof plane is often not clearly understood. When the small conducting disc is covering a small area of the surface of the charged conductor, the charge does not remain on the surface of the conductor but moves to the outside, i.e. it covers the outer surface of the disc which now virtually becomes part of the surface. On lifting the disc, the charge still remains on its outer surface: it does not flow underneath, as this is the inside of a 'hollow' conductor. Thus the proof plane is removed with its charge, the charge on the conductor spreading over the area from which the charge has been removed.

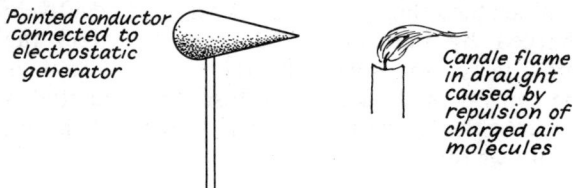

Fig. 32.7 Charge streams off a pointed conductor

rounded surfaces. Point 'collectors' of charge are used in collecting the charges induced in electrostatic generators (Fig. 32.7). A lightning conductor (Fig. 32.8) is made of copper and has

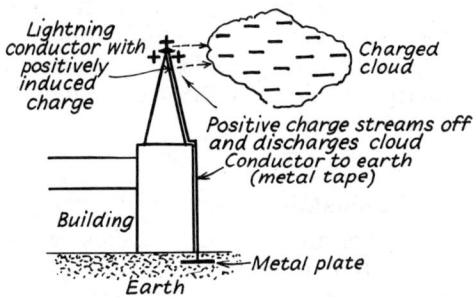

Fig. 32.8 Action of lightning conductor

sharp points of platinum. It is connected to earth by means of a metal band running down the building and terminating in a buried metal plate. During a storm, when charged clouds come near to the conductor, induced* charge of opposite sign streams off the points, so discharging the cloud and avoiding an intense spark from the cloud passing to earth through the building.

Charging by induction

We have seen earlier that a charged rod brought near to an insulated conductor affects the distribution of charge and raises its potential. Fig. 32.9a shows a positively charged rod held near to a conductor. Below it the potential diagram is shown. The potential of the conductor, because it is the same all over, is represented by the horizontal line *AB*. The potential diagram as it was before the introduction of the conductor is shown dotted. The changes in potential at the ends *A* and

* Refer to next paragraph. Fig. 32.9b shows the effect, but note the charges are reversed.

B are produced by the piling up of the charge. It thus appears that the charges induced equalize the potential all over the surface of the conductor. Fig. 32.9b shows the change when the conductor

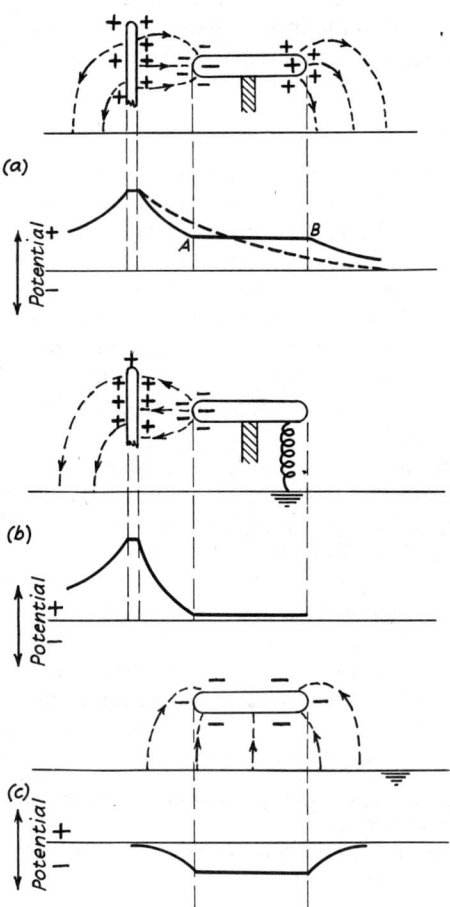

Fig. 32.9 Charging by induction

is earthed. Note that the positive charge and the field lines have gone. Field lines must have a potential difference between their ends, so the lines from the conductor to earth are destroyed. The potential of the conductor is zero when under the influence of the charged rod. When this is removed, the result is as shown in Fig. 32.9c. The conductor is now left negatively charged and, being isolated, its potential is also negative. The removal of the dominating charged rod shows this.

The electroscope can be charged in this way: in fact it is usual to charge it positively, using a negatively charged ebonite or Alkathene rod.

Method

1. Hold a charged rod near to the insulated conductor (or electroscope) to be charged.

2. Earth the conductor by touching.

3. Remove the earth connection.

4. Withdraw the inducing charge: the conductor will be charged oppositely to the charge on the rod.

Faraday's ice-pail experiment

1. An open can (Faraday used an ice-pail) is connected to an electroscope, and a charged, insulated, spherical conductor* is lowered into it without touching it. The leaves of the electroscope diverge but collapse again when the sphere is removed. If the sphere is now inserted and allowed to touch the inside of the can, the divergence remains unchanged. Furthermore, on removal the sphere is found to be completely discharged (Fig. 32.10).

When the charged sphere is inserted into the can, *all* the field lines from it end on the inside surface, inducing an equal and opposite charge on the inside of the can: an equal like charge is at the same time induced on the outside of the can. When the sphere touches the can and assumes the same potential, the field lines between them are destroyed and the equal opposite charges are neutralized. Thus, the can is charged, but none of the charges resides on the inside. Faraday's first ice-pail experiment shows that

(a) the induced charge is equal and opposite to the inducing charge.

(b) no charge resides on the inside of a hollow charged conductor.

2. The hollow can is now insulated by standing it on a slab of suitable insulating material and charged, say, positively. (This can be done with a negatively charged rod by induction, with an E.H.T. unit, or with an electrophorus.)† Into it is lowered an insulated sphere, which is then earthed. On removal, this sphere is found to be negatively charged. This experiment is repeated several times, and each time the sphere is found to carry an equal charge (Fig. 32.11).

---

* The sphere should either be suspended by a nylon thread or mounted on the end of a thin polythene rod. The thread or rod should be half a metre long in order to eliminate the effect of the hand.

† *See* electrical machines below.

When the sphere is earthed, it becomes negatively charged by induction. This charge lowers the potential of the sphere by the same amount as the can raises it, so that the resultant potential is zero. As the sphere always carries an equal charge, it follows that

the potential is the same at all points inside a hollow, charged conductor, and equal to the potential of the conductor itself.

Not only are the results of the above experiments of theoretical importance, they also have practical application. By the method of the second experiment we can obtain charge in exactly equal amounts. We can also completely discharge a conductor by bringing it into contact with the inside of a hollow conductor. Simply touching the electroscope with a charged body does not cause *all* the charge to flow to the electroscope. The charge is only shared.

Electrostatic shielding

A conductor electrostatically shields whatever is enclosed within it from the effects of charges outside. This fact is utilized in shielding components within radio sets and other electronic devices. High-voltage generators, which are very dangerous, are usually surrounded by earthed wire-netting screens. These protect any observers from electric shocks.

The research laboratory of the Central Electricity Generating Board has recently produced a suit for the use of engineers engaged in repairing faults on the new 400 000 V transmission lines. The suit is made of cotton and contains steel wires. The engineer is hoisted to the line on an insulating plastic chain and connection is then made between the suit and the transmission line. The suit, which covers the whole body including the head, is provided with a peak which gives protection to the body round an observation slit. This has solved an international problem of providing protection to workers engaged on line maintenance without the need to disconnect the supply.

Equal and opposite charges produced by rubbing

We can utilize the experimental principles of the ice-pail experiments to show that the fur with which an Alkathene rod is rubbed is given a positive charge of the same size as the negative

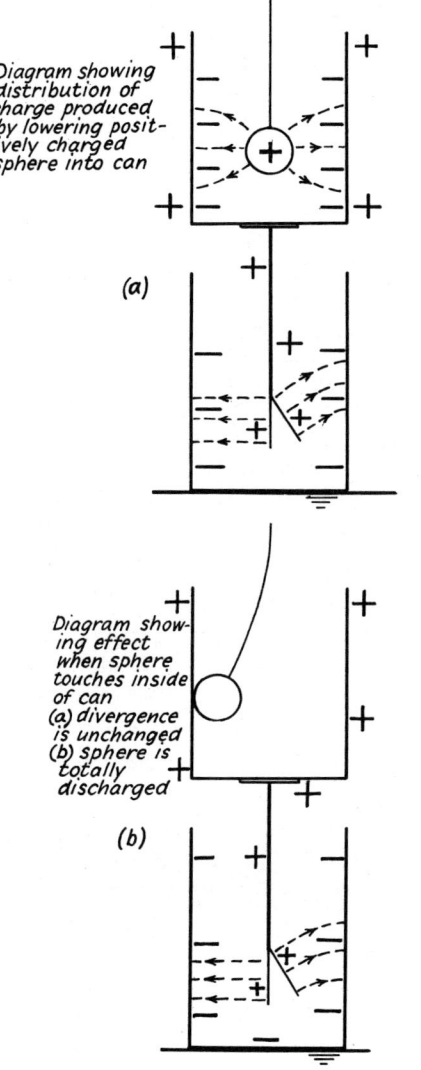

Diagram showing distribution of charge produced by lowering positively charged sphere into can

(a)

Diagram showing effect when sphere touches inside of can (a) divergence is unchanged (b) sphere is totally discharged

(b)

Fig. 32.10 Faraday's ice-pail experiment (1)

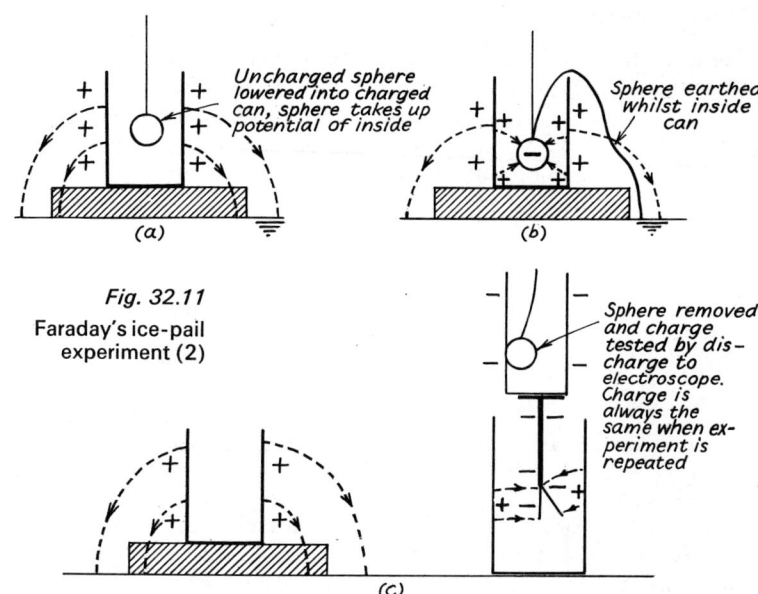

Uncharged sphere lowered into charged can, sphere takes up potential of inside

Sphere earthed whilst inside can

(a)

(b)

Fig. 32.11
Faraday's ice-pail experiment (2)

Sphere removed and charge tested by discharge to electroscope. Charge is always the same when experiment is repeated

(c)

by putting mercury in the can and stirring it with an Alkathene rod.

Another modification of the experiment is to rub the ends of uncharged acetate and Alkathene strips while holding them in a can or on the plate of an electroscope. No divergence results. When one strip is removed, however, a divergence results. This is positive or negative according to which strip is left on the plate.

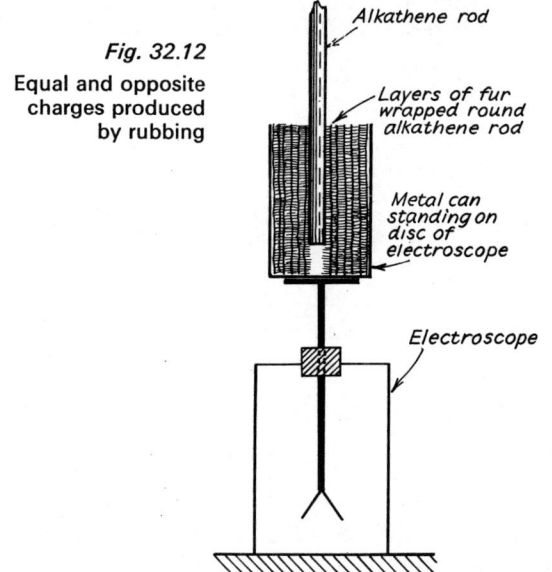

Fig. 32.12
Equal and opposite charges produced by rubbing

Alkathene rod

Layers of fur wrapped round alkathene rod

Metal can standing on disc of electroscope

Electroscope

charge given to the Alkathene. A suitable piece of fur is wrapped round the end of an Alkathene rod and packed loosely into a can resting on the disc of an electroscope (Fig. 32.12). The rod is now moved up and down and so charged by the fur. The leaves of the electroscope are not diverged. When, however, the rod is removed, the leaves show a positive divergence, which again is brought to zero by reinserting the rod.

The experiment can conveniently be performed

Capacitance

When a charge is given to a conductor, its potential is increased. The increase in potential depends upon the size of the body to which the charge is given. For example, a fixed charge raises the potential of a small sphere by a greater amount than it would do that of a large sphere.

The charge required to raise the potential of a body by unity is called the 'capacitance' of the body.

If Q units of charge raise the potential by V units, then $C = \dfrac{Q}{V}$. If Q is measured in coulombs and V in volts, then C is measured in farads.

A conductor has a capacitance of one farad when a charge of one coulomb raises its potential by one volt.

The farad is a very large unit, and capacitances are usually measured in microfarads, μF. (1 farad $= 10^6 \, \mu$F.)

An earthed conductor brought near to a charged conductor lowers its potential. The divergence of the leaves of an electroscope is reduced by bringing the hand near to the cap of the instrument. This is due to the capacitance of the conductor (the inside system) being increased by the presence of the earthed conductor. The arrangement is called a *condenser* or *capacitor*.

A capacitor is an arrangement by which the capacitance of a conductor is increased by bringing near to it another conductor (usually earthed).

In an ideal capacitor, all the field lines from one conductor terminate on the other.

Factors influencing the capacitance of a capacitor

To investigate the factors influencing capacitance, the apparatus represented diagrammatically in Fig. 32.13 is used. X and Y are two metal plates. X is earthed, and Y is insulated and connected to an electroscope. Y is given a charge. When the plate X is brought nearer to Y, the divergence of the electroscope is decreased. This indicates a fall in potential and, because the charge is constant, it indicates a rise in capacitance $\left(C = \dfrac{Q}{V}\right)$. On inserting a slab of glass, ebonite or other insulator between the plates, the divergence decreases, showing again an increase in capacitance. If, now,

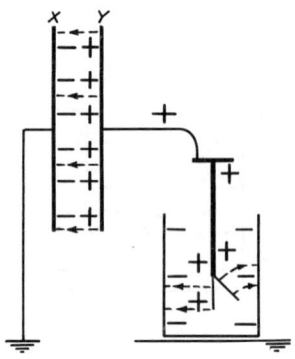

Fig. 32.13 Apparatus for investigating factors affecting capacitance

plate X is moved across plate Y so that there is a reduction in the common area between the plates, the divergence increases, showing a decrease in the capacitance. With more precise experiments we can show that:

Capacitance $\propto$ area of plates, A

$\propto$ medium between plates, ε

$\propto \dfrac{1}{\text{distance between plates, } d}$

i.e. $C \propto \dfrac{\varepsilon A}{d}$

The medium between the plates is known as the *dielectric* and ε is the dielectric constant or relative permittivity. Some approximate values for it are: vacuum 1.00, ebonite 2.8, glass 6, rubber 2.2, mica 6, quartz 4.5.

Capacitors

A capacitor (or condenser) stores electric charge. One with a large enough capacitance may store enough charge to give a large spark when the plates are joined. Capacitors are classified, not only according to their capacitance, but also according to what potential difference can be put across their plates without causing a breakdown of the insulation. A dielectric between the plates increases the capacitance, and so capacitors are made smaller by using as separators thin sheets of an insulating material, such as mica.

A Leyden jar is a well-known form of capacitor (Fig. 32.14). It consists of a glass jar with inside and

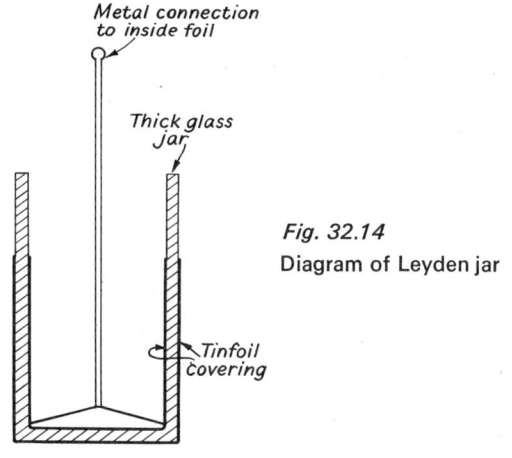

Fig. 32.14

Diagram of Leyden jar

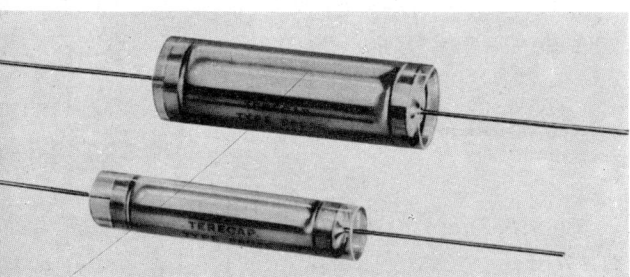

Fig. 32.15 Fixed capacitors

The larger capacitor is only 3.5 cm long and has a capacitance of 1 μF (375 V d.c. test). The dielectric is a polyethylene terephthalate plastic

chemical method for effectively producing 'plates' very near together, and enables the size of a high-value capacitor to be kept small. These capacitors are 'polarized', i.e. they must be connected the correct way round in a circuit.

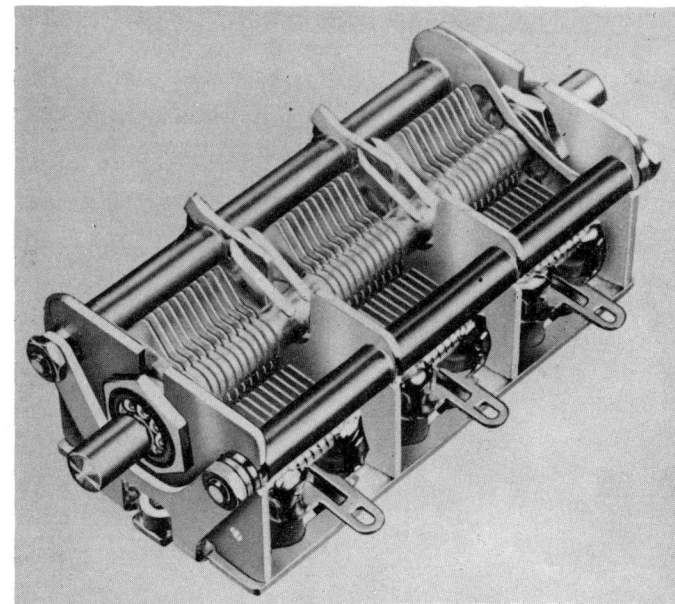

Fig. 32.16 Variable air capacitor

The photograph shows a high-quality 3-gang capacitor. The fixed and movable plates are clearly seen. The total capacitance is 0.0003 μF

outside coverings of metal foil. These coverings do not extend to the edge of the jar. The inside covering forms one plate of the capacitor and the outside the other. As glass has a high permittivity, the potential difference across the plates may be very high before a breakdown occurs.

The capacitor is a component of all electronic circuits. It will be seen in Chapter 40 that, in effect, a capacitor offers no barrier to an alternating current, whereas it stops a direct current. It is also a vital component of a tuned circuit. As such it has often to be of variable capacitance. Figs. 32.15 and 32.16 show examples of fixed and variable capacitors. Often the plates of fixed capacitors are long strips of tinfoil which are rolled up with thin, waxed paper between them as the dielectric. Another type of capacitor called *electrolytic* uses a

Electrical machines

The electrophorus

This is a simple device for obtaining an unlimited quantity of charge. It consists of a cake of insulator (or dielectric) such as ebonite,* the undersurface of which may be covered with a metal 'sole'. (This is not essential as the bench surface acts as the conducting sole when the cake is placed on it.) A metal plate provided with an insulating handle may be placed on the cake. The cake is laid on the bench, sole downwards, and the upper surface is negatively charged by rubbing it with fur. The plate is then placed on the charged

---

* Perspex, Alkathene or any modern dielectric is excellent.

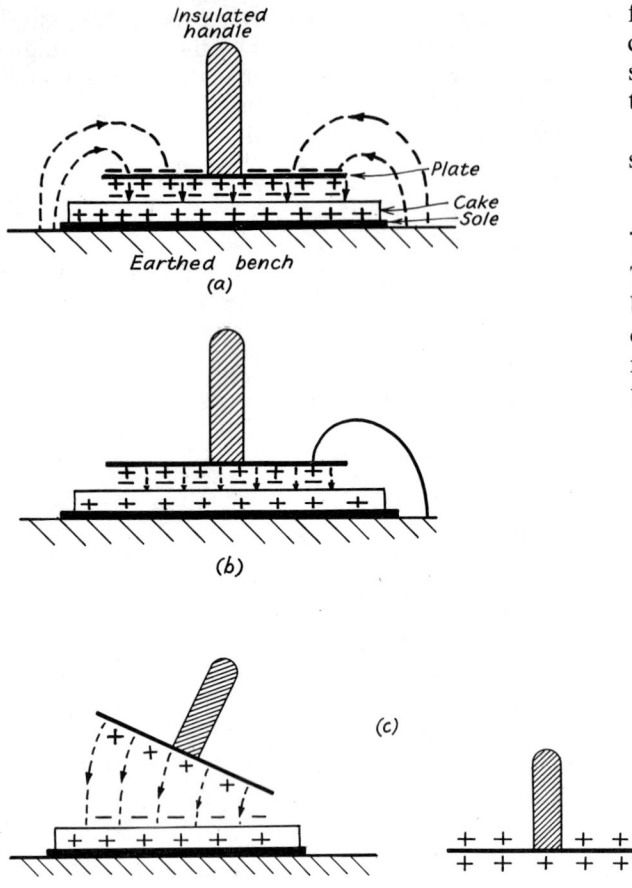

Fig. 32.17 The electrophorus
(Positive charge on upper surface due to induction;
induced negative charge goes to earth)

for no charge is removed from the cake. The charge on the plate is always the same size if the same procedure is followed. The diagrams illustrate the sequence of events.

The source of energy is the work done in separating the plate from the cake (Fig. 32.17*c*).

The van de Graaff generator

This is a modern electrostatic generator which has been used to provide the very high potential differences, of the order of three million volts, required in nuclear research. One of these generators (Fig. 32.18) is in use in the Cavendish labora-

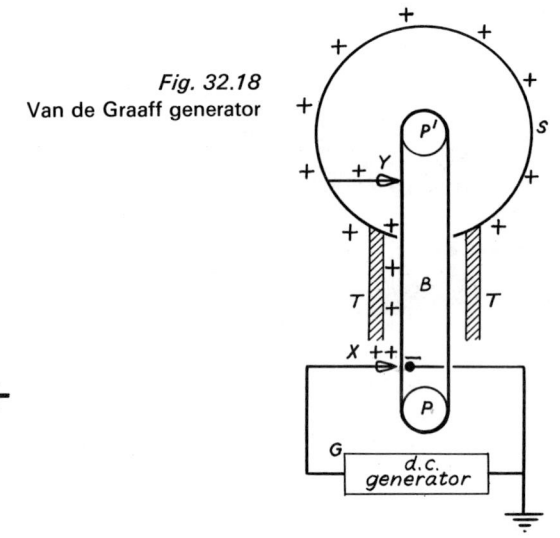

Fig. 32.18
Van de Graaff generator

face of the cake. The arrangement is shown in Fig. 32.17*a*. (If the plate is now removed, it will be found to carry very little charge. It makes contact with the cake at very few points and the negative charge is held by the attraction of the induced positive charge on the sole.)

When the plate is resting on the cake a positive charge is induced on its near surface, and a negative charge on the upper surface.

The plate is now earthed, and so the negative charge is conveyed to earth. The field lines from the plate to earth are destroyed (Fig. 32.17*b*). The earth connection is removed and then the plate is taken from the cake (Fig. 32.17*c*). At a great distance the positive charge spreads over the plate. This process may be repeated indefinitely

tory in Cambridge. It consists of a large sphere *S* supported on an insulating pillar or cylinder *T*. An endless belt of rubber, plastic material or other good insulator, passes round a driving pulley *P* and an idling pulley *P'* within the upper sphere. Charge from a d.c. generator *G* is sprayed on the belt by a row of pointed conductors *X*, and is collected off the belt by a row of similar conductors *Y*, within the sphere. The collected charge passes to the outside of the sphere and raises its potential.

These generators have been made in sizes ranging from miniature toys to the huge American generator which is housed in an airship hangar. Recent models are reduced in size by enclosing them in an atmosphere of nitrogen under pressure.

atoms. They are not vibrating about fixed positions like the metal atoms, but move freely, resembling to some extent, the molecules in a gas. It is the drift of these electrons which constitute a current in a metallic conductor.

Electrons cannot move freely in an insulator, nor over its surface. If a charge is put on an insulator, it remains where it is put.

When the plates of a capacitor are charged, the electric field between the plates affects the distribution of charges in the molecules of the dielectric and we say they become 'polarized'. Each molecule has a positive charge on one side and a negative on the other (Fig. 32.20); it becomes

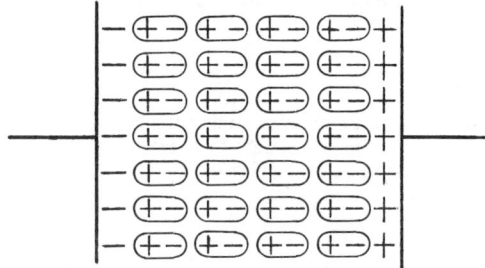

Fig. 32.20 Polarization in dielectric

an electric dipole. This is the form in which the energy is stored in a capacitor.

We can demonstrate this with a Leyden jar which has removable metal cylinders on the outside and inside of a glass vessel. The jar is first charged and then the three parts are separated. This must be done using insulating tongs. The metal vessels, when separated from the glass dielectric, are found to be uncharged. When the capacitor is reassembled (again using tongs), it is found to be still charged. This shows clearly that the charge (and hence the energy) is associated with the glass. The molecules are polarized. If the glass vessel is passed through a flame before reassembling, all the charge will be removed and the capacitor will now be uncharged once more.

Fig. 32.19 Van de Graaff generator at Oak Ridge National Laboratory, USA
The driving motor can be seen and the bottom of the belt is just visible

Conductors: dielectrics (insulators)

We must get quite clear the difference between these two types of material. Metals contain atoms which are arranged in an orderly pattern, but among the crystal lattice are free electrons which are unattached or loosely attached to

QUESTIONS

1. Describe the gold-leaf electroscope, and explain how you would use it to detect the sign of an electric charge.

How would you demonstrate that the charges produced on an ebonite rod and on a piece of fur by rubbing them together are equal in size and opposite in sign?   (O)

2. Describe the leaf electroscope and explain how you would use it to test the sign of an electric charge.

A hollow can is placed on the cap of an electroscope. An insulated charged conductor is lowered into the can, allowed to touch the inside, and then withdrawn. State and explain what is observed at each stage.   (O)

3. Describe the process of charging by induction, and explain how this process is used in charging an electrophorus.

State and explain what happens to the leaves of a charged electroscope when a metal plate held in the hand is brought up towards the cap, without touching it, and is then removed again.   (O)

4. Describe the electrophorus, and explain what happens at the various stages of its operation.

How would you demonstrate that the charge carried away on the plate is opposite in sign to that on the cake?

What is the source of the electrical energy possessed by the charged plate?   (O)

5. What do you understand by the statement that a body is electrified positively? Describe experiments to show:

(a) that some bodies are electrified positively and others negatively by friction;

(b) that equal quantities of positive and negative charge are produced by both friction and induction.   (L)

6. Describe the structure of a simple form of electroscope.

Explain briefly how you would use it to show (a) that the charge on a hollow conductor is entirely on the outside surface, (b) that the capacity of a charged conductor is increased when an earthed conductor is brought near, (c) that 'points' on a conductor facilitate its discharge.   (L)

7. When a strip of polythene is rubbed against an insulated metal plate the polythene becomes negatively charged, the metal positively charged. Explain this by reference to the movement of electrons.

If you touch these two charged bodies with your finger, i.e. you 'earth' them, only one of them becomes discharged. Which one becomes discharged, and why not the other?

If these two charged bodies are held near a flame, both become discharged. Explain how this happens.   (C)

8. Describe how you would investigate the distribution of charge over the surface of an isolated negatively electrified spherical conductor. State the result you would obtain.

Explain why this distribution would be affected by the presence of a large metal plate which is 'earthed' and held near to and facing the sphere.

Draw a diagram showing the distribution of charge and the electric field between the sphere and the plate.   (JMB)

9. Describe, explain and compare what happens when the following three experiments are performed separately:

(a) a charged metal sphere is lowered slowly into a metal can on the cap of a leaf electroscope until it touches the bottom;

(b) a similar metal sphere is brought up slowly to touch the outside of a similar metal can on the cap of an electroscope;

(c) the first experiment is repeated with an equally charged similar sphere but a larger metal can.

Comment on the state of electrification of the sphere on removal in each case.   (JMB)

10. What is meant by *electrostatic induction*?

Given a positively charged insulated conductor describe and explain how you would proceed to charge another insulated conductor negatively by induction.

How would you decide whether or not the induced charge was as great as the inducing charge, and what would you expect to find?

How might you arrange that the induced charge is numerically equal to the inducing charge? How could you check whether you had achieved this result?   (O & C)

11. (i) A charged insulated metal sphere, X, shares its charge with a smaller insulated metal sphere, Y, originally uncharged, by touching it. Describe an experiment to show that the charge which remains on X is greater than that given to Y.

(ii) Describe and explain the effect produced upon the divergence of the leaves of a charged gold-leaf electroscope by slowly bringing a large earthed metal plate near to, but not touching, the cap.   (JMB)

12. (i) Describe the distribution of charge over a charged pear-shaped conductor. (ii) Describe and explain the effect produced by a steel needle projecting from the surface of a charged conductor.   (JMB)

13. Describe and explain experiments, ONE in each case, to show how the capacity of a parallel plate condenser depends on TWO of the following factors:

(i) the area of the plates, (ii) the distance between the plates, (iii) the nature of the dielectric. Given a supply of tin foil and waxed paper explain how you would make a compact condenser of large fixed capacity.   (JMB)

14. (a) Describe *two* experiments to demonstrate that the whole of an electric charge on a charged hollow conductor resides on the outside.

Explain how a charge is transferred completely from a charged body to a hollow conductor.

(b) Describe *two* experiments designed to show the discharging action of points and mention *two* practical uses of the phenomenon.   (L)

15. Draw and describe a gold-leaf electroscope and explain how (a) you would charge it by induction, (b) you would attempt to use it to measure a potential difference of about 100 V.

When a 100 V battery is suitably connected to an electroscope a deflection of the leaves is produced. When the electroscope is disconnected, discharged, and the connections to the battery interchanged the same deflection of the leaves is produced. Explain this and describe a simple experiment to support your explanation.   (O & C)

16. Define *capacitance*.

Describe some simple form of capacitor, such as a Leyden jar. State the factors on which its capacitance depends.

A capacitor of capacitance 8 μF is charged to a potential of 500 V in 1 min. Find the charge on the capacitor, and also the average value of the charging current. (O)

17. Describe and explain an experiment to show that the divergence of the leaves of an electroscope is due to a difference of potential between the cap and the case.

The cap of an electroscope is connected to a large insulated metal plate and the system is charged. A similar plate, connected to earth, is placed opposite to and parallel with the first one. State and explain what happens when (a) the earthed plate is moved nearer to the charged one without touching it, (b) a sheet of glass is placed between the plates.

Name the practical device which is based on the above experiment and state what is achieved by its use. (L)

18. Draw a labelled diagram to show the structure of any one type of moving-leaf or moving-fibre electroscope.

When such an electroscope has been charged by momen-tary connection to a high-voltage supply, the following effects are observed:

(a) There is an extremely slow decrease in the reading of the electroscope.

(b) When the electroscope has had a capacitor, consisting of two parallel metal plates, connected to its terminals the rate of decrease of the reading is even less.

(c) When a beam of X-rays is passed between the plates of the capacitor the rate of decrease of the reading is quite rapid. Give an explanation of these effects and describe how you would use this, or any other apparatus, to compare the intensity of the X-ray beam with that obtained after passing the beam through 1 millimetre of lead. (C)

19. Two parallel horizontal metal plates are connected to a high-voltage d.c. supply so that the upper plate is positive. A charged oil drop remains stationary between the two plates.

(a) What is the sign of the charge on the oil drop?

(b) What would be the result of switching the supply off?

(c) What would be the result of increasing the potential difference between the plates? (JMB)

33 The magnetic effect of an electric current

Electrostatics and current electricity

We have seen in the previous chapters that an electric current consists of moving electric charges. Static electricity and current electricity must not be regarded as different. We study them separately because we use different instruments for their measurement. In static electricity we deal with very small charges and with large potential differences, be used, but it must be connected the correct way round in the circuit.) The vibrating switch, which touches first a and then b, rapidly connects the accumulator to the capacitor and then discharges it through the microammeter.

With $V=2$ V and $C=4$ μF, each charge on the capacitor $=2\times4\times10^{-6}$ C. If this is discharged through the microammeter 50 times each second,

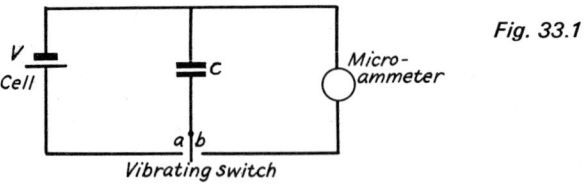

Fig. 33.1

whereas in current electricity we deal with large charges and with relatively small potential differences. A sensitive electroscope will not respond when a single, dry cell is connected to it, but one or two high-tension batteries will cause a divergence.

A rapid succession of small charges may constitute a measurable current. The following experiment illustrates this. In Fig. 33.1, C is a capacitor preferably with a large capacitance and V is an accumulator. (An electrolytic capacitor may

current $=$ charge $\times 50$ A
              $=2\times4\times10^{-6}\times50$ A
              $=400$ μA

This experiment is usually carried out to determine capacitance by employing a calibrated microammeter.

A modification of the experiment is to discharge the capacitor through a water voltameter, introduced in the circuit in place of the microammeter. The gaseous products can be collected and measured.

Another experiment to show that static electricity and current electricity are identical in nature is to connect an induction machine to a water voltameter. Although the charge generated is small, it is sufficient to liberate a small volume of gas if the machine is operated for some time.

If a number of charges from an electrophorus are put on a capacitor, they will cause a throw of the coil when discharged through a suitable galvanometer.*

Circuits

Fig. 33.2 shows an electric circuit, but before we can pursue the study of current electricity, we must understand certain elementary facts about circuits.

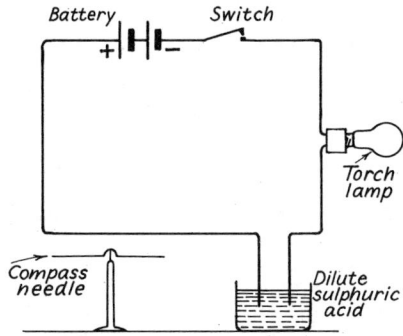

Fig. 33.2 An electric circuit

First, the current must have a complete path from one terminal (positive) of the battery to the other (negative). This path requires a conductor of electricity throughout. Copper wires are usually employed because copper is a good conductor of electricity. All metals are good conductors, but copper and silver are the best: silver is not used simply because of its high cost. In the circuit we are considering, we have a liquid conductor at one point. Solutions of acids and metallic salts in water are conductors. In connecting up a circuit, the ends of the copper wire must be clean, for oxides are non-conductors or insulators. Insulated wire, with bared ends, is used for connecting up.

---

* A galvanometer designed for measuring charge is called a 'ballistic' galvanometer. As the passage of the charge through the galvanometer is of very short duration, the throw of the coil is measured: the charge is not sustained to give a steady current. The charge is proportional to the throw of the coil.

Examples of the best insulators, as we have seen in the previous chapters, are rubber, silk, glass, ebonite, certain oils and certain plastics.

What is an electric current? When we speak of the current in a stream or in a pipe, we refer to the rate at which the liquid is flowing. The word is used in a similar sense in electricity but, in this case it is the flow of electric charge. In metals and other solid conductors, there are some electrons which are loosely attached to the atoms. When a battery is placed across the ends of a conductor, the electrons drift along the conductor from molecule to molecule. It is a good thing always to visualize an electric current as charge flowing in a wire.

The current is the same at every point in a circuit. If this were not so, then there would be an accumulation of electric charge at some point. A fundamental fact in current electricity is that there can be no accumulation of electric charge at any point in a circuit in which the current is constant.

Let us now close the switch in the circuit (Fig. 33.2) and observe what happens. Note the following three effects:

1. the compass needle is deflected,
2. the bulb lights, and
3. bubbles of gas are evolved from the wires in the acid.

This experiment illustrates the three effects of a current, namely, the magnetic, heating and chemical effects.

We shall now investigate these effects in detail.

The magnetic effect of a current

The production of a current of electricity was first recorded in 1800 by Volta, who invented a simple form of cell. Six weeks later, the chemical effect was discovered, but it was not until 1819 that Oersted discovered the magnetic effect. He was lecturing at the University of Copenhagen on electricity and probably, while talking of its possibilities, tried the effect of a current on a compass needle. He held the wire carrying the current over a compass needle, and obtained a deflection of the needle. This showed that a current must have a magnetic field associated with it.

Consider a long wire stretched across the room

and carrying an electric current. We should expect the field to be the same all along the wire at the same distance from it, and the same at an equal distance from it in every direction. Thus, we deduce that the field should be symmetrical about the wire. In a plane at right angles to the wire, the field lines should be circles.

Pass a thick piece of wire or a brass rod through a hole in a piece of wood supported horizontally. Sprinkle iron filings over the board and pass a large current through the wire.*

(a)

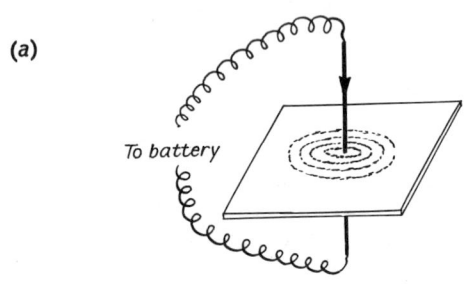

(b)

Fig. 33.3 Magnetic field produced by a straight wire carrying a current

(*a*) experimental arrangement
(*b*) photograph of field produced with iron filings

Tapping will set the iron filings in circles about the wire (Fig. 33.3). We cannot deduce the direction of the field lines. This has to be found experimentally and is easily done with a compass needle (Fig. 33.4). If a sufficiently large current cannot be obtained to set the iron filings, a smaller current will show the circular field lines with a number of

* With a single wire a current of 30–50 amperes is necessary to produce a good field.

small compass needles. The direction of the field is shown by the compass needles. It may be remembered by the following rule:

Maxwell's corkscrew rule. Imagine a corkscrew being driven in the direction of the current: the motion of the thumb indicates the direction of the field (Fig. 33.4).

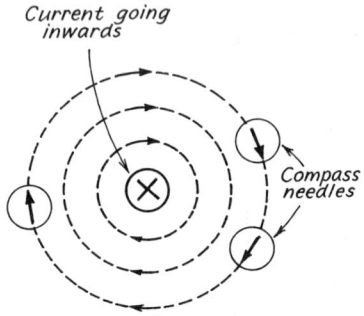

Fig. 33.4 Magnetic field produced by a straight wire carrying a current (in section)

There are other rules among which the 'GRIP' rule (Fig. 33.5) is the best known. If the wire carrying the current is grasped by the *right* hand with the

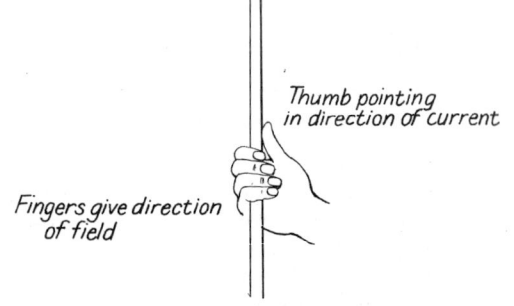

Fig. 33.5 The *right*-hand GRIP rule

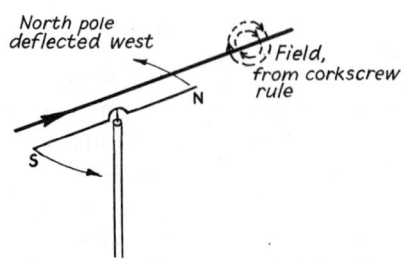

Fig. 33.6 The SNOW rule
Current S to N over needle: north pole deflected to the west

thumb pointing in the direction of the current, then the fingers indicate the direction of the field lines.

In applying this rule, it must be remembered that the right hand has to be used. The advantage of the corkscrew rule is that either hand can be used. (*See also* Fig. 33.6.)

To return to Oersted's experiment, the deflection of the needle is found by applying the above rule. The current tries to turn the compass needle at right angles to it, but the earth's field prevents this.

Fig. 33.7 Photograph of Faraday's apparatus

Faraday devised a simple piece of apparatus to show the direction of the field about a straight wire. A modified form is shown in Fig. 33.7 and Fig. 33.8.

Two bar magnets *NS* are joined by the cross-piece *AB* and suspended by thread *C*. The current is led up a brass rod provided with a mercury cup at *Q*. Into this dips one end of a wire connected to the cross-piece *AB* at *R* and continued at right angles

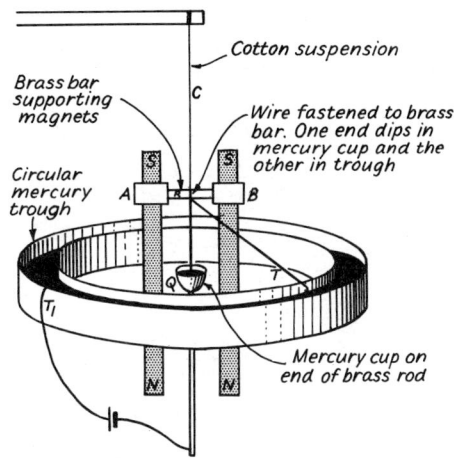

Fig. 33.8 Faraday's apparatus

Rotation of a pole round a current. Current flows up brass rod, along horizontal wire and into circular mercury trough. Magnetic field of current acts only on N-poles, as current does not pass near to S-poles

as shown, with the other end dipping into the circular mercury trough TT_1. The current is conveyed by this path into the mercury trough. By this means, only the N-poles of the magnets are near to the current and the magnet system rotates showing the rotation of a free N-pole round a current.

Field due to a coil

Next consider the magnetic field produced by a current passing round a circular coil. Applying Maxwell's rule, the field lines due to each little section of the current will be threaded through the coil, all in the same direction. But the field lines cannot cross, so they crowd together and become flattened on the inside. Fig. 33.9*b* shows a cross-section of the resulting field, which can easily be demonstrated experimentally. If several turns of wire are used in the coil, a current of a few amperes will produce a very good field which can be revealed

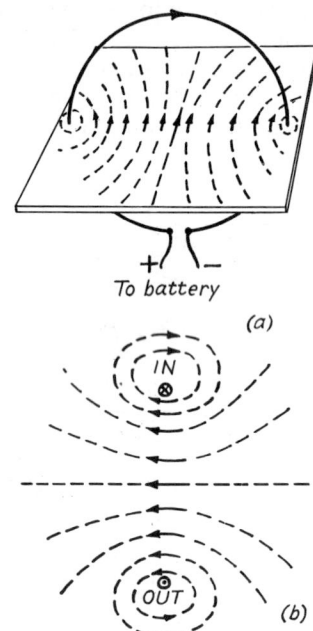

To battery

(a)

(b)

Fig. 33.9 Field due to a circular coil carrying a current

by iron filings (Fig. 33.10). Note that the field is perpendicular to the plane of the coil at its centre.

The field of a circular coil is like that produced by an iron disc with the same diameter as the coil, magnetized, with one face north and the other south.

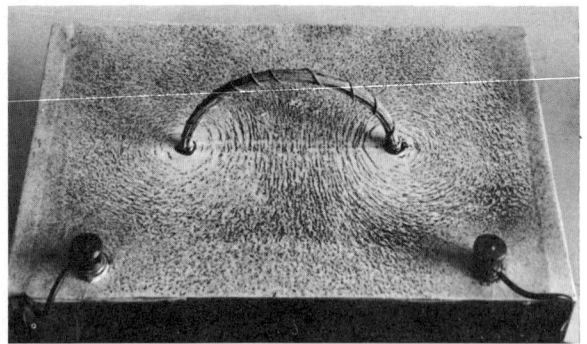

Fig. 33.10 Photograph of field produced by a circular coil carrying a current

Field due to a solenoid

A solenoid is a spiral or helix of wire through which a current is passing. The field due to a solenoid can be regarded as the resultant field of a number

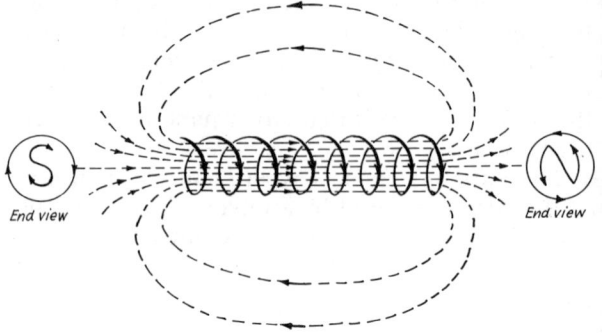

Fig. 33.11 Field due to a solenoid

of circular coils placed close together (Fig. 33.11). This will be like that of a number of magnetized discs arranged in a pile with the N-pole of one touching the S-pole of the next. This arrangement forms a bar magnet.

Note:

(*a*) The field inside a solenoid is uniform—the lines of force are parallel (Fig. 33.12).

(*b*) The external field of a solenoid resembles that of a bar magnet (Fig. 33.13).

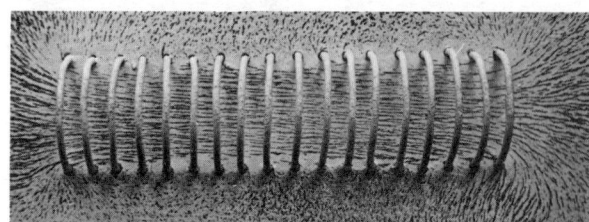

Fig. 33.12 Photograph of internal field of a solenoid

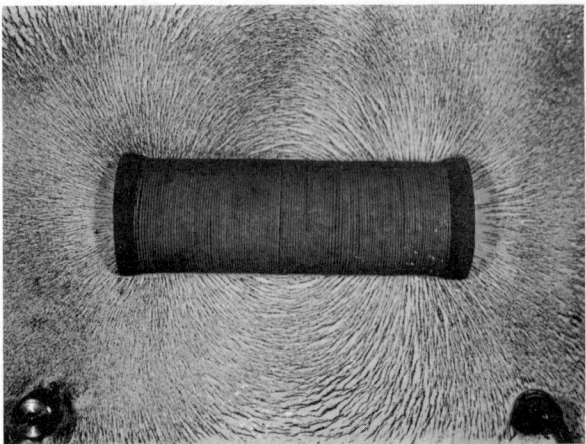

Fig. 33.13 Photograph of external field of a solenoid

The field can be considerably increased by the addition of an iron core. The iron becomes a powerfully induced magnet in the strong internal field. (Fig. 33.11 indicates how to find the polarity of the ends knowing the direction of the current.) This arrangement is an electromagnet. If the straight solenoid is bent into a horseshoe shape, we have an electromagnet of the usual design (Fig. 33.14). Note that the current must pass in opposite directions round the coils in order to get opposite poles. Soft-iron pole-pieces are usually placed on the top of the poles. As the field lines keep their path in air as short as possible, they crowd into the small area of the pole-pieces and they produce a very strong field in the gap (Fig. 35.15).

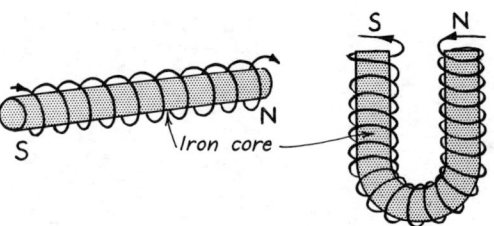

Fig. 33.14 Electromagnets

Electromagnets

Electromagnets of the type described above are used mainly in laboratories for demonstration purposes. Those for commercial use are nearly always circular. A short solenoid of many turns is embedded in a circular groove machined in a short, iron cylinder (Fig. 33.16). This provides a path in the

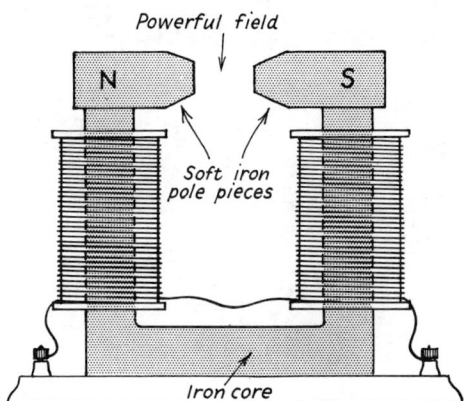

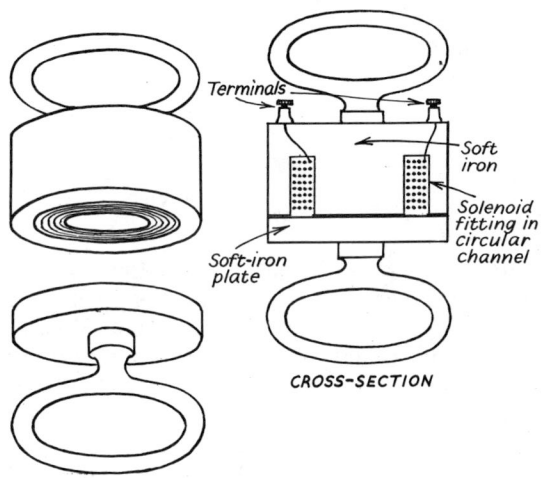

Fig. 33.16 Lifting electromagnet

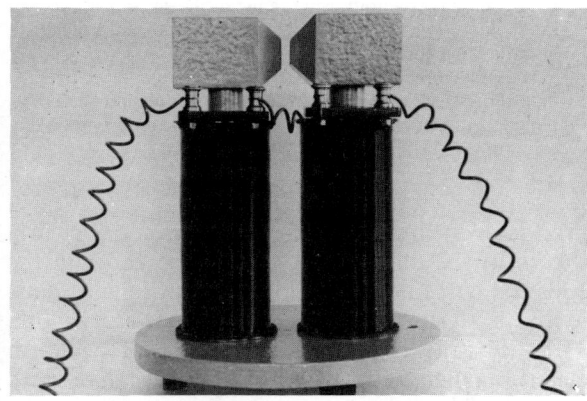

Fig. 33.15 Laboratory form of electromagnet

iron for the field lines outside, as well as inside, the solenoid. If a circular plate of iron is put over the open end, the path of the field lines is wholly within the iron. This considerably strengthens the force of attraction between the magnet and the plate.

Apart from the use of an electromagnet for lifting large iron objects or scrap iron, many instruments employ electromagnets of special form and for special purposes. (*See* page 275.)

The electric bell

The principal component of an electric bell (Fig. 33.17) is an electromagnet. When a current passes through the coils of the electromagnet, the soft-iron core is magnetized and attracts the soft-iron bar or *armature*. When this happens, the hammer which

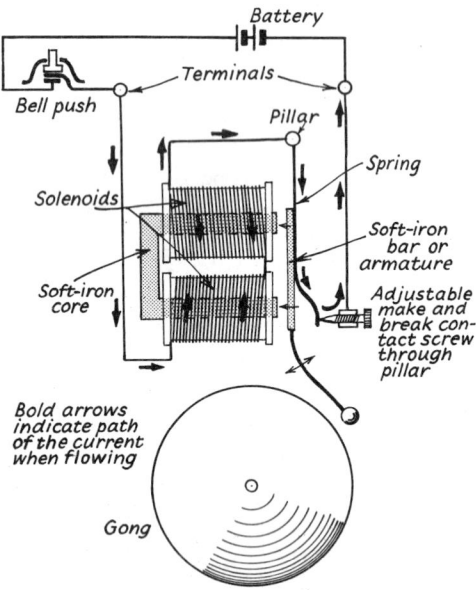

Fig. 33.17 The electric bell

is attached to the armature hits the gong. The armature is supported on a strip of springy steel which is firmly fixed at its other end. When the attraction takes place, contact between the armature and contact-screw is broken. This breaks the electric circuit and so stops the attraction of the electromagnet. The action of the steel spring returns the armature and re-establishes contact at the screw. The cycle of operations is repeated to

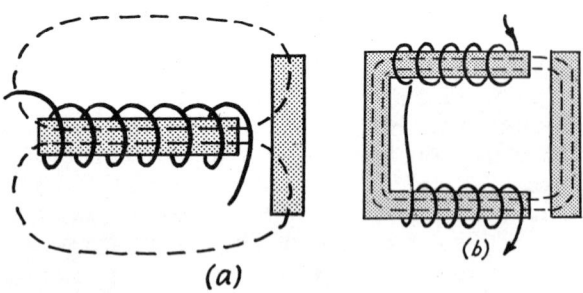

Fig. 33.18 Diagrams to explain the function of double coils

Fig. 33.19 Circuit breaker

When an overload current passes through the solenoid, a short movable iron slug is drawn from one end into the centre of the solenoid. This reduces the air gap in the magnetic circuit, thus causing an increased pull on the armature just above the solenoid.

The circuit breaker, or cut-out, has many advantages over an ordinary fuse. It is re-set by a simple switch and can be adjusted to withstand a large momentary surge without tripping

produce continuous ringing. The circuit is clearly shown in the diagram. After passing through the coils the current passes along the spring and back to the battery by way of the contact-screw. Fig. 33.18 illustrates the great advantage of using an electromagnet with two poles rather than with one pole. The arrangement provides a path for the field lines which is almost entirely in iron.

The relay

The relay (Fig. 33.21) is a simple and useful device incorporating an electromagnet. It is used where one circuit has to influence or control another. Relays are used extensively in communications and the telephone network involves many in each circuit. They are useful where an automatic device, using

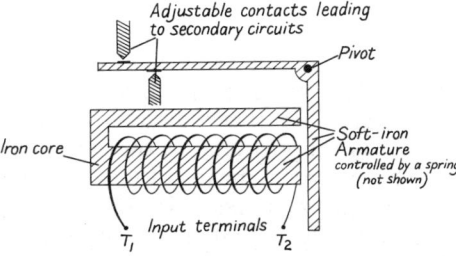

Fig. 33.21 The relay

a small current, has to control a circuit carrying a large current. The principle involved is that when a comparatively small current is made to flow through a magnetizing solenoid, the iron core attracts the armature which closes a gap between two points in the second circuit.

Detection and measurement of a current using the magnetic effect

A simple current detector or galvanometer

We have seen that a circular coil carrying a current produces a magnetic field at its centre at right angles to the plane of the coil. Thus, a compass needle placed at the centre will tend to set in the direction of the magnetic field. If the coil is arranged in the magnetic meridian, its field will be at right angles to the earth's field and, when the current is

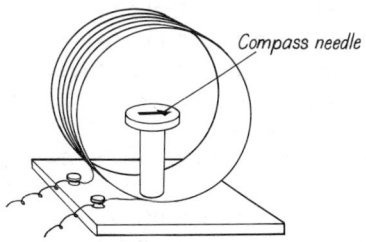

Fig. 33.22 A simple current detector

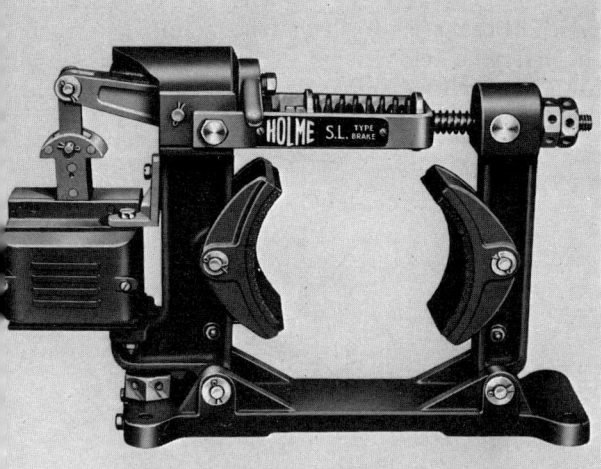

Fig. 33.20a Electromagnetic brake

While the electromagnet seen at the left of the photograph is energized, the brake-shoes are held apart allowing the controlled machine to operate. On pressing the 'stop' button or on electrical failure, the electromagnet becomes de-energized and the brake-shoes automatically close by means of the heavy duty spring, which can be seen behind the name-plate. This type of equipment is classed as 'fail safe' because the controlled machine will automatically stop when the electrical circuit is broken

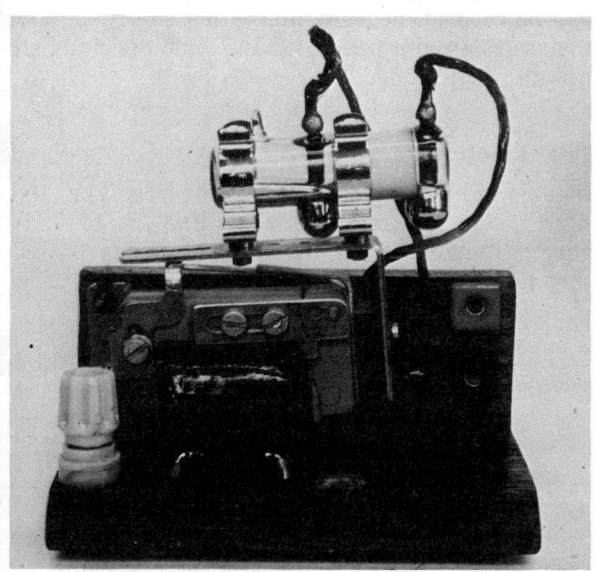

Fig. 33.20b Mercury switch

The sealed glass tube contains mercury which moves from end to end as the tube is rocked, and so makes and breaks the circuit between two contacts fused through the glass. The tilting of the tube is operated by the electromagnet which can be seen in the picture

switched on, a compass needle at the centre will be deflected from the meridian and set along the direction of the resultant of the two fields. This simple arrangement can be used for detecting very small currents.

The model shown in Fig. 33.22 is easy to make.

Ammeters

An ammeter is an instrument used to measure current in amperes. the ampere being the unit of current (*see* page 347). The magnetic field produced by a current is proportional to the strength of the current. This affords a means of measuring it.

Repulsion type of ammeter. Two curved soft-iron strips *A* and *B* (Fig. 33.23) are arranged in a small

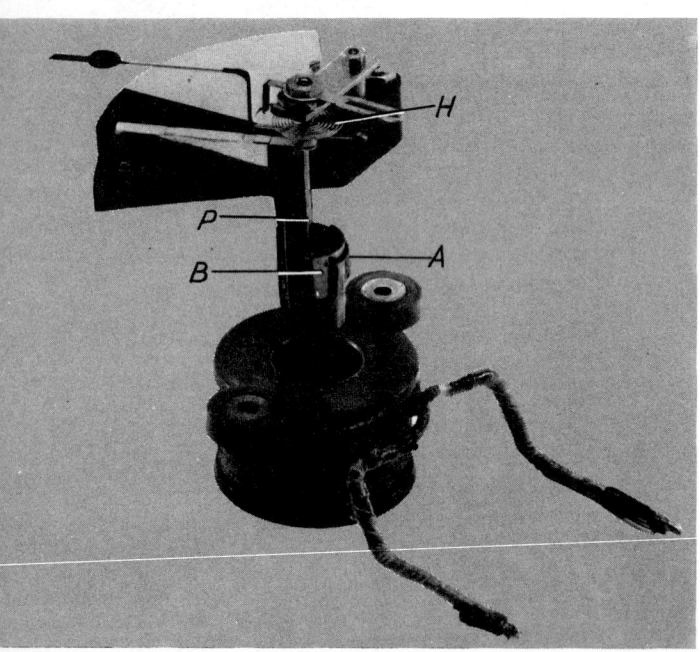

Fig. 33.23 Repulsion-type ammeter (moving system has been removed from coil)

coil. *A* is fixed and *B* is attached to a movable lever pivoted at *P*. When a current is passed through the coil the strips become induced magnets with like poles adjacent. The repulsion between them causes *B* to move away, and this is indicated by the movement of the pointer. The moment of the repulsive force is balanced by that produced by the coiling of the hair-spring *H*. The stronger the current, the greater is the force of repulsion, resulting in a higher reading on the scale.

The moving-iron type ammeter. We have seen on page 281 that a piece of iron tends to move to where the magnetic field is stronger. This principle is used in the design of a simple ammeter (Fig. 33.24). The

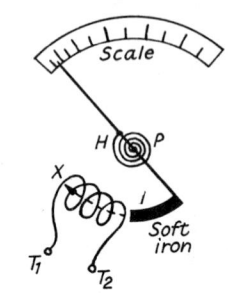

Fig. 33.24 Moving-iron ammeter

current is passed through a solenoid near to which is a small piece of iron *I* attached to a pointer pivoted at *P*. The iron is drawn into the solenoid, and as it enters, the pointer moves round the scale. Equilibrium is produced either by the hair-spring *H* or simply by gravity acting on the piece of iron.

Both the above ammeters respond in the same way to currents sent in either direction. Hence they can be used to measure alternating currents. Note the scales are not linear, i.e. the scale divisions are not all of equal size.

The telephone earpiece

Fig. 33.25 shows a telephone earpiece. The first such instrument was made by Bell, in 1877. A varying current from the microphone is passed through the coils, making an electromagnet of vary-

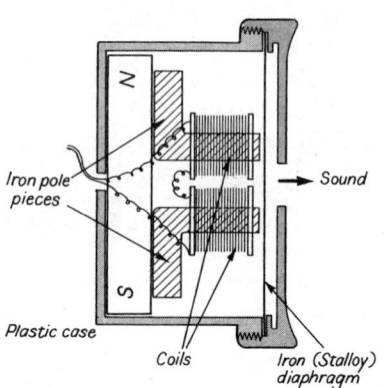

Fig. 33.25 Telephone earpiece

ing strength. This attracts the soft-iron diaphragm, setting up vibrations in it, similar to those of the microphone diaphragm, and thereby reproducing the sound waves. The iron pole-pieces within the coils are attached to the poles of a horseshoe magnet. The incorporation of this magnet makes the earpiece reproduce the true frequency and not double the input frequency, e.g. without it an alternating current of 50 Hz would be reproduced as a note of frequency 100 Hz. It can also be used as a microphone but as such is not very powerful.

The commonest type of microphone is represented diagrammatically in Fig. 33.26. A diaphragm of mica or thin metal has a carbon disc attached. This forms one end of a chamber packed with carbon granules. At the other end is another carbon disc. The vibrations of the diaphragm caused by sound waves alter the density of the packing of the granules and so the electrical resistance. Thus, when a cell is connected in series

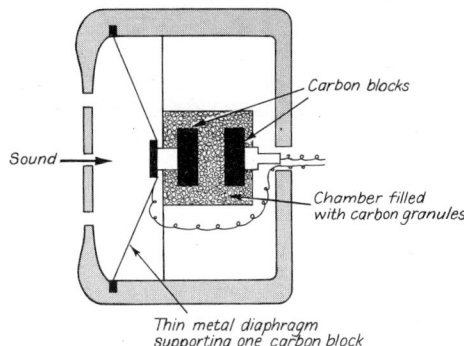

Fig. 33.26 Carbon microphone

with the microphone and an earpiece, the vibrations of the microphone diaphragm cause fluctuations in the current, and these, in turn, cause the earpiece diaphragm to vibrate and reproduce the sound waves.

QUESTIONS

1. Draw diagrams showing the pattern of the magnetic field due to (*a*) a long straight wire carrying an electric current, (*b*) a circular wire carrying a current. In each case mark the directions of the current and the field.

Give a labelled diagram of an electric bell operated by a battery, and explain how it works. (O)

2. How would you use an electric current to magnetize a piece of iron or steel?

What are the chief differences in the magnetic behaviour of soft iron and steel?

Describe a telephone receiver, and explain how it works. (O)

3. A low-voltage battery is sometimes used, in conjunction with an electromagnet, to switch on a large current in another circuit. Draw a labelled diagram to show the structures of the electromagnet and other fittings which make them particularly suitable for the purpose. (C)

4. Describe an experiment to investigate the magnetic effect of a current flowing in a long straight wire. Draw a diagram showing the nature of the field produced.

Sketch the field due to the same current flowing through two straight parallel wires (*a*) when the directions of the current are the same, (*b*) when they are opposite. Neglect the earth's magnetic field. (O & C)

5. Draw diagrams showing the lines of magnetic force in a horizontal plane associated with a current flowing in (*a*) a long, straight, vertical wire, (*b*) a solenoid with its axis horizontal. In each diagram show the direction of the current

and in (*b*) state a rule which gives the polarity of the coil. Describe the method you would use to plot the lines of force in either (*a*) or (*b*). (JMB)

6. Explain how you would investigate the magnetic field due to a current flowing in a long solenoid. Sketch the type of result you would expect to find for a particular direction of the current.

A magnetic needle free to pivot in any direction about its centre of gravity is placed inside a horizontal solenoid the axis of which is in the magnetic meridian. For a given strength and direction of current the needle is observed to set vertically. Explain this. (O & C)

7. How would you magnetize a rod of steel *AB*, by electrical means, so that the end *A* becomes a N pole? Illustrate your answer by a suitable diagram giving the direction of the current and the polarity of the magnet.

Describe a moving iron ammeter and explain its action, pointing out why iron is preferred to steel. Explain why this instrument can be used to measure both alternating and direct current. (L)

8. Explain the construction and mode of action of (*a*) the carbon microphone used in telephones, (*b*) the telephone receiver. Illustrate your answer by means of diagrams.

A telephone receiver contains, among other things, a diaphragm of magnetic material and a permanent magnet. State and explain carefully what would be the effect on the frequency of a musical note reproduced by the receiver if the magnet were to be replaced by a piece of soft iron of similar size and shape. (O & C)

34 Current, potential difference, resistance

An electric current

The magnitude of a current is measured by the rate of flow of charge. But we have seen that there are two kinds of electric charge, and so a current could be positive charge flowing in one direction or negative charge flowing in the other. All our observations to date show that charge is carried on particles. The association of electric charge with atoms was envisaged by Faraday and supported by Weber and Helmholtz.

The simple experiment described on page 305 shows an electric current passing through a liquid conductor and also through the metal wire connecting the circuit.

The mechanics of conduction

In Chapter 32 we mentioned that the atoms of some substances (in particular metals) have one or more outer orbiting electrons which become detached and move in the interatomic space. The atoms are left with one or more electrons short and so they are positively charged—they are ionized. They are not free to move about like the free electrons. When an electric field is created, e.g. by connecting a battery across the ends of the metal, the free electrons drift in the electric field towards the

electrode of the cell at the higher potential. This drift of electrons is the current.

In a liquid conductor there are both positive and negative moving charges, but this will be explained in Chapter 36.

Again a different method of conduction occurs in a gas and this is dealt with in Chapter 41.

When the switch is closed in a simple circuit such as that on page 305, the lamp lights up at once. We are inclined, therefore, to think that the charge from the cell passes round the circuit extremely quickly. This is not so for it can be compared with a conveyor belt delivering packages to a higher level. If the belt is started when full of packages, delivery at the top starts at once. Those packages loaded at the bottom take some time to reach the top. When a switch is closed charge is delivered at once—not the charge directly from the cell, but charge which was in the connecting wires.

The speed with which charge flows through a metal depends upon the potential difference applied, but it is never more than a few centimetres per minute. The speed with which the charged atoms flow through a liquid conductor can be measured directly by employing coloured chemical indicators.

We have explained a current in a metal as a steady flow of electrons in the field produced by the battery. As a charge is acted upon by a force when in a field, the electrons should have an acceleration due to the constant force acting on them. A constant current infers charge moving with a constant speed. Thus Ohm's law indicates that the electrons must be subjected to some retarding action which causes them to attain this constant speed, an effect which can be produced, for example, in a falling raindrop (*see* page 112).

We shall see later that when electrons in a vacuum are subjected to a constant force they move with a constant acceleration.

The measurement of current and charge

An electric current is the rate of flow of charge. If we define the unit of either current or charge, we can define the other from it. We have seen in Chapter 1 that, in devising the SI system of units, we have coupled the basic units of mass, length and time to electricity by adopting the ampere, the unit of current, as a basic unit. We are not yet able to understand the definition of the ampere: it is given on page 347.

> **The unit of charge—the coulomb. If the current in a wire is one ampere, then the charge which flows past a point in the wire in one second is one couloumb.**

Thus   **coulombs = amperes × seconds**

charge = It

We could define the unit of current in terms of the number of electrons passing a point in a wire in one second. In other words, we could define the coulomb as equal to the charge on so many electrons and define the ampere from the coulomb. It is, however, found advantageous to define the ampere first and from it to define the coulomb: the charge on the electron is then expressed in coulombs.

The water analogy of current

Consider the flow of water in a pipe. Water will not flow from *A* to *B* (Fig. 34.1) along a horizontal pipe such as *AB* unless it is forced. It can be forced by providing a head of water in a reservoir *R* or by attaching a pump *P*. In both of these, and in any other method which may be used, the pressure at *A* must be greater than that at *B*. Or in other words, a pressure difference must be created across the ends of the pipe. The pump provides the motive force which produces the pressure difference.

Electric charge is very similar. The free electrons

in wire *XY* will not flow unless they are forced along. To do this, a battery or generator is applied across the ends of *XY* to provide the electromotive

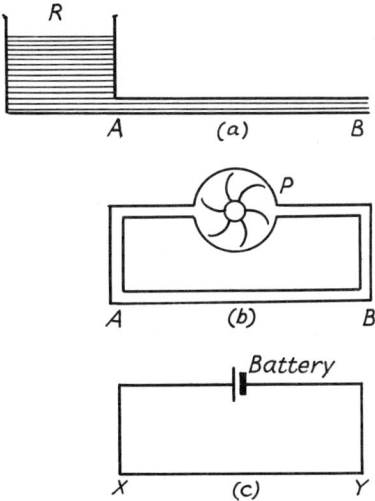

Fig. 34.1 e.m.f. and potential difference

force (e.m.f.) and to create the electrical pressure difference across the ends. This electrical pressure difference is the *potential difference* (p.d.), and we measure it in volts.* We conclude, then, that a potential difference must be applied across the ends of a conductor to produce a current in it.

We have already mentioned potential difference in the chapter on electrostatics and have seen that when a p.d. exists across two charged conductors and they are connected, charge passes from one to the other until the two potentials become equal. In electrostatics only relatively small charges are involved and the flow of these charges causes a very small current which only persists until all the charge has flowed across, which is usually for a very short interval of time. In current electricity we are involved with much smaller potential differences but with much larger charges. A battery or a generator can replenish the charge as fast as it is conveyed through the conducting wire and so a constant p.d. can be maintained.

The idea of a water circuit can be developed further and can be demonstrated as illustrated in

---

* For definition of the volt, see page 322.

Figs. 34.2a and b.* The apparatus is filled with water and circulation is produced by a pump P driven by a small electric motor. The flowmeter† F

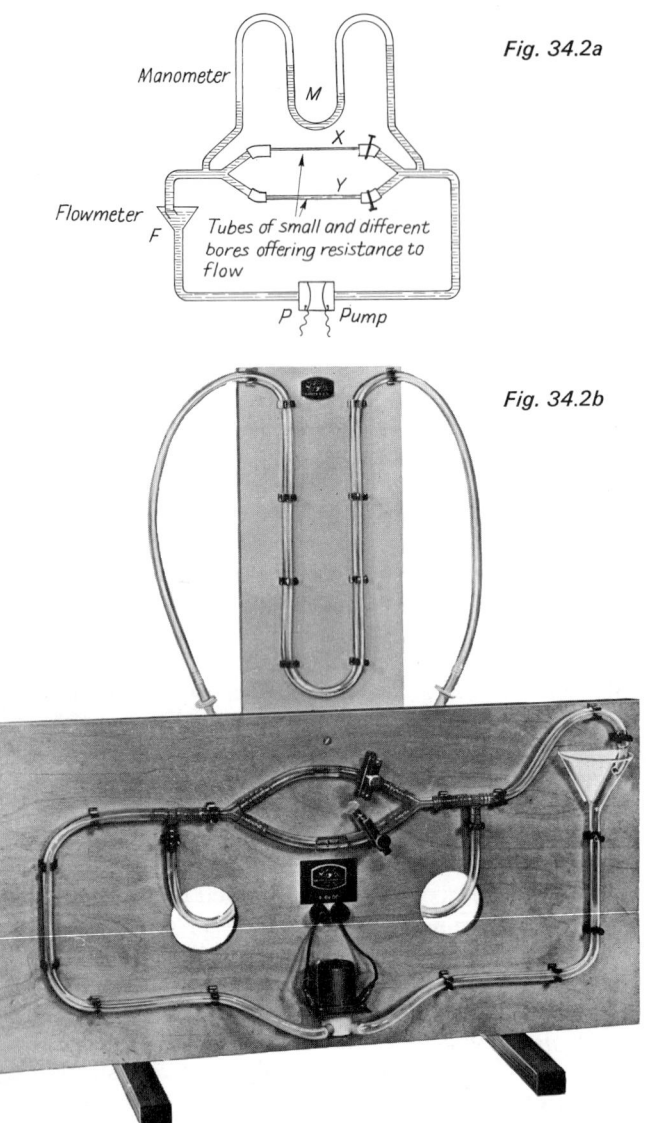

Fig. 34.2a

Fig. 34.2b

gives the same reading when placed anywhere in the main flow. The gauge M measures the pressure differences across the ends of the parallel tubes X

and Y. These have small bores compared with the other tubes in the circuit.

Certain elementary facts concerning currents and circuits can be appreciated by considering the water circuit.

1. There can be no accumulation of charge at any point in the circuit.

2. When the current is started by a switch at one point in the circuit, the current commences at all points at the same time.

3. The current is the same at all points in a simple circuit.

4. When the e.m.f. is increased (the pump worked harder) the current is increased and so is the p.d. across the resistors* (the reading of M is increased).

Ohm's law

The relation between the current and the potential difference was first studied by G. S. Ohm in 1826. He applied a cell† across the ends of a metallic

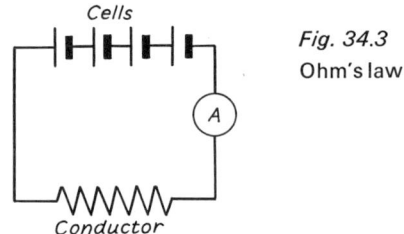

Fig. 34.3
Ohm's law

conductor and measured the current produced in it (Fig. 34.3). (Ohm had to measure the current by observing its effect upon a compass needle, but we can put an ammeter in the circuit and read off the current directly.) He then doubled the p.d. by using two similar cells, and again found the current. Continuing in this way, by adding more cells he found how the current was affected by changing the p.d. His conclusion is contained in the following law, which is of the greatest importance.

---

* Nuffield *G. to E.*, IV, No. 116.

† This interesting device is described in the Nuffield reference above. The rate of flow is judged from the swirling of the water in the funnel.

---

* The terms 'resistor' and 'conductor' are opposites. We usually employ resistor for a body which offers a big resistance to the flow and conductor to one which offers a comparatively small resistance. An insulator has a *very* high resistance.

† Ohm found the cells he used unsatisfactory and he resorted to the use of copper-bismuth thermocouples. Accumulators can be used for this demonstration.

Provided the temperature and other physical conditions of a metallic conductor are kept constant, the current is proportional to the potential difference applied across its ends.

$$\text{Current} \propto \text{p.d.}$$
$$I \propto V$$
$$I = \text{constant} \times V = kV$$

or if
$$k = \frac{1}{R}$$

$$\underline{I = \frac{V}{R}}$$

The constant of proportionality k is called the *conductance* and represents the ability of the conductor to pass the charge. It is more usual to employ the reciprocal R of the conductance, called the *resistance*,[*] which represents the ability of a conductor to resist the flow of charge. From Ohm's law we deduce:

$$I = \frac{V}{R} \quad ; \quad V = IR \quad ; \quad R = \frac{V}{I}$$

From any law, such as the one above, we can define one new quantity and one only. Thus, we can define the unit of current, p.d., or resistance, provided we define the other two by means not connected with Ohm's law. We decide to define the unit of resistance from the law, and define those of current and p.d. by other methods.

The unit of resistance—the ohm. A conductor[†] has a resistance of one ohm when a p.d. of one volt, across its ends, produces a current of one ampere through it.

The resistance of a wire depends upon its length and inversely upon its cross-section.

$$R = \rho \frac{l}{a}$$

where l = length, a = cross-section

ρ is the constant depending upon the material and is called the *specific resistance* or *resistivity*.

The resistivity is the resistance of a piece of the material of uniform cross-section of 1 m^2 and of length 1 m. It is measured in ohm metre (Ωm).

ρ for copper is very small, but certain alloys have much larger values, e.g. constantan (*see* table).

---

[*] For measurement of resistance, *see* Chapter 38.
[†] Or resistor.

RESISTIVITIES/ohm metre	
GOOD CONDUCTORS	
Aluminium	3.2×10^{-8}
Copper	1.8×10^{-8}
Iron	12.0×10^{-8}
Mercury	96.0×10^{-8}
Brass	6.6×10^{-8}
Constantan	49.0×10^{-8}
INSULATORS	
Glass	10^{12}
Paraffin wax	10^{14}
Perspex	10^{16}
SEMI-CONDUCTORS	
Selenium	10^{-4}
Germanium	10^{-3}

approximate values

Wire of such alloys is used for making standard high-resistance coils.

Further consideration of the relationship between p.d. and current

Ohm's law is the basic law for calculating the current in an ordinary circuit. Although its importance must not be underrated, its limitations must be appreciated. It is only applicable to conductors, e.g. metals, in which the charge carriers quickly attain a steady velocity in an electric field. We shall be studying other examples of conduction where this simple relationship does not apply. Ohm's law infers that the graph of V against I is a straight line through the origin. Liquid conductors give approximately straight lines but not usually through the origin. The graphs for semi-conductors and gases have characteristics peculiar to themselves. We shall study conduction in liquids and in gases in a little more detail in Chapters 36 and 41.

The distinctly different behaviour of some substances to the passage of charge through them has led to their use for special purposes. In this category we put the many semi-conductor devices among which are transistors. The resistance of semi-conductors is affected by temperature and also by the action of light.

Circuits

Ohm's law is used to find the current in a circuit or part of a circuit. When conductors are connected together to make up a circuit, they may all be strung together chain-like, or some may be connected as are X and Y in Fig. 34.4. Resistors P and Q are connected in *series*, and X and Y in *parallel*.

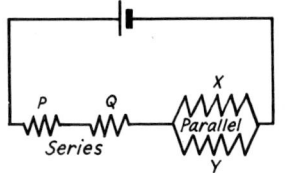

Fig. 34.4
Resistors in series and parallel

Resistors in series

We shall now deduce the effective resistance of two resistors R_1 and R_2 connected in series.

In Fig. 34.5*a* the current is the same through both resistors. The potential is falling from X along the wire.

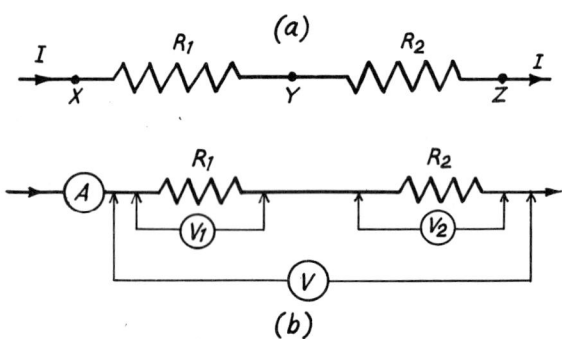

Fig. 34.5 Resistors in series

We can write:

p.d. across XZ=p.d. across XY+p.d. across YZ.

Using Ohm's law, $V=IR$
this is $IR=IR_1+IR_2$
where R is the effective resistance.

Hence, and for any number of resistors,

$$R=R_1+R_2+R_3+\ldots$$

This can be shown experimentally by using meters, as in Fig. 34.5*b*.

Read the voltmeters and note:

reading of V=reading of V_1+reading of V_2

But $R=\dfrac{\text{reading of } V}{\text{reading of } A}$

$$R_1=\frac{\text{reading of } V_1}{\text{reading of } A}$$

$$R_2=\frac{\text{reading of } V_2}{\text{reading of } A}$$

$$R=R_1+R_2$$

Resistors in parallel

Consider now two resistors R_1 and R_2 in parallel (Fig. 34.6). The main current divides into two parts I_1 and I_2, just as a flow of water may divide and pass along two pipes. In the case of water in pipes,

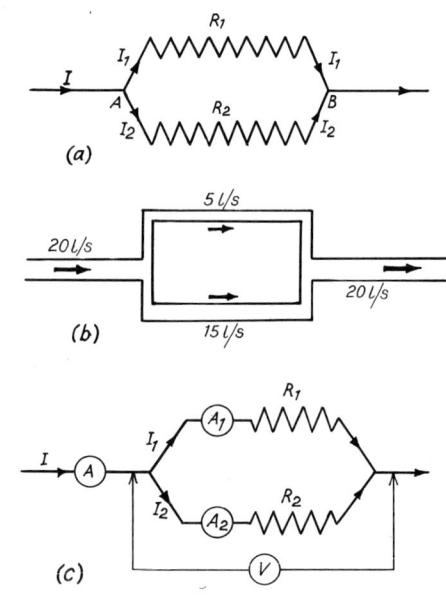

Fig. 34.6 Resistors in parallel

we immediately appreciate that the sum of the currents in the two pipes must add up to that in the main pipe.

This applies also for a flow of electricity.

We can write

$$I=I_1+I_2$$

But the p.d. along R_1 must be equal to that along R_2 as both their ends are connected. Let this p.d. be V.

p.d. across AB=p.d. across R_1=p.d. across R_2

i.e. $V=IR=I_1R_1=I_2R_2$ from Ohm's law, where R is the effective resistance.

$$I=\frac{V}{R}; \quad I_1=\frac{V}{R_1}; \quad I_2=\frac{V}{R_2}$$

Substituting in the above equation for the current:

$$\frac{V}{R}=\frac{V}{R_1}+\frac{V}{R_2}$$

Hence, and for any number of resistors,

$$\frac{1}{R}=\frac{1}{R_1}+\frac{1}{R_2}+\frac{1}{R_3}+\cdots$$

The two resistors R_1 and R_2 can be replaced by one resistor R which will not alter the p.d. across AB, nor the main current I.

This result can be verified experimentally using the circuit shown in Fig. 34.6c.

Note:
   reading of A=reading of A_1+reading of A_2
Effective resistance of R_1 and R_2 in parallel

$$=\frac{\text{reading of }V}{\text{reading of }A}$$

Calculations

Ohm's law is the most important law concerning electric circuits. It can be applied to a single conductor or to a complete circuit, as is illustrated in Examples 1 and 2.

Example 1

In the circuit shown in Fig. 34.7, find the current and p.d. across the 15-ohm resistor.

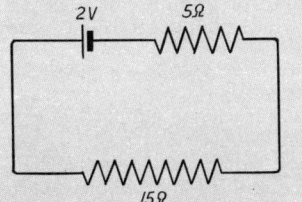

Fig. 34.7

Applying Ohm's law to the whole circuit:

$$\text{Current}=\frac{\text{total e.m.f.}}{\text{total resistance}}$$

$$=\frac{2}{5+15}=0.1 \text{ A}$$

To find the p.d. across the 15-ohm resistor, apply Ohm's law to it:

$$\begin{aligned} V&=IR \\ &=0.1\times15 \\ &=1.5 \text{ V} \end{aligned}$$

The p.d. is 1.5 V.

Example 2

In the circuit shown in Fig. 34.8, find the currents I, I_1, and I_2.

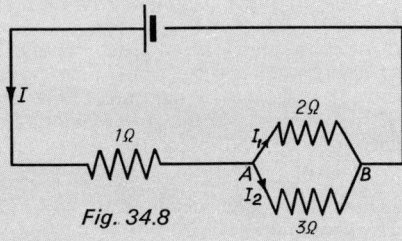

Fig. 34.8

To find the effective resistance of 2 and 3 ohms in parallel, we have

$$\frac{1}{R}=\frac{1}{R_1}+\frac{1}{R_2}$$

$$=\frac{1}{2}+\frac{1}{3}=\frac{5}{6}$$

R (in ohms)$=1.2$

$$\text{Main current } I \text{ (in amperes)}=\frac{\text{total e.m.f.}}{\text{total effective resistance}}$$

$$=\frac{2}{1+1.2}=0.91$$

To find I_1 and I_2:

$$\begin{aligned} \text{p.d. across } AB&=I_1R_1=I_2R_2 \\ 2I_1&=3I_2 \\ \therefore \ I_1&=\frac{3}{2}I_2 \end{aligned}$$

Also
$$I=I_1+I_2$$

Substituting for I_1 we have:

$$\frac{3}{2}I_2+I_2=\frac{10}{11}$$

$$\frac{5}{2}I_2=\frac{10}{11}$$

$$\therefore \begin{cases} I_2=0.36 \\ I_1=0.55 \end{cases}$$

The currents are 0.91A, 0.36A and 0.55A.

Effect of temperature on resistance

In stating Ohm's law we have added the necessary condition of keeping the temperature constant. When the temperature of a solid is increased, the molecules vibrate more quickly. This apparently increases the forces opposing the flow of charge in a metal and so the ratio of V/I increases.

The change in resistance of a metal with temperature is almost a linear relationship and is used to measure temperatures over a very wide range. The platinum resistance thermometer (page 130) employs this principle and is a very accurate thermometer.

The resistance of liquid conductors usually decreases with rise of temperature: so also does that of carbon.

QUESTIONS

1. A metal rod of 1 mm diameter has a resistance of 0.01 Ω. What would be the resistance of a rod of the same metal and length if the diameter were **2** mm?

2. If the resistivity of copper is 2×10^{-8} Ωm, what length of wire of diameter 0.1 mm will have a resistance of 1 Ω?

3. Two resistors, 3 and 5 Ω, are connected to a battery of e.m.f. 10 V, first in series and then in parallel. What current in each case is taken from the battery?

4. Find the p.d. across each of two resistors 4 and 5 Ω connected in series across a battery of e.m.f. 12 V.

5. In a stage lighting circuit, 10 lamps each of resistance 480 Ω, are wired in parallel to work off a 240 V main. What resistance must a dimmer have so that, when introduced in series with the mains, it can reduce the current to half its maximum value?

6. Accumulators are charged from a 200 V d.c. supply through a lamp resistor. If each lamp has a resistance of 500 Ω, how many such lamps must be used in parallel to obtain a charging current of about 2 A?

7. State Ohm's law, and define the *resistance* of a conductor.
Describe how you would measure a resistance by the method of substitution.
Calculate the length of wire of diameter 0.5 mm and resistivity 44×10^{-8} Ωm needed to make a 5 Ω resistance coil. (O)

8. Define *resistance* and *specific resistance* (or *resistivity*).
The formula for the effective resistance R of two resistors R_1 and R_2 in parallel is

$$\frac{1}{R} = \frac{1}{R_1} + \frac{1}{R_2}$$

Describe how you would test this formula experimentally.
Two lamps, each rated as '12 V, 36 W', are connected in parallel, and joined in series with a variable resistance across a 110 V supply. Find the value to which the resistance must be adjusted in order that the lamps can be run under their rated conditions. (O)

9. Obtain a formula for the effective resistance of two resistors R_1 and R_2 in parallel.
Describe how you would determine the resistivity of the material of a wire, with the help of a Wheatstone bridge.
Two equal lengths of wire, made of the same material but of different diameters, have an effective resistance of 5 Ω in series and of 0.8 Ω in parallel. Find the resistance of each wire, and calculate the ratio of their diameters. (O)

10. State Ohm's law, and describe an experiment to test it.
A laboratory draws its low-voltage supply from an accumulator of negligible internal resistance, placed 100 m away from the laboratory terminals. When a current of 2.0 A is taken, a voltmeter across the terminals reads 8.0 V; when the current is 2.5 A, the terminal voltage is 7.0 V. Calculate the e.m.f. of the accumulator, and the resistance per m length of the cable used to wire the terminals to the accumulator. (O)

11. State Ohm's law and describe an experiment you would perform in order to verify it.
Ten accumulators each of e.m.f. 2.2 V and internal resistance 0.1 Ω are connected in series to form a battery. The terminals of the battery are connected by a 10-Ω resistance. Find the current in the circuit and the potential difference between the terminals of the battery. (D)

12. A cell of e.m.f. 2.00 V and negligible internal resistance is connected in series with a resistance of 3.50 Ω and an ammeter of resistance 0.50 Ω. Calculate the current in the circuit. When a resistor of 0.100 Ω is connected in parallel with the ammeter, the current changes. Calculate the current in the cell. (C)

13. State Ohm's law and explain the terms *potential difference*, *current*, *resistance* and *resistivity*.
A length of constantan wire of cross-sectional radius (r) 0.5 mm is to have a uniform layer of copper deposited on it so as to halve its resistance. What must be the thickness of the copper? Assume the layer is a strip of width $2\pi r$.
[Resistivity of copper=1.72×10^{-8} Ωm; resistivity of constantan=49×10^{-8} Ωm.] (O & C)

14. State Ohm's law and define *resistance*. Distinguish between *resistance* and *resistivity*.
Compare the resistances per unit length of aluminium and copper wires of the same mass per unit length.
[Resistivity of copper=1.7×10^{-8} Ωm. Density of copper $=8.9 \times 10^3$ kg/m^3. Resistivity of aluminium=2.8×10^{-8} Ωm. Density of aluminium=2.7×10^3 kg/m^3.] (O & C)

15. State Ohm's law and define *resistance*, *resistivity*.
A filament lamp of 2 Ω resistance is run from a battery of e.m.f. 6 V and 0.5 Ω internal resistance. Assuming there are no other resistances in the circuit, calculate the current passing through the lamp and the p.d. across the lamp terminals. (O & C)

16. State and prove the formulae for the combined resistance of two wires connected (*a*) in series, (*b*) in parallel.

Two resistors of 10 Ω and two of 5 Ω, a 2 V battery and a galvanometer are connected to form a balanced Wheatstone bridge. What current is taken from the battery in the two positions in which it can be connected? Neglect its internal resistance.   (O & C)

17. A lamp filament glows when connected to a cell of negligible resistance. An ammeter measures the current as 1.0 A and a voltmeter across the lamp measures the p.d. as 2.0 V. If by mistake the instruments were interchanged, what would be the effect on the lamp, and approximately what would you expect them to read?   (O & C)

18 Discuss the conditions under which Ohm's law is obeyed. Why is it that the passage of a current through a gas does not obey the law?

19. Draw a labelled circuit diagram to show how you could investigate the variation of the current through a diode valve with variation of the applied p.d. for a constant filament current. Sketch a graph of the result you would expect to obtain and comment on this result in terms of the resistance of the valve.   (JMB)

35 Energy in a circuit; heating effect of a current

Electricity is a form of energy, i.e. it can do work. It can be converted into other forms, for example, by the electric motor and the electric radiator. It requires the expenditure of other forms of energy to generate it.

Consider the energy consumed in a circuit. Work has to be done to maintain a current in a conductor or, what is exactly the same thing, to transport a charge through a conductor.

The electric charges—free electrons in the case of metallic conduction—wander in their haphazard motion among the ions of the metal until they are made to drift by the electric field produced by the applied e.m.f. Above we have visualized the drift of charges to be like packages being lifted on a conveyor belt. A constant flow of charges means a constant current. The energy used in the process depends upon the mass of the packages and the height through which they are lifted. The height resembles the potential difference. Thus the work done in maintaining an electric current is the product of the p.d. and the charge conveyed. This gives us the means of defining the unit of potential difference, the *volt*.

If we make the p.d. unity when we carry unit charge, i.e. 1 coulomb, then we do unit work.

The volt: the p.d. across two points of a conducting wire is one volt when one joule of work is done in conveying one coulomb from one point to the other.

Or: The volt is the p.d. across two points of a conducting wire carrying a constant current of one ampere when the power dissipated between the two points is one watt (page 326).

Hence we have:

$$\text{Work or energy} = \text{p.d.} \times \text{charge}$$
$$\text{joules} = \text{volts} \times \text{coulombs}$$

$$\begin{aligned} \text{Energy consumed} &= V \times Q \\ \text{(in joules)} \qquad &= V \times I \times t. \end{aligned}$$

By using Ohm's law this can be expressed as

$$I^2 \times R \times t \quad \text{or} \quad \frac{V^2 \times t}{R}.$$

Joule's law

Joule stated that the heat developed in an electric circuit is

$$\propto (\text{current})^2$$
$$\propto \text{resistance}$$
$$\propto \text{time}.$$

The simplest way of verifying this law is to arrange a coil of resistance wire in a measured mass of water contained in a vacuum flask and pass a current through it for a measured time. (Fig. 35.1). The rise in temperature can be observed for different periods of current flow and for different currents flowing for equal times. To verify the dependence of the heat generated on the resistance, it is easier to use two flasks possessing heating coils of different resistances connected in series.

Many modern forms of apparatus for measur-

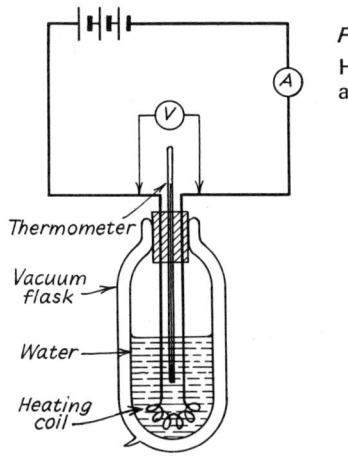

Fig. 35.1
Heating effect of
a current

Thermometer

Vacuum
flask

Water

Heating
coil

whole of the electric energy is converted into heat
if none goes into other forms of energy. In the
case of an electric motor, most of the energy is
converted into mechanical work, but some goes
to raising the temperature of the coils of the
motor. In an electric radiator all the energy goes
into heat, if we neglect the small amount of visual
radiation produced.

The hot-wire ammeter

The heating effect may be used to measure current.
A hot-wire ammeter is shown in Figs. 35.3 and
35.4. Such ammeters can be used to measure

ing specific heat capacities and specific latent
heats are electrical. The heat supplied by a coil
of wire to a solid or liquid is very conveniently
measured, with reasonable accuracy, using an
ammeter and a voltmeter. The vacuum-flask
apparatus described above can be used to measure
the specific heat capacity of a liquid.

The heat developed is proportional to the
square of the current. This shows mathematically
that heat is always developed, whichever way the
current is flowing.

The dependence of the generation of heat upon
the resistance is demonstrated effectively by passing
a current through similar-gauge copper and iron
wires mounted on an asbestos board as shown in
Fig. 35.2. The iron wire, having a much higher

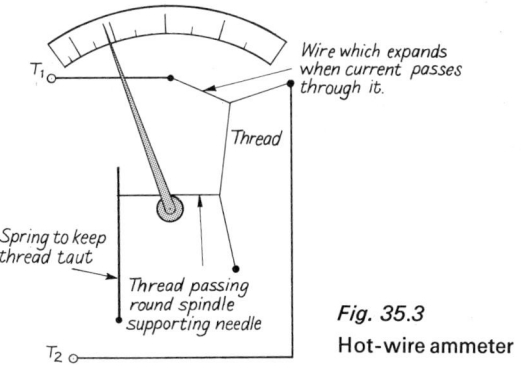

Wire which expands
when current passes
through it.

T_1

Thread

Spring to keep
thread taut

Thread passing
round spindle
supporting needle

T_2

Fig. 35.3
Hot-wire ammeter

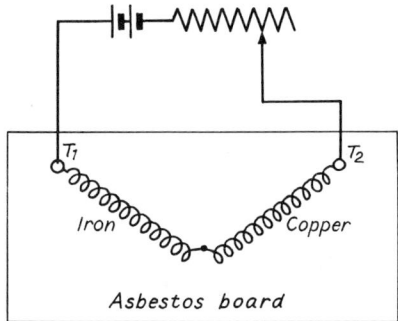

T_1 T_2

Iron   Copper

Asbestos board

Fig. 35.2 To show heat developed is proportional to
the resistance

resistance than that of the copper, can be made to
glow while the copper wire remains quite cold.

When an electric current passes through a
solid, liquid or gas, some heat is generated. The

Fig. 35.4 Hot-wire ammeter

alternating currents, but they are sluggish and easily burnt out by overloading.

Uses of the heating effect

The heating effect of a current is used to advantage in many appliances such as radiators, irons, cookers, etc. found in the home. The ordinary filament lamp consists of a very fine wire of tungsten, which glows white hot when a current is passed through it.

Fuses

All electrical installations incorporate fuses in the circuits (Fig. 35.5) to protect the wiring in the event of failure of an appliance. A fuse is a weak

Fig. 35.5 Double-pole main switch and fuses

spot in the circuit, specially arranged so that it breaks down if an abnormally high current passes. It consists of a short length of thin tinned copper wire mounted in a porcelain holder, which can easily be removed from a porcelain base into which it clips. The need for a fuse can be best appreciated by considering a simple case. Suppose an electric fire, taking 10 A, is used on a wall plug which is wired with cable capable of carrying,

safely, a load of 20 A. This cable will be of fairly thick copper and will have a very low resistance. If the supply is 200 V, the resistance of the fire must be 20 Ω $\left(R=\dfrac{V}{I}\right)$ Should the fire develop a fault, such as would result from a section of the element being cross-connected, the resistance may drop to about 5 Ω or less. The current then would increase to $\dfrac{200}{5}=40$ A. This excessive current would damage the permanent wiring to the plug by overheating. It is to safeguard against such an occurrence, which may easily cause a serious fire, that the fuse is inserted. A circuit, such as the one mentioned above, would include a fuse capable of carrying a current of about 15 A without getting warm. If this current were exceeded, the heat generated in the fuse wire would melt it and, in so doing, break the circuit.

The mains supply

The majority of consumers receive their electrical supply through their local authority, from the grid (*see* Chapter 40). This is supplied at a voltage of 240 and is alternating current. The street cables usually contain four wires for a three-phase supply. One wire (neutral) is maintained at earth potential and the other three wires (live) have an alternating potential varying between ± 325 V. These are separately paired with the neutral to give three independent circuits, which are led into neighbouring houses in order to balance the loads on the three phases.

For an a.c. supply to a house, the leads from one phase are taken through sealed fuses and the meter. This part of the installation is the property of the supply company and the sealing of the fuses prevents any unauthorized person changing them in any way. To do so might result in damage to the supply cables or to the meter.

The wiring diagram (Fig. 35.6) is largely self-explanatory, but certain comments may be desirable. The p.d. at the plug outlet socket is the same as that at a lampholder, i.e. the same p.d. is used for power and lighting. The reason there are two separate circuits is simply a matter of expense. The thick cable necessary for carrying the large currents used by radiators and other domestic appliances is expensive and only comparatively

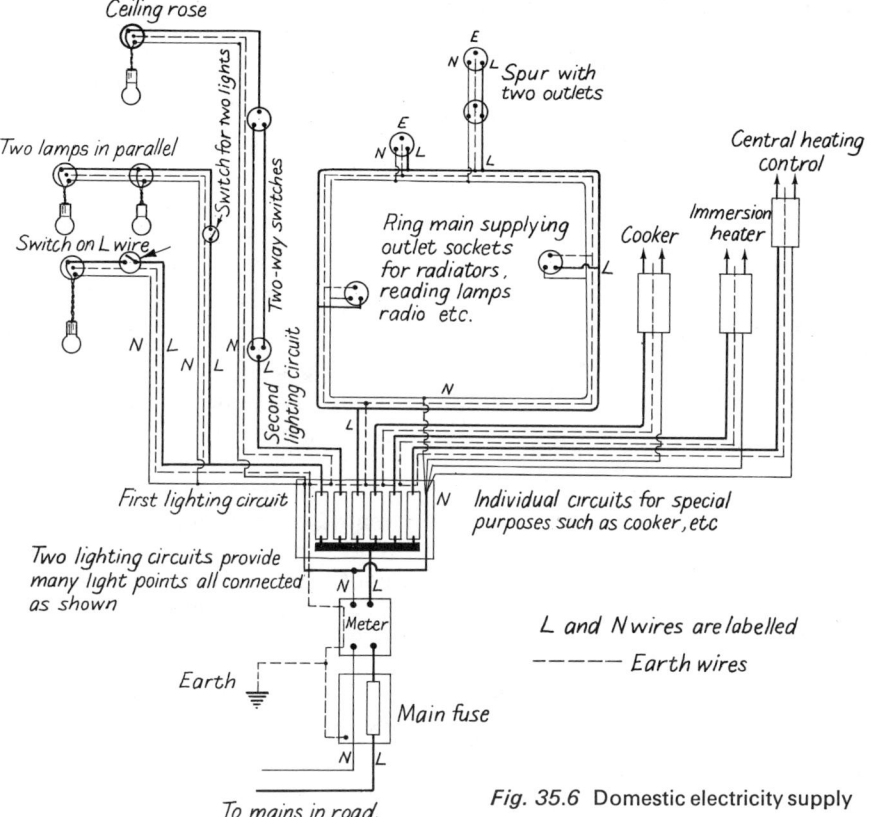

Fig. 35.6 Domestic electricity supply

thin cable is required for the small currents in the lighting circuit.

The cables can be run in a steel tube (conduit), but are now more usually moulded in rubber or in PVC (polyvinyl chloride). The latter forms excellent protection against dampness and does not deteriorate. A steel conduit must be continuous and well earthed at one or more points. PVC cable encloses two insulated wires (live and neutral) and also a bare copper wire (earth). The third wire is connected to a good 'earth' such as the main water pipe. This earth wire is terminated at the third terminal of an outlet socket or at a plate in the ceiling from which the flexible wire carrying the bulb hangs. All switches break the live L wire and not the neutral N wire. Otherwise the live wire in a lamp-holder, when the switch is at off, could complete a circuit with any conductor connected to the earth, e.g. a water tap or a concrete floor. This also accounts for the advisability of having no metal fittings near to earthed conductors. Ordinary light switches are

not permitted in a bathroom where they can be touched by a person standing in the bath or in contact with the water pipes. A fault in the switch could lead to the person receiving a fatal shock.

The ring main

This method of wiring is now in common use for outlet sockets for appliances. Until a few years ago all such outlets had to be separately wired from the distribution box and provided with a pair of fuses (one on each of the L and N wires). The modern method is much more economical of cable. As will be seen from the diagram, the cable is in the form of a ring and so the current to each outlet has two paths. This reduces the current and so permits lighter cable to be used. Fuses are now inserted in the live L wire only. One main fuse is incorporated in the complete ring, but the plugs used for the appliances are individually fused to suit the requirement of the appliance. For example, a 2 A fuse is adequate for a standard lamp, but a 13 A fuse is necessary for a 3 kW radiator. The wires to an appliance are colour coded. The plug terminals are marked L, N and E. The BROWN wire goes to the L terminal; the BLUE to the N; the GREEN/YELLOW striped to the E terminal.*

Power in a circuit

At the beginning of this chapter we saw that the energy, in joules, used in a circuit $= V \times I \times t$

Power is rate of working.

The unit of power is obtained by considering unit work to be done in unit time. The unit of power is the watt, defined as follows:

one watt = one joule per second

therefore   **power, in watts, in a circuit = V × I**

watts = volts × amperes

The power of an appliance for use on the mains is usually quoted in watts. This enables us quickly to calculate how much it will cost per hour in use. Common powers of lamps are 60, 100, and 150 W and fires 1, 2, and 3 kW. The approximate powers of other common appliances are, clock, 2 W; iron, 300–750 W, vacuum cleaner, 300 W; radio, 60–100 W; kettle, 1500 W; cooker, 8 kW.

The horsepower, which is a unit of power based on British gravitational units, will eventually be superseded by the kW.

The British horsepower = 746 W
The metric horsepower = 735.5 W

The unit of electrical energy

Electrical energy can easily be measured and we buy it at a fixed price per unit of energy. The unit of energy used above is the joule. But this is a very small unit: a 60-W lamp uses 60 J per second or 3600 J per minute. We have, therefore, to devise a larger unit.

---

* The old colours were live—red; neutral—black; earth—green.

The joule is the energy used in one second by a circuit working at the rate of one watt.

To increase the size of the unit we increase the time to one hour and the power to 1000 W, and the unit becomes:

the Board of Trade unit or kilowatt-hour.

The kilowatt-hour is the energy consumed in one hour by a circuit working at the rate of one kilowatt.

Note.   1 kW h $= 1000 \times 3600$ W s
$= 3\ 600\ 000$ J

This is the unit used in measuring electrical energy commercially (*see* Example). The electric meter measures the energy consumed in kilowatt-hours.

It should be noted that the kWh, although a metric unit, is not a derived SI unit of energy. The Central Electricity Board are proposing to continue the use of this unit for domestic purposes. The SI unit which should be used is the joule or, more suitably, the MJ (10^6 J).

Example

If electrical energy costs 1p per kilowatt-hour, calculate the cost per week of the electrical service to a house which uses the following appliances for the estimated times shown.

4	60-W lamps	2 hours
6	100-W lamps	8 hours
1	750-W iron	6 hours
2	2000-W radiators	6 hours
1	250-W vacuum cleaner	2 hours

Total energy consumed

$= 4 \times 60 \times 2 + 6 \times 100 \times 8 + 1 \times 750 \times 6$
$+ 2 \times 2000 \times 6 + 1 \times 250 \times 2$

$= 480 + 4800 + 4500 + 24\ 000 + 500$ Wh
$= 34.28$ kWh

Cost $= 34.28 \times 1$p
$= 34.28$p

The cost is 34.28p per week.

QUESTIONS

[$g = 10 \text{ m/s}^2$; specific heat capacity of water $= 4.2 \times 10^3$ J/kg K]

1. Compare the amounts of heat generated in two copper wires of equal lengths connected in series, one having twice the diameter of the other.

2. An electric radiator is rated at 2000 W. How much heat does it develop per hour?

3. Define the terms *watt* and *kilowatt-hour*.
    An electric iron is rated at 600 W when used on a 240-V main. Find its resistance.

4. A current of 4 A is passed through a heating coil of resistance 5 Ω. How long will it take when placed in 200 g of water to raise its temperature from 10 °C to 90 °C?

5. An electric kettle is rated at 800 W and is used on a 200-V supply. Find (*a*) what current it takes, (*b*) how long it will take to boil 0.5 dm^3 of water at 20 °C (assume the kettle to be 80% efficient), and (*c*) how much it will cost in performing (*b*) if electricity costs 1½p per kilowatt-hour.

6. A private consumer uses on the average each day two 100 W lamps for 5 hours and one 2000 W radiator for 2 hours. What will be the weekly cost at 1p per kilowatt-hour?

7. How would you investigate experimentally the way in which the rate of production of heat in a wire carrying an electric current depends on the strength of that current?
    A 3 kW electric heater is immersed in a storage tank containing 180 dm^3 of water at 16 °C. Calculate, to the nearest minute, how long it will take to raise the temperature to 50 °C, and how much it will cost if the Electricity Board's tariff is 1½p a unit.   (O)

8. Describe an experiment to investigate the rate at which heat is generated when an electric current is passed through a wire. Draw a diagram of the apparatus and suggest a suitable material for the wire. How would you expect the rate of heat generation to depend on the current?
    Water is flowing at a steady rate of 1 dm^3/min through a pipe in which there is an electrical heating element connected to the 200 V main supply by wires of negligible resistance. If the rise in temperature of the water after passing the heater is 60 °C, calculate (*a*) the current in the heater, (*b*) the resistance of the heater.   (O & C)

9. Describe an experiment to show that the rate of production of heat in a wire by a current is directly proportional to the square of the current.
    An electric iron is rated at 350 W when used on a 230-V supply. Find: (*a*) the current flowing in the heating coil, (*b*) the heat developed in the iron per hour, (*c*) the cost of using the iron for 4 hours, given that the cost of a kilowatt-hour is 1½p.   (JMB)

10. Describe an experiment to show how the quantity of heat produced by an electric current flowing in a conductor depends upon the strength of the current. State the conclusion to which the experiment would lead.
    Two wires, *P* and *Q*, of 4 and 5 Ω resistance respectively, are connected first in series and then in parallel with a battery of e.m.f. 8 V and of negligible internal resistance. In each

instance, (*a*) draw the circuit, (*b*) calculate the value of the current through each wire, (*c*) compare the quantities of heat produced per second in the two wires, stating clearly in which of the two wires (*P* or *Q*) the quantity is the greater.   (L)

11. An electric immersion heater is marked 12 V, 72 W. Describe with the aid of a circuit diagram, an electrical method of checking this marking, given an accumulator battery of e.m.f. 13 V as the source of supply. Give *one* reason why the heater should be immersed in water during the experiment. Calculate the rise in temperature which the heater can cause per minute in 0.5 kg of water when the heater is connected to (*a*) a 12-V supply, (*b*) a 13-V supply.   (C)

12. An electric motor, supplied by an accumulator of electromotive force 12.0 V and of negligible internal resistance, is geared to a mechanism which does 132 J of work in 6.0 s. The motor takes a steady current of 4.0 A and its armature has a resistance of 0.50 Ω. Calculate (*a*) the power supplied by the battery, (*b*) the power utilized by the mechanism, (*c*) the power lost in heating the armature.   (C)

13. State Ohm's law.
    An accumulator of electromotive force 12.0 V and internal resistance 0.05 Ω is connected, by copper cables having a total resistance of 0.55 Ω, to an electric heater of resistance 2.52 Ω.
    Calculate (*a*) the potential difference between the terminals of the heater, (*b*) the power taken by the heater, (*c*) the power losses in the cables.   (C)

14. Describe in detail how you would show experimentally that when electrical energy is completely transformed into heat energy the same amount of heat energy is always generated when a fixed amount of electrical energy disappears. Give a diagram of your apparatus and state the precautions you would take.
    How long would a 2 kW electrical immersion heater take to heat a tank containing 150 kg of water, from 15 °C to 75 °C?   (O & C)

15. Define *joule* and *volt*. What is the numerical relationship between the kilowatt-hour and the joule?
    An electric lamp, the power of which is between 50 and 100 W, is connected to 220 V a.c. mains. A variety of volt-meters and ammeters, of different types and ranges, is available. State what types of instrument you would select, of what approximate ranges, and how you would use them, in order to measure the exact wattage of the lamp. Draw a circuit diagram.
    1 kg of water, in a vessel of negligible thermal capacity, are heated by an immersion heater. Assuming that the process is 100% efficient, and that electrical energy costs 2p a unit, calculate the cost of raising the water from 20 °C to boiling-point.   (C)

16. Write down an expression for the rate of heat production in a circuit of resistance *R* Ω when connected to a source of potential difference *E* volts.
    Calculate the working resistance of a 0.75 kW, 230 V electric fire. How many joules are given out by the fire in 30 minutes?   (O & C)

17. Define *volt*, *joule*, *kilowatt*.

A 3 kW electric fire is intended to be used on a 200 V supply. What is the value of the current and of the resistance?

If the fire is connected to a 240 V supply, what power will it consume on the assumption that its resistance is unchanged?

(O & C)

18. State and prove the expressions for the combined resistance of two wires connected (*a*) in series, (*b*) in parallel, and state the law on which your proofs are based.

Two resistances maintained at constant temperature dissipate 100 W each when connected in parallel to a 230 V supply. Calculate (*a*) the value of each resistance, (*b*) the power dissipated in each when they are connected in series to the same supply. Would you expect to find the same or a different result for (*b*) when two 230 V, 100 W tungsten filament lamps are used? Justify your answer.   (O & C)

19. State Ohm's law and define *potential difference*, *resistance* and *resistivity*.

Water flows through a pipe past an electrical heating element at the rate of 3 dm^3/min. If the potential difference across the element is 230 V and the rise in temperature of the water is 60 °C, calculate (*a*) the power which must be supplied to the heater, (*b*) its resistance.   (O & C)

20. Explain the meanings of the following units: *joule*, *watt*, *kilowatt-hour*.

State the law which determines the rate of generation of heat in an electric circuit.

A current of 0.5 A passes through a filament lamp and the p.d. across its terminals is 200 V. Calculate (*a*) the resistance of the lamp, (*b*) the power consumed in running it and (*c*) the heat developed in it per hour.   (O & C)

21. An electric coil-heater *A* connected to a 20 V supply, and an immersion heater *B* connected to a 250 V supply each take 1 kW of power from their respective supplies. Calculate (*a*) the current in each circuit, (*b*) the resistance of the connecting cables in each case if only 10 W of the power is to be lost in the cables.

Explain the fact that the cables to *A* are likely to be thick, but need not be so well insulated as those to *B*.

Draw a wiring diagram to show how the heating coil and metal casing of *B* should be connected to a 3-hole socket of a household system, and explain the purpose of this form of connection.   (C)

22. An electric blanket for use on a 200 V supply has two identical heating elements (resistors), each of resistance 1000 Ω which may be connected to the mains with the help of a special switch, in three different ways:

   (i) one element, only, is used, the other being disconnected;

  (ii) both elements are used in series;

 (iii) both elements are used in parallel.

(*a*) Draw diagrams to show the elements connected in series, as in (ii) and in parallel, as in (iii). Label the diagrams clearly.

(*b*) Determine the current *in each element* when used as in (i), (ii) and (iii).

(*c*) Determine the maximum power needed to operate the blanket from the 200 V supply.

(*d*) If the cost of supplying electrical energy were 2p a kilowatt-hour, for how many hours could the blanket be used at maximum power for 1p?

(*e*) If one of the elements 'burnt out' so that no current could flow through it, state, giving your reason, in which *one* of the three arrangements of the elements the blanket would still be certain to give out some heat.   (L)

23. Define *specific latent heat of fusion* and *specific heat of vaporization*.

One method of determining the specific latent heat of fusion of ice electrically is as follows. A known mass of ice at 0 °C is added to a quantity of warm water at some temperature θ °C. When all the ice has melted, cooling the water in the process, heat energy is supplied electrically until the temperature θ °C is reached again. Since the water has now been warmed to its original temperature, the heat energy supplied must have been just sufficient to melt the ice at 0 °C and then raise the temperature of the resulting ice-water up to θ °C.

In such an experiment, 30 g of ice was added to warm water at 40 °C. After all the ice had melted, a steady current of 4 A at 12.5 V was passed through a coil immersed in the water, and after 5 minutes the temperature reached 40 °C again. Calculate, in joules,

(*a*) the quantity of energy supplied by the heater,

(*b*) the quantity of energy used to raise the ice-water from 0 °C to 40 °C,

(*c*) the quantity of energy used to melt the ice at 0 °C.

Hence find the specific latent heat of fusion of ice in J/kg. What are the chief sources of error in such an experiment?

[Take the specific heat capacity of water to be 4.2×10^3 J/kg K.]   (O)

36 The chemical effect of a current; cells

Conduction through liquids

We now come to the passage of an electric current through a liquid. First, we must find out if all liquids are conductors. Use the circuit shown in Fig. 36.1. Put distilled water in the tank and

metallic salts. Distilled water is found to be a very poor conductor of electricity, but it is rendered conducting by the addition of an acid, an alkali or a metallic salt.

If this experiment is carried out using a very

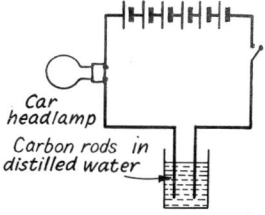

Car headlamp

Carbon rods in distilled water

Fig. 36.1
Conduction through liquids

employ carbon rods or clean steel plates for leads into the water. Switch on and note there is no effect produced. Add to the water a little sulphuric acid and note that the lamp glows brightly. Repeat the experiment, adding other acids, alkalis and

sensitive galvanometer or microammeter in place of the lamp no current will be indicated with such substances as ether, alcohol, benzene, turpentine and certain oils, but there will be a very small reading with water.

When a current is passed through a liquid, e.g. dilute sulphuric acid, contained in a glass tube, the liquid gets hot: further, a compass needle, suitably placed, is deflected. In addition to these two effects a chemical effect may be observed. The chemical action in many cases results in the discharge of products at the plates.

A further series of similar experiments should be performed using fused metallic salts in place of the solutions. The salts are strongly heated in a hard glass tube. For example, if the ends of the carbon rods are submerged in molten sodium chloride and a current is passed, the products liberated are sodium and chlorine. The effect is better demonstrated, however, using heated lead chloride, for a bead of lead can be observed to form on the negative rod. With fused potassium bromide and potassium iodide, the brown bromine vapour and the purple iodine vapour can be observed coming off the positive carbon rod.*

The chemical effect of a current was discovered in 1800 by Carlisle and Nicholson. In the years which followed many experimenters investigated the effects using various solutions, but their results were unconnected until Faraday completed his famous researches on the subject and produced order out of chaos. Most of the terms we use are those first used by Faraday.

The production of a chemical action by an electric current is called *electrolysis*. A compound which, when dissolved in water, or when molten, conducts an electric current, is termed an *electrolyte*. The current is led into and from the liquid by way of plates or *electrodes*. The electrode where the current enters is called the *anode* and that where the current leaves, the *cathode*. The whole cell comprising electrodes, electrolyte and container (which in many cases is of special shape) is called an *electrolytic cell*.†

The following facts are worth noting:

(*a*) The products are liberated only at the electrodes.

(*b*) The chemical action begins immediately the current flows.

(*c*) A metal or hydrogen is liberated at the cathode while a non-metal or oxygen appears at the anode.

(*d*) The final products depend not only upon the electrolyte, but also upon the electrodes.

(*e*) Acids, alkalis and salts are electrolytes.

The ionic theory

Chemical action arises from action between electric charges in atoms. From the study of chemical actions in which metals take part, an 'electro-chemical series' has been compiled which places metals in the order of their chemical activity. The following gives the order of a few metals in the series:

$+$ K Ca Na Al Zn Pb (H) Cu Hg Ag Au $-$

The higher (more positive) metal in the series will replace the lower in an appropriate chemical action, e.g. zinc will replace copper in copper sulphate solution and form zinc sulphate and copper. Hydrogen has been given a place in the series because it behaves, in many ways, like a metal and has a basic function in the series. The chemical activity of a substance depends upon its ability to lose electrons and to become ionized.

The action of metals on water and on dilute acid causing the liberation of hydrogen, illustrates their order in the series. K, Ca and Na will react with water and liberate hydrogen, but Zn and Pb require a weak acid from which to release the hydrogen.

Non-metals are given a place in the series. They are electro-negative and come after Au in the order:

C   N   P   S   O   Cl   F

Electrolytes

An electrolyte contains ions. These are atoms or groups of atoms and are formed when the solution is made or when the substance is fused. An ion carries a number of charges equal to its valency. Thus a copper ion carries twice the charge carried on a hydrogen ion. Ions do not display the properties possessed by the uncharged atom or radical group. The hydrogen ion is usually associated with a molecule of water forming an H_3O^+ ion. The copper ion is associated with four molecules of water forming a $Cu(H_2O)_4^{2+}$ ion.

The degree of ionization in an electrolyte varies

---

* *See* Nuffield *Chemistry Basic Course*, page 160.

† An electrolytic cell designed for measuring the mass of a product is termed a *voltameter*.

considerably. For example, ionization in water*
is very small indeed, while that in strong acids,
alkalis and salts is 100 per cent. These form strong
electrolytes. The weaker acids, such as carbonic
and acetic acids, are weak electrolytes. Non-
electrolytes are mostly organic compounds such
as chloroform, alcohol, sugar and oils.

The mechanics of electrolysis

It must be emphasized that ionization occurs
when an aqueous solution of an electrolyte is
made or when an electrolyte is fused. No electric
current is necessary and hence, no electrical
energy is used in ionizing the electrolyte.

When two metal plates are put into a solution
of an electrolyte and these are connected to the
poles of a battery giving an e.m.f. of a few volts
(4–6 V), effects as mentioned above are observed.

The positive ions drift to the negatively charged
electrode (cathode) and the negative ions to the
positively charged electrode (anode).

At the electrodes the ions become neutralized
by accepting or giving electrons. In effect the
cathode supplies electrons to the positive ions and
so acts as a reducing agent: the anode accepts
electrons from the negative ions thereby producing
oxidation.

When the ions have their charges neutralized,
the atoms or radical groups so formed display
their ordinary chemical properties. They are
liberated at the electrodes, being deposited on the
electrodes when solid, or freely evolved when
gaseous.

Preferential discharge of ions

In an aqueous solution of an electrolyte there are
present, not only the ions of the electrolyte, but
also those of water, viz. OH^- and H_3O^+. At the
electrodes not all the ions are neutralized and
liberated. The particular ones liberated are those
which require the least energy to liberate. Ions
can be arranged in an electrochemical series. The
following shows some members of this series:

$$K^+ \quad Na^+ \quad H^+ \quad Cu^{2+} \quad Ag^+ \quad OH^- \quad Cl^- \quad SO_4^{2-}$$

---

* About 1 molecule in 5.5×10^8 is ionized. This small ionization is what caused the small deflection of the galvanometer
mentioned in the experiment at the beginning of the chapter.

In general a higher ion in the series is liberated
in preference to a lower one. This statement may
not hold if there is a large difference in the number
of ions present. For example, in a solution of a
chloride, such as NaCl, the number of Cl^- ions
present may be much greater than the number of
OH^- ions. In this case the Cl^- ions will overwhelm
the OH^- ions and chlorine will be liberated.

The nature of the electrodes may influence the
liberation of a particular ion. It is to eliminate
this that platinum is very often used for electrodes.

The examples given below serve to illustrate
preferential discharge.

Dilute sulphuric acid (platinum electrodes)

This is usually regarded as the electrolysis of
water as the acid content remains constant. The
traditional apparatus is shown in Fig. 36.2 and it
allows the collection of the liberated gases.

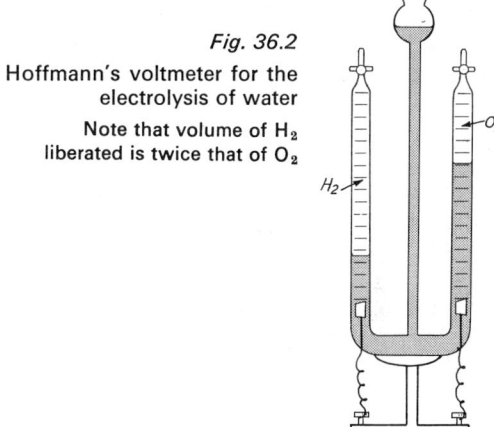

Fig. 36.2

Hoffmann's voltmeter for the
electrolysis of water

Note that volume of H_2
liberated is twice that of O_2

In dilute sulphuric acid there are H_3O^+ and
SO_4^{2+} ions from the sulphuric acid, together
with a very small number of H_3O^+ and OH^-
ions from the water. A possible result of electrolysis
is that H_3O^+ ions are liberated at the cathode
and an equal number of OH^- ions liberated at
the anode. This latter will be in preference to the
liberation of SO_4^{2-} ions. The OH^- ions at the
anode form oxygen and water and release electrons.

At the cathode hydrogen ions gain electrons and
hydrogen gas is evolved:

$$4(H_3O)^+ + 4e^- \rightarrow 4H_2O + 2H_2$$

At the anode OH^- ions lose electrons and form
water and oxygen:

$$4OH^- \rightarrow O_2 + 2H_2O + 4e^-$$

Thus, twice the number of hydrogen molecules as oxygen molecules is liberated, giving the volume ratio of 2:1, which agrees with the observed result.

Copper sulphate solution (platinum electrodes)

Ions present are $Cu(H_2O)_4^{2+}$, H_3O^+, OH^-, SO_4^{2-}. The $Cu(H_2O)_4^{2+}$ ions are liberated at the cathode in preference to the H_3O^+ ions as they require less energy for their discharge. At the anode the OH^- ions are liberated. This is in preference to both the liberation of the SO_4^{2-} ions and the passing into solution of platinum ions.

At the cathode copper is deposited:

$$Cu(H_2O)_4^{2+} + 2e^- \rightarrow 4H_2O + Cu$$

At the anode oxygen gas is evolved:

$$4OH^- \rightarrow 2H_2O + O_2 + 4e^-$$

Copper sulphate solution (copper electrodes)

Here, the same action takes place at the cathode, but at the anode Cu atoms from the electrode give up electrons and pass into solution as $Cu(H_2O)_4^{2+}$ ions thereby maintaining their concentration in the electrolyte.

At the cathode copper is deposited:

$$Cu(H_2O)_4^{2+} + 2e^- \rightarrow 4H_2O + Cu$$

At the anode copper is removed:

$$Cu + 4H_2O \rightarrow Cu(H_2O)_4^{2+} + 2e^-$$

The net result of this electrolysis is the transference of copper from the anode to the cathode. This is the principle employed in electroplating. The article to be plated is made the cathode, while the electrolyte is a solution of the metallic salt; the anode is a plate of the pure metal. The metal deposited comes from the solution and the deficiency is made up from the anode.*

Sodium chloride solution (carbon electrodes)

Ions present are Na^+, H_3O^+, Cl^-, OH^-. At the cathode H_3O^+ ions are liberated in preference to Na^+ ions. (The OH^- ions make the solution alkaline—sodium hydroxide.) Chlorine gas is evolved at the anode.

---

* In some cases an anode of a different metal is used: the electrolyte has then to be replenished.

At the cathode hydrogen gas is evolved:

$$2H_3O^+ + 2e^- \rightarrow 2H_2O + H_2$$

At the anode chlorine gas is evolved:

$$2Cl^- \rightarrow Cl_2 + 2e^-$$

If this electrolysis is carried out with fused sodium chloride instead of with an aqueous solution, the Na^+ ions are liberated at the anode in the absence of hydrogen ions and metallic sodium is deposited.

$$Na^+ + e^- \rightarrow Na$$

Faraday's work

In his famous researches on the chemical effect of an electric current, Faraday passed the same current for different times through a voltameter and weighed the products liberated. He then varied the current, keeping the time constant. The masses of the products liberated were found to be proportional to the time and to the current. But the product, *current × time*, is equal to the quantity of electricity and Faraday expressed this result in the form of a law.

First law of electrolysis

The mass of a product liberated in electrolysis is proportional to the current *I* and to the time *t* for which the current is flowing, i.e. to the quantity *Q* of electric charge passed. ($Q = I \times t$.)

Faraday next passed the same current for the same time through a number of voltameters

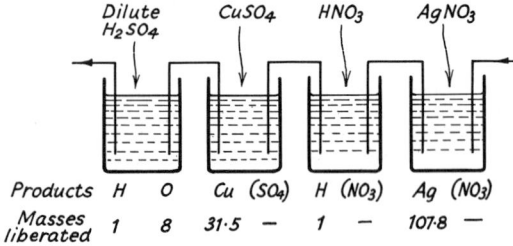

Fig. 36.3 Faraday's experiment

Liberated products and their masses, taking mass of hydrogen as 1

containing different electrolytes. Where possible, the masses of the products obtained were noted (Fig. 36.3).

The mass of a product liberated was equal to a constant times the quantity of electric charge passed. This constant z was called the *electrochemical equivalent* (e.c.e.).

The electrochemical equivalent of a substance is the mass liberated by one coulomb of charge, i.e. by one ampere flowing for one second.

It is measured in kg/C or g/C

$$\text{Mass liberated} = zQ \text{ or } zIt$$

Specific charge

The specific charge is the charge associated with unit mass (1 kg).

The specific charge is easily obtained from the above, for it is equal to $1/z$:

$$\text{specific charge of hydrogen} = \frac{1}{z} = \frac{1}{1.09 \times 10^{-8}}$$
$$= 9.65 \times 10^7 \text{ C/kg}$$
$$\text{or } 9.65 \times 10^4 \text{ C/g}$$

In electrolysis the charge is carried by the ions or charged atoms. This charge can be calculated* (*see* Chapter 41).

From the results of Faraday's second experiment, the same number of coulombs liberates 108 kg of silver as liberates 1 kg of hydrogen. Hence, because the mass of the silver atom is 108 times that of the hydrogen atom and, assuming an even distribution of the charge among the atoms, the same charge is associated with each silver atom as with each hydrogen atom. In the case of copper, however, there is double this charge on each atom. This is so for all divalent substances. Hence, the copper ion Cu^{2+} carries a double positive charge. This is interpreted as indicating that the ionized copper atom has lost two electrons. Likewise the oxygen ion O^{2-} has gained two electrons. The valency expresses the number of elementary charges an ion possesses.

Second law of electrolysis.

The mass of any ion liberated by a fixed charge is proportional to the $\frac{\text{atomic weight}}{\text{valency}}$.

This quantity when multiplied by the specific charge gives the universal Faraday constant, 9.65×10^7 C/kg mol or 9.65×10^4 C/mol.

---

* In Chapter 41 the identification of the charges associated with ionization is discussed. It is established that the charges concerned are electrons. These negatively charged particles are fundamental constituents of atoms and their drift in a metal conductor constitutes an electric current.

The ratio atomic weight/valency is known as the *chemical equivalent E* of the substance. (This term is not now used as frequently as it used to be.) Faraday expressed his second law: for the same quantity of charge passed through any electrolyte,

$$\text{mass liberated} \propto \text{chemical equivalent}$$

Hence, $\qquad m = kEQ$

k being a universal constant.

When $m = E$, i.e. for the liberation of a gramme-equivalent (a mass equal to the chemical equivalent in grammes),

$$Q = \frac{1}{k} = 96\ 500$$

This holds for all substances and means that 96 500 C will liberate a gramme-equivalent of any ion.

Faraday's first law can be verified experimentally by measuring the masses of copper deposited on the cathode of a copper voltameter by different quantities of charge and showing them to be proportional. A quicker method is to use a water voltameter, Hoffman's apparatus (Fig. 36.2) in which dilute sulphuric acid is electrolysed into oxygen and hydrogen. It is found that on doubling the value of It, the volume of each gas, measured at the same pressure and temperature, is doubled: hence the mass is doubled.

The second law is verified by passing the same quantity of electricity through a copper and a silver voltameter in series, and showing that the masses of copper and of silver deposited are 31.5 and 108 times the mass of H_2 liberated. In a water voltameter, the volume of hydrogen liberated is twice that of oxygen, but as oxygen is 16 times as dense as hydrogen, the mass of oxygen liberated is 8 times that of hydrogen, again in the same proportion as their chemical equivalents (*see* Fig. 36.3).

Determination of the e.c.e. of copper

This experiment is typical of that used for determining the e.c.e. of many metals. A simple voltameter is used, consisting of a copper plate for the cathode flanked by two copper plates coupled together for the anode. The electrolyte is copper sulphate solution. The simple circuit comprises battery (6 volts), ammeter, rheostat, voltameter and switch in series. The cathode is first cleaned with emery paper, washed,* dried and then accurately weighed. The cleaning is very essential, for the copper deposited during electrolysis will not adhere if the cathode is even slightly contaminated. The weighing must be carried out carefully for the increase in mass will probably be less than one gramme. To get this increase accurate to 1 per cent demands that the two weighings be taken to better than 10 milligrammes. For convenience, the current in the simple circuit can be adjusted by using a dummy cathode, to avoid wetting the one to be cleaned and weighed. The current should not be too great† or the deposit will not be hard. A good quality ammeter is essential, for the result depends directly on its accuracy.‡

The rheostat having been adjusted by a trial run, allows the suitable current to be attained at once, and so the duration of the experiment can be determined precisely. This is usually about thirty minutes for a laboratory determination, during which time the current should be constant. The cathode is then removed, carefully washed, dried and re-weighed. The e.c.e. is found by:

e.c.e. [grammes per coulomb]

$$= \frac{\text{mass deposited [grammes]}}{\text{current [amperes]} \times \text{time [seconds]}}$$

Uses of electrolysis

Chemical

1. Extraction of metals.
2. Purification of metals.
3. Electrotyping.
4. Electroplating.

The above uses are fully discussed in Chemistry texts.

Physical

1. *Measurement of electrical energy.* An early form of meter made use of electrolysis. A small fraction of the total current used was passed through a water voltameter. The volume of gas liberated was proportional to the quantity of electricity passed. With a constant p.d. this was also proportional to the energy consumed.

2. *Measurement of current.* The unit of current, the ampere, is defined in terms of the attraction between neighbouring current-carrying conductors (Chapter 39). The measurement of current absolutely requires such precise and elaborate apparatus that the determination is carried out in only a few laboratories such as the National Physics Laboratory. The e.c.e. of silver has been determined very accurately so that we may use the value (0.001 118 23 g/C) to measure a current.* The cathode of a silver voltameter placed in the circuit is weighed before and after passing the constant current for a measured time. The increase in mass (grammes) divided by the product of the e.c.e. (grammes/coulomb) and the time (seconds) gives the current in amperes.

Cells†

The generation of electricity by chemical means was recorded by Volta in 1800. He invented what is known as a *voltaic pile,* consisting of a number of plates of two different metals separated by felt impregnated with electrolytes, as shown in Fig. 36.4. A modified form of this apparatus, which we

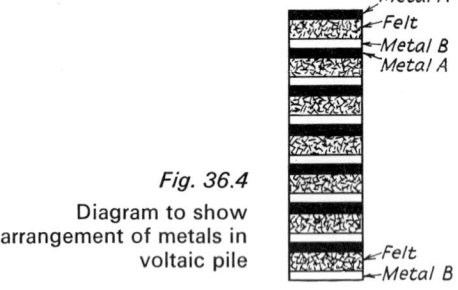

Fig. 36.4
Diagram to show arrangement of metals in voltaic pile

call the *simple cell,* consists of a plate of copper and one of zinc immersed in dilute sulphuric acid.

---

* It is effective to wash the cathode in water and then in methylated spirit: drying is quick with little oxidation.

† The current should not exceed about 1 A for 50 cm² of cathode area.

‡ Care should be taken to read the instrument accurately, e.g. the zero error should be noted and the needle observed by viewing the scale normally.

---

* The International Ampere used to be defined as that unvarying current which will liberate 0.001 118 grammes of silver in a silver voltmeter in 1 second.

† See *On Teaching Voltaic Cells and Electrolysis,* H. G. Andrew, Modern Science Memoir 41.

If the zinc is pure,* nothing is observed until the two plates are connected internally or externally by bringing them together or by connecting them with a wire. Then, bubbles of gas will be seen coming from the copper plate. This gas is hydrogen. Furthermore, a current flows in the connecting wire, which can easily be detected with a compass needle. The action causes a decrease in the mass of the zinc, but the copper remains unchanged.

Copper and zinc are not the only metals which can be used to produce a simple cell. Any two metals can be chosen from the electrochemical series given earlier in this chapter.

Theory

For an electric current to flow in the wire connecting the copper and zinc plates of the simple cell, it is necessary that there be a p.d. across the plates. Indeed, the electrochemical series is compiled by measuring this p.d. The more electropositive the metal is (i.e. the higher in the series) the more ready is the metal to give up one or more electrons.

In the dilute sulphuric acid of the simple $Cu-Zn$ cell are H_3O^+, SO_4^{2-} and OH^- ions, as mentioned earlier in this chapter.

The following actions occur:

(*a*) At the surface of the zinc plate, atoms go into solution as Zn^{2+} ions, i.e.

$$\underset{atom}{Zn} \rightarrow \underset{ion}{Zn^{2+}} + \underset{electrons}{2e^-}$$

(*b*) The negative charge (electrons) is released to the external circuit and passes to the copper plate.

(*c*) The negative charge neutralizes the positive charge on the H_3O^+ ions in the vicinity of the copper plate and the resulting hydrogen is evolved as a gas.

$$\underset{electrons}{2e^-} + \underset{ions}{2H_3O^+} \rightarrow \underset{hydrogen\ gas}{H_2} + \underset{water}{2H_2O}$$

The distribution of charge on the zinc and copper plates builds up a potential difference across the plates. It is this p.d. which causes the drift of electrons round the external circuit. When the circuit is broken the flow ceases, but the p.d. across the plates is maintained. Zn^{2+} ions do not pass into the solution because there is accumu-

---

* Pure zinc does not dissolve readily in sulphuric acid.

lation of positive ions round the zinc plate and an accumulation of electrons on the plate.

The action at the negative electrode where electrons are produced is *oxidation*: that at the positive electrode where electrons are absorbed is *reduction*.

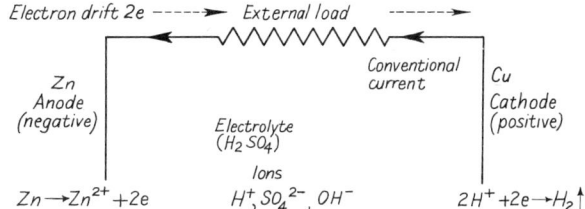

Fig. 36.5 The simple cell

It should be noted that in the case of cells, the positive electrode is the cathode (current enters) and the negative the anode (current leaves). In all cells oxidation occurs at the anode and reduction at the cathode.

Defects of the simple cell

1. *Local action.* If an ordinary commercial zinc plate is used, this goes on dissolving in the acid when the cell is not being used. This action is due to the impurities in the zinc forming with it a host of tiny simple cells. If one could see the details of the action, the hydrogen would be found coming off from the impurities and not from the zinc. A cheaper remedy than using pure zinc is to amalgamate the plate. To do this the plate is dipped in dilute sulphuric acid and then a small quantity of mercury is rubbed over the surface. This is quite easy to do and gives the surface a silvery appearance. The amalgam is a solution of pure zinc in mercury. The mercury has no effect on the action of the cell, and so there is virtually a layer of pure zinc on the surface. As this is used up in working the cell, more dissolves from underneath to take its place.

2. *Polarization.* The bubbles of hydrogen which collect on the copper plate form a barrier and so increase the resistance of the cell. Further, a back e.m.f. is set up because the reaction (*c*) above tends to reverse.

$$2H_2O + H_2 \rightarrow 2H_3O^+ + 2e^-$$

To correct this defect, the layer of hydrogen has to

be removed. Brushing it away illustrates what is required, but it is not a satisfactory remedy. The best method is to remove it chemically, i.e. to combine it with something with which it forms a liquid compound. The hydrogen, on liberation, is in the nascent form and will combine readily. A substance used for this purpose is called a *depolarizer*. Obviously, most depolarizers are oxidizing agents. Polarization and the effect of a depolarizer can easily be demonstrated. Make up a simple cell and connect it as in the circuit shown in Fig. 36.6.

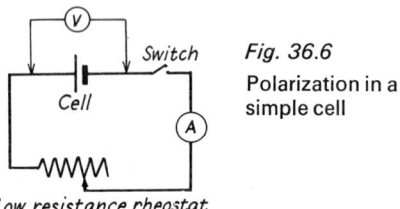

Fig. 36.6
Polarization in a simple cell

Close the switch and note the ammeter and voltmeter readings. Quite soon the voltmeter reading will drop and cause a corresponding drop in the current. When the voltmeter ceases to fall rapidly, add some potassium dichromate solution to the acid of the cell. The reading of the voltmeter will shoot up to its original value and remain there.

The Leclanché cell

This is a well-known cell and its components are shown in Fig. 36.7. The electrolyte is sal-ammoniac

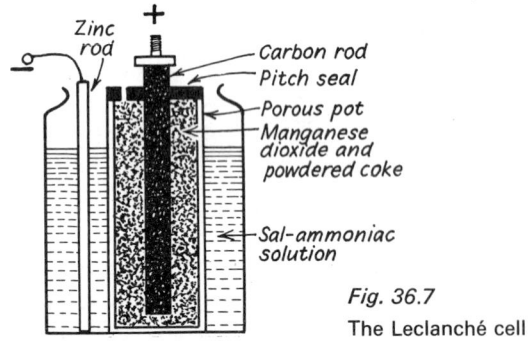

Fig. 36.7
The Leclanché cell

solution (NH_4Cl). The positive electrode is manganese dioxide. The carbon plate acts as a collector of the charge from the manganese dioxide, the action being aided by adding powdered carbon to the manganese dioxide, which increases its poor

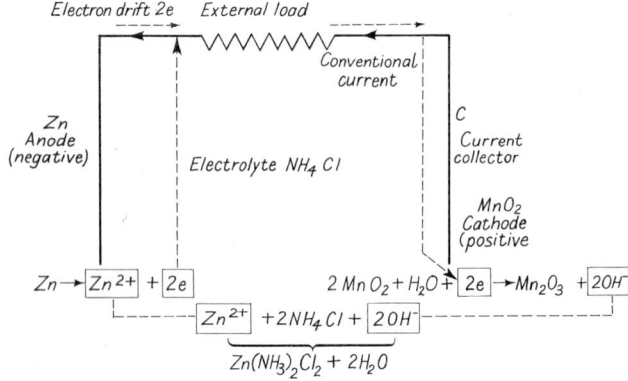

Fig. 36.8 The Leclanché cell (after Ever Ready)

conductivity. (The carbon plate is often called the positive electrode because the terminal is attached to it.) The carbon plate and its surrounding powder are contained in a porous pot. This allows the free passage of ions, but polarization soon sets in. The zinc is oxidized to zinc ions.

$$Zn \rightarrow Zn^{2+} + 2e^-$$

and the manganese dioxide reduced to a lower oxide.

$$2MnO_2 + H_2O + 2e^- \rightarrow Mn_2O_3 + 2OH^-$$

From this action the cell obtains its energy. The popularity of this cell is on account of its constancy for occasional or intermittent use over long periods. It is useful for such purposes as ringing electric bells and, when once set up, will last for many months. The only attention it requires is filling up with water and, perhaps, the addition of a little more sal-ammoniac. Its e.m.f. is about 1.5 V.

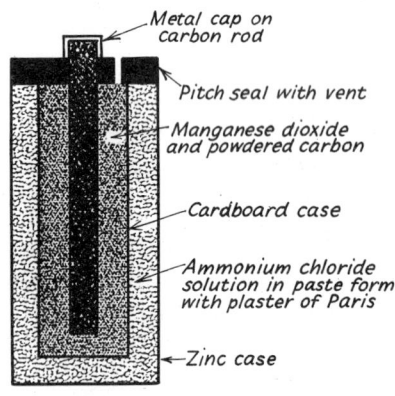

Fig. 36.9 The dry cell

The dry cell

The dry cell, used for torch batteries, is a modified form of Leclanché cell. The electrolyte is prevented from running by forming it into a paste with plaster of Paris, Fig. 36.9. The performance of the cell is much improved by the addition of certain substances which are trade secrets of the manufacturers.

Other dry cells

The zinc-carbon or modified Leclanché dry cell is still by far the most common type but, during recent years, new types such as the *alkaline manganese*, the *mercury* and the *silver oxide* cells have been introduced. These have definite advantages over the old ones, but they are much more expensive and so are used only for special purposes, such as for hearing aids and watches and also in connection with the space programme.

The Daniell cell

The construction of this cell is shown in Fig. 36.11. It is not much used, but it can serve as a cheap standard cell. Its e.m.f. remains very constant.

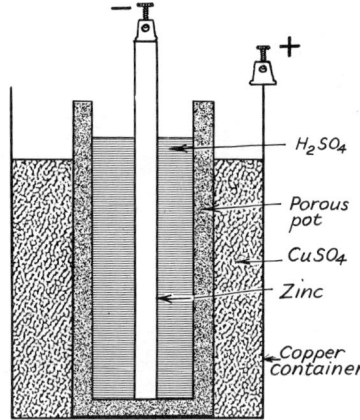

Fig. 36.11 The Daniell cell

The Weston cadmium standard cell

This cell, Fig. 36.12, was adopted in 1908 as a standard of e.m.f. Its performance remains very

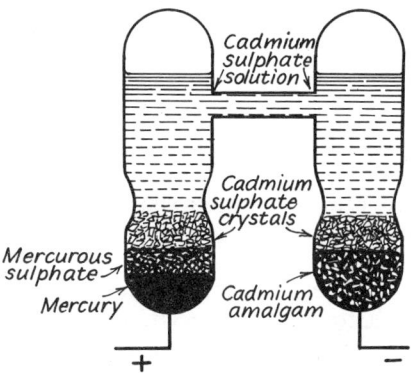

Fig. 36.12 Weston cadmium cell

constant (1.018 V at 20 °C), but it must be treated carefully and used only to standardize a potentiometer. It must never be used to provide an appreciable current in a circuit.

Fig. 36.10 Watch powered by tiny electric battery
The battery maintains a small tuning-fork in vibration with a frequency of 360 Hz

Fig. 36.13
This is an experimental tractor manufactured by the American firm of Allis-Chalmers. It develops 15 kW from 1008 fuel cells

Fuel cells

The cells described above obtain their electrical energy by consuming certain contents of the cells. When one of these is used up, the cell is either thrown away or the deficiency made good. It should, then, be possible for a cell with suitable consumable constituents to be continuously replenished and so to provide a continuous output. The first *fuel* cell, as these are called, was built in Cambridge by F. T. Bacon in 1959. Oxygen and hydrogen were the fuels and about 150 W of power were developed. At the present time much work is being done on these cells and there is a possibility that they will be developed to produce electrical energy in commercial quantity at an economic price.

Accumulators

The accumulator or secondary cell has now largely superseded the primary cell. In the latter, the electrical energy is generated by the chemical action within the cell and the constituents are used up. The accumulator is a storage cell: the electric charge has first to be put into it, and this is stored by chemical means. The chemical action is reversible and so the charge can be recovered from the cell at will.

The lead-acid accumulator

Arrange two lead plates in slightly diluted sulphuric acid and pass a current through the cell so formed.

After a few minutes, stop the current and connect a small torch bulb across the lead plates. The bulb will glow for a few moments, showing that some of the electrical energy supplied to the cell is recoverable.

The modern accumulator consists of lead plates in sulphuric acid. In order to increase the capacity for storing electric charge, several plates are used which are alternately connected and are kept apart by wooden separators.

The plates are manufactured in grid form so that the products of the action will adhere more easily. The accumulator is usually supplied charged with electricity, the plates having deposits of lead and lead dioxide. During discharge the action at the anode (negative electrode) is

$$Pb \rightarrow Pb^{2+} + 2e^- \text{ (oxidation)}$$

and at the cathode (positive electrode)

$$Pb^{4+} + 2e^- \rightarrow Pb^{2+} \text{ (reduction)}$$

At the anode, Pb^{2+} and SO_4^{2-} ions come together leaving in excess two H_3O^+ ions.

At the cathode a similar action takes place and the total of four H_3O^+ ions react with the two O^{2-} ions left over from the reduction of the lead dioxide.

$$Pb^{4+}(O^{2-})_2 + 2e^- \rightarrow Pb^{2+} + 2O^{2-}$$

This action involves the removal of four H_3O^+ and two SO_4^{2-} ions from the solution, which is equivalent to consuming two sulphuric acid molecules. Thus during discharge the specific gravity of the acid falls. In the charging process the reverse chemical changes take place and the specific gravity of the acid rises.

Charging. Accumulators are conveniently charged from the a.c. or d.c. mains by using a motor generator. This consists of a mains motor coupled directly to a low-voltage d.c. generator. Charging from the a.c. mains is usually done by first transforming the voltage down to a low value and then rectifying it.

The e.m.f. of an acid accumulator is about 2.1 V when fully charged. In use, this e.m.f. is maintained until, when nearly discharged, it drops rapidly. The e.m.f. should not be allowed to fall below 1.8 V. If this happens the cell is damaged, for the chemical action is reversible only over a limited range. The relative density of the acid is a good indication of the state of the charge. When fully charged it should be about 1.25, and recharging is indicated when it drops to 1.11, or to the value stated on the accumulator. Unlike the e.m.f., the relative density drops regularly during discharge. Distilled water must be added occasionally to make up for the loss due to electrolysis in the charging process.

The alkali accumulator

There are two types of alkali accumulator in common use. They both have positive electrodes consisting of nickel hydroxide packed in perforated steel containers. The negative electrodes are, in one case, finely divided iron packed into perforated steel tubes and, in the other, a mixture of iron and cadmium in flat perforated steel containers. The electrolyte is a solution of potassium hydroxide in distilled water (relative density 1.15 to 1.20).

The second type has properties very near to those of a lead plate accumulator, while the first type is better for traction purposes and stands up to use in hot climates.

On discharge, the chemical action is as follows.

(*a*) At the anode the iron becomes bound up with the OH^- ions,

$$Fe + 2OH^- \rightarrow Fe^{2+}(OH^-)_2 + 2e^-$$

(*b*) At the cathode the nickelic hydroxide is reduced

$$2Ni^{3+}(OH^-)_3 + 2e^- \rightarrow 2Ni^{2+}(OH^-)_2 + 2OH^-$$

The reverse action takes place on charging. There is no change in the electrolyte on charging or discharging. The e.m.f. of each cell is 1.2 V. The

great advantage of alkali accumulators is that they can be charged and discharged at a heavy rate without damage. They are relatively lighter than lead-acid accumulators.

Internal resistance of a cell

In a simple circuit containing a cell and resistors, the current is the same at all points. This applies even to points within the cell. Thus, a cell provides the e.m.f. to drive the current through itself as well as through the external resistors. The p.d. required is opposite in direction to the e.m.f. of the cell. Hence, when the cell is working, the p.d. across the electrodes is the e.m.f. of the cell less the p.d. required to maintain the current through the cell. This latter p.d. is greater, the greater the current the cell is maintaining: it becomes zero when the current is zero. Hence,

The e.m.f. of a cell is the p.d. across its terminals when on open circuit.

The opposing p.d. is equal to Ir, where I is the current and r the *internal resistance* of the cell.

If E is the e.m.f. of a cell and V the p.d. across its terminals when it is working, we have:

$$V = E - Ir$$

and

$$V = IR$$

where R is the external resistance in the circuit (Fig. 36.14).

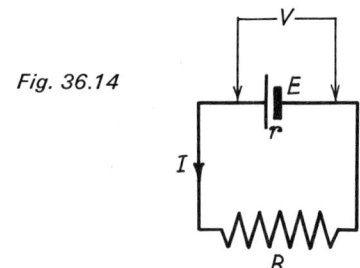

Fig. 36.14

The internal resistance of a cell is determined by the size of the electrodes, their distance apart, and the conductivity of the electrolyte. The inclusion of a porous pot in a cell considerably increases the internal resistance.

The maximum current a cell can give is governed by its internal resistance. It is equal to E/r (*see* Example overleaf).

Example

A high-resistance voltmeter reads 1.56 V when connected across a cell which is not on load. When a 5.0 Ω resistor is connected across the cell terminals, the volmeter reading drops to 1.20 V. Find the internal resistance of the cell.

Let internal resistance in ohms $=r$
Applying Ohm's law to the whole circuit, Fig. 36.14, we have:

$$I = \frac{\text{total effective e.m.f.}}{\text{total effective resistance}}$$

$$= \frac{E}{R+r} \tag{1}$$

Applying Ohm's law to the external resistor, we have:

$$V = IR$$

or to the cell, when we have:

$$V = E - Ir$$

Combining eqn. (1) with either of the above, to eliminate I,

$$\frac{V}{R} = \frac{E}{R+r}$$

$$\frac{R+r}{R} = \frac{E}{V}$$

$$r = \frac{E-V}{V} . R$$

Substituting the values given,

$$r = \frac{1.56 - 1.20}{1.20} \times 5.0$$

$$= 1.5 \ \Omega$$

The internal resistance of the cell is 1.5 Ω.

Experimental determination of the internal resistance of a cell

The principle of the method is to determine the p.d. across the cell terminals when it is working. A resistor R is connected across the cell (Fig. 36.14) and the p.d. across its terminals determined by using a good quality high-resistance voltmeter, or better, by a potentiometer (*see* page 357). To obtain a series of readings, R is given different known values. The readings are tabulated as shown.

Resistance R/Ω	Reading of voltmeter V/V	$r = \dfrac{E-V}{V} . R/\Omega$
20		
18		
.		
.		
.		
.		
6		

E is the value of V when $R = \infty$

The internal resistance varies also with the current the cell is giving and so the value of r in the above table will not be constant.

The e.m.f. of a cell is independent of its size and depends only on the constituents. The larger the cell, the smaller the internal resistance.

If cells are connected in series, the total e.m.f. is found by adding the individual e.m.f.s. The total internal resistance is also the sum of the separate resistances (Fig. 36.15).

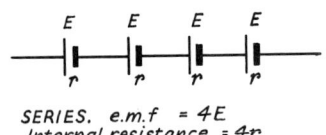

SERIES. e.m.f = 4E
Internal resistance = 4r

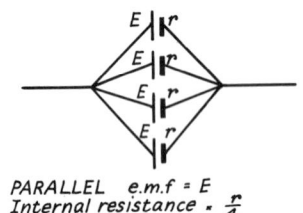

PARALLEL   e.m.f = E
Internal resistance = $\frac{r}{4}$

Fig. 36.15 Connecting cells

The e.m.f. of a number of similar cells in parallel is the same as that of each separately, but the internal resistance is reduced in proportion to the number of cells used.

QUESTIONS

1. State Faraday's laws of electrolysis and describe an experiment to illustrate *one* of them.

Calculate the volume of mercury liberated in 40 minutes by a current of 2 A passing through a solution of a mercury salt.

[Electrochemical equivalent of hydrogen=0.00001044 g/C, chemical equivalent of mercury=200.6, density of mercury =13.6×10³ kg/m³.] (L)

2. Explain the terms *electrolysis, electrolyte, ion, anode, cathode.*

What experiments would you perform in order to find out the factors upon which the mass of a substance deposited during electrolysis depends? What would you expect to find?

A current of electricity passes for 2 hours through a copper voltameter in series with a water voltameter. If the mass of copper deposited is 1.1 g calculate (a) the average value of the current, (b) the volume of hydrogen at s.t.p. evolved in the water voltameter.

[e.c.e. of copper, 0.00033 g/C; equivalent weight copper, 31.8; density of hydrogen at s.t.p. 0.09 kg/m³.] (O & C)

3. Explain as fully as you can how an electric current can pass through an electrolyte, and state the laws of electrolysis.

A current of 10 A between copper electrodes is passed for 30 minutes through a solution of copper sulphate.

Calculate the average thickness of copper deposited on the cathode if the area submerged is 300 cm².

[e.c.e. of copper, 3.3×10⁻⁴ g/C. Density of copper, 9.0×10³ kg/m³.] (O & C)

4. How would you find the electrochemical equivalent of copper by experiment? Draw a diagram of the circuit that you would use, explain the procedure and show how you would calculate the result.

In an electrolysis experiment the ammeter records a steady current of 1 A. The mass of copper deposited is 0.66 g in 30 minutes. What is the error in the ammeter reading?

[Electrochemical equivalent of copper is 0.00033 g/C.] (JMB)

5. Define *coulomb* and *electrochemical equivalent.*

A 200 V, 100 W lamp is joined in series with a copper voltameter of negligible resistance and the circuit is connected to a 200 V direct current supply. Draw a circuit diagram to illustrate the arrangement and indicate clearly the electrode on which copper is deposited. Calculate (a) the current through the voltameter, (b) the resistance of the lamp, (c) the mass of copper deposited per hour, assuming the electrochemical equivalent of copper to be 0.00033 g/C. (L)

6. Draw a diagram showing the circuit you would use to deposit electrolytically a thin film of silver on a copper object. State any *two* precautions you would take to obtain a satisfactory deposit.

It is required to electroplate a copper object of total surface area 400 cm² with a layer of silver 0.002 cm thick. If a current of 2 A is used, how long will the process take?

[The electrochemical equivalent of silver=0.001118 g/C and the density of silver=10.5×10³ kg/m³.] (L)

7. State the laws of electrolysis.

Two cells, each of e.m.f. 1.10 V and internal resistance 1.50 Ω, joined in series, supply current to a copper voltameter of resistance 1.40 Ω. Draw a circuit diagram of the arrangement, mark the polarity of the cells, and indicate clearly the electrode on which copper is deposited in the voltameter. Calculate (a) the current through the voltameter, (b) the electrochemical equivalent of copper, if 0.60 g of copper is deposited per hour. (L)

8. State the laws of electrolysis.

A current is passed through two voltameters connected in series. One is designed for copper plating an object, the other for electrolysing water. Draw a diagram of the voltameters indicating the complete circuit. State what happens at each of the four electrodes. If 0.12 g of copper is liberated in the one voltameter, calculate the volume at s.t.p. of the gases liberated in the other.

[Chemical equivalent of copper=31.5; density of hydrogen at s.t.p.=0.09 kg/m³.] (L)

9. State Faraday's laws of electroysis.

A current is passed through two voltameters in series. Copper is deposited on the cathode of the first, while hydrogen is liberated from the cathode of the second. If 1.36 g of copper are deposited in one hour, calculate (a) the current, (b) the volume of hydrogen liberated (measured at a temperature of 17 °C and a pressure of 750 mmHg.

[The atomic weight of copper=63.6; the density of hydrogen at s.t.p.=0.090 kg/m³; 96 500 C will liberate one gramme equivalent.] (C)

10. Describe a method of measuring electric current by measuring the amount of chemical action it produces.

A current is passed through a silver voltameter and the p.d. between the plates of the voltameter is maintained at 2 V. If 4.026 g of silver are deposited on the cathode in 1 hour, calculate (a) the resistance of the voltameter, (b) the energy consumption in kilowatt-hours.

[Electrochemical equivalent of silver=0.001118 g/C.] (O & C)

11. Describe and explain the effects produced at the electrodes when a steady current is passed in series through solutions of (a) dilute sulphuric acid with platinum electrodes, (b) copper sulphate with copper electrodes.

If in 45 minutes the mass of one of the copper electrodes is found to have increased by 0.75 g, what is (a) the value of the current, (b) the volume of hydrogen measured at s.t.p. liberated at one of the platinum electrodes?

[Electrochemical equivalent of copper: 0.00033 g/C, atomic weight of copper: 63.5, valency 2; atomic weight of hydrogen: 1, valency 1; density of hydrogen at s.t.p. 0.09 kg/m³.] (O & C]

12. State Faraday's laws of electrolysis and describe *two* experiments to illustrate them.

A steady electric current passes through a silver voltameter and a 10 Ω resistor arranged in parallel. A high-resistance voltmeter connected across the terminals of the 10 Ω resistor reads 6 V, and 1.5 g of silver are deposited on the cathode of the voltameter in 20 minutes. Calculate the resistance of the voltameter.

[The electrochemical equivalent of silver=0.001118 g/C.] (O & C)

342 Electrical energy

13. When a current passes through an electrolyte, effects may be observed similar to those produced by a current in a wire. What are these effects?

A current of 2.0 A is passed through a copper voltameter for 180 minutes. What is the change in mass of the cathode?
[e.c.e. of copper 3.3×10^{-4} g/C.] (O & C)

14. State Faraday's laws of electrolysis and explain the meaning of a *faraday*.

A battery, a copper voltameter and a water voltameter are connected in series. After 30.0 minutes 0.50 g of copper is deposited. Calculate the average current and the volume of hydrogen, measured at s.t.p., liberated in the water voltameter.
[e.c.e. of copper is 33×10^{-5} g/C. Atomic weight of copper $=63.5$; valency 2. Density of hydrogen at s.t.p.$=0.090$ kg/m^3.] (O & C)

15. In electrolysis 1 gramme equivalent of hydrogen is liberated by approximately 9.6×10^4 C. Taking the charge on the electron as 1.6×10^{-19} and the density of liquid hydrogen as 0.07×10^3 kg/m^3 estimate the volume of a hydrogen atom. (O & C)

16. Give an account of the construction and mode of operation of the simple cell.

From what defects does the simple cell suffer and how are they minimized in *either* (a) the Leclanché cell *or* (b) the Daniell cell? (O & C)

17. Describe the simple cell and explain how it works. What are the chief practical disadvantages of this cell as a source of intermittent small currents? How have these disadvantages been reduced in the 'dry cell'? (O & C)

18. Given a voltmeter, an ammeter and a resistance box, how would you determine the internal resistance of a Daniell cell? Draw a circuit diagram. Explain what readings you would take and how you would calculate the result.

A potentiometer wire of length 100 cm and resistance 2 Ω is connected in series with a 2 V accumulator of negligible internal resistance and a variable resistance box. What value must be given to the variable resistance in order that the fall of potential along the potentiometer wire may be 0.001 V/cm? (C)

19. (a) A resistance coil R is connected in series with a cell of e.m.f. 1.56 V and an ammeter of resistance 0.100 Ω. If the p.d. across the cell terminals is 1.40 V and the current is 0.20 A, find the value of R and the internal resistance of the cell.

(b) A high resistance voltmeter reads 1.18 V when connected across the terminals of a cell. When a resistor of 6 Ω is also connected across the cell terminals, the reading of the voltmeter drops to 1.03 V. Find the internal resistance of the cell.

20. Define the *electromotive force (e.m.f.)* of a cell.

Describe how you would use a potentiometer to compare the e.m.f.s of two cells.

A wire of resistance 8 Ω is connected across the terminals of a cell of e.m.f. 1.1 V and internal resistance 3 Ω. Find the current passing, and also the p.d. between the terminals of the cell. (O)

21. Give an account of the Lechanché cell. What is polarization, and how is it overcome in this cell?

A battery is made up of 80 Leclanché cells in series; each has e.m.f. 1.5 V and internal resistance 2 Ω. What is the e.m.f. of the battery? What is the p.d. between its terminals when it is passing a current of 20 mA? (O)

22. Explain the terms *electromotive force, internal resistance* of a cell.

A battery of storage cells is to be charged in series across the 200 V d.c. mains. The e.m.f. of each cell is 2.5 V and its internal resistance 0.1 Ω and the charging current is to be 8 A. What is the greatest number of cells which can be charged in this way and what additional resistance must be connected? (O & C)

23. Explain what is meant by the *internal resistance* of a cell. Describe an experiment to show that a Daniell cell has internal resistance.

The e.m.f. of a cell is 1.1 V and its internal resistance is 2 Ω. When its terminals are joined by a wire a current of 0.22 A flows. Find the resistance of the wire and the potential difference between the terminals of the cell.

Three such cells in series are connected to two coils of resistances 3 Ω and 2 Ω respectively. Calculate the current through the cells when the coils are (a) in series, (b) in parallel. (L)

24. What are the defects of a simple cell? How are they overcome in the Leclanché cell?

A battery of ten accumulators is to be charged from a d.c. source of 50 V. Each accumulator has an internal resistance of 0.05 Ω, and an e.m.f. of 2.5 V. The charging current is to be 2 A. Calculate the resistance that must be included in the circuit. (C)

25. Draw a labelled diagram to show the structure of any one type of accumulator in a fully charged condition.

An accumulator is labelled '2 volts, 50 ampere-hours, normal charging rate 3 A'. Draw a fully labelled circuit diagram to show how you would re-charge it after discharge, using a 6 V d.c. supply. Estimate the cost of the re-charge in this circuit if the supply costs 2p per kilowatt-hour. (C)

26. Give short explanations of the terms *electromotive force, internal resistance* of a cell.

A battery can supply a current of 0.20 A when a 20 Ω resistor is connected across it. When a second 20 Ω resistor is connected in parallel with the first the current through *each* resistance becomes 0.15 A. Suggest a reason for this and calculate (a) the e.m.f., (b) the internal resistance of the battery. (O & C)

27. State the meaning of *electromotive force, internal resistance* of a chemical cell.

Why is the potential difference between the terminals of a cell not always equal to its electromotive force?

The terminals of a cell of 2 V e.m.f. and of negligible resistance are connected by two resistors R_1 and R_2 arranged in series. The resistance R_2 consists of two resistance coils of 4 Ω and 6 Ω respectively arranged in parallel. If a current of 0.5 A flows through the circuit, calculate the value of the resistance R_1, and deduce the p.d. across the 4 Ω coil. (O & C)

28. Why is the internal resistance of an accumulator much less than that of a Leclanché cell?

Three Leclanché cells, each of e.m.f. 1.5 V and 3 Ω internal resistance, are connected (a) in series, (b) in parallel, and placed in a circuit of resistance 9 Ω. Find the current in each case.

29. Describe how you would set up a simple cell.

It is observed that, when a small lamp bulb is connected across the terminals of such a cell, the lamp glows brightly at first, but soon ceases to glow at all. Explain the reason for this.

A cell of e.m.f. 1.5 V and internal resistance 5 Ω is connected to a lamp of resistance 25 Ω. Find (a) the current taken by the lamp; (b) the p.d. across it; and (c) its power consumption. (O)

37 Force acting on a current in a magnetic field

A current flowing in a straight wire produces a symmetrical magnetic field about it. A N-pole placed at the point P (Fig. 37.1) will be acted upon by a force F, as shown. Newton's third law of motion states that, '*to every action there is an equal and opposite reaction*'. Hence, the N-pole at P must act on the current with an equal force R in the opposite direction. If the N-pole at P were stationary, and the conductor movable, the latter would move in the direction of R. Now the only effect the N-pole can have on the current is to produce a magnetic field about it. The field lines will be outwards from P, and so the field at the current will be as shown. Hence, we conclude that when a current is in a magnetic field it is acted upon by a force. But the directions are important. In the diagram, the current is inwards, the field due to the pole is from right to left and the force in the direction of R. The three directions are mutually at right angles. The following rule enables the direction of the force (and the motion of the conductor which may result) to be found quickly.

> Fleming's left-hand rule:
> Arrange the first two fingers and the thumb of the LEFT hand mutually at right angles:
> Point the First finger along the Field:
> Point the seCond finger along the Current (conventional):
> then the thuMb gives the direction of the Motion.

Demonstrations of Fleming's left-hand rule or the motor effect

1. N and S are the poles of an electromagnet providing a strong field (Fig. 37.2). A length of flexible

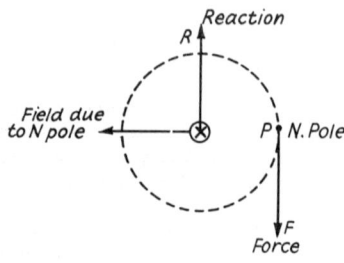

Fig. 37.1 Force on a wire carrying a current in a magnetic field

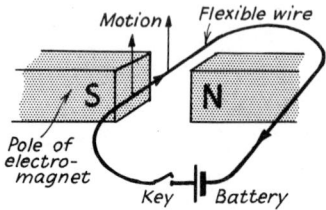

Fig. 37.2 Force acting on a wire carrying a current when in a magnetic field

When key is pressed, wire jumps out of gap between poles

wire is supported on a piece of paper or cardboard in the field. Closing the spring key sends a current through the conductor which jumps out of the field.

This effect upon a current in a magnetic field can be understood by considering the combined field. Fig. 37.3 shows the combined field with the

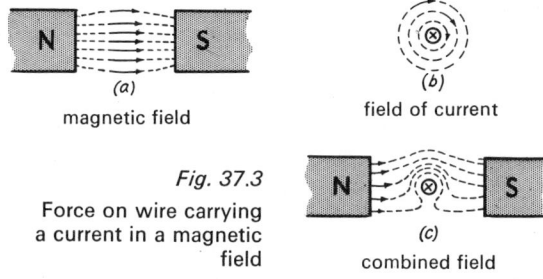

(a)

magnetic field

(b)

field of current

Fig. 37.3

Force on wire carrying
a current in a magnetic
field

(c)

combined field

distorted field lines. As these are in tension, they act like a catapult on the wire carrying the current.

2. Two metal rails, mounted on wood, are arranged one on each side of the horseshoe magnet (Fig. 37.4). The short rod *AB* completes the circuit and the force resulting from the effect rolls the rod along the rails.

3. *Barlow's wheel*. A stellate wheel (Fig. 37.5) is mounted, as shown, so that the tips make contact with mercury in a small trough in the baseboard. Magnets are arranged on each side of the trough so that the tip dipping in the mercury is in a magnetic field. The force produced causes the rotation of the wheel.

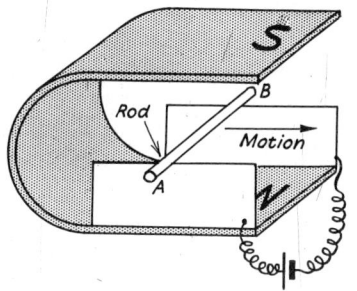

Fig. 37.4 Motion of conductor in magnetic field

The rod completes the circuit across metal rails in the field of a horseshoe magnet. The motion of rod is reversed by reversing the current or field

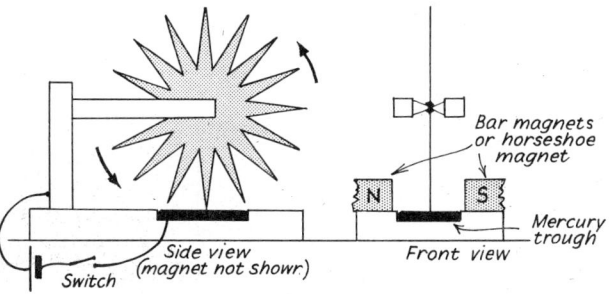

Fig. 37.5 Barolow's wheel

A disc will do, but the stellate wheel is more effective and does not make such a great demand on the accumulator.

4. *Faraday's experiment*. Along with the experiment on page 307, Faraday exhibited this apparatus (Fig. 37.6) to his wife on Christmas Day, 1821, at the Royal Institution.

The upper pole of a vertical bar magnet (Fig. 37.6*b*) is surrounded by a circular mercury trough.

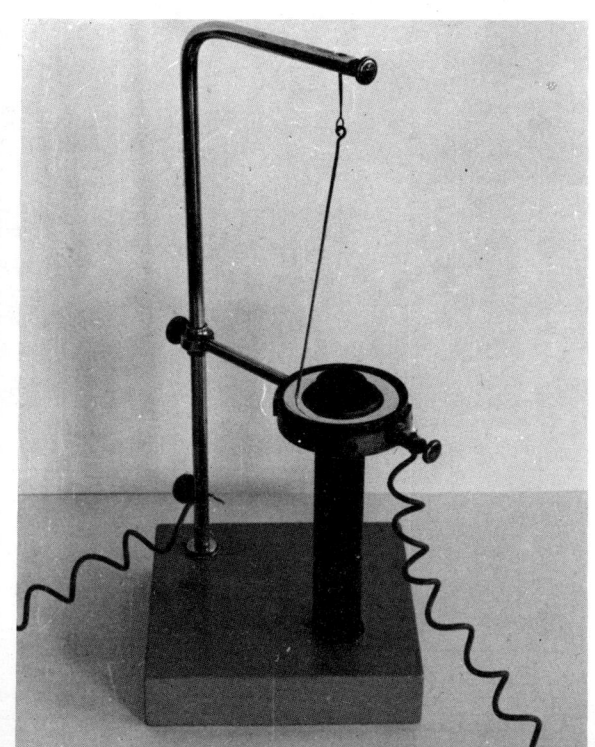

(a)

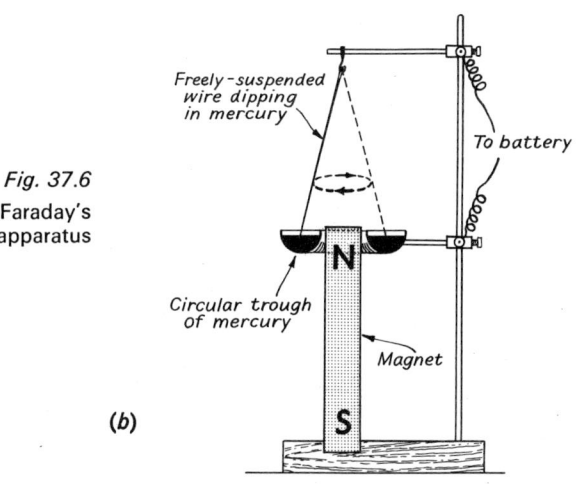

Fig. 37.6
Faraday's
apparatus

Freely-suspended
wire dipping
in mercury

To battery

Circular trough
of mercury

Magnet

(b)

A short straight wire hangs loosely from a loop supported above the magnet. The lower end of the wire dips into the mercury trough. Connections from the battery are made to the upper wire and the mercury. A tangential force on the wire is produced and causes rotation of the wire round the pole.

5. A wire carrying a current produces a field which will cause a force on a neighbouring wire carrying a current. There is mutual attraction between adjacent wires carrying currents in the same direction and repulsion when the current directions are opposite (Fig. 37.7).

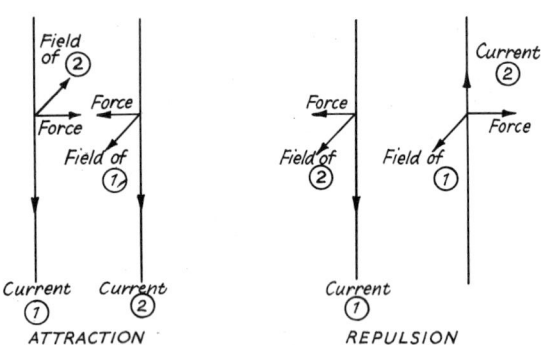

Fig. 37.7 Forces between neighbouring currents

A development of this is Roget's jumping spiral. The turns of a solenoid attract each other and, if the solenoid is loosely coiled, it contracts when a current is passed through it. If an open solenoid is supported vertically with its lower end just dipping

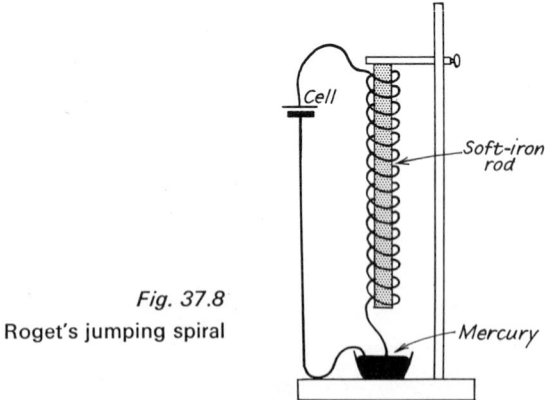

Fig. 37.8
Roget's jumping spiral

into a mercury trough, it contracts when a current is passed and breaks the circuit (Fig. 37.8). Being springy, it regains its original length and re-makes

the circuit. The repetition of this action provides the 'jumping'. The effect is much increased in vigour by thrusting a soft-iron rod down the solenoid.

6. The following experiment illustrates the force acting on a current passing through a liquid. A crystallizing dish of about 10 cm diameter, is wrapped with about 100 turns of insulated wire which forms a circular coil. Within the dish a strip of copper about 5 cm by 30 cm is bent round the inside wall. A copper rod is clamped vertically in the centre of the dish which is filled with copper sulphate solution. A current is passed through the coil which produces a vertical magnetic field. A second circuit produces a current through the solution from the rod to the strip. The direction of this current is radial and the interaction with the vertical magnetic field causes a tangential force on the liquid. The resulting rotation of the liquid can be seen by putting sawdust on the surface. The experiment lends itself to vertical projection. (*See* Fig. 37.9.)

In Chapter 41, the deflection of the cathode rays

Fig. 37.9 Apparatus to show the force acting on a current-carrying liquid conductor in a magnetic field

The coil round the glass vessel provides a vertical magnetic field and the current passes through the electrolyte ($CuSO_4$) from the centre electrode to the copper ring

by a bar magnet provides yet another illustration of the force acting on a current in a magnetic field.

The ampere

In number 5 above, it is shown that there is a force of attraction between two parallel current-carrying conductors when their currents flow in the same direction. The magnitude of the force can be shown to depend upon the product of the currents, upon their lengths and, inversely, upon the distance they are apart. By making the magnitude of each of these quantities unity, we obtain the SI unit of current— the *ampere*.

> **The ampere is that constant current which, when flowing in two straight and parallel conductors of infinite length and placed one metre apart, produces between the conductors a force of 2×10^{-7} newton per metre length.**

Note on the ampere

This number of 2×10^{-7} N was not chosen in adopting SI units. The ampere is defined in this way in order to retain a unit of the same size as in the past. Previously it was related in the cgs system to the magnetic field produced by a current.

An alternative method of defining unit current would have been by fixing the number of electrons which flowed past a point in a conductor in one second. It follows from the above definition that a current of one ampere means a rate of flow of 6.25×10^{18} electrons per second.

The current balance

This instrument, shown in Fig. 37.10, is used for the absolute measurement of current at the National Physical Laboratory. It is shown diagrammatically in Fig. 37.11. The two smaller coils are suspended from the arms of a balance and hang symmetrically between two pairs of larger coils. The same current is passed through all six coils in such directions that the forces all act to produce a deflection of the balance beam in the same direction. The total force on the coils is measured by the load required to restore a balance. The following relation holds,

$$mg = kI^2.$$

k is an expression which involves only the dimensions of the coils, which can be measured

Fig. 37.10 Current balance at the National Physical Laboratory

accurately. Thus, an expression for the current is obtained which depends only on a mass, the acceleration due to gravity and the statistics of the apparatus.

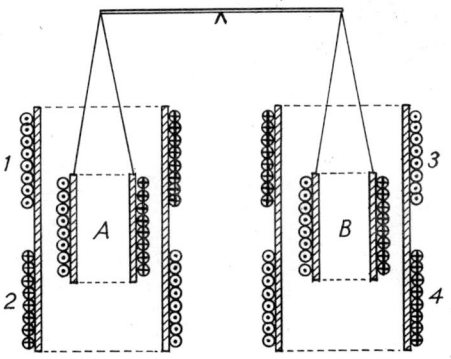

Fig. 37.11 Current balance

Diagram shows coils in section. ⊕ represents wire with current going in and ⊙ wire with current coming out. Coils 1, 2, 3 and 4 are fixed : coils *A* and *B* are suspended from the ends of a balance beam. With the current as shown, coil *A* will be acted on by an upward force and coil *B* by a downward force

Forces acting on a coil in a magnetic field

Consider a rectangular coil *ABCD* mounted on an axis *XY* in a magnetic field (Fig. 37.12). (For clarity, the leads to the coil are omitted.) The

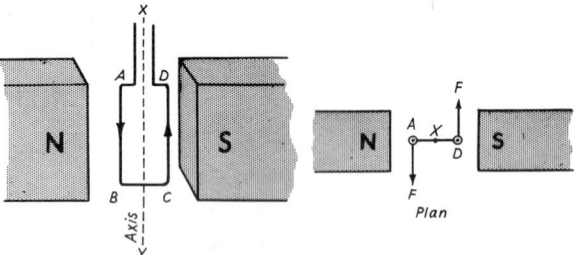

Fig. 37.12 Rotation of a coil in a field

sides *AB* and *CD* are acted on by forces *F* as shown. The top *AD* and the bottom *BC* are acted on by no force when in the position shown, but with the coil turned through 90°, the forces will be upward and downward respectively. The only forces effective in producing rotation, however, are those on *AB* and *CD*. These twist the coil until it is at right angles to the magnetic field, i.e. until the maximum number of field lines pass through it.

This is the principle employed in the moving-coil galvanometer and the electric motor.

The moving-coil galvanometer

This instrument consists of a suspended or pivoted coil in the field of a permanent magnet.

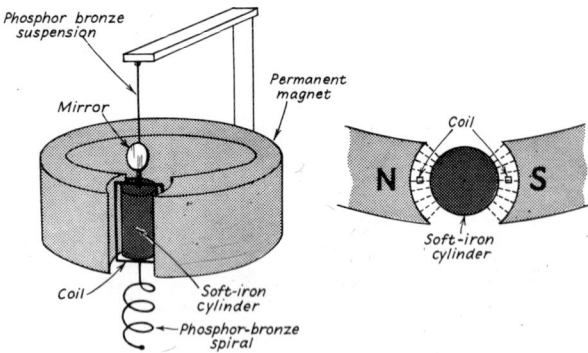

Fig. 37.13 The moving-coil galvanometer

The essential parts are shown in Figs. 37.13 and 14. In the sensitive suspended-coil instrument, the current is led into the coil by a fine phosphor-bronze

Fig. 37.14 Moving-coil galvanometer

strip, and out by a phosphor-bronze spiral. When a current is passed, the coil tries to turn at right angles to the field, but this turning effect is balanced by torsion in the suspension.

A soft-iron cylinder is supported inside the coil. This concentrates the field and makes it radial. The radial field ensures that the turning effect produced by the same increase in current is the same for all positions of the coil. This results in the instrument having a linear scale.

The deflection of the coil is measured by observing the image produced by a beam of light reflected from the mirror on to a scale.

The portable type of moving-coil galvanometer has a pivoted coil mounted on jewelled bearings. The current leads are two hair-springs, one on each end of the coil, which also serve to produce the mechanical moment required to balance the electromagnetic moment. A pointer is attached to the coil.

Many galvanometers now have a taut suspension. The spiral is replaced by a second strip, which is kept in tension by a spring, and the coil 'floats' in the field.

Ammeters and voltmeters

The moving-coil galvanometer is very suitable for incorporation in an ammeter or a voltmeter.

A meter should not affect the circuit into which it is inserted. Examination of the simple circuit shown in Fig. 37.15 illustrating the use of an ammeter and a voltmeter reveals:

An *ammeter* is placed *in* the circuit and must have a *low* resistance.

A *voltmeter* is placed *across* two points of the circuit and must have a *high* resistance (Fig. 37.16).

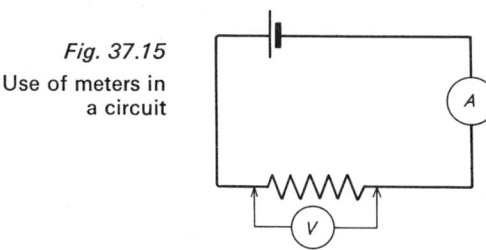

Fig. 37.15
Use of meters in
a circuit

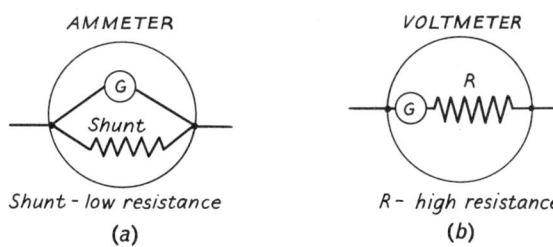

Fig. 37.16

A standard galvanometer now manufactured for meters has a resistance of 5 Ω and gives one scale division deflection for one milliampere. Such an instrument can be modified to read amperes by by-passing a known fraction of the main current through an easy alternative route. We call this 'shunting' the galvanometer: a low resistance is connected in parallel with it and, thus, the resistance of the instrument as a whole is greatly reduced.

To produce a voltmeter, a high resistance is put in series with the galvanometer. The resistance of the whole instrument is then high.

An ammeter is a shunted galvanometer.
A voltmeter is a galvanometer with a high resistance in series.

(*See* Fig. 37.17 and the example overleaf.)

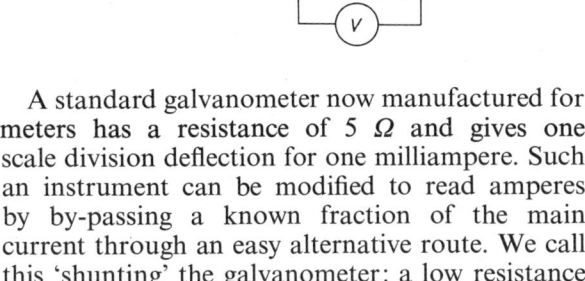

Fig. 37.17 Universal indicator
The photograph shows a milliammeter (0–15 mA : 5 Ω resistance). *Left*: a series resistor is attached converting it into a voltmeter reading to 3 V. *Right*: a shunt is attached converting it into an ammeter reading to 30 A

Example

Find (*a*) the value of the shunt necessary to make the standard galvanometer (5 Ω) read amperes instead of milliamperes, and (*b*) the value of the series resistance to make it read in volts (Fig. 37.18).

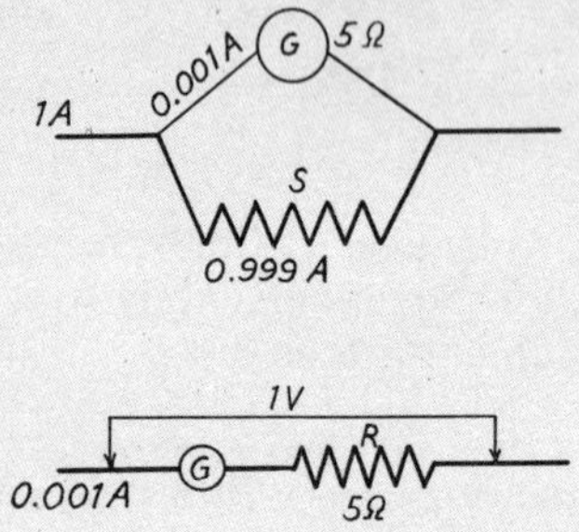

Fig. 37.18

(*a*) One division deflection is produced by 0.001 A. Only 0.001 A of each ampere must pass through the galvanometer.

Therefore,

$$\text{current in shunt} = 0.999 \text{ A.}$$

But

$$\text{p.d. across galvanometer} = \text{p.d. across shunt}$$

therefore

$$0.001 \times 5 = 0.999 \times S$$

where *S*, in ohms = resistance of shunt.

$$S = \frac{5}{999} = 0.005\ 005.$$

The resistance of the shunt is 0.005 005 Ω.

(*b*) A p.d. of 1 V across the total resistance 5 + *R* must produce a current of 0.001 A.

Substituting in Ohm's law:

$$I = \frac{V}{R}$$

$$0.001 = \frac{1}{5 + R}$$

$$5 + R = \frac{1}{0.001} = 1000$$

$$R = 995.$$

The series resistance required is 995 Ω.

Moving-coil instruments are normally used to measure direct current. They can, however, be provided with a rectifier which enables them to be adapted for alternating current measurement.

The electric motor

Briefly, the electric motor consists of a coil mounted on an axle free to rotate in a powerful magnetic field (Fig. 37.19).

We have already seen that when a current is passed through a coil in a magnetic field, the coil always tends to set at right angles to the field. If the coil is mounted on a shaft, and continuous rotation is desired, the moment must be reversed every time the coil is at right angles to the field. To reverse the moment, either the field or the current must be reversed. It is more convenient to reverse the latter, and this is accomplished by means of a commutator. This consists of a copper tube over an insulated cylinder fitted on the shaft (Fig. 37.20). The tube is fastened to the cylinder and is cut into two parts by two diametrically-opposite saw-cuts along its length. The two ends of the coil are fastened one to each of the two

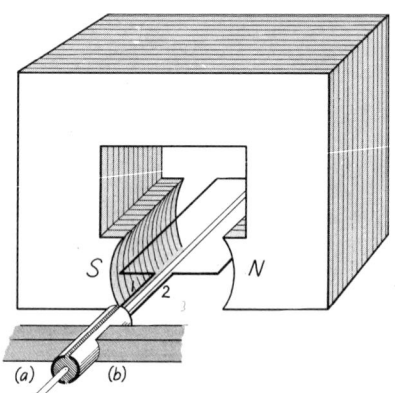

Fig. 37.19 The electric motor

Reversal of current by commutator

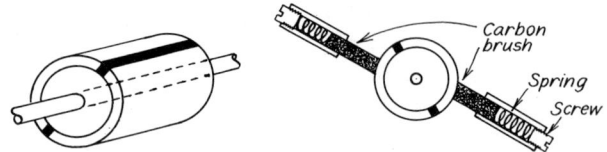

Fig. 37.20 Commutator. *Right:* commutator showing brushes

sections of the commutator. The current is fed into the coil by brushes made of springy brass or, in a large motor, carbon. If a source of d.c. is connected across the brushes, as in Fig. 37.19, the current will always enter the coil by brush (*a*), and leave by brush (*b*). When section (1) of the commutator is in contact with brush (*a*), the current will flow round the coil in one direction: but this direction is reversed when section (2) comes round to brush (*a*). The gaps in the commutator cause a reversal every time they cross the brushes, and the position of the brushes is so adjusted that this takes place when the coil is at right angles to the field.

The forces acting on the sides of the coil, and so the moment producing rotation, is considerably increased by winding the coil on iron. This core, called the *armature,* is usually built up of thin circular sheet-iron stampings which are electrically insulated from each other by thin paper stuck on one of their faces. The advantage of a laminated core is explained in Chapter 39.

The torque (twisting moment) produced by the current and field is greatest when the coil is in position (*a*) (Fig. 37.21) and becomes zero in

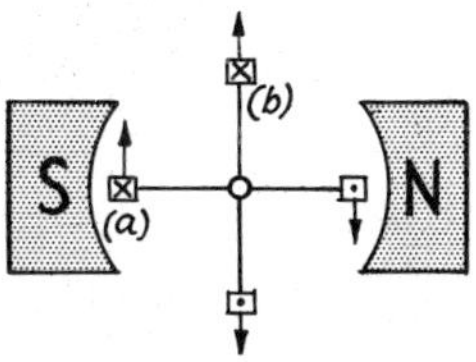

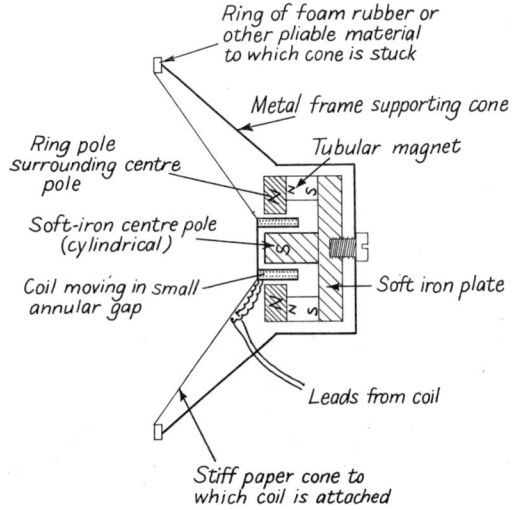

Fig. 37.21

Forces acting on armature coil

position (*b*). This leads to the device of winding several coils on the armature, each with two opposite sections in the commutator. By this means, the current can be sent through the particular coil which is in the region (*a*) of maximum torque, and as this coil passes on, the commutator switches the current through the next.

This is a simple form of armature winding, but there are many more designed to increase efficiency for special purposes.

The magnetic field is always provided by an electromagnet except in toy motors. The coils producing the field are called *field coils*, and they may be connected to the armature in series or parallel. When in series, the motor is series-wound, and when in parallel, shunt-wound (Fig. 37.22). A compound-wound motor is a mixture of the two. A series-wound motor is most suitable for traction, and a shunt-wound for machine

driving or other purposes requiring a constant speed under a varying load.

A back e.m.f. is developed in the armature of a

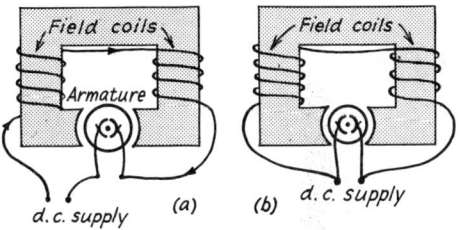

Fig. 37.22

(a) series-wound motor (b) shunt-wound motor

motor when it is in use. This is discussed in Chapter 39.

The moving-coil loudspeaker

The loudspeaker in common use today employs the principles outlined above. The early speakers were modelled on the telephone earpiece. The modern speaker reverses the moving and stationary components. The coil is situated in a radial magnetic field formed between a ring pole-piece and a central pole. These are energised by a tubular permanent magnet. (Some speakers employ electromagnets.) The arrangement is shown in section in Fig. 37.23. By applying Fleming's

Ring of foam rubber or other pliable material to which cone is stuck

Metal frame supporting cone

Tubular magnet

Ring pole surrounding centre pole

Soft-iron centre pole (cylindrical)

Soft iron plate

Coil moving in small annular gap

Leads from coil

Stiff paper cone to which coil is attached

Fig. 37.23 Moving-coil loudspeaker

left-hand rule, it will be found that the coil moves inwards and outwards according to the direction of the current. A varying current, such as is created in a wireless receiver, when passed through the coil causes vibrations which are conveyed to the cone, so setting up sound waves. Such systems can be made to respond to a range of frequencies from 20 Hz to about 40 000 Hz.

QUESTIONS

1. State Fleming's left-hand rule and describe two experiments to illustrate it.

2. Explain the mutual attraction between two neighbouring conductors carrying currents in the same directions.

3. Draw a diagram of a moving-coil galvanometer. How can its sensitivity be increased?

4. Describe an experiment which shows that a wire carrying a current in a magnetic field experiences a mechnical force. Draw a diagram indicating the relative directions of force, field, and current.

Give a labelled diagram of a moving-coil galvanometer, and explain how, by use of a low resistance in parallel, the galvanometer may be made to read as an ammeter.   (O)

5. Describe an experiment to show that a wire carrying an electric current at right angles to a magnetic field experiences a mechanical force, and state the rule for finding the direction of the force, when the directions of current and field are known.

Explain how this principle is applied in the electric motor, and give a clear labelled diagram showing the construction of a series-wound direct-current motor. Explain briefly the function of each part.   (O)

6. Show by a diagram the shape and direction of the magnetic field due to an electric current in a straight wire and by reference to this diagram (or another) explain how a straight wire carrying a current would be forced to move in a magnetic field which is at right angles to the wire.

Describe briefly the structure of a simple electric motor and draw a diagram to show the directions of the field, the current and the resulting motion.   (JMB)

7. Describe a simple experiment to show that a force is exerted on a current-carrying conductor placed at right angles to a magnetic field. Use this information to explain the action of a moving coil galvanometer and show in a simple diagram the directions of the forces acting on the coil. A moving coil galvanometer of resistance 5 ohms gives a full-scale deflection when a current of 10 milli-amperes passes through it. How would you convert it to an ammeter reading up to 1.5 amperes?   (JMB)

8. Explain the action of a direct current motor by considering a single turn of wire free to rotate between the poles of a permanent magnet. Draw a diagram to show the commutator, the polarity of the magnet, the direction of the current, and the way the wire will rotate.

Say briefly how your motor would have to be modified to make it of practical use.   (JMB)

9. Describe an experiment you have seen which shows that a magnetic field exerts a force upon a wire carrying a current.

Explain clearly how this force is produced.

Describe how this effect is utilized in the construction of an electric motor.   (D)

10. Describe an experiment to show that when a current is passed through a conductor situated in a strong magnetic field, it will experience a mechanical force. State a rule connecting the direction of motion with that of the field and of the current, and show how the principle is applied to the construction of a moving coil galvanometer. What determines the magnitude of the deflection produced by a given current?   (L)

11. Explain, with the help of diagrams, the construction, and give the underlying principle of the moving-coil galvanometer.

How is the deflection arranged to be proportional to the current?

An ammeter of resistance 8 Ω gives its full-scale reading for a current of 0.001 A. How would you adapt it for use (a) as a voltmeter reading up to 1 V, (b) as an ammeter reading up to 1 A?   (O & C)

12. Sketch and describe a moving-coil galvanometer and explain how it works. Show on a diagram the direction of the deflection for a given direction of current.

Explain the features which are responsible for (a) making the deflection proportional to the current, (b) producing a big deflection for a small current.

A galvanometer gives its full-scale deflection for a current of 0.5 mA. Its resistance is 1 Ω. Explain how you would adapt it for use as a voltmeter reading up to 5 V.   (O & C)

13. Describe a form of d.c. motor which does not use a permanent magnet and explain how it works. Why may the core of the magnet be made in a single piece while that of the rotating armature is always laminated?

As the motor gains speed the current falls from the value it had when the motor is first switched on. Why is this?   (O & C)

14. Draw a diagram to show the structure of a simple form of d.c. motor. Explain the action of the motor.

A d.c. motor takes a current of 6 A under a potential difference of 25 V when it is on load and doing 7200 J of work per minute. Calculate the power which is not transformed to useful work and give an account to show where the power loss can occur. If the motor is taken off load it runs faster and takes much less current. Explain this.   (C)

15. An ammeter is to be calibrated by means of a current balance. The current balance is an instrument in which two horizontal straight wires, one vertically above the other (the lower wire fixed), are connected in series in such a way that the current in them also flows through the ammeter to be calibrated. The force required to maintain the upper wire at a constant distance from the lower wire is then measured for different currents in the circuit.

In an experiment using a particular current balance, for

each value of the current the ammeter is read, and the correct value of the current is calculated from the measured value of the force, using the formula

$$I^2 = 0.625F,$$

where I is the correct current in amperes and F is the force in μN (10^{-6} N).

The following readings are obtained:

Ammeter readings

(X) in scale divisions	2.0	2.5	3.0	3.5	4.0
Force (F) in μN	60	95	137	185	245

Plot the graph of F against X^2.

From the graph find the force, F', required when the ammeter reading is 4.0 scale divisions. Calculate the correct value of the current flowing, which is equal to $\sqrt{(0.625F')}$.

When using the current balance, explain what would be the effect of:

(a) reversing the current, so that it flows in the reverse direction in both wires;

(b) reversing the current in one of the wires?

Give your reasons. (C)

38 Electrical measurements

In the ordinary electrical laboratory, we cannot measure current, potential difference and resistance absolutely: we compare their values with those of fixed standards. Standards of resistance and potential difference are comparatively cheap. For a few pounds we can buy resistors known to a high degree of accuracy, also standard cells of accurately known e.m.f. We cannot have a similar current standard, but by using a silver voltameter

current, but again, it must be emphasized that these are really comparisons.

Measurement of resistance

Ammeter and voltmeter method

The simplest and most direct way of determining resistance is by measurement of current and

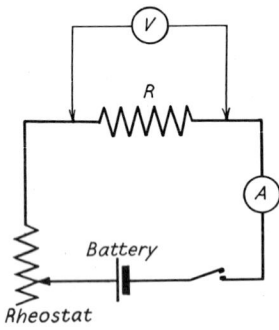

Fig. 38.1
Measurement of resistance

(*see* Chapter 36) we can measure a steady current very accurately by determining the mass of silver deposited on the cathode.

Ammeters and voltmeters are not used where the highest accuracy is demanded. Methods of measuring current and potential differences are employed which make use of standard resistors and of standard cells.

We shall deal with a few laboratory methods of measuring resistance, potential difference and

potential difference using an ammeter and a voltmeter. In the circuit in Fig. 38.1 R is the unknown resistor, A the ammeter and V the voltmeter. Then,

$$R \text{ (ohms)} = \frac{\text{reading of voltmeter } (V)}{\text{reading of ammeter } (A)}$$

If a rheostat is included in the circuit, a series of readings can be taken. By plotting the readings a more accurate value for R can be obtained. It is the slope of the V/I graph.

The ohmmeter

This is a meter for finding the value of a resistor directly. Referring to Fig. 38.1, if the p.d. is constant, only the ammeter reading will change when the resistor R is changed. The ohmmeter (Fig. 38.2) consists of an ammeter connected to a small dry cell enclosed in the instrument. The

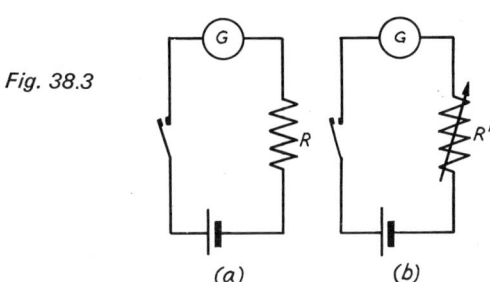

Fig. 38.3

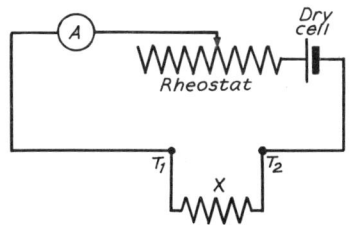

Fig. 38.2 The ohmmeter

R' (Fig. 38.3b) the value of which is changed until the same deflection is registered. When this is so, the currents, hence the resistances, must be equal, i.e. $R = R'$. A common type of plug resistance box is shown in Fig. 38.4. A dial type is now becoming popular.

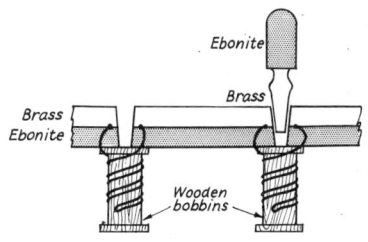

Fig. 38.4 Plug resistance box

Note: (1) resistance is introduced when plug is withdrawn;
(2) wire is wrapped double on bobbins (non-
inductive winding, *see* Chapter 39)

unknown resistor X is connected across the terminals $T_1 T_2$ of the instrument. This completes the circuit, and a current will flow which will be inversely proportional to the resistor X. This means that a resistor of small value will produce a large current, and vice versa. The ammeter is not calibrated to read the current, but to read directly the resistance of X in ohms. This resistance scale will be the inverse of a current scale, since the pointer will be fully deflected for zero value of X and only slightly deflected when X is large. A variable resistor is included to adjust the readings when the p.d. produced by the dry battery de-creases. Before using the instrument, the terminals $T_1 T_2$ are joined by a short wire of negligible resistance. If the pointer is not deflected to the zero of the scale, the variable resistor is altered until it is.

Method of substitution

Another simple method of determining resistance is by substitution.

A circuit as shown in Fig. 38.3a is used. R, the unknown resistor, is in series with a cell, and a galvanometer that gives a suitable deflection. This deflection is noted, and then the unknown resistor R is replaced by a known variable resistor

The Wheatstone bridge

When two resistors such as R_1 and R_2 in Fig. 38.5 are connected in parallel, the p.d. across the ends of R_1 is equal to the p.d. across the ends of R_2.

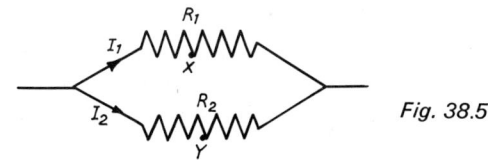

Fig. 38.5

Thus, for any point X on R_1 there must be a point Y on R_2 at the same potential. If X and Y are joined by a wire, then no current will flow in it. X and Y will divide R_1 and R_2 into parts whose resistances will be in direct proportion.

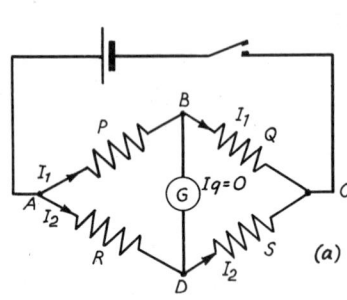

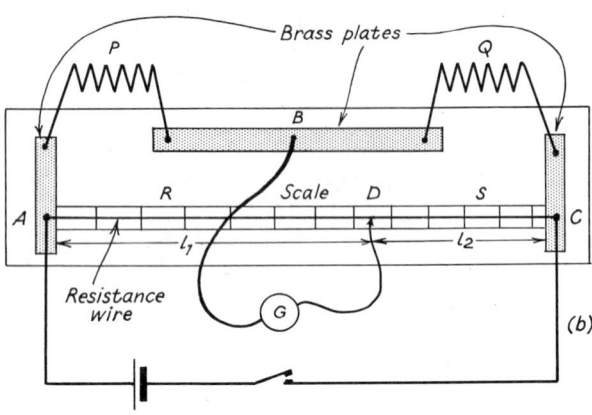

Fig. 38.6 The Wheatstone bridge

The Wheatstone bridge circuit is shown in Fig. 38.6*a*. It consists of four resistors connected as shown. One of them, say *P*, is unknown. The value of one or more of the other resistors is changed until there is no deflection of the galvanometer. When this adjustment is made, the bridge is said to be balanced.

For a balanced bridge,
current through galvanometer = 0

hence,

potential at *B* = potential at *D*

and   {current in *P* = current in *Q*
         {current in *R* = current in *S*

also   {p.d. across *P* = p.d. across *R*
         {p.d. across *Q* = p.d. across *S*

which is
$$I_1 P = I_2 R$$
$$I_1 Q = I_2 S$$

hence,
$$\frac{P}{Q} = \frac{R}{S}$$

The values of *R* and *S* need not be known absolutely.

An elementary form of Wheatstone bridge for use in the laboratory is shown in Fig. 38.6*b*. A metre or half-metre scale is mounted on a baseboard and over it is stretched a length of uniform resistance wire, the ends of which are soldered on the brass plates at *A* and *C*. The wire replaces *R* and *S* and, being uniform, its resistance is proportional to its length. Across *AC* is connected a cell—a dry cell is excellent and convenient. The resistors *P* (unknown) and *Q* (known) are connected across *AB* and *BC* respectively. The galvanometer *G* is connected to *B* and to a movable

contact *D*, which is moved along the wire until a balance is obtained. A sensitive galvanometer suitable for detecting rather than measuring a current is employed.

If $AD = l_1$, and $DC = l_2$, we can write

$$\frac{P}{Q} = \frac{R}{S} = \frac{l_1}{l_2}$$

Error is least when a standard resistor of approximately the same value as the unknown is employed. This brings the point *D* near to the mid-point of *AC*. Also the known and unknown resistors *P* and *Q* should be interchanged and a second result obtained. If the new readings do not agree closely with the first readings reversed, the results are not reliable. The error is probably due to bad connections or to bad soldering of the resistance wire. The mean value of *P* should be found.

Measurement of resistivity

If the resistivity of the material of a wire is required, the resistance of a length of the wire is first determined using the method above. The dimensions of the wire are then obtained, first the precise length used (not including that wrapped round the terminals), and then the diameter. The diameter is measured with a micrometer screw-gauge and the average of several readings taken at different points along the wire and across diameters at right angles is found. (The zero error of the micrometer should be noted.) The resistivity is given by:

$$R = \rho \cdot \frac{l}{a}$$

or
$$\rho = \frac{R \times \pi \, (\text{diameter})^2}{\text{length} \times 4} \, \Omega \, \text{m}$$

For the potentiometer method of measuring resistance, *see* below: for the measurement of the internal resistance of a cell, *see* page 340.

Measurement of potential difference —the potentiometer

The simplest method of measuring a potential difference is by use of a voltmeter. The ideal voltmeter should have an infinite resistance so that it does not take current from the main circuit. Electrostatic and valve voltmeters which satisfy this condition are not suitable for general use, and voltmeters commonly used are modified moving-coil galvanometers (*see* page 348). These are made with as high a resistance as is practicable.

In the laboratory measurement of potential difference we use an instrument called a *potentiometer*. In its simplest form* it consists of a straight length of uniform resistance wire mounted above a scale. For school use, the wire is stretched

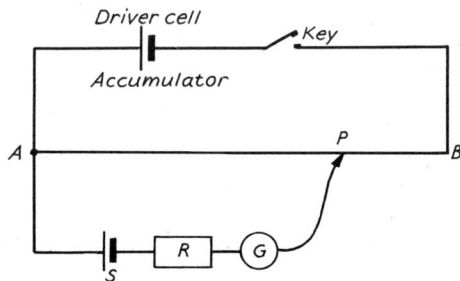

Fig. 38.7 The potentiometer

tightly over a metre scale mounted on a baseboard. Across the ends is connected a constant electromotive force, e.g. a well-charged accumulator. In Fig. 38.7, AB is the potentiometer wire. To calibrate the instrument, a standard cell†

S is connected at A, as shown. Suppose the e.m.f. of S is 1.5 V and the p.d. across AB is 2.0 V. Because both positive terminals are connected at A, the potential of the negative plate of S will be 1.5 V below that at A. This will be the same as that at P, where $AP = \frac{3}{4} AB$. Hence, if the negative plate of S be joined to the point P, no current will pass along the connection.

In carrying out the experiment, a galvanometer is inserted in the lead SP in order to indicate that there is no current flowing. To the free end of the wire a jockey* is attached which is momentarily applied to points along AB until the position of P is located. This can be accomplished with extreme accuracy. In the simple experiments described, the position of P is easily determined to 10^{-3} m. Hence, an accuracy of 1 part in 1000 can be approached.

It should be noted also that the resistance of the connecting leads does not influence the setting of P. As no current flows through the galvanometer for a correct setting, the arrangement is equivalent to a voltmeter of infinite resistance.

The potentiometer may be modified in order to measure a very small p.d. by inserting a high resistance at B.

The potential gradient along the wire is given by $\frac{\text{e.m.f. of } S}{AP}$ V/cm.

R, in the circuit, is a resistor (about 1000 Ω) inserted in order to protect the standard cell. A good standard cell, such as a Weston cadmium cell, is never used as a source of power, i.e. to supply a current. Its use is confined strictly to the calibration of a potentiometer as just described above.

Comparison of the e.m.f. of two cells

A potentiometer is used to measure the e.m.f. of each cell in turn. The circuit diagram is given in Fig. 38.8. With cell E_1 in the circuit, P_1 is the position of the movable contact for no deflection of the galvanometer. Repeating the experiment with cell E_2 in circuit, the new position of the jockey is P_2.

---

* For industrial purposes potentiometers are made up with dial controls for specific measurements.

† The best cell to use is a Weston cadmium cell, but a Daniell cell is a good substitute.

---

* This is a brass rod with one end sharpened to an edge and the other provided with a terminal. The edge is applied to the wire AB, so giving a good electrical connection. It is also used with the Wheatstone bridge.

Then,
$$E_1 = kAP_1$$
$$E_2 = kAP_2$$

where k is the voltage drop per cm along the wire.

$$\frac{E_1}{E_2} = \frac{AP_1}{AP_2}$$

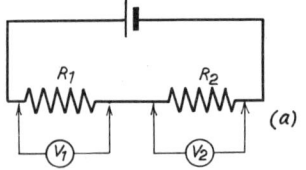

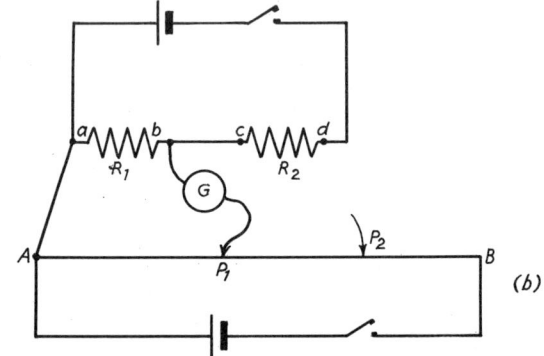

Fig. 38.9

Comparison of resistances by potentiometer

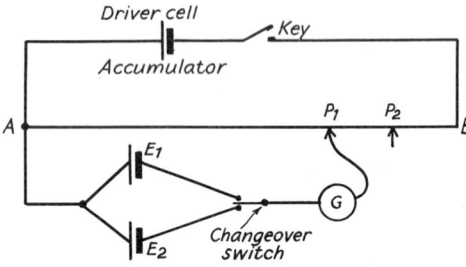

Fig. 38.8 Comparison of electromotive forces

Note that this method of comparing electromotive forces is a good one, for AP is a measure of the p.d. across the cell terminals when the cell is not supplying a current, which is the true e.m.f. of the cell.

In using a potentiometer, if the galvanometer deflection is always in the same direction for P anywhere along AB, then:

(a) the e.m.f. being measured may be greater than that of the driver cell, or

(b) opposite poles instead of like poles of the two cells may be connected at A, or

(c) the driver cell circuit may be faulty.

Comparison of resistances

R_1 and R_2 are two resistors to be compared, and are connected in the circuit in Fig. 38.9. If we put two voltmeters V_1 and V_2 across the resistors, then we shall have:

$$V_1 = IR_1 \quad ; \quad V_2 = IR_2$$

and
$$\frac{R_1}{R_2} = \frac{V_1}{V_2}$$

The method is improved by using a potentiometer to find V_1/V_2. Leads from the ends a and b of R_1, Fig. 38.9b, are taken to the potentiometer wire AB, and P_1, the position of the jockey for

no current, is found. Likewise, P_2 is found for R_2 with leads taken from c and d, and we get:

$$\frac{R_1}{R_2} = \frac{V_1}{V_2} = \frac{AP_1}{AP_2}$$

Measurement of current

The measurement of current by using a silver voltameter has been referred to on page 334. This method is accurate, but it has two practical draw-

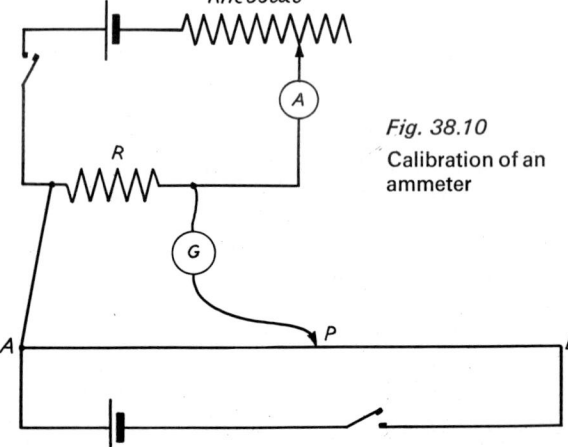

Fig. 38.10

Calibration of an ammeter

backs: (1) it takes a long time to make the determination, and (2) the current has to be maintained constant throughout.

A more flexible method for the laboratory is one in which the p.d. across a standard resistor is determined. The method is highly suitable in calibrating an ammeter, for numerous values of the current can be found to cover the whole range of the meter.

The circuit employed is shown in Fig. 38.10. R is the standard resistor and A the ammeter to be standardized. The potentiometer is first stan-

dardized, and the voltage drop per cm along AB is found by using a standard cell in a preliminary experiment.

The balance point P is found for a given current through R. The value of the current is given by

$$\frac{\text{p.d. across } AP}{R}, \text{ e.g. if } R=1 \ \Omega, \text{ p.d. in volts}=$$

current in amperes. This is repeated for different currents through R obtained by using the rheostat. The calculated values are compared with the currents shown by the ammeter.

QUESTIONS

1. Describe how you would find as accurately as possible the resistance of a piece of wire by using a Wheatstone (metre) bridge.

A piece of manganin wire length 200 cm and cross-section $0.001 \ cm^2$ has a resistance of $8.8 \ \Omega$. Calculate the resistivity of manganin.

If the ends of the wire are soldered together to form a circle, what is its effective resistance between any two points which are at the opposite ends of a diameter? (JMB)

2. State Ohm's law and define the terms *potential difference*, *electric current* and *resistance*.

Show how to calculate the combined resistance of two conductors connected in parallel.

Resistances of 20 000 Ω and 10 000 Ω are connected in series across a battery of constant voltage V. A voltmeter of resistance 10 000 Ω connected across the 20 000 Ω resistance reads 15 V. Find V. (O & C)

3. What is meant by the *electromotive force* and the *internal resistance* of a cell?

Describe and explain a method of measuring the internal resistance of a cell.

A cell can supply a current of 0.50 A through a resistance of 4 Ω and 0.92 A through a 2 Ω resistance. Calculate the e.m.f. and internal resistance. (O & C)

4. Describe (*a*) how you would determine the resistance of a given coil of resistance wire, (*b*) how you would determine the power used by a given electric lamp when the potential difference between its terminals is 6.0 V. Draw a circuit diagram in each case.

A metal wire, 80.0 cm long and 1.0 mm^2 in cross-section, has a resistance of 0.92 Ω. Calculate the resistivity of the metal. (C)

5. Describe how you would compare the e.m.f.s of two cells using a slide-wire potentiometer.

Explain the theory of the experiment.

What are *two* of the possible causes of a failure to find a null point in this experiment? (JMB)

6. What do you understand by 'the e.m.f. of a cell'?

How would you use a potentiometer to compare the e.m.f. of a Leclanché with that of a Daniell cell?

Why is it better to use a potentiometer rather than a voltmeter for this experiment? (JMB)

7. Prove the relation between the resistances in the arms of a balanced Wheatstone (metre) bridge. How would you use the bridge to find the resistance of a length of wire as accurately as possible?

If the wire was 120 cm long, of cross-section $0.12 \ mm^2$ and of resistance 4 Ω, what would be the resistivity of the material of which the wire is made? (JMB)

8. State Ohm's law and show how it leads to a definition of electrical resistance.

Describe in detail a method for determining the resistance of a length of wire.

If the wire has a resistance of 1.32 Ω, a length of 110 cm and an area of cross-section of $0.00415 \ cm^2$, find the resistivity of the material of which it is made. (L)

9. Describe how to use a Wheatstone bridge to find the resistance of a piece of wire. Give the theory of the method and the practical details of the experiment.

The resistance of 100 cm of a thin strip of metal is found to be 2.50 Ω. The cross-section of the strip is a rectangle 2.0 mm by 0.50 mm. Calculate the resistivity of the metal. (C)

10. Draw the circuit which you would use to compare the e.m.f.s of two cells by means of a potentiometer. What are the advantages of this instrument over an ordinary voltmeter?

A cell of e.m.f. 2.0 V and negligible internal resistance is connected to two resistances in series of values 500 Ω and 1000 Ω respectively. Calculate the p.d. across each resistance.

If a voltmeter of 500 Ω resistance is placed first across one resistance and then across the other, calculate the two readings it will give.

11. Find an expression for the resistance of two conductors when they are connected (*a*) in series, (*b*) in parallel.

An ammeter gives its full-scale reading for a current of 0.1 A and its resistance is 0.5 Ω. Explain how you would adapt it (*a*) to give a full-scale reading of 2 A, (*b*) for use as a voltmeter to read up to 100 V.

If the voltmeter (*b*) reads 100 V when connected to the

terminals of a battery of internal resistance 80 Ω, calculate the voltage of the battery on open circuit.   (O & C)

12. Explain the terms *electromotive force* and *internal resistance* as applied to a cell.

Describe how you would attempt to measure accurately these quantities for a dry battery for which they are known to be roughly 100 V and 1000 Ω. You are given a meter reading 0–10 mA, and resistances of any value you require.

A voltmeter of resistance 10 000 Ω is connected across a battery of internal resistance 1000 Ω. If the voltmeter reading is 100, calculate the e.m.f. of the cell.   (O & C)

13. What is meant by (*a*) the *electromotive force* (*b*) the *internal resistance*, of a cell?

Describe a method you have used to find the internal resistance of a cell. Explain how the result is calculated and give a diagram of the circuit.

A cell can supply a current of 0.80 A through a 2 Ω resistance and 0.25 A through a 7.5 Ω resistance. Calculate its e.m.f. and internal resistance.   (O & C)

14. What is meant by (*a*) the *electromotive force*, (*b*) the *internal resistance* of a cell? How would you attempt to measure them?

A cell can supply a current of 0.2 A through a resistance of 3.3 Ω. The 'open-circuit' voltage of the cell is 1.05. Calculate the internal resistance.   (O & C)

15. Explain the principle of the potentiometer and describe how you would use a potentiometer to measure accurately a potential difference of about 1 V. Mention two features of the complete instrument which might influence the accuracy of your measurements.

A potentiometer for small p.d.s consists, in part, of a uniform wire of length 1 m and of resistance 0.1 Ω in series with a fixed resistance of value 9.9 Ω, the whole being connected across a 2 V cell of negligible internal resistance. The p.d. to be measured is balanced by that across a length of 15 cm. Calculate its value.   (O & C)

16. Describe a form of moving-coil meter and explain the principles which underlie its operation. Illustrate your answer by a diagram.

A milliammeter is connected in series with a high resistance and a 12 V accumulator, and the resistance is altered until the full-scale deflection of 1 mA is reached. A wire of resistance per unit length 1 Ω/m is now connected across the milliammeter terminals and its length is varied until the deflection of the milliammeter is reduced to 0.50 mA. The length of wire between the terminals is measured and found to be 12.5 cm.

(*a*) Calculate the approximate value of the series resistance.

(*b*) Making any reasonable approximation, find the resistance of the milliammeter.

What value of shunt would be needed to change the full-scale deflection of the meter to 6 mA?   (O & C)

17. (*a*) An electric lamp is marked 200 V, 150 W. What would you expect the resistance of the lamp to be when it is operating? When the lamp is connected across a 1.5 V cell of negligible internal resistance a current of 0.1 A flows. What is now the resistance of the lamp? What can be concluded from these figures?   (O & C)

18. For a metal wire at constant temperature, the readings of an ammeter and a voltmeter suitably connected to it were as follows:

Voltmeter (volts)	0.90	1.60	2.00	2.50	3.00	3.50	3.75
Ammeter (amperes)	0.40	0.65	0.82	0.98	1.20	1.40	1.45

Plot the graph of V against A. Given that there was no zero error in the ammeter, show that there must be a zero error in the voltmeter and find its value. Explain why, in practice, it is more important to set such meters to zero rather than attempt to correct any errors from a graph.

Find the resistance of the metal wire.

Draw a circuit diagram to show the arrangement of apparatus you would use to obtain results such as are shown in the above table. Write brief notes on your procedure. (C)

39 Induced currents

In 1831, Faraday made a momentous discovery. He showed how an electric current can be produced in a circuit without the aid of a battery. A similar effect had been produced a little earlier by Henry, in America, though he did not investigate it fully at the time.

Faraday's was not a chance discovery, but a reward for many years of careful and thoughtful work. His interest in electricity was first aroused from reading of Oersted's discovery and, during the ten years which followed, he sought after a full understanding of the connection between magnetism and electricity.

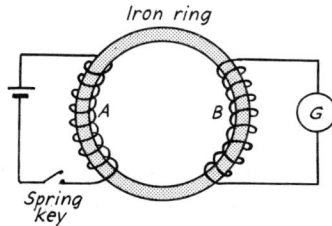

Fig. 39.1 Faraday's apparatus with which he discovered electromagnetic induction

In his famous experiment a soft-iron ring was wound with two coils *A* and *B* (*see* Fig. 39.1). *A* was in a circuit containing a battery and key,

and *B* was joined directly to a galvanometer. When the key was closed, a momentary current was induced in coil *B*, which was indicated by a deflection of the galvanometer. On opening the key, another momentary deflection was observed, but this time in the opposite direction.

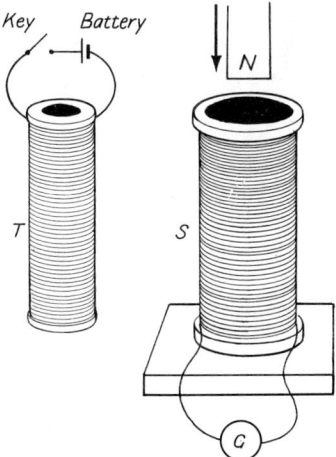

Fig. 39.2 Apparatus for investigating laws of electromagnetic induction

Coil *T* can be inserted into coil *S*

The fundamentals concerning the production of an induced current can best be studied using the simple apparatus shown in Fig. 39.2 and 39.3

Fig. 39.3 Apparatus for demonstrating induction

S is a solenoid of many turns coupled directly to a galvanometer G.

change in the magnetic field about S. Thus, we conclude:

Whenever the number of field lines threaded through a circuit is changing, an e.m.f. is induced across its ends. If the circuit is closed, this e.m.f. produces a current.

We shall now investigate the magnitude and direction of the induced current.*

Repeat Experiment 1, inserting the magnet, first slowly and then quickly. The first deflection will be small, and the second large, but the former will persist for a longer time than the latter. Thus, we can, by threading a certain number of field lines, either induced a small current for a long time or a large current for a short time. This suggests that the product current × time is a constant. Hence, a definite change in magnetic field or flux will induce a definite charge which can be either flooded quickly through the closed circuit or allowed to trickle through it.

The direction of the induced current can be deduced by consideration of the source of the electrical energy. To generate a current, work

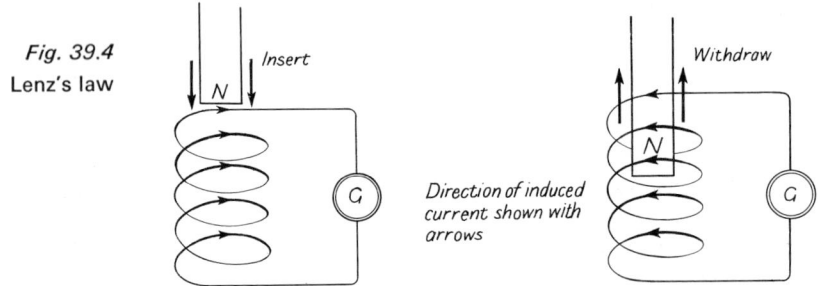

Fig. 39.4
Lenz's law

Insert

Withdraw

Direction of induced current shown with arrows

Experiment 1. Insert the N-pole of a bar magnet into the solenoid. Note a momentary deflection of the galvanometer.

Experiment 2. Remove the magnet and note an opposite momentary deflection.

Experiment 3. Repeat experiments 1 and 2, using the S-pole of the magnet and note that the deflections are reversed.

Experiment 4. Close the key in the second circuit and thrust the solenoid T into S. Note a deflection of the galvanometer.

Experiment 5. Keeping the smaller solenoid T within the larger solenoid S, open and close the spring key. Note opposite deflections.

All the above experiments involve similar changes to the solenoid S—they all cause a

must be done. The induced current in solenoid S flows in such a direction as to produce a field which will oppose the motion of the magnet. The magnetic field of a solenoid resembles that of a bar magnet, hence, the top of the solenoid must act as a N-pole. Thus, the induced current will be anticlockwise, looked at from above when the N-pole is inserted. On removing the magnet the induced current opposes the action by making the top end of the solenoid a S-pole, i.e. the induced current is clockwise (Fig. 39.4). This result is contained in Lenz's law, *see* next page.

---

* It should be clearly understood that it is the e.m.f. which is induced, and this, if the circuit is closed, produces the current. With an open circuit, the e.m.f. is produced across the open ends but there is no current.

Summary of laws of electromagnetic induction

1. **Whenever the magnetic flux through a circuit is changing, an e.m.f. is induced in it.**

2. **The magnitude of the induced e.m.f. depends on the rate at which the magnetic flux is changed— or the rate at which the field lines are cut. (Neumann's law.)**
The induced charge is constant for a fixed change of flux.

3. **The direction of the induced e.m.f. is such that its effect opposes the change producing it. (Lenz's law.)**

Lenz's law can be demonstrated by the experiment shown in Fig. 39.5. A coil of wire of many turns and with its ends joined is suspended by two lengths of thread. When one pole of a magnet

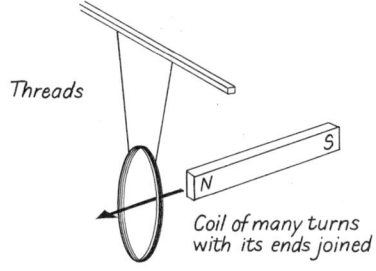

Fig. 39.5 Experiment to demonstrate Lenz's law

is quickly threaded through the coil, it swings away from the pole. If the pole of the magnet is removed, the coil will swing in the same direction as the movement of the magnet.

Faraday's experiment can now be better understood. The link between the two coils is a magnetic one through the iron ring. The current in coil *A* (Fig. 39.1) sets up field lines which pass round

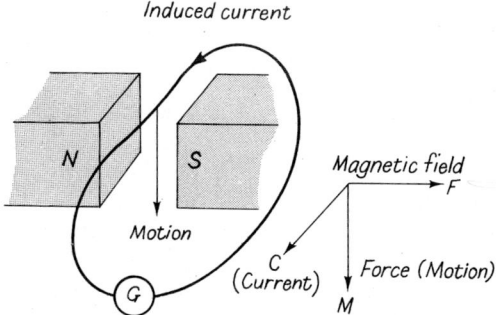

Fig. 39.6 Induction of a current in a wire

the iron ring and so thread the coil *B*. The threading of these field lines induces the current.

If a straight wire connected to a galvanometer is swept between the poles of an electromagnet, a current is induced in it: the wire cuts the field lines. The direction of the current is found by applying Lenz's law. If the motion is downward, the opposing force must be upward. Thus, with the magnetic field from left to right, the current must be outward to make the opposing force upward (Fig. 39.6) (Fleming's left-hand rule).

To find the direction of the induced current rapidly in such a case, we use the following rule:

Fleming's right-hand rule:
Arrange the first two fingers and the thumb of the RIGHT hand mutually perpendicular:
Point the thuMb along the direction of the Motion:
Point the First finger along the direction of the Field:
Then the seCond finger gives the direction of the induced Current.

The generator or dynamo

The above experiment illustrates the fundamental principle of a generator. If a pivoted coil is rotated in a magnetic field (Fig. 39.7), an inward current will be induced in side *A* as it moves upward through the field, and an outward current in *B* as it moves downward. As the ends of the coil will not cut field lines, no induction will take place in them. The currents in *A* and *B* will be the same way round the coil, but their directions

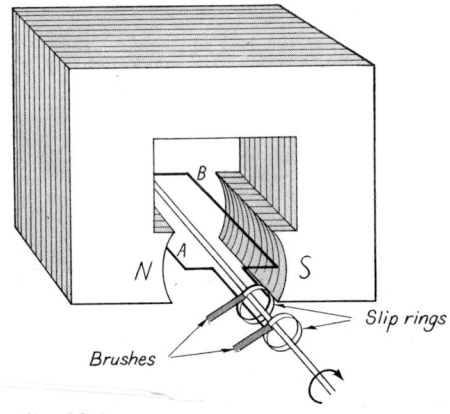

Fig. 39.7 Alternating-current generator

will be reversed each time the plane of the coil is at right angles to the field. To collect the induced current from the coil, the ends are led to copper rings (slip rings) on the rotating shaft. The current is taken from these by brushes in the manner described when dealing with the motor (page 350). The output e.m.f. from this simple generator is alternating. It is a maximum when the coil has its plane along the magnetic field, that is, in the region (1) (Fig. 39.8), and gradually falls to zero

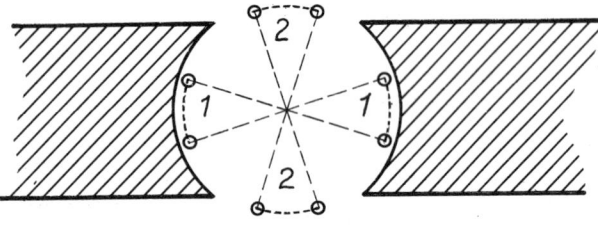

Fig. 39.8

and then reverses in region (2). The graph of the output is shown in Fig. 39.9. The section

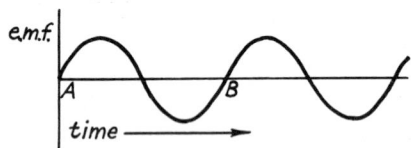

Fig. 39.9 Alternating-current output

AB represents the output for one complete revolution of the coil, and is called a *cycle*.

The d.c. generator or dynamo

The design of a simple d.c. generator is like that of the electric motor. The magnetic field is provided by an electromagnet and the armature is wound on a cylindrical iron core.

The output current must be in one direction and as constant as possible. By using a com-

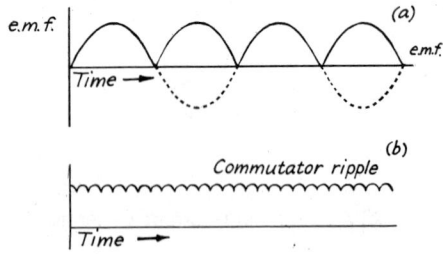

Fig. 39.10

(*a*) pulsating d.c. output (one coil); (*b*) output (many coils)

mutator, instead of the slip-rings, the negative half of each cycle is reversed and the result is a unidirectional pulsating output (Fig. 39.10*a*).

To remove the pulsations, and so produce a steady output, more coils are introduced into the armature. The final output shows only small rapid fluctuations (commutator ripple) (Fig. 39.10*b*).

In a d.c. generator the output current from the armature passes through the field coils. There is sufficient residual magnetism in the cores of the field coils to generate the initial current which builds up until the cores are magnetically saturated.

The a.c. generator

The idea of an a.c. generator has already been given. It is uneconomical to use one coil only. If there are three equally spaced coils in the armature, and the three circuits are kept separate by having one common slip-ring and three others, the outputs will be as shown in Fig. 39.11. The e.m.f.s are similar, but reach their maxima at different times. We say there is a phase difference between them. This is called 3-phase output and illustrates the character of the a.c. supply from the Grid (*see* page 378).

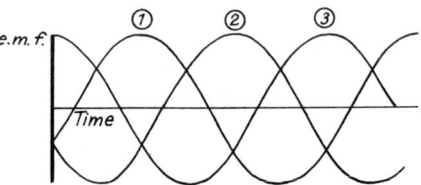

Fig. 39.11 Three-phase a.c. output

The a.c. generator is not built like the d.c. generator. The stationary (stator) and rotating (rotor) parts are interchanged. The field magnets rotate inside the stator containing the coils in which the currents are induced. Obviously this change is permissible, as the necessary condition for induction of a current is a change in magnetic flux which occurs in both cases. Also the e.m.f. supplied to the field coils is small and, therefore, there is little sparking at the brushes.

The a.c. output cannot be sent through the field coils, which require d.c. A large a.c. generator at a power station has a small d.c. generator on the same main shaft from the turbine, to supply direct current to the rotor coils (*see* Fig. 39.14).

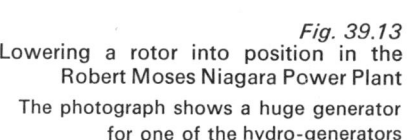

Fig. 39.12 The Robert Moses Niagara Power Plant
The capacity of the plant is 2190 MW

Fig. 39.13
Lowering a rotor into position in the
Robert Moses Niagara Power Plant
The photograph shows a huge generator
for one of the hydro-generators

Fig. 39.14 A modern generator
The 60 MW turbo-generators in the Ince Generating Station of the Central Electricity Generating Board

Eddy currents

If a block of metal is in a changing magnetic field, currents are induced in it which have no definite paths but which are all in directions obeying Lenz's law, i.e. they all oppose the main current. They are called *eddy currents* and represent lost energy. This is transformed into heat. To reduce these eddy currents, the armatures of motors, and the cores of transformers, are built up of thin iron sheets which are stamped out in their correct shape and then insulated by a covering of thin paper or varnish. This prevents the passage of current across the laminate, but does not seriously affect the magnetic field.

When a thick piece of sheet metal is moved in a strong magnetic field, the resistance to its movement gives an effect like moving it in treacle. Metal objects fall very slowly in the exceptionally powerful magnetic fields produced by the large electromagnets made for research work, such as those in cyclotrons (page 420).

An interesting demonstration is provided by Waldenhofen's pendulum. A piece of copper plate is attached to a wire or rod, the other end of which is pivoted so that the plate swings across the field of an electromagnet (Fig. 39.15). With no current flowing through the coils of the electromagnet the plate swings freely, but when the

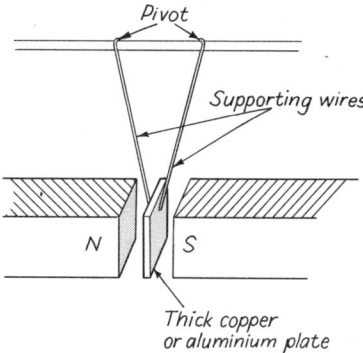

Fig. 39.15 Waldenhofen's pendulum

The induction furnace

The heat developed by eddy currents is utilized in the induction furnace. To increase the effect, a high-frequency alternating supply is sent through a solenoid, which is sometimes made from copper tubing. This permits the circulation of cold water through the solenoid and so prevents its temperature becoming excessive. The rapidly changing magnetic field within the coil quickly brings to a high temperature any conducting material placed in it. It is useful for melting small quantities of precious metals, which are contained in a suitable crucible. It is possible to heat metal components within sealed glass containers and so to carry out certain processes without destroying the containers.

current is switched on and the field created, the plate is braked and brought to rest quickly. This is due to the production, in the plate, of eddy currents which work against the motion causing them. If the plate is replaced by one which has vertical slots cut in it, then eddy currents are greatly reduced, and there is little braking.

Damping in a galvanometer

To reduce the oscillations of a moving-coil galvanometer, the coil is wound on a closed metal frame. This constitutes a closed circuit of low resistance and, when the coil is deflected, there is a current induced in the frame, the effect of which works against the deflection of the coil. It does not alter the final deflection, but causes the coil to come to rest quickly. When damping is not required (as in the case of the ballistic galvanometer) the coil is wound on a frame of non-conducting material.

Fig. 39.16 Zone refining of silicon

The impure silicon bar is passed slowly through the energized coil of the induction furnace. It is melted locally and the impurities are carried to the end of the bar. 'Superpure' metals are required for withstanding the high temperatures produced in a spaceflight project, and also in the production of transistors (*see* Chapter 41)

Back-e.m.f. in a motor

The armature coil of a motor is a coil through which the magnetic flux is changing rapidly. This induces an e.m.f. which works against the e.m.f. applied to the motor. It is termed the *back-e.m.f.* and, when the motor is running at full speed, this back-e.m.f. attains a value nearly equal to that applied across the armature coils. The effect of this back-e.m.f. is easier to understand if we take a definite example.

Suppose an e.m.f. of 100 V is applied to the armature and the back-e.m.f., at full speed, is 95 V. The effective e.m.f. across the armature is then only 5 V and, if an appreciable current, say 5 A, is required in it, its resistance must be only 1 Ω. But, in starting up from rest there will be no back-e.m.f. and the applied 100 V will produce a current of 100 A, which will probably be large enough to burn out the armature coils.

Thus, for large motors, which do not very quickly attain full speed, a starting resistance is introduced. This consists of a variable resistance in series with the armature. As the armature gains speed, this resistance is slowly cut down until it is zero at maximum speed.

The induction coil

In Expt. 5, described on page 362, an e.m.f. was induced in a circuit by a change of current in a neighbouring circuit. This effect is called *mutual induction*. We have also seen that the induced e.m.f. depends on the rate of threading the field lines. This will therefore depend on the number

hammer then recoils and once more completes the circuit. This cycle is repeated rapidly and at each 'make' and 'break' e.m.f.s are induced, but in opposite directions.

A capacitor is put across the contacts of the make-and-break where arcing occurs. This suppresses the arc at the break and makes it less intense, the effect being to produce a much greater induced e.m.f. at the break than at the make.

The induction coil is used when a very high voltage is required, for example, to produce a spark between two points such as in a sparking-plug of a motor-car engine, or to work electric discharge tubes.

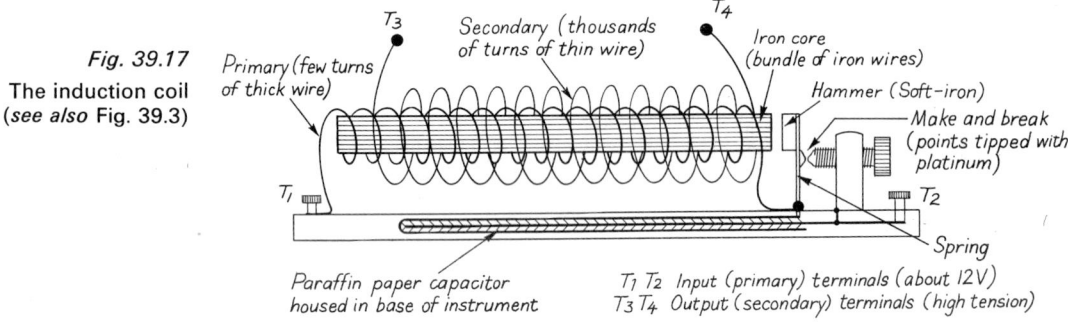

Fig. 39.17

The induction coil (*see also* Fig. 39.3)

Primary (few turns of thick wire) T_3 Secondary (thousands of turns of thin wire) T_4 Iron core (bundle of iron wires)   Hammer (Soft-iron)   Make and break (points tipped with platinum) T_1 T_2 Spring

Paraffin paper capacitor housed in base of instrument

T_1 T_2 Input (primary) terminals (about 12V)
T_3 T_4 Output (secondary) terminals (high tension)

of turns on the large solenoid, and the induced e.m.f. may be made very large by using a great number of turns. In the experiment referred to, the inner solenoid carrying the inducing current is called the '*primary*', and the outer solenoid, across which the e.m.f. is induced, is called the '*secondary*' coil:

$$\frac{\text{output e.m.f.}}{\text{applied e.m.f.}} = \frac{\text{number of turns on secondary}}{\text{number of turns on primary}}$$

A specially designed instrument for obtaining a high voltage by such a method is called a *transformer*. An induction coil is a transfromer specially adapted for working off d.c.

It must be emphasized that an induced e.m.f. is produced across the secondary only when the primary current is changing. To change this current, the induction coil uses a hammer make-and-break such as is used in the electric bell. In Fig. 39.17 the primary current magnetizes the soft-iron core which attracts the hammer and breaks the circuit. The spring supporting the

Ignition system of a car

Fig. 39.18 shows the ignition system of a car. The petrol-air mixture, fed into the cylinders by way of the carburettor, is exploded in each cylinder in turn. The succession of the sparks is produced

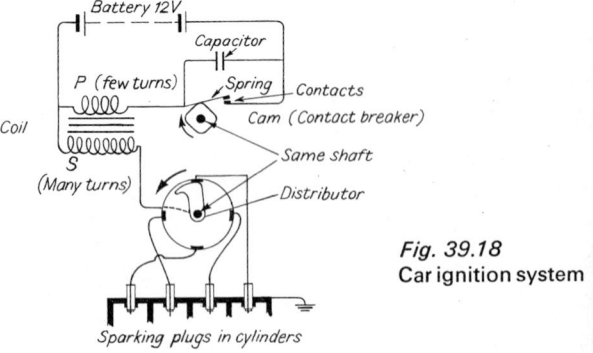

Battery 12V   Capacitor P (few turns)   Spring   Contacts   Cam (Contact breaker)   Coil   Same shaft S (Many turns)   Distributor

Fig. 39.18
Car ignition system

Sparking plugs in cylinders

by means of the distributor which closes connections to the spark plugs. The distributor contains a rotating arm which conveys the electrical pulse from the induction coil to the four cylinders. On the same shaft as the arm are cams which break the primary circuit of the coil regularly. At each break a pulse of charge is produced in the coil and is conveyed to the spark plug in the appropriate cylinder.

The transformer

When a source of a.c. is available, a transformer is used instead of an induction coil. There is no make-and-break in a transformer, for the primary supply of a.c. forms the constantly varying current necessary to induce the e.m.f. in the secondary. The iron core of the primary is made continuous,

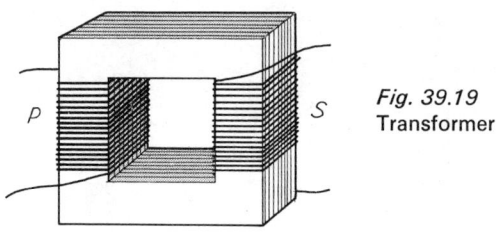

Fig. 39.19
Transformer

so that the field lines are contained wholly in iron. This considerably increases the magnetic flux through the secondary. Various forms of closed iron rings are used for cores, and they are built up of laminated stampings to minimize energy losses due to eddy currents. The use of the transformer is described in the next chapter.

The generation of electrical energy

The conventional method of generating electrical energy is to employ a turbine (steam or water) to rotate the rotor of a dynamo. With a steam turbine the energy is generated by burning solid fuel to create the steam. The nuclear power stations employ nuclear energy instead of coal to generate the steam.

Several quite different methods of generating electricity are being investigated. A great economic advance would follow if the energy created in a nuclear reactor could be directly converted into electrical energy. Experiments to this end are now being performed.

The magnetohydrodynamic (MHD) generator (Fig. 39.20) sends a stream of very hot gases

Fig. 39.20 Experimental MHD generator at the Central Electricity Research Laboratories, Leatherhead

This generator may be in limited use within ten years. The present difficulty is the finding of materials which will stand up to temperatures of 3000 °C. It may be incorporated in conventional generators and so increase greatly the over-all efficiency

through a magnetic field. The hot gas is ionized; hence, the arrangement is virtually a conductor moving in a magnetic field. The generated current is collected from two grids placed on the two sides of the magnetic field. It has been adapted to work from the exhaust of a rocket to contribute, it is hoped, to the electrical energy available to the rocket.

The thermocouple, described in Chapter 15, provides a minute amount of electrical energy. New types of thermocouple, made with semiconductors,* are found to give a much increased thermoelectric effect, so that when a large number of couples is arranged in series, an appreciable amount of energy is generated. In this case, again, it is hoped that further research will enable much larger quantities of electricity to be generated. The process is not efficient, but is certainly convenient when heat is available. Transistor wireless sets can be operated in remote areas not supplied

---

* Non-conducting material to which a small amount of a good conducting material has been added.

FLUE OUTLET

ECONOMIZER SECTION

THERMOELECTRIC MODULES
WIRED IN SERIES

DIRECT CURRENT MOTOR

BELT DRIVE
CENTRIFUGAL BLOWER

BURNER POUCH

Fig. 39.21 Thermoelectric furnace
Thermoelectricity is generated from the heat of the furnace and used to drive the blower motor

Fig. 39.22 Relay communications satellite
The outside surface is covered with 8215 solar cells which convert sunlight into electricity to power the receivers and transmitters

with electrical power, by heating the junctions of many couples with any type of flame.

Fuel cells, mentioned in Chapter 36, are another type of electrical generator which promises to become useful in certain cases.

Semiconductors are now being used as photoelectric cells which generate appreciable quantities of electrical energy. These *solar* cells absorb sunlight and transform it into electricity. Fig. 39.22 shows many of these cells on the surface of a satellite from which it will absorb energy to generate the necessary electrical power for its instruments.

QUESTIONS

1. What are *eddy* currents? How are they minimized in armature and transformer cores?

2. What is the function of the starting resistor for use with a large electric motor?

3. Explain the difference between an induction coil and a transformer. Which is the more efficient and why?

4. Show how an electric current can be produced in a close loop of copper wire without touching it or joining it to a source of current. Give the conditions to obtain a maximum current.

Give a diagram of a simple dynamo producing alternating current. Indicate the direction of rotation of the armature, the polarity of the field magnets, and the direction of the current in the armature winding. How does it differ from a dynamo producing direct current?   (S)

5. State the laws which determine (*a*) the direction, (*b*) the magnitude, of the e.m.f. produced by electromagnetic induction, and show how these laws may be verified given a coil of wire, a galvanometer, a cell to calibrate the galvanometer for direction of current and a strong bar magnet.

Describe *either* (i) an induction coil *or* (ii) a transformer. In either case explain its mode of action.   (L)

6. State the laws of electromagnetic induction and explain how these laws are applied in the induction coil.

Why is a condenser usually connected across the contact breaker of an induction coil?   (O & C)

7. Describe a simple form of a.c. generator and explain the underlying principle.

A ring-shaped coil consisting of a few turns of insulated wire is steadily rotated about a fixed diameter as axis. The diameter is at right angles to a steady, uniform magnetic field.

Sketch, using the same axes, graphs of the variation with time of the p.d. between the ends of the coil (as observed, for example on a suitably connected cathode-ray oscilloscope) so as to show clearly the effects of:

(a) doubling the number of turns in the coil keeping the rotation speed and the magnetic field constant;

(b) doubling the speed of rotation keeping the number of turns and the magnetic field constant.   (O & C)

8. Explain briefly what is meant by *electromagnetic induction*, illustrating your answer by reference to at least one application of the effect.

State the factors which determine the size of an induced electromotive force. Describe carefully how you would investigate these factors experimentally.

When the terminals of a freely swinging moving-coil galvanometer are short-circuited the oscillations quickly die away. Make clear the reasons for this effect, and hence explain the law which determines the direction of an induced current produced in a conducting circuit.   (C)

9. State the laws of electromagnetic induction.

Draw a labelled diagram to show the structure of an a.c. transformer. Give an explanation of the characteristic features of the structure.

A step-up a.c. transformer has an input of 6 V and a turns-ratio of 51:2. Calculate the theoretical voltage output.   (C)

10. Describe, with the aid of a diagram, the structure of an a.c. transformer suitable for lighting a 12 V lamp from a 240 V supply. Explain the action of the transformer.

A transformer is used to light a 24 V, 60 W lamp from a 240 V a.c. main. The current in the main cables is 0.30 A. Calculate (a) the efficiency of the transformer, (b) the cost of running the lamp for 500 hours at a cost of 2p per unit, (c) the current in the lamp.   (C)

11. Explain the existence of a small electromotive force between the ends of each axle of a moving railway train.

State the factors on which the magnitude of the e.m.f. depends and draw a diagram showing the direction in which it acts when the train is travelling northwards in the magnetic meridian.   (O & C)

12. Explain why a small potential difference can be measured between two railway lines when a train is passing along them. The rails can be assumed to be insulated from each other except where connected by the wheels and axles of the train.

How would you expect the measured p.d. to depend on (a) the speed of the train, (b) the direction of travel, assumed horizontal, (c) the distance apart of the rails?·State the laws on which you base your answers.   (O & C)

13. Draw a diagram of a telephone receiver and transmitter and explain the principles underlying their operation.

Sketch and explain a circuit suitable for telephone communication between two points a short distance apart.   (O & C)

14. State the two laws of electromagnetic induction, and describe an experiment to illustrate one of them.

Explain the principle of the transformer. Why is a transformer with a solid iron core less efficient than one with a core made of a number of thin iron stampings?   (C)

15. Describe experiments to illustrate the chief features of electromagnetic induction. State the laws which determine (a) the direction of the induced current in a conducting circuit, (b) the magnitude of the induced electromotive force.

Explain with the help of a diagram *one* practical application of electromagnetic induction.   (O & C)

16. State the laws of electromagnetic induction and briefly describe *two* experiments to verify them.

Give brief explanations of the following observations:

(a) The secondary winding of an induction coil consists of many turns of thin wire while the primary winding contains fewer turns of thick wire.

(b) The current taken by a direct current electric motor when running is much less than when starting.   (O & C)

17. Draw a section through a simple alternating current transformer and explain how the laws of electromagnetic induction are applied in its construction and operation.

The secondary windings of a transformer in which the voltage is 'stepped up' are usually made of thinner wire than the primary. Why is this?   (O & C)

18. How would you demonstrate that a momentary current can be obtained by a suitable use of a magnet and a coil of wire? What is the source of the energy associated with the current so obtained?   (C)

19. State the laws of electromagnetic induction and describe *three* experiments to illustrate them.

Give an account of the construction and action of an alternating current transformer, and mention one of its important uses.   (O & C)

20. State the two laws of electromagnetic induction.

A transformer is designed to work from 240 V a.c. mains and to give a supply at 8 V to ring house bells. The primary coil has 4800 turns.

(a) Would you expect the secondary coil to be of thicker or of thinner wire than the primary?

(b) About how many turns would you expect it to have?

(c) Why is the iron core made of laminations (or sheets) of iron instead of being in one solid piece?

(d) What would happen if the transformer were connected to 240 V d.c. mains?

(e) Do you think the primary current will increase or decrease when a bell is being rung?

Give reasons for your answers.   (O & C)

40 Alternating current*

The magnetic, heating and chemical effects of an alternating current

In the previous chapter we have seen how an alternating current is produced. We must now study its properties and contrast it with direct current.

First, does an alternating current display magnetic, heating and chemical effects? It does, but in most cases the experiments used to demonstrate the effects with direct current have to be modified.

a rheostat. Try Oersted's experiment, that is, bring the wire carrying the current over a compass needle. Nothing happens! The explanation is simple. Because the current is reversed 100 times each second (assuming the mains have a frequency of 50 Hz) the compass should be deflected first in one direction and then in the other, 50 times each second. The compass needle is too sluggish or its inertia is too great to allow it to vibrate so quickly.

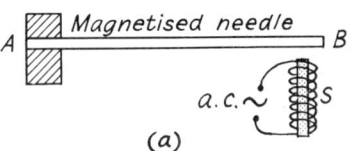
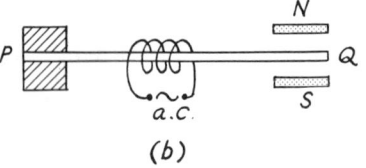

Fig. 40.1 Alternating-current vibrators

Obtain an alternating e.m.f. of a few volts, using a transformer on the a.c. mains, and connect a length of wire across the secondary, in series with

The field round the wire can be shown using iron filings (*see* page 306). The magnetism induced in the iron filings changes its direction with every change in the direction of the current, but this does not affect the behaviour of the filings.

The magnetic effect of an alternating current is often demonstrated by the production of a vibration. In Fig. 40.1a *AB* is a magnetized knitting-needle clamped at *A*. *S* is a solenoid with

---

* For fuller experimental details and further experiments on this subject *see*,
(1) *Alternating Currents*, Modern Science Memoirs 21, Dance, Savage and Ghey (John Murray).
(2) *School Experiments with Alternating Current*, Pearce (Bell).

a soft-iron core. When an alternating current is passed through it, the polarity of the core changes rapidly and the end *B* of the magnet is attracted and repelled 50 times each second. If the natural frequency of *AB* (which can be changed by altering the position of the clamp) is equal to that of the a.c. passed through *S*, the amplitude of the vibrations may be quite large.

A slight modification of this apparatus is more common (Fig. 40.1*b*). *PQ* is a steel bar (a knitting-needle is often used) passing through a solenoid carrying the a.c. If a small horseshoe magnet is now placed so that the end *Q* is in its field, the bar vibrates. Can you explain this?

When a powerful magnet is brought near to a carbon filament lamp working off the d.c. mains, the large spiral filament is either attracted or repelled. The carbon filament bends sufficiently to show this without breaking. With a.c. the filament vibrates to the alternations in the magnetic field of the spiral filament, and appears much broader (due to persistence of vision).

The heating effect of a current depends on the square of the current (Joule's law). Hence, heat is generated whichever way the current is flowing. An electric lamp glows on the a.c. mains as it does on the d.c. mains of the same voltage. But there is a difference. As an a.c. supply rises to a maximum value in one direction and then decreases and reverses to a maximum in the other direction, the heat developed rises to a maximum and falls to zero 100 times a second. This is understood better from the graph of the square of the current against the time (Fig. 40.2). The filament of an ordinary

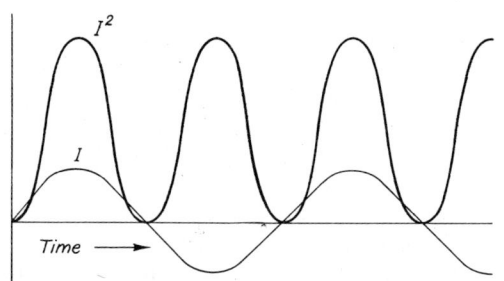

Fig. 40.2 Graph to explain heat developed in an a.c. circuit

lamp does not show distinctly the rapid fluctuations in the heat developed, for it has not sufficient time to cool down when the current becomes zero.

The modern discharge lamps, however, show the current fluctuations clearly. If a neon lamp is swung pendulum-like by the flexible wire connected to it, a continuous arc of light is not seen, but instead, the spiral and plate are seen to flash alternately. When held still, both appear to glow continuously, due to persistence of vision.

Another way of showing the intermittent flashing is by illuminating a quickly moving stick. It will be seen in several distinct positions, which are farther apart the quicker the stick is waved. A development of this experiment is the stroboscope. A disc is divided into equal sectors, alternately black and white, and rotated by fitting it on a variable-speed electric motor. If the disc is now illuminated by an a.c. discharge lamp, a speed can be attained for which the sectors on the disc appear stationary. During the time the lamp is out, the disc rotates through an angle equal to twice that of each sector, so that each black sector moves into the position previously occupied by the one before. The sectors on the disc appear stationary again when the speed is doubled. This stroboscopic effect* is frequently observed at the cinema when the wheels of a moving vehicle appear still, or even going slowly backwards. Try to explain this latter effect.

An a.c. produces a chemical effect, but with a water voltameter equal volumes of gas are obtained at the electrodes. The gas is a mixture of oxygen and hydrogen in the ratio of one to two parts by volume.

The rapid reversal of anode and cathode can be shown in the following interesting manner. A sheet of white blotting-paper soaked in potassium iodide solution is laid on a copper plate which forms one electrode. The other electrode is in the form of a piece of copper wire bent over at the end to make it smooth. If this is drawn over the blotting-paper, a broken line is produced. When the wire is the anode, iodine is produced and this causes a brown stain. (If starch is added to the solution, a blue mark is obtained.) The line is broken and not continuous because of the changes in polarity. Each space and dash indicates that the alternating current has gone through one complete cycle. This experiment can be adapted to measure the frequency of an alternating current.

---

* The same effect has been employed in measuring velocities etc. in Chapter 8 and also in viewing progressive waves in Chapter 20.

Measuring instruments

We have already described instruments which can be used for a.c. measurements of current and p.d. These are illustrated in Figs. 33.23, 33.24 and 35.3. The response of each of these instruments depends on the square of the current flowing through it. For example, the heat developed in the hot-wire ammeter depends upon the energy passed through it, that is, upon I^2. The recorded reading of the meter is not I^2, but the square root of the mean value of I^2. The inertia of the moving part in each of these meters is so great that the variations of I^2 cannot be shown, but only the mean value of I^2. It is the value of the direct current

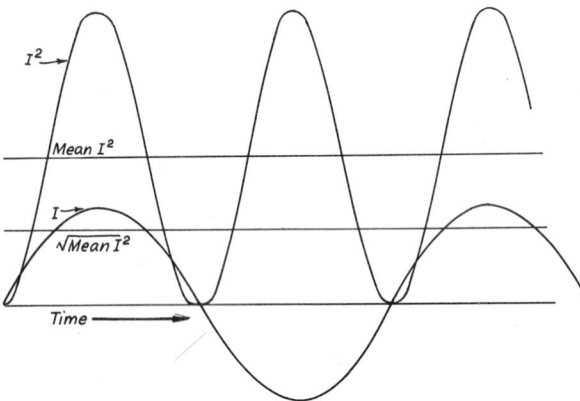

Fig. 40.3 Root mean square
The scale used shows:
$I_{max} = 3$ units, $I^2_{max} = 9$ units. Mean $I^2 = 4.5$ units
$\sqrt{}$ (mean I^2) = r.m.s. = 2.1 units

which would develop the same energy as the alternating current. It is usually referred to as the root mean square (r.m.s.) value. The graph (Fig. 40.3) shows its significance clearly.

Transformers

One advantage of alternating current over direct current is its flexibility in use. This is effected by means of a transformer (*see* page 369) which permits of a change in p.d. without any appreciable loss of energy. To take a simple example, Suppose a car lamp (12 V, 24 W) is used on the d.c. mains, 240 V. A rheostat is necessary across which the p.d. will be 228 V when a current of 2 A is flowing.

Thus, only $\dfrac{12}{240}$ of the energy used is being usefully employed, the remainder being wasted in the form of heat developed in the resistor. With a.c., the lamp could be connected directly to the output of a suitable step-down transformer, and there would be practically no wastage of energy.

For laboratory use the pattern shown in Fig. 40.4 is convenient. The core is laminated and ⎵ shaped and is closed by a yoke.* The coils are easy to wind and, using the most common size

Fig. 40.4 A laboratory pattern transformer

of core for this experimental type of transformer, viz. 2.5 cm², we allow 7 turns for 1 V. Hence, for a 240 V supply, the primary should have $240 \times 7 = 1680$ turns. For a secondary to yield 12 V, 84 turns are required. Such a secondary is capable of giving a much greater current than that in the primary, hence, thicker wire should be used for the secondary coil. Any output, from a few volts to several thousand, can be provided by winding a suitable secondary. It is essential that the core is earthed when a step-up transformer is used on the mains.

If a transformer were 100 per cent efficient, we should have:

---

* The term used for the iron bar which closes the ends of the core and so makes it a closed iron 'ring'.

input power=output power

(primary voltage) × (primary current)

=(secondary voltage) × (secondary current)

$$V_p I_p = V_s I_s$$

But

$$\frac{V_s}{V_p} = \frac{\text{number of turns on secondary}}{\text{number of turns on primary}} = \frac{I_p}{I_s}$$

The very large value of I_s which is produced when V_s is very small, can be shown by using a secondary coil of 6 turns of thick copper rod. If an iron nail is connected across the output terminals it will get red-hot, and finally melt and break the circuit. This very large current, which can be produced with a secondary of few turns, can be utilized in welding.

The output voltage of a transformer drops when the iron yoke closing the limbs of the ⊔ shaped core is removed. This is due to a reduction in the number of field lines from the primary threaded through the secondary.

If a solid iron core is used instead of the usual laminated one, there is again a reduction in output energy. This is due to the production of eddy currents which cause a warming up of the core.

Domestic chime or bell circuit

The circuit used is shown in Fig. 40.5. The secondary of the mains step-down transformer has an

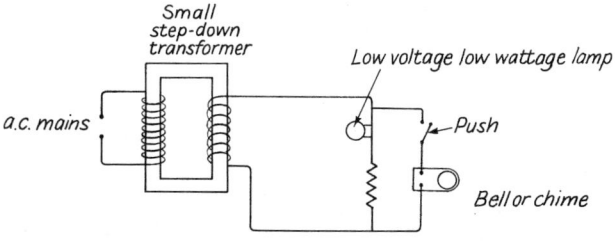

Fig. 40.5 Mains bell or chime circuit

output of a few volts. This provides current for the lamp which illuminates the push button. The bell or chime mechanism is in parallel with the lamp and functions when the push closes the circuit. The lamp circuit has a higher resistance than the bell circuit and so little current passes through the lamp loop when the push is depressed and the lamp goes out. The primary of the trans-

former, which is permanently connected across the mains supply, has a very large resistance (impedance) and so takes very little current.

Effect on primary power by a change in secondary power

Arrange the circuit shown in Fig. 40.6.

When the switch is closed in the secondary, the current in the primary is increased. This is shown by an increase in the reading of the ammeter.

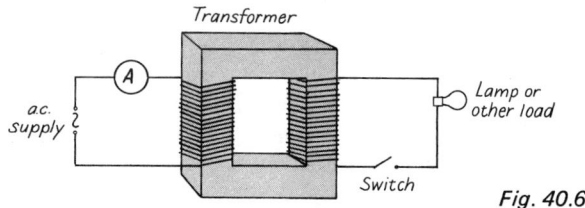

Fig. 40.6

The circuit is similar to that used for an electric bell. The bell replaces the lamp, and the bell-push, the switch. The primary of the bell transformer takes little current when the secondary is not delivering current, and it is left permanently connected to the a.c. mains supply.

Self-inductance

Chokes

The transformer and the induction coil illustrate the principle of mutual induction—the induction in one circuit of a current due to the effect of a neighbouring circuit.

If we now consider a simple circuit comprising a solenoid, switch, and battery, when the switch is closed, the current will flow and establish a magnetic field about the solenoid. But we know that when field lines are threaded through a circuit, an e.m.f. is produced which opposes the change producing it. Hence, during the setting up of the magnetic field of the solenoid, there is present a back-e.m.f. which prevents the current rising to its maximum value immediately. Likewise the current does not drop immediately to zero when the switch is opened (Fig. 40.7). This back-e.m.f. becomes zero very quickly in most circuits, but where a circuit has a strong magnetic field round it, the back-e.m.f. may persist for an appreciable time. This effect is called *self-induction,* and the property possessed by a circuit, *self-*

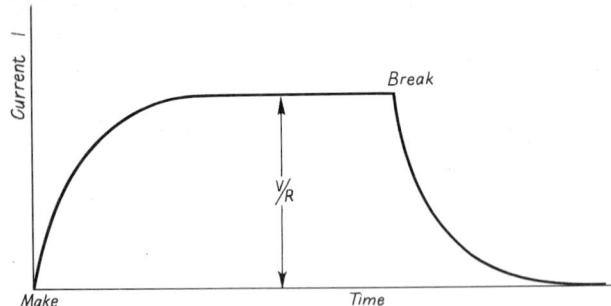

Fig. 40.7 Graph showing current variation with time at make and break of an inductive circuit

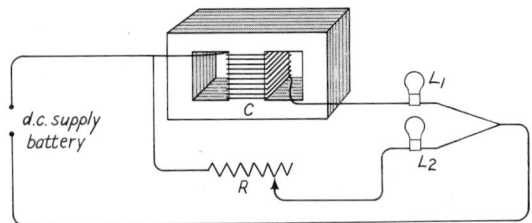

Fig. 40.9 Experiment to demonstrate effect of a choke

inductance. An iron core increases the field associated with a solenoid. Fig. 40.8 shows a typical self-inductance; such is called a *choke*.

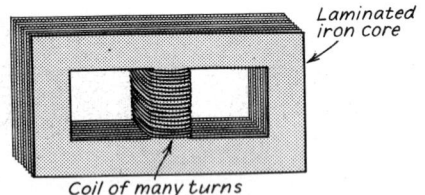

Fig. 40.8 Common design of choke

The circuit shown in Fig. 40.9 presents a method of demonstrating self-induction. L_1 and L_2 are two similar lamps suitable for the low voltage supply. C is a coil of 1200 turns. When the switch is closed the rheostat is adjusted so that both lamps are equally bright. The switch is then opened and, after a few moments, again closed. It will be noticed that lamp L_1 takes a little time to attain equal brightness with L_2. If a very large choke is available, L_1 may take several seconds to reach its normal brightness.

When used with alternating current of mains frequency, the magnetic field has no time to establish itself before the current is reversed, thus there is always a back-e.m.f. working against the applied e.m.f. The current then never rises to as high a maximum as it would otherwise do, and so the choke acts as a resistor. The perfect choke has self-inductance but no resistance. In practice, however, a coil of wire must have some resistance. The working of a choke can be seen using the circuit in Fig. 40.9, but this time using an a.c. supply. It will now be found that the lamp L_1 will not glow, whereas L_2 glows brightly. A choke

then offers resistance to an a.c. which is increased with an increase in the frequency.

Non-inductive resistance coils

Self-inductance is effective in d.c. circuits when the current is switched on or off. For example, in a Wheatstone bridge circuit the throw of the galvanometer needle when the galvanometer key is pressed, may be due to inductance in one or more of the resistors. In d.c. bridge circuits the battery key should always be closed a few moments before the galvanometer key is closed. This gives time for the current to become established.

Coils in resistance boxes (Fig. 38.4) are always wound non-inductively. The ends of the wire are secured and then the wire is wound double on the bobbin. With such an arrangement the resultant field of the coil is very small and hence the coil has no inductance. The employment of non-inductive resistors in a.c. bridge work is most important.

Capacitors (condensers)

We have seen in the section on electrostatics that a capacitor stores electric charge. An interesting and instructive experiment can be carried out using d.c. mains, or alternatively two h.t. batteries to give, together, about 200 V. The circuit in Fig. 40.10 shows a neon lamp N, a capacitor

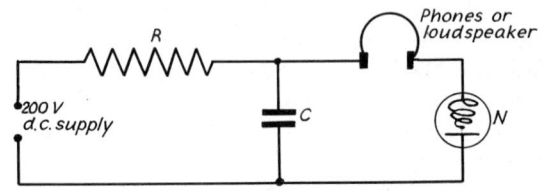

Fig. 40.10 Flashing neon tube

C, a resistor R, and headphones. The neon lamp will not pass a current and glow until a p.d. of more than about 160 V is put across it. The capacitor C is charged through the resistor R. Because R has a high value (about 500 000 Ω) it takes a little time for sufficient charge to pass through it to raise the p.d. across C to 160 V. When this value is just exceeded, the capacitor discharges through N. This is indicated by a flash in the lamp. The rebuilding of the p.d. starts again and the process is repeated. The frequency of the flashes is controlled by the values of R and C, the smaller they are, the more rapid are the flashes. If headphones are also incorporated each flash is accompanied by a click, and by suitable choice of R and C a musical note can be produced.

Effect of a capacitor in an a.c. circuit

If the switch S_1 in the circuit shown in Fig. 40.11 be closed, the capacitor C will be charged and the p.d. across its plates will be that of the d.c. supply.

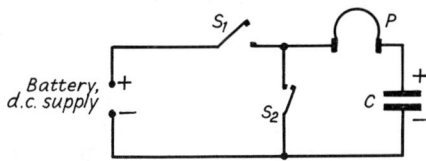

Fig. 40.11 Capacitor in d.c. circuit

A click in the phones P will indicate the passage of the charge. If S_1 is opened and S_2 closed, C will be discharged and there will be another click in the phones.

With an a.c. supply, this charging and discharging is carried out rapidly and there is a rapid clicking in the phones indicating the passage of charge. When terminal A (Fig. 40.12) is positive, positive charge collects on plate X of the capacitor which induces a negative charge on plate Y by

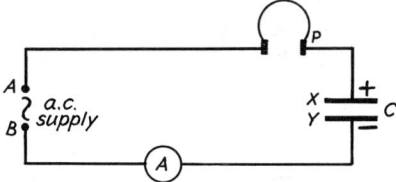

Fig. 40.12 Capacitor in a.c. circuit

driving the positive charge round to B. When A becomes negative, during the next half cycle of

the supply, X becomes negatively charged and positive charge is induced on C, being drawn round the circuit from B. Hence, as the terminals A and B change polarity, the induction of charges on the opposite plates of capacitor C produces the effect of C conducting the a.c. through it.

So we conclude that a capacitor stops the flow of a direct current but it does not act as a barrier to an alternating current. The insertion of a capacitor in a circuit carrying an alternating current reduces the current: the greater the capacitance and the greater the frequency of the alternating current, the smaller is the reduction in the current.

Phase

The phase of an alternating current has been mentioned in connection with 3-phase generators. Although very little more can be included at this stage, the following simple cases of phase change should indicate that its consideration is most important in all a.c. circuits, and particularly in electronics.

When a choke is present in an a.c. circuit, we have seen that its effect is to produce a back-e.m.f. as the alternating p.d. grows. The current is, consequently, not only reduced, but retarded, and so reaches a maximum after the applied p.d. attains its maximum. This is called *lag*.

The effect of capacitance in an a.c. circuit is just the opposite to that of inductance. The current *leads* the applied p.d.

The graph (Fig. 40.13) shows the variations clearly.

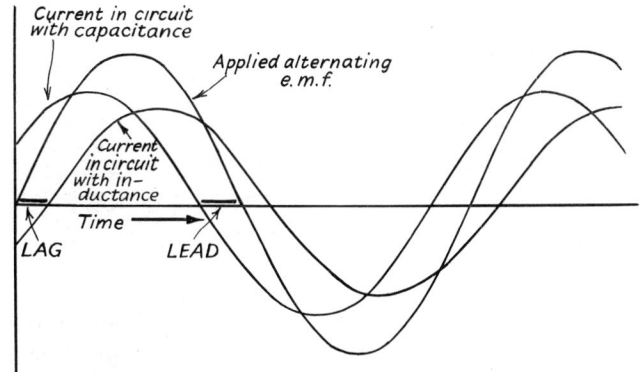

Fig. 40.13 Graph of currents in circuits containing inductance and capacitance showing lag and lead

Distribution of electrical energy

The transfer of electrical energy at low voltage, for the supply of a large district, requires the use of very thick cables, which are costly. But, if the transmission is carried out at a high voltage with the consequent reduction in the current, much thinner cables can be used.

The modern method of transmission involves the use of high voltages. Electrical energy is generated to about 4000 V a.c. and is then transformed up to 132 000 V.* At this voltage it is fed into a network of transmission cables which covers the whole country. For use in any locality, the supply from the overhead cables is transformed down in stages to 240 V. This network is called the Grid. It enables the electricity to be generated at the most convenient places in the country, i.e. where coal is plentiful or where there is a natural source of energy. With this scheme only a few large generating stations are required. These sell the electricity they generate to the controllers of the Grid, and the Local Authorities buy their supply from the controllers of the Grid.

The standard frequency of the a.c. mains is 50 Hz. This is maintained with exceptional constancy. A.C. mains clocks contain a small motor, the speed of which is determined by the frequency of the mains. The accuracy of these clocks testifies to the constancy of the mains frequency.

---

* Work is now proceeding on the erection of a 'Super Grid' which will connect the main industrial areas of Great Britain and which will supply power at 275 000 V.

QUESTIONS

1. Draw a section through a simple alternating current transformer and explain how the laws of electromagnetic induction are applied in its construction and operation.

100 kW of power are being supplied to a factory through wires of resistance 0.1 Ω. What power is lost in the leads if the voltage at the factory end of the wires is (a) 230, (b) 10 000?   (O & C)

2. Describe and explain the effect of a choke on the current in an a.c. circuit.   (JMB)

3. (a) In this country electricity is distributed from the power stations in the form of a.c. at a high voltage. Discuss the advantages of this system.

(b) Explain:

Either how a.c. can be transformed from a low to a high voltage. *Or* how a.c. can be converted to d.c. by the use of a diode valve.

In both processes there is a loss of energy. Why does this occur in the process you have explained?   (JMB)

4. Distinguish between alternating current and direct current.

Give a diagram of a transformer suitable for working an electric bell on an a.c. circuit and explain its mode of action.

Why is the grid system for the distribution of electric power not possible with d.c.?   (S)

5. A simple transformer consists of two coils A and B wound on a bar of iron; coil A carries an alternating current. Explain the following facts. (i) The terminals of B have a potential difference which is alternating. (ii) The bar of iron becomes hot. (iii) If the transformer is placed on a sheet of iron, a humming sound is heard.   (C)

6. Describe with the aid of a diagram, a simple a.c. generator, and explain how it acts.

An a.c. voltmeter connected to a 50 Hz supply registers 240 V (which is the r.m.s. voltage, not the peak voltage). Also connected to the supply is a resistance of 480 Ω. Draw a diagram to show how the current in the resistance varies with time over an interval of six hundredths of a second. Show clearly, on the diagram, what you consider to be a likely value for the peak current.   (C)

7. What is meant by the *frequency* and the *peak voltage* of an alternating electrical supply?

Draw a labelled diagram of a simple a.c. generator: explain how the generator acts. At what stages in its action will the peak voltage occur, and what factors affect the magnitude of the peak voltage?   (C)

8. Show, by means of a labelled *sketch graph*, what you understand by 'an alternating current of frequency 50 Hz and peak value 1 A'. If this current is passed through a hot-wire ammeter, properly calibrated, the ammeter reads less than 1 A; give a reason for this.   (C)

Part Ten MODERN PHYSICS

A beam of electrons can be focused to concentrate sufficient power to weld and drill metals. The photograph shows such a beam of power 500 watts drilling through a 0.18 cm sheet of stainless steel in less than one second. Droplets of metal are shown being ejected

41 Electrons, ions, electrical discharge through gases

The elementary particle of electricity

Evidence for the particulate nature of charge

In Chapter 2 we built up the idea that matter is composed of atoms. Later, in Chapter 32, we saw that, by rubbing, we can charge some bodies with electricity. This process involves the removal of one type of electric charge from a body to leave it deficient in that type or, as we more often put it, to leave it charged with the opposite type of electric charge. But what is the nature of electric charge? Does it have separate existence? Is it a continuous fluid or does it resemble matter by being constituted of discrete particles? From our study of electricity to date, evidence points to its having a particulate nature. In electrolysis we have seen that the same charge is required for the liberation of a mole of any monovalent substance. But a mole is the same number of atoms whatever the substance. It does not follow definitely from this that the charge is equally distributed among the atoms, but it seems a reasonable assumption to make provisionally until we can pursue our investigations further and obtain more definite proof.

Making this assumption, the charge on each ion would be

$$\frac{\text{Faraday's constant*}}{\text{Avogadro's number}}$$

$$= \frac{96\,500}{6.0 \times 10^{23}} = 1.6 \times 10^{-19}\ \text{C}$$

* *See* Chapter 36.

But to return to charging by friction: in this process charge passes across from body to rubber or vice versa. We now ask if it is possible for charge to leave a body by any other means. We have seen how molecules leave a liquid by the process of evaporation. Can electric charge likewise be made to 'evaporate' from a body?

Thermionic emission

About 1900 Richardson investigated more fully what was known as the Edison effect: 20 years previously, Edison had shown that a current passed between the heated filament of an electric lamp and a metal plate mounted near to it within the glass envelope, when the plate was at a positive potential with respect to the filament. Richardson explained the effect as an emission of negatively charged particles from the heated wire. By surrounding the heated wire with a metal cylinder raised to a potential above that of the wire, the charges emitted from the wire were collected. The wire and cylinder were contained in an evacuated enclosure as shown in Fig. 41.1. It was found that

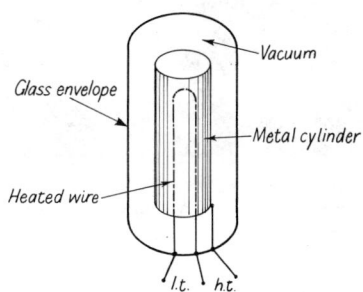

Fig. 41.1 Emission of negatively charged particles from a hot wire

a current was obtained only when the potential of the cylinder was positive with respect to the wire. Thus the charge emitted must have been negative.

Furthermore, this electric current consisted of a stream of negatively charged particles passing through a vacuum and was not associated with atoms. (As previously pointed out the *conventional* current is taken as opposite to the drift of the negatively charged particles.)

The name *electron* was proposed for the fundamental negatively charged particle by Dr Stoney in 1891 before its nature was definitely established.

The magnitude of the elementary charge and its mass

The ratio of the charge to the mass of the electron can be determined experimentally by observing the effect of electric and magnetic fields upon a beam of the moving charges. Such a beam constitutes an electric current, and hence it can be deflected by a magnetic field (Fleming's left-hand rule). The magnitude of the deflection can be worked out theoretically and shown to be dependent upon the ratio e/m, where m is the mass and e the charge (assuming all the particles to have the same velocity).

The charges generated at a heated filament are drawn away by a plate at a high positive potential. The accelerated charges pass through a small hole in the positively charged plate and then on to a zinc sulphide screen, which fluoresces when the charges fall on it. By subjecting this beam to magnetic and electric fields of known strengths and measuring the deflections produced, a value for e/m can be calculated.

This experiment was first performed by Sir J. J. Thomson in 1897: his apparatus is described later. A short description of a laboratory method using a 'fine beam' tube is also given later in this chapter.

The charge on the electron

An accurate determination of the charge associated with an electron was carried out by Millikan in 1906. It is a model example of experimental physics and illustrates the painstaking care necessary to achieve the highest accuracy with the available means. The apparatus shown in Fig. 41.2 is the same in principle as that which Millikan used. *A* and *B* are circular metal plates about 5 cm in diameter and separated by a ring of insulating material about 1 cm thick. There is a small hole in the upper plate through which drops of oil, produced by an atomizing spray, may fall. This hole is closed during observations in order to cut down convection currents. These falling drops are viewed through a window in the insulating ring, using a reading microscope.

It is found that most oil drops pick up electrons from the air and, when a suitable potential difference is applied across the plates, the drops are acted upon by a force. This force is upward when the upper plate is at a positive potential. The observations are carried out on one drop which,

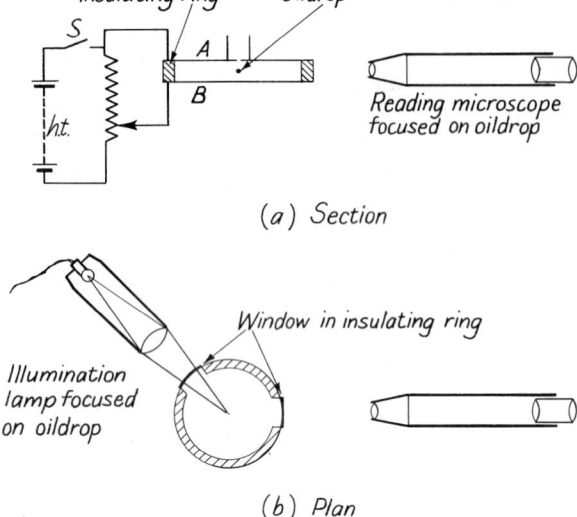

(a) Section

(b) Plan

Fig. 41.2 Millikan's apparatus to determine the charge on an electron

by adjustment of the electric field, is held stationary in the field of view of the microscope. This occurs when the upward force on the drop, due to its charge and the field, is equal to its weight. The field is then removed and the drop allowed to fall freely under gravity. Its terminal velocity is quickly attained and is measured by readings on the eyepiece scale. From the terminal velocity, the radius and the density, the mass of the drop can be calculated. Observations are made, in turn, on several oil drops and the charge which each carries is determined. When all these charges are compared, they are found to have a common factor. This is due to each drop carrying, not one, but several electrons and the common factor is the charge on a single electron. This is found to be 1.6×10^{-19} C. By combining this with the value of e/m, the mass of the electron is found to be 9.1×10^{-28} g or 9.1×10^{-31} kg.

Discharge of electricity through gases

Much modern research work is based upon the fundamental experiment of the discharge of electricity through a gas. The advances in this section of physics have been coupled with improvements in vacuum pumps and in vacuum techniques.

The French scientist Masson, in 1853, was the first to investigate electrical discharge through

gases at low pressure. At atmospheric pressure, air is an insulator; a very high potential difference is required to produce a discharge between two electrodes, even when they are only a few millimetres apart. When the electrodes are contained within a glass tube in which the pressure is reduced,

Fig. 41.3 Discharge tube attached to rotary pump for reduction of pressure. An induction coil provides the h.t. supply

a discharge will occur with a much smaller potential difference. A tube provided with electrodes fused through its ends, as shown in Fig. 41.3, is called a *discharge tube*.* If a high potential difference is applied across the electrodes while the pressure is progressively reduced, the form of discharge passes through a series of changes. When the pressure is 1 or 2 cm of mercury, a potential difference of about 10 kV produces a noisy discharge which takes the form of thin, pink streamers. At a lower pressure, the discharge broadens into a wide band and becomes quiet.

The colour of the discharge is different for

---

* If a heated filament is employed as the cathode, the discharge is more easily started and is more intense.

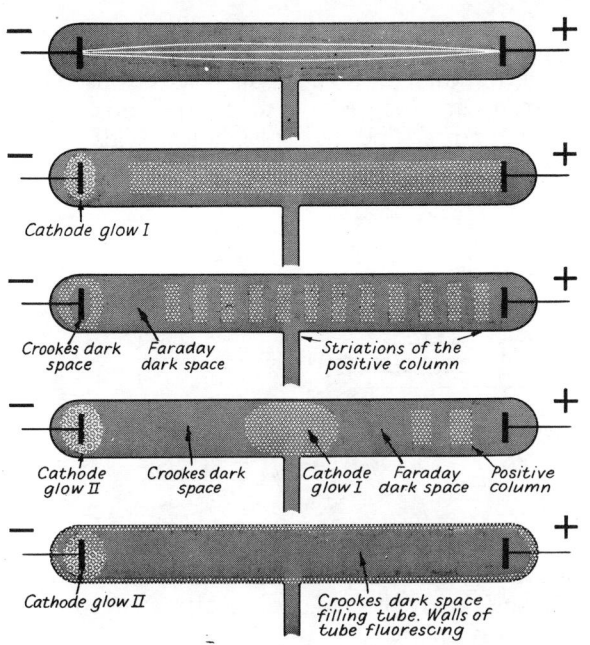

Cathode glow I

Crookes dark space   Faraday dark space   Striations of the positive column

Cathode glow II   Crookes dark space   Cathode glow I   Faraday dark space   Positive column

Cathode glow II   Crookes dark space filling tube. Walls of tube fluorescing

Fig. 41.4 The character of the discharge

different gases, e.g. red for neon, blue-green for mercury. The modern advertising signs and the strip lamps used for lighting are discharge tubes with coatings of fluorescent powders on their inside surfaces.

If the pressure is steadily reduced beyond the above stage, the band of light breaks up. The successive changes in the character of the discharge are shown in Figs. 41.4 and 41.5. Dark spaces and striations (alternate bright and dark bands) develop and then retract into the anode until, finally, when the pressure is less than 1 mm of mercury, the Crooke's dark space fills the whole tube and the glass fluoresces.

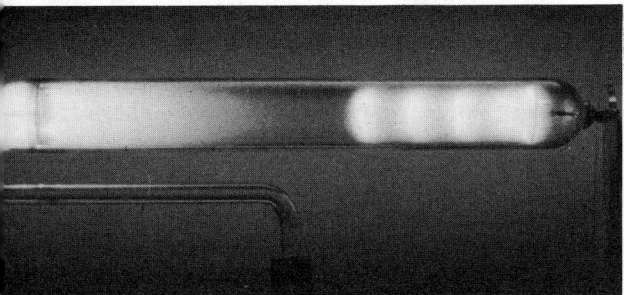

Fig. 41.5 Photograph of discharge showing striations

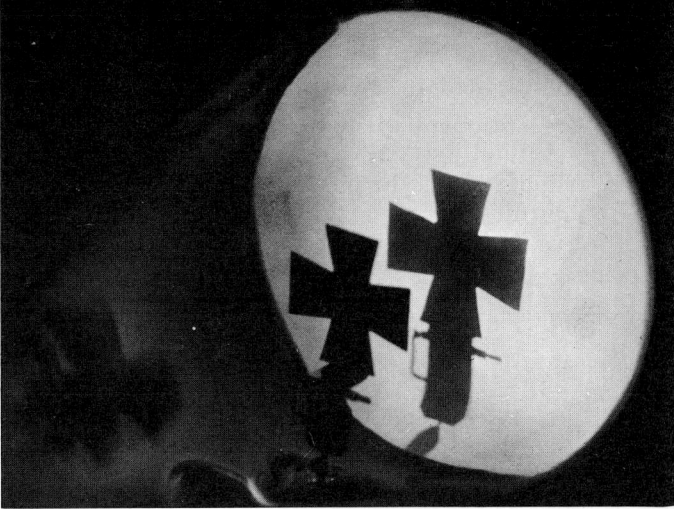

Fig. 41.6 Photograph of shadow of Maltese cross case by cathode rays

Cathode rays

The anode casts a shadow which indicates that the fluorescence is caused by rays sent out from the cathode. This is shown clearly by the apparatus in Fig. 41.6. The end of the tube may be coated with fluorescent paint, and the anode, in the shape of a maltese cross, casts a sharp shadow upon it. This experiment also illustrates that the cathode emits rays in straight lines normal to its surface. These are called *cathode rays*, and a narrow beam of them is obtained in the tube shown in Fig. 41.7. The anode has a hole or slit cut in

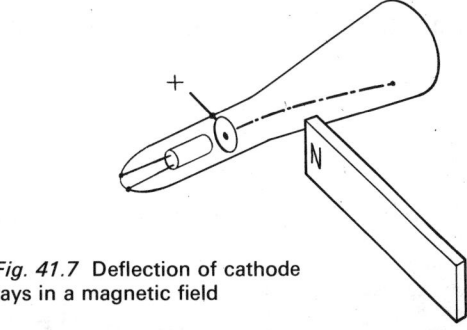

Fig. 41.7 Deflection of cathode rays in a magnetic field

it, and is placed near the cathode so that the rays which pass through the hole can be investigated. The rays are deflected by a magnet brought near the tube, and the direction of this deflection shows that they carry a negative charge; the fact that all the rays are similarly deflected shows that they all possess the same speed. This is a demonstration of the force acting on a current in a magnetic field.

Measurement of charge/mass (e/m) for cathode rays

Thomson was the first to determine the value of e/m for cathode rays. His apparatus is shown in Fig. 41.8. The rays were passed through a small hole to produce a fine pencil. This passed between the charged plates and on to a fluorescent screen at the end of the tube. The magnetic field was provided by a small electromagnet outside the tube. The direction of this field was perpendicular to the plane of the diagram. The deflection of the spot on the screen was noted with one field acting alone. The other field was then made to oppose the effect of the first field and was varied until the forces on the particles produced by the two fields were equal and opposite, i.e. the deflection of the spot on the screen was brought back to zero. From the dimensions of the apparatus and the strength of the fields, the value of e/m was found.

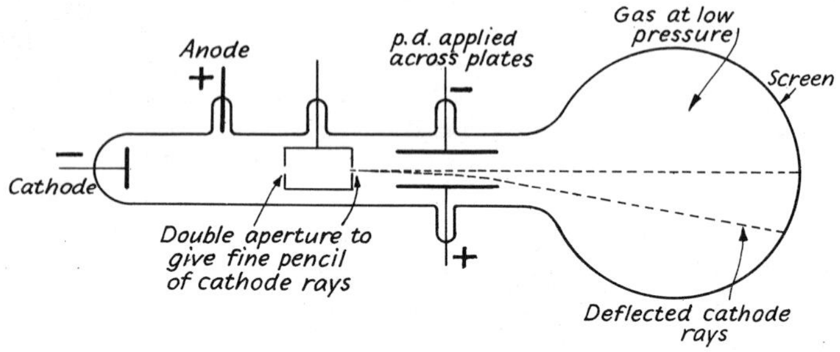

Fig. 41.9 shows a modern tube for determining

e/m. The cathode is a heated filament and provides a plentiful and variable supply of electrons. These electrons are focused on to a small hole in the anode and a fine beam emerges. This passes between two parallel metal plates before reaching the screen at the end of the tube. A p.d. of a few thousand volts is applied across the plates and the electrons are deflected, being attracted towards the positive plate.

The magnetic field is provided by two coils outside the tube and can be controlled by varying the current in them. The deflection produced is in a direction mutually perpendicular to the magnetic field and to the electron beam.

By adjusting the values of the fields, compensating deflections can be produced to give an undeflected cathode beam.

Again, from the values of the fields, etc., e/m can be found.

Fine beam tube

An interesting form of discharge tube for investigating cathode rays is shown in Fig. 41.10. The cathode rays are generated by a heated filament and are accelerated by an anode near to the filament. Two large coils, one on each side of the tube provide the magnetic field. The tube contains hydrogen gas at low pressure. The cathode rays ionize the hydrogen atoms and in consequence they emit a blue glow. The rays are emitted from the gun in an upward direction, but they are deflected by the magnetic field produced by the current through the coils. By varying this current the path of the rays can be made circular. From the

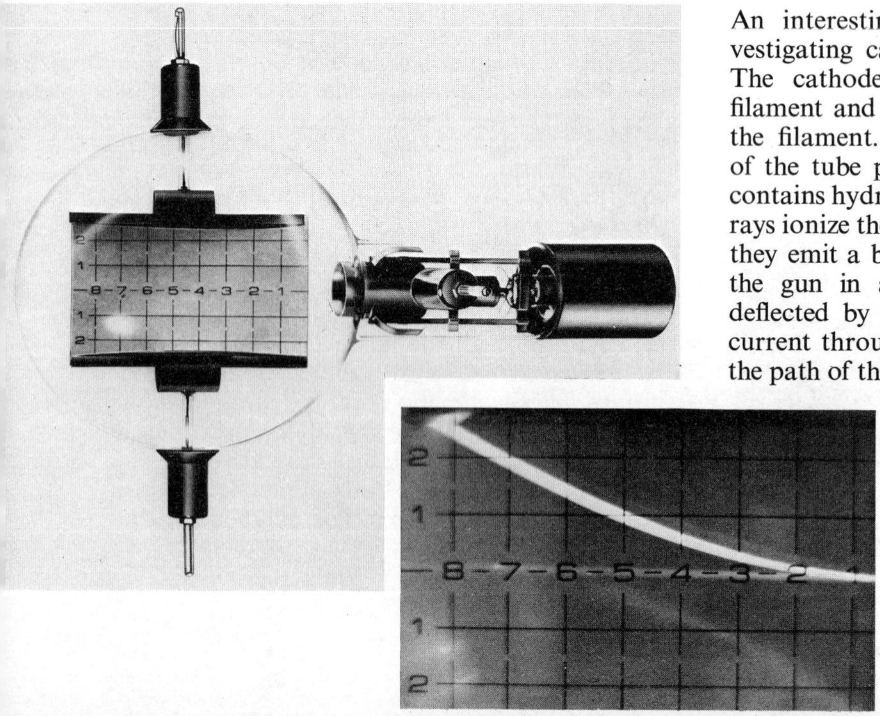

Fig. 41.8

Discharge tube for deflection of cathode rays

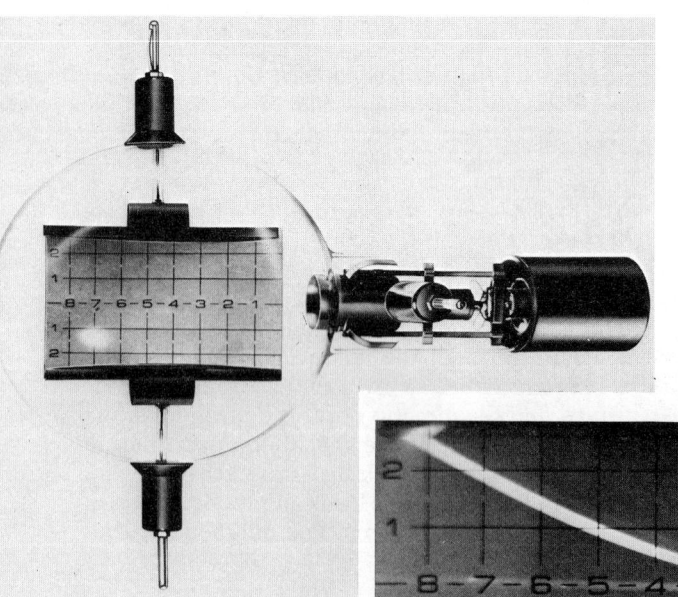

Fig. 41.9

Cathode rays

41.11). Cathode rays pass along the tube and strike the fluorescent screen. To one side of this path is a Faraday cylinder which is a metal cylinder with one closed end. It is mounted on a metal rod which passes through the tube to the outside. By manipulating a magnet near to the tube, the rays can be deflected into the open end of the cylinder. When this is connected to an electroscope, a divergence is obtained which is shown to be due to negative charge collecting in the cylinder.

Positive rays

Thomson found that rays were also emitted from the anode. He called these *positive rays* and, investigating them by methods similar to those he employed for cathode rays, found them to be positively charged atoms (positive ions) of the gas in the tube. The apparatus used is shown in Fig. 41.12.

The positive rays for the anode (not shown in the diagram) pass through the very small hole

Fig. 41.10 Fine beam tube

diameter of this circular path, the value of e/m can be calculated.

The velocity of the rays is governed by the accelerating p.d. across the cathode and anode. The kinetic energy of the rays is shown by the heat they generate when they fall on a solid surface.

Perrin's tube

A direct proof that the particles carry a negative charge is demonstrated using a Perrin tube (Fig.

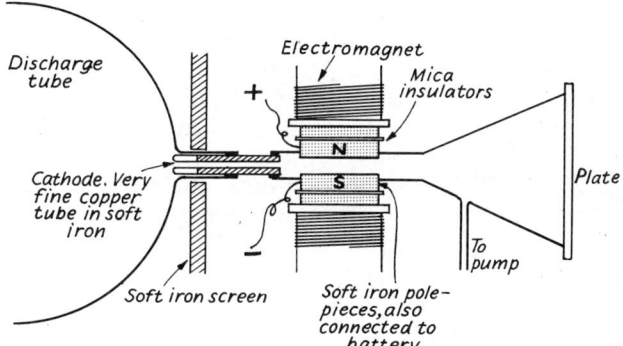

Fig. 41.12 Thomson's apparatus for magnetic and electric deflection of positive rays

Fig. 41.11 Perrin tube

in the cathode. The electric and magnetic fields are in the same direction and hence, the deflections produced are at right angles to each other. The positive ions, unlike the electrons forming the cathode rays, are formed all along the discharge tube from the anode to the cathode and consequently arrive at the cathode with different velocities. The slower moving ions are deflected more than those moving faster so that, instead of a single spot on the screen, a curved trace is produced. It can be shown mathematically that the trace for ions of the same mass is a parabola.

Thomson's plates, however, showed several parabolas which corresponded not only to the atoms known to be present in the tube, but also to atoms of unknown atomic mass.

The periodic table—atomic number —mass number

The advance of chemistry with increasing knowledge of the properties of the elements led Mendeleeff in 1869 to point out that the elements could be arranged so that they fell into specific groups which coupled them together by virtue of their properties. In this way the rare gases, the alkali metals. the halogens, etc. were brought together into groups. The elements were numbered in order of their atomic masses. This classification, however, revealed one or two discrepancies: the sequence of atomic masses did not always fit in with the Mendeleeff sequences.

The work of Thomson showed that an element can have more than one atomic mass. This means that a quantity of an element may consist of a mixture of atoms of different atomic masses. For example, Thomson found neon to contain atoms of mass 20, 22 and 23 units. These different forms of one element are called *isotopes*. The masses of isotopes, taking hydrogen as 1 (or, better, carbon as 12, which we now use as standard) are all whole numbers. These are called *atomic mass numbers*.

The full significance of atomic number and mass number will be better appreciated after a study of atomic structure which comes later in Chapter 43.

Thompson's technique of separating isotopes by means of magnetic and electric deflections was developed by Aston and is now well advanced. The instrument devised for this purpose, called a *mass spectrograph*, brings particles of the same mass to a line focus. The photograph produced resembles an ordinary line spectrum, each line corresponding to a particular isotope.

X-rays

It has been mentioned that when the Crookes's dark space fills the whole of a discharge tube, the walls of the tube fluoresce. This is due to the production of X-rays, which are generated when cathode rays of sufficient speed strike a target. The existence of X-rays was first proved in 1895

by Röntgen, who observed that the fluorescence of certain compounds was produced by the radiation from a discharge tube even when it was wrapped in brown paper.

A modern form of X-ray tube is shown in Fig. 41.13. It differs greatly from the simple tube of Röntgen in which the cathode rays were allowed to strike a heavy anode (or anti-cathode). The

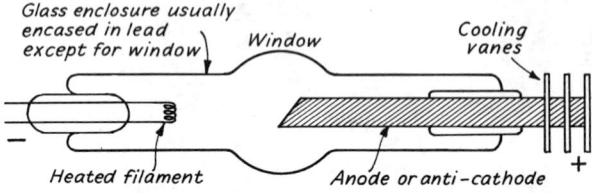

Fig. 41.13 Heated cathode X-ray tube

modern tube uses a heated filament as cathode which supplies an abundance of electrons and so produces a more intense beam of X-rays. The high p.d. across the electrodes accelerates the electrons to the speed necessary to excite the X-rays. Only a small fraction of the kinetic energy of the electrons becomes X-radiation; the majority is absorbed by the target which becomes hot. This necessitates the use of some form of cooling device. The large units use water cooling.

Properties of X-rays

X-rays falling on a plate coated with barium platinocyanide cause it to glow. This effect is made use of in visual investigations employing X-rays. The coating is usually put on a thick plate of lead glass which absorbs the rays and gives protection. When an object is supported between the X-ray tube and the plate, a shadow-graph of the object is seen on the plate which is viewed from behind. The shadow-graph is produced by different parts of an object absorbing the rays to a different degree (*see* below). As X-rays affect a photographic plate, a record may be made by replacing the barium platinocyanide screen by a photographic plate.

X-rays are unaffected by magnetic and electric fields. This property immediately differentiates them from cathode and positive rays and so suggests that they consist of waves and not particles. But it was not until 1912, however, that this was definitely established when Laue was able to show that X-rays could be diffracted. He found that the lattice structure formed by the regular arrangement of atoms in a crystal acted as a diffraction grating for the X-rays. He used the simple arrangement shown in Fig. 41.14. The photographic

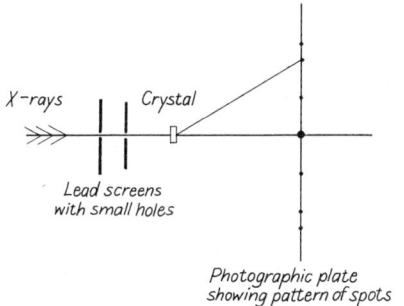

Fig. 41.14 Lane diffraction patterns

plate showed a symmetrical pattern of spots about the central spot. These were produced by diffrac-

tion. The symmetrical arrangement of the atoms in the crystal acted in a manner which can be compared with the results which would be obtained using two diffraction gratings at right angles. The effect is produced by the layers of atoms which are separated by distances of the order of the wavelength of X-rays.

X-ray diffraction and crystal structure

In the early days of X-ray diffraction the Laue patterns were found to be too difficult to interpret. In 1913, Sir William and W. L. Bragg discovered a new method of X-ray diffraction which enabled them to find the arrangements of atoms in many crystals.

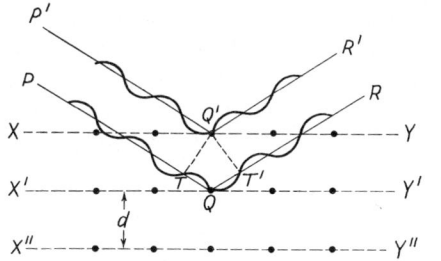

Fig. 41.15 Crystal diffraction
In the diagram $TQ + QT'$ is just one wavelength. The wave scattered by Q is in phase with that scattered by Q'

The principle of the Bragg method can be seen from Fig. 41.15. X-rays are incident upon the face

Fig. 41.16

The Bragg spectrometer

The photograph shows the mounting for the crystal and the ionization chamber. The chamber is connected to the Wilson tilted electroscope and the leaf system is observed through the telescope

XY of the crystal. The radiation is scattered by the individual atoms (or ions). The effect resembles the scattering of visible light by dust particles and each atom becomes a centre of a new disturbance. The total effect produced is as if the planes containing the atoms were mirrors. For a particular angle of incidence, the radiation scattered by the atoms in layer *X'Y'* will be in phase with that scattered by the layer *XY*. This will also hold for radiation reflected from other layers *X"Y"*, etc., if these are equally spaced. This condition will be obeyed when the path difference between *PQR* and *P'Q'R'* is a whole number of wavelengths. This condition is given by

$$(TQ+QT')=2d \sin \theta = n\lambda$$

where *d* is the grating space: *θ* the grazing angle: *n* is an integer.

There will be reinforcement of the intensity of the reflected beam when the above equation is satisfied. There will be several angles at which this will occur, which are found by putting values of 1, 2, 3, etc. for *n* in the above equation.

The essential parts of the Bragg spectrometer are shown in Fig. 41.17. X-rays are passed through screens having narrow slits. This fine beam falls upon the crystal which is mounted on a table capable of rotation as in the ordinary spectrometer. The ionization chamber is rotated about the crystal and

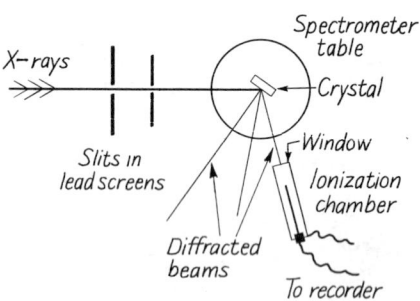

Fig. 41.17 Principle of the Bragg spectrometer

the response measured by the instrument connected to the chamber. This gives a measure of the of intensity as well as the direction of the diffracted X-rays.

Monochromatic X-rays have to be used. From the formula either *d* or *λ* can be found if the other is known. In the first experiments a value for *d* for rocksalt was calculated from the known value of

Avogadro's number. The wavelength of the X-rays was then found.

It is now possible to determine the wavelength of X-rays using ruled gratings similar to those used for visible light. The gratings have a very large number of lines per centimetre and the incident beam is arranged to have a large angle of incidence (grazing incidence).

X-ray crystallography has progressed at a fast rate since 1913 and the equipment used today is ingenious and complicated compared with the Bragg spectrometer.

It is possible, from the measurements made by X-ray diffraction, to build up a picture of the arrangement of the atoms in a molecule. (*See* Fig. 41.18.) This work has had tremendous success in the biological sciences where knowledge of the structure of the complex molecules has led to their greater understanding and even their synthesis.

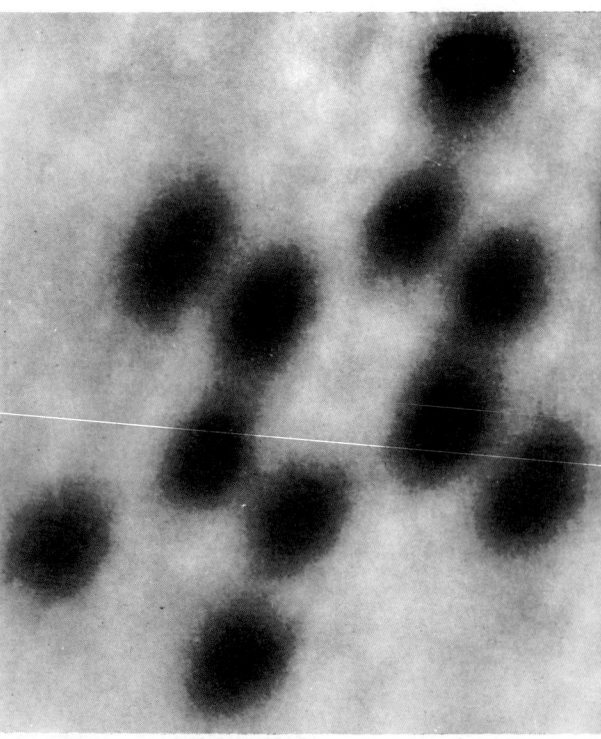

Fig. 41.18 Hexamethylbenzene molecule showing arrangement of atoms

Photograph synthesized by M. L. Huggins from X-ray diffraction data supplied by L. O. Brockway and J. M. Robertson. Magnification is about 200 000 000 times

Fig. 41.19 X-rays reveal hidden face in painting

A shows ordinary photograph of *'Girl with Rabbit'* by Sir William Beechey. *B* is an X-ray photograph of the same picture, which reveals a face that has been painted over

Absorption of X-rays

We have already placed X-rays in our complete spectrum of electromagnetic waves (*see* page 252). Their wavelength is the region of 10^{-10} m.

The shorter waves are termed 'hard' X-rays and the longer, 'soft' X-rays. These terms, 'hard' and 'soft', indicate their penetrative powers. Röntgen discovered that X-rays would pass through matter and he quickly appreciated their medical possibilities. The penetration of X-rays into a substance is inversely proportional to its density. For this reason lead is used as a protection against X-rays. This is also why X-ray photographs of the bones and organs of the body are possible. Different organs have different absorptive powers and so the the transmitted radiation produces a shadow picture (Fig. 41.19).

Ionization produced by X-rays

In addition to their photographic effect, X-rays can be detected by their ionizing effect. The use of this effect in measuring the intensity of X-rays was proposed by Thomson and, as we have just recorded, was used in the Bragg spectrometer. A charged electroscope subjected to X-radiation becomes discharged because the molecules of the air become ionized and thus conduct the charge away from the leaf system to the case. The ionizing effect is measured with an ionization chamber, a simple form of which is described later (page 402).

Precautions to be taken when using X-rays

X-radiation has a disastrous effect upon living tissue. Small amounts of radiation, such as one receives in the taking of an X-ray photograph of the body, have negligible effect, but prolonged or repeated exposure is harmful and in human beings may produce permanent illness. The effect is cumulative, that is, the blood does not recover from the damage caused by the radiation and, therefore, the effects of repeated doses add up. All apparatus generating radiations of this type should be enclosed in lead casing of sufficient thickness to absorb completely all stray radiation.

Generation of X-rays and X-ray spectra

When an electric charge is accelerated, radiation is produced. When very fast-moving electrons are stopped or slowed down, radiation of a very short wavelength is created. This is the method by which the 'background' X-radiation is generated. The electrons in the cathode rays are slowed down when they pass near to the positive nuclei of the metal atoms of the anode. This background radiation has a minimum wavelength corresponding to electrons being stopped by a head-on collision and losing all their kinetic energy. If the potential difference across the X-ray tube is increased, the electrons have a greater speed and so the minimum wavelength of the X-rays is reduced. Thus we see that the 'hard' X-rays (short wavelength) are produced by a very high potential difference across the tube. In addition to this background radiation, some of the high-speed electrons may penetrate into the innermost orbitals of the metal atoms and eject the electrons. In the readjustment of the electrons in the orbitals, short-wave radiation of a definite wavelength is emitted. More detail of the generation of such monochromatic radiation is given in Chapter 43.

Thus X-radiation consists of a background radiation covering a range of wavelengths down to a minimum value, along with certain lines of specific wavelength. The graph (Fig. 41.20) shows

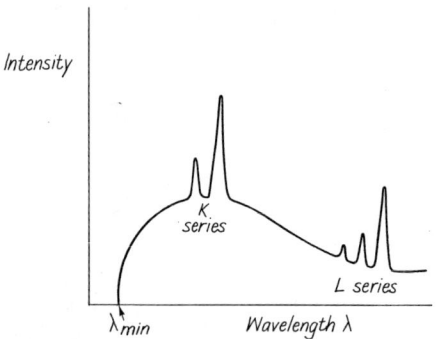

Fig. 41.20 Intensity/wavelength curve for X-radiation

the distribution of the intensity of the radiation among the wavelengths.

The lines in the X-ray spectrum mentioned above are characteristic of the metal forming the anode (or anti-cathode) of the X-ray tube. An investigation of the characteristic X-ray spectra was made by Moseley* in 1913. He found that the spectra

---

* Moseley became famous for his work under Rutherford at an early age. His outstanding work was on X-ray spectra. He was killed in Gallipoli during the First World War.

fitted into a sequence which denoted the order of the elements in the periodic table and assigned to each element a number. This is now called the atomic number *Z* and, as we shall see, it gives the number of protons in the nucleus of an atom.

Thermionic valves

The diode

Earlier in this chapter we mentioned Richardson's discovery of the emission of electrons from a heated wire. Fleming, in 1904, put the heated filament in an evacuated enclosure with a second electrode (plate or anode). The arrangement is represented diagrammatically in Fig. 41.21. It was found that, when the

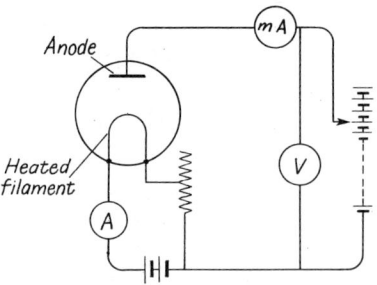

Fig. 41.21 Diode circuit

second electrode was given a positive potential with respect to the filament, a current passed round the circuit. The arrangement was called a *thermionic valve* or *diode*.

If a p.d. of different values is applied to the diode, the current increases with the p.d., but the curve flattens out, i.e. the current attains a maximum value. If the filament current is increased, a higher maximum results (Fig. 41.22).

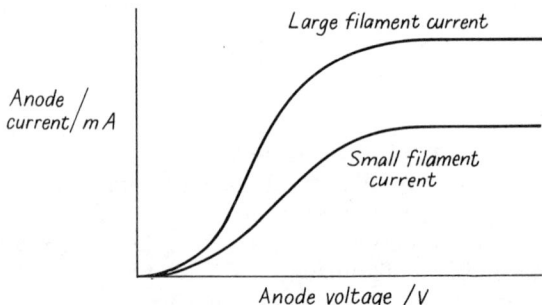

Fig. 41.22 Current/p.d. graph for diode

Electrons are emitted from the filament (cathode). These remain as a cloud round the cathode until a p.d. is applied across the diode making the filament negative. Electrons are attracted to the anode and their flow through the diode creates the current in the external circuit. Increasing the p.d. takes electrons from the space charge at an increasing rate until the electrons are removed at the same rate as they are emitted. This produces the maximum current through the tube. An increase in the filament current increases the rate of emission of electrons and so results in a higher maximum or saturation current. It should be noted that the graph of current against p.d. is not a straight line and hence Ohm's law is not obeyed.

If the battery is now replaced by an alternating voltage supply, as in Fig. 41.23, the anode is given

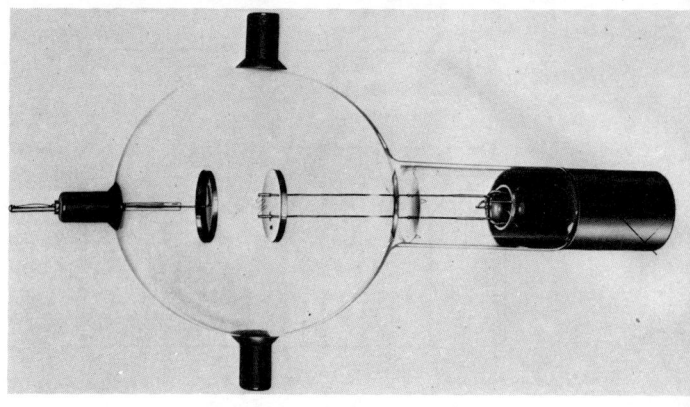

Fig. 41.24 Experimental diode

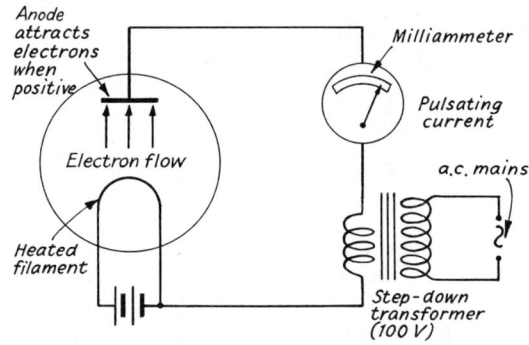

Fig. 41.23 The diode—rectification

alternately a positive and a negative potential. In this case, electrons flow to the anode only when it is positive, none travelling when it is negative. So the current is pulsating, flowing only when the anode has a positive potential. The milli-

ammeter registers a mean value of these pulsations, as was explained in the previous chapter. This is an example of rectification.

It has been found that certain oxides, e.g. those of barium and strontium, emit a plentiful supply of electrons when slightly warmed. Many modern valves are 'indirectly heated', the active oxide being on a cylinder (cathode), which is gently heated by a centrally placed wire carrying a current. This arrangement permits the use of a.c. through the 'heater'.

The triode

In 1906 De Forest added a third electrode to the diode and thereby greatly increased its usefulness. The third electrode is a grid, interposed between the filament and the plate, which influences the electron flow between them. In construction it often takes the form of a spiral of wire surrounding the cathode. The effect of a change in the p.d.

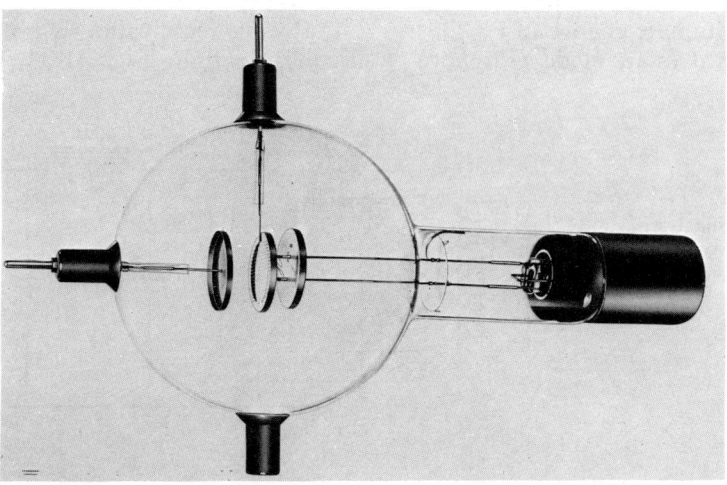

Fig. 41.25
Experimental triode

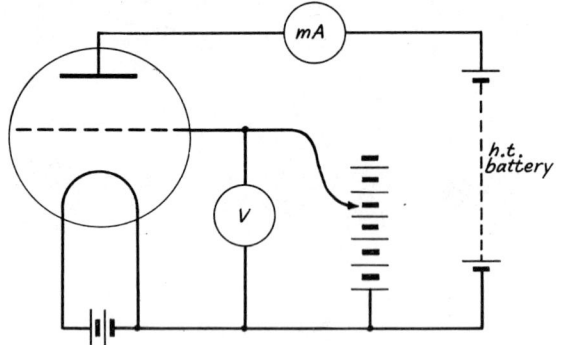

Fig. 41.26 The triode

Circuit for investigating variation of anode current with
grid voltage

(V_G) across the grid and filament on the anode
current (I_A) is investigated, using the circuit shown
in Fig. 41.26. As V_G is varied, I_A is measured and

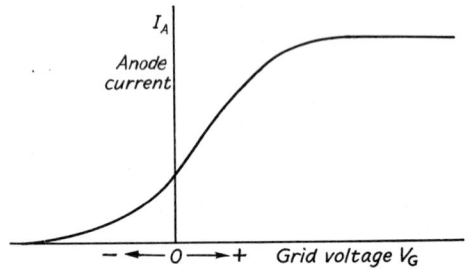

Fig. 41.27 Characteristic curve for triode

a graph drawn (Fig. 41.27). In the region where the
graph is straight and steep, a small change in V_G
gives a large and proportionate change in I_A. This
is amplification. Today valves are made with more

than three electrodes, the pentode (five electrodes)
being very common. Until the advent of the
transistor, about 1948, these valves formed the
basic components of radio sets and a vast range
of other electronic devices.

Rectification: power units

The power unit in a mains wireless set provides a
good example of the use of the diode as a rectifier.
The triodes in the set require a direct p.d. of 100
to 200 V, and the power unit supplies this from
the a.c. mains. In Fig. 41.28, T is a transformer

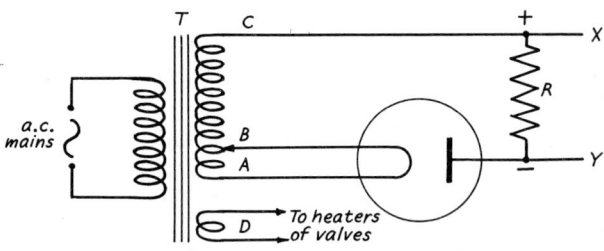

Fig. 41.28 Half-wave rectification

with a secondary BC of the right number of turns
to provide the working p.d. required by the
valves. AB is another secondary of few turns to
provide the heating current to the filament of the
diode. Another secondary D, also of few turns,
provides the heater current to the other valves.
Only when C is positive will any current flow
through the diode: the flow of electrons ceases
when the plate is negative. This unit is called a
half-wave rectifier, because it completely stops
the flow of electrons in the negative half of the
cycle of the a.c. supply.

Full-wave rectification is obtained by using the
circuit shown in Fig. 41.29, which accepts the

Fig. 41.29

Full-wave rectification
circuit for power unit
giving smoothed output

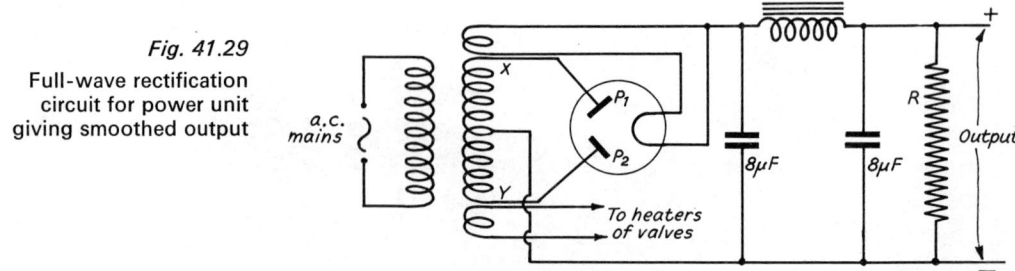

negative half of the a.c. wave and makes it positive. When the end X of the secondary is positive, a current flows through the valve via anode P_1, but no current passes via anode P_2. The reverse happens when end Y is positive. The pulsating output (Fig. 41.30) is smoothed by the use of chokes and capacitors. The high-value (8μF) capacitors provide an easy path for a.c. but shut off d.c. The chokes, on the other hand, provide a difficult path for a.c., but an easy one for a steady

are not the only ones which can be used, but they serve as examples. Antimony (valency 5) with germanium (valency 4) provides 'superflous' electrons; in this respect it differs from aluminium (valency 3) with germanium in which there is a 'deficiency' of electrons. The former produces negative (N-type) and the latter positive (P-type) germanium. When small pieces of these two types of germanium are joined together to make what is termed a *P-N junction diode,* it will allow the

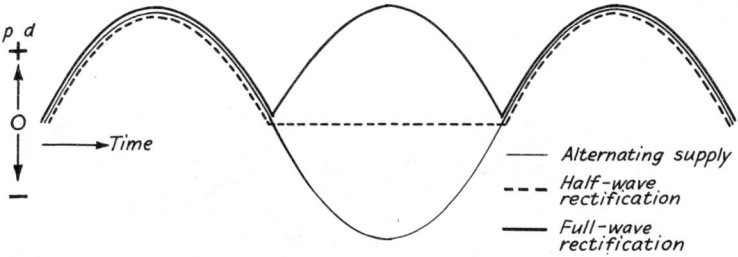

Fig. 41.30

Diagram showing half- and full-wave rectification (unsmoothed)

— — — Alternating supply

— — — Half-wave rectification

——— Full-wave rectification

d.c. R is a high resistance across the output, which may consist of several resistors in series so as to make available various output potential differences of lower values.

passage of electrons more easily from the N- to the P-type than in the opposite direction. Such an arrangement, with a low resistance to current in one direction and a high resistance to current in the other, can be developed as a rectifier.

Semiconductors

Diode

A semiconductor is a non-conducting crystalline material into which has been introduced a small amount of 'impurity' which renders it slightly conducting. The usual 'semiconducting device' is a piece of very pure germanium into which has been introduced a small amount of antimony or aluminium. The manufacture of these devices is very exacting (*see* Fig. 41.31). The substances named

The transistor

The transistor is a development from the junction diode. A simple form consists of three layers of germanium of different types arranged alternately, i.e. P-N-P, or N-P-N. External connections are soldered to the three regions, called *emitter, base* and *collector*. A detailed description is not possible here, but suffice it to say that the transistor can be made to function as a triode valve and can amplify a signal 10 000 times. The great advan-

Fig. 41.31

Germanium zone-refining unit

The germanium required for transistors is purified to 1 part in 10⁸ by this method

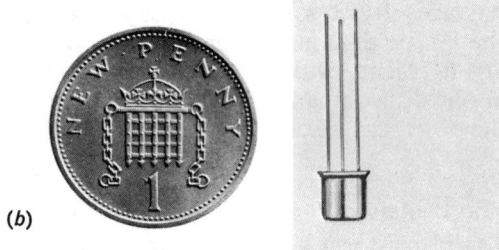

Fig. 41.32 Transistors

(*a*) a general-purpose power transistor
(*b*) a germanium alloy transistor for computer circuits

The tunnel diode

This is one of the most recent developments in semiconductor devices. Conditions across a P-N boundary are so arranged that electrons should not pass across, but it is not impossible for one to do so. As there are many millions of electrons present, there are some—a relative few—which can make the crossing, and these are said to 'tunnel' through the boundary. These devices have several advantages: they are exceedingly small; their power consumption is fantastically low— about one-hundreth that of a transistor; their action is rapid—one hundred times faster than

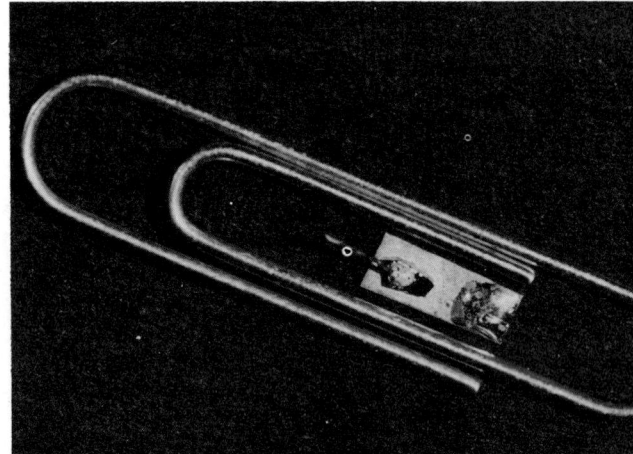

Fig. 41.33 Tunnel diode
An early form of tunnel diode. The paper-clip is for comparison of size

tages of these semiconducting devices is that they are so very small and that they function with relatively small potentials. There is also no 'warming-up' delay on switching-on for they do not contain filaments such as those in valves which consume so much energy. It must be pointed out that semiconductors are still a comparatively recent discovery, so that novel applications and developments are constantly being found. They can be made to function under the action of light and can therefore be adapted for light measurements. They form the basic constituents of solar cells and they have an important future in refrigeration. In the early days of their development they could handle only small power, but this difficulty has been overcome and their power-handling capacity is now remarkable.

that of a transistor, which enables them to be used in very high frequency (VHF) circuits; they can function at high and at very low temperatures; and they can be employed in low-noise amplifiers. These features make it highly probable that they will be used extensively in the immediate future.

The cathode-ray oscilloscope

The properties of the cathode-ray tube are made use of in the cathode-ray oscilloscope, which is one of the most important scientific instruments of the twentieth century. It is a development of the discharge tube, as designed by Thomson for measuring the electrostatic deflection of cathode rays.

Fig. 41.34

Cathode-ray oscilloscope

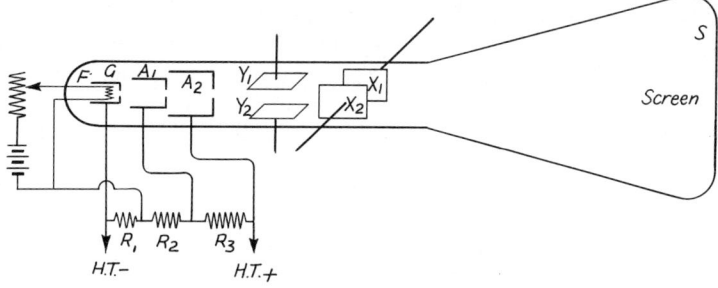

A plentiful supply of electrons is generated by the heated filament F (Fig. 41.34). This is surrounded by a metal cylinder G maintained at a negative potential with respect to the filament. The electron stream is focused upon a hole at the end of this cylinder and then proceeds to two anodes A_1 and A_2. The positive potential of A_2 is larger than that of A_1 and is about 2000 V. The high potential accelerates the electrons so that they travel with a large speed to the fluorescent screen S, on which they produce a spot of light. The tube is evacuated hence the beam is not slowed by collisions with residual air molecules. The potentials of G, A_1 and A_2 are supplied by a potential divider as shown. R_1 adjusts the brightness of the trace on the screen and R_2 provides adjustment to the focusing. The unit comprising the filament and anodes is often called the 'gun'. The electron beam passes through the electric fields produced by two pairs of metal plates, Y_1Y_2 and X_1X_2. These are known as the *X- and Y-deflection plates*.

If a steady potential difference is put across the Y-plates, an electric field proportional to the p.d. is created between the plates which deflects the electron beam, and the spot on the screen is displaced either upwards or downwards according to the polarity of the plates. Likewise a p.d. across the X-plates causes a sideways displacement. An alternating voltage across the plates causes the displacement to alternate and to produce a line trace on the screen. The length of the line will be determined by the magnitude of the alternating voltage: its periodic time will be determined by the mains frequency.

If alternating voltages are applied to both the X- and the Y-plates, the resultant motion of the spot will be that produced by the combination of two simple harmonic motions. Two simple harmonic motions of the same frequency and amplitude will produce a trace in the form of (*a*) a straight line inclined at 45° when the phase difference is 0° or 180°, and (*b*) a circle when the phase difference is 90° or 270° (Fig. 41.35).

The ratio of the frequencies of two alternating sources can be compared by applying them across the X- and Y-plates and changing the frequency

Fig. 41.35

Cathode-ray oscilloscope

Some traces that can be produced on the screen by the electron beam

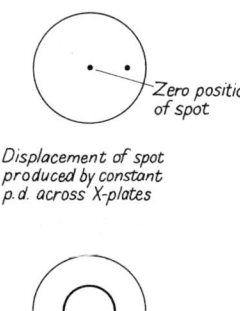

Displacement of spot produced by constant p.d. across X-plates

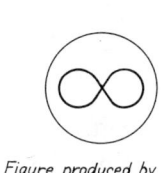

Vertical line produced by alternating p.d. across Y-plates

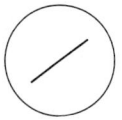

Inclined line produced by alternating p.d.'s (in the same phase) applied to both X-and Y- plates

Circle or ellipse produced by alternating p.d.'s with phase difference of 90° across X-and Y-plates

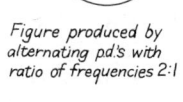

Figure produced by alternating p.d.'s with ratio of frequencies 2:1

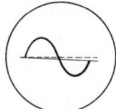

Waveform of single cycle of mains supply. Note faint 'fly-back' of spot to re-start trace, and that the cycle is not quite complete

of one until a steady pattern is produced upon the screen. The form of this pattern determines the simple ratio between the two frequencies. The scale of frequencies on the oscillator used for this purpose can first be calibrated by comparing it with the a.c. mains frequency which, in England, is maintained constant at 50 Hz.

The cathode-ray oscilloscope can be used to show the form of any voltage fluctuation which may be of very short duration, such as the voltage changes when a switch is operated. The form of the displacement produced by the Y-plates is not revealed unless an X-shift is also produced. For this purpose a 'time base' is employed. This device produces a p.d. across the X-plates which increases with time and so displaces the spot across the screen with a constant speed. When the spot reaches the edge of the screen, the p.d. must rapidly drop so that the spot *very* quickly flies back to its starting-point. The details of the time base are too involved to record here. The effect is that, when an alternating p.d. is applied across the Y-plates, the time base will at the same time draw the spot across the screen from left to right. The path described by the spot will be a sine curve. The repetition of the action of the time base can be adjusted to have the frequency of the voltage fluctuation under investigation. In the case of the mains voltage across the Y-plates, if the frequency of the time base is made 50 Hz, the sinusoidal trace on the screen can be repeated 50 times each second in exactly the same position, thereby producing a fixed and bright sinusoidal trace.

Most oscilloscopes incorporate an amplifier so that a small fluctuating p.d. can be increased sufficiently to produce an appropriate displacement.

Deflection of the cathode rays can be produced by magnetic fields. Such can result from current-carrying coils suitably placed outside the tube.

The experiments which can be performed with the cathode-ray tube are numerous. A varying current is investigated by passing it through a resistor and taking leads from the ends of the resistor to the Y-plates.

The importance of the cathode-ray oscilloscope is due particularly to the fact that the electron beam has virtually no inertia. An a.c. of frequency 50 Hz sent through a moving-coil ammeter with centre zero will not register because the alternations are too quick: the sluggish coil

cannot move as quickly as the current changes its direction. The electron beam in a cathode-ray tube, however, will oscillate when subjected to an alternating electric field of frequency greater than a million hertz.

The television tube is a cathode-ray oscilloscope with large screen. The picture is built up by use of a time base which sends the spot across the screen from left to right and also causes it to progress downwards. The brightness of the spot varies and it makes altogether 605 journeys across the screen to cover the picture once. When the movement is repeated many times a second a steady image is produced.

Photoelectricity

Another way that electrons are freed from a substance is by the *photoelectric effect*. When ultraviolet light falls on certain metallic surfaces, e.g. sodium, potassium or caesium, electrons are emitted. The electrons behave like molecules released from a liquid by evaporation. They form a cloud round the surface of the metal unless they are moved away by the attraction of a positively charged plate in the vicinity.

The effect can be demonstrated very simply using a similar method to that employed by Hallwacks who first showed the effect in 1888.

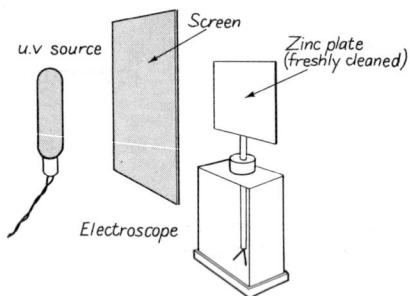

Fig. 41.36 Demonstration of photoelectricity

In Fig. 41.36, a zinc plate is mounted on an electroscope. This plate must be freshly cleaned with glass-paper just before performing the experiment. The electroscope and plate are given a negative charge: the screen is then removed and the radiation from the ultraviolet lamp allowed to fall on the plate. The divergence of the electroscope is soon reduced to zero. The experiment is then repeated with the electroscope positively charged and this time there is no decrease in the divergence. The results are explained by the emission of photoelectrons when the plate is negatively charged. The positive charge on the plate prevents the escape of the negatively charged electrons.

The explanation of this effect is based on the quantum theory and is given later in Chapter 43.

Applications of the photoelectric effect —photocells

The effect is adapted in the design of photocells. These are electric cells which are activated by the action of light. There are three types in common use. They all depend upon the release of electrons by the action of light, although their functioning differs.

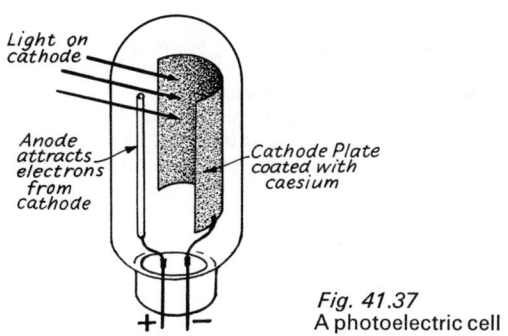

Fig. 41.37
A photoelectric cell

Photoemission gas-filled cells—A form of this type of cell is shown in Fig. 41.37. The plate is coated with caesium (or other active metal) and is enclosed along with the anode in a glass vessel which contains an inert gas (argon or neon) at low pressure. The cell responds to ultraviolet and to visible light. A steady p.d. of about 100 V is applied across the cell and the current through the cell is proportional to the intensity of the radiation falling on it.

Barrier layer cells—This type of cell is used in lightmeters and exposure-meters and has been mentioned in Chapter 21. This cell creates an e.m.f. by the action of light. A plate of semi-conductor, e.g. cuprous oxide, is covered by a thin metal layer through which the light can pass (Fig. 41.38). Electrons are released at the boundary and an e.m.f. is created.

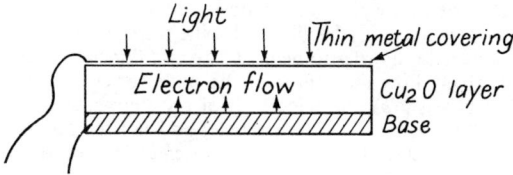

Fig. 41.38 Barrier layer photoelectric cell

Selenium cells—Selenium has the property of releasing electrons when light falls upon it. When illuminated, selenium releases electrons which cause a relatively large drop in its electrical resistance. Thus, when included in a suitable circuit, there is a large increase in the current. Selenium cells are slow acting compared with the gas-filled cells, but they have their uses. They are very good in circuits for controlling burglar alarms, automatic counters, automatic doors, etc.

QUESTIONS

1. Discuss the evidence leading to the conclusion that electric charge exists in fundamental units.

2. Describe, with diagram, how you would show that the cathode in a discharge tube emits some form of rays.

3. What experiments could you use to show that cathode rays are negatively charged particles.

4. Describe the effect of (*a*) a magnetic field, (*b*) an electric field on a stream of cathode rays. What conclusions can you draw from the experiments?

5. Compare and contrast the properties of the positive and cathode rays produced in a discharge tube.

6. Describe how X-rays are produced and give some of their properties.

7. What is the photoelectric effect? Describe one practical application of it.

8. Explain the application of thermionic emission to the valve. Describe how a simple diode functions. Draw the circuit for full-wave rectification. Explain the function of chokes and capacitors in the circuit.

9. What are cathode rays? Describe briefly, with the aid of a diagram, any one way in which they can be produced.
    How would you demonstrate the deflection of cathode rays by a magnetic field? By reference to the direction of the

deflection in a particular case, explain how it gives evidence as to the electrical nature of the rays.

Describe briefly any one other experiment which leads to the same conclusion.   (C)

10. Describe and explain how a narrow pencil of electrons of constant velocity can be produced.

The electrons in a cathode-ray tube beam are accelerated through a potential difference of 8000 V. The beam current is 0.25 mA. What power is dissipated in the screen of the tube?   (O & C)

11. Describe the diode valve and give details of the cathode circuit. State the essential condition, in each instance, to produce an anode current which is (i) zero, (ii) a maximum, assuming that the cathode is at a constant temperature suitable for the normal working of the valve.   (L)

12. Assuming that 9.6×10^4 C liberate 1.0 g of hydrogen in electrolysis, and 1.0 g of hydrogen contains 6.0×10^{23} atoms, calculate a value for the charge on the electron.   (O & C)

13. Describe how charged particles moving in a vacuum at constant speed must be directed with respect to a uniform magnetic field in order to be bent into circular paths. Explain qualitatively how the radius of the path varies with (a) the sign of charge, (b) the magnitude of the charge, (c) the mass and (d) the velocity of the particle, and (e) the strength of the field. Discuss whether a uniform *electric* field gives rise to a circular path.   (O & C)

14. Draw a labelled diagram of an a.c. transformer for transforming an input of 240 V down to an output of 24 V. Explain how the step-down occurs.

Draw a circuit diagram to show the output of the transformer connected to a diode valve and an accumulator so as to charge the accumulator. Explain the rectifying action of the diode in this circuit.   (C)

15. What is *thermionic emission*? Mention *one* resemblance to and *two* differences from the emission of β-rays.

Describe how you would investigate experimentally the way in which the current passed by a thermionic diode varies with the magnitude and sign of the applied voltage when the cathode temperature is fixed. Sketch the kind of curve you would expect your results to follow.   (O & C)

16. If the current in the beam of a cathode-ray tube (as used in a television receiver) is 0.12 mA, how many electrons strike the screen each second?

If the accelerating voltage is 3000 what is the kinetic energy in joules of an electron as it strikes the screen?

Take the charge on an electron as 1.6×10^{-19} C.   (O & C)

17. Draw a labelled diagram to show the structure of any *one* simple type of cathode-ray tube, using a heated cathode, designed to demonstrate the fluorescent effect of cathode rays. Show clearly on your diagram the electrical connections needed for working the tube and the position of the resulting fluorescence. At what stages in their motion are the cathode rays (a) accelerated, (b) retarded? What becomes of the kinetic energy lost during the retardation?

Describe, with an illustrative diagram, any *one* experiment which indicates that cathode rays are negatively charged. (C)

18. Cathode rays which consist of a *beam* of free electrons can cause fluorescence, or can otherwise be made visible, when given a high velocity. Describe briefly:

(a) how the free electrons may be obtained;

(b) how a high velocity beam is obtained;

(c) an apparatus used to show that the electrons travel in straight lines and what will be observed when so used;

(d) a simple means of deflecting the beam. Indicate the direction of the deflection in relation to the means you have chosen.   (L)

42 Radioactivity

Discovery

While investigating the connection between X-rays and fluorescence in 1896, Becquerel discovered that salts of uranium emit a radiation similar to X-rays, which will fog a photographic plate and also ionize the air. This new discovery did not attract much attention at the time, but further experiments were carried out by Becquerel and his assistants, among whom was Madame Curie. In 1898 she discovered that pitchblende* contained a more powerfully active substance than uranium. This she named 'polonium'. Working with her husband, Madame Curie succeeded in separating a third, and even more powerful, constituent of the pitchblende. Scientists of the day were sceptical about the discovery of two new elements, so the Curies set about the mammoth task of extracting them from several megagrammes of pitchblende. After three years' work they produced about 1/30 g of the new and very active substance, which they called *radium*.

---

* Pitchblende is the natural ore from which uranium is obtained.

A young New Zealander, Ernest Rutherford (later to become Lord Rutherford), working in Cambridge, became interested in these new discoveries and joined in the investigation. By 1899, he had proved the existence of two distinct types of radiation emitted by the active elements. He called them α- and β-radiations. Later, a third radiation, γ, was identified by Villiard in France. The phenomenon of the emission of radiation by active substances was called *radioactivity*. For nearly forty years, until his death in 1937, Rutherford pursued this subject. His researches are an epic in the annals of experimental physics. He is regarded by many as the greatest of all experimental physicists. His intuitive ability was unique. Unlike his great counterpart of the nineteenth century, Michael Faraday, he was not an individual worker, but a leader. He infused into all those with whom he collaborated, a burning desire to pursue the mysteries of the structure of the atom. The rapidity of the progress of twentieth-century physics, is, in no small measure due to the genius of this great man.

The early measurements in radioactivity were

made using the property of ionization. Madame Curie had been working on the measurement of very small currents with a sensitive galvanometer, and she employed the knowledge so gained to measure the radiation. The radioactive specimen was placed on a plate of a charged air capacitor. The ionization of the air caused a very small current across the capacitor plates.

Rutherford and his co-workers developed sensitive forms of electroscopes and with these the fundamental principles of radioactive change were discovered. Today we have many instruments for investigation of the radiation but, before we describe them, we must learn more about the properties of the radiations they have to measure.

Properties of the radiations

All three types of radiation produce ionization and are able to penetrate matter. The relative magnitudes of these properties differ considerably for the different radiations.

α-radiation

The radiation from a small source of radium produces strong ionization. A thin sheet of aluminium or even a visiting-card is found to cut off the radiation. Any source of α-particles, when brought close to the plate of a charged electroscope, will discharge it by ionizing the air round the plate.

α-particles have a range of about 3 cm in air: their penetrative power is not great. This range is increased by reducing the pressure of the surrounding air. In 1903, Rutherford was able to show that a beam of α-particles could be deflected by a very strong magnetic field. Using apparatus similar in principle to that employed by Thomson for cathode rays, Rutherford obtained a value for e/m for the radiation. The magnetic deflection of the particles showed also (i) that the rays consisted of particles and were not waves* and (ii) that, as could be deduced from the direction of the deflection, the particles carried a positive charge. The charge on these particles was not measured until a few years later, when it was found to be twice the size of the charge on an electron. The

---

* Waves are unaffected by electric and magnetic fields.

mass of the particles obtained from the value of e/m was about four times that of the hydrogen atom, i.e. the particle appeared to be a doubly charged helium ion. Direct proof of this came in 1909 when Rutherford and Royds performed what Rutherford himself considered to be the most elegant experiment of his lifetime. A source of α-particles, S, was contained in a tube (Fig. 42.1)

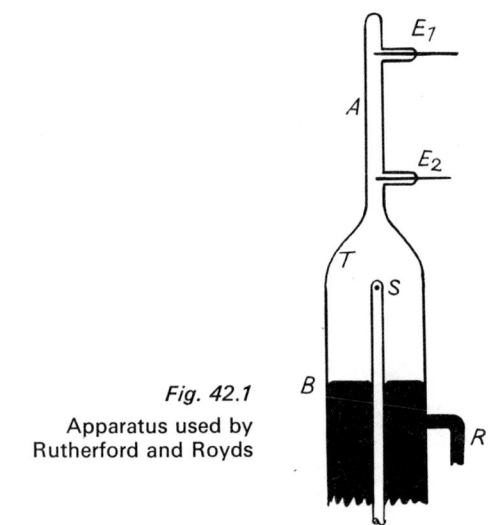

Fig. 42.1

Apparatus used by
Rutherford and Royds

which had very thin glass walls through which α-particles could pass. The surrounding space in the tube T was evacuated through the tube R and then sealed off by raising the mercury level to B. The apparatus was left for two days to allow the α-particles to collect in the space T. The collected particles were driven into the capillary tube A by raising the level of the mercury. A discharge across electrodes E_1 and E_2 was produced by applying a suitable p.d. and the spectrum of the glow examined. This contained the strong yellow line in the helium spectrum. After leaving the apparatus for six days, the whole helium spectrum was produced. Thus the identification of α-particles as helium nuclei was established.

β-radiation

In 1899 Rutherford observed that a thin sheet of aluminium did not absorb all the radiation from a uranium source. Another radiation which penetrated the first aluminium sheet had a penetrative power about 100 times that of the first radiation.

So the two radiations α- and β- were discovered. About the same time Becquerel and others showed that the radiation from uranium could be deflected by electric and magnetic fields. Their observations were on the β-radiation and so they demonstrated its particulate nature and also the nature of the charge it carries. The Curies also showed the charge to be negative by collecting the particles in a hollow conductor (Faraday cylinder). The mass and charge of the particles were found to be identical with those of the electron and so β-radiation was identified as a stream of electrons. The velocity of emission was found to vary over a range with maximum values approaching that of light.

The simple properties of β-particles can be demonstrated easily using modern equipment. A suitable counter (*see* page 402) will record the rate at which β-particles enter the counting tube. Thus it is only necessary to set up a counter near to a source and the effect of interposing absorbing screens can be investigated. A fine pencil of β-particles can be produced by passing it through slits in lead plates and its deflection can be shown by bringing up a magnet to deflect it on to, or off, the counter.

Fig. 42.2 Radiograph of watch

γ-rays

The third type of radiation was not discovered until 1900. The radiation was found to be very penetrative and resisted all attempts to deflect it by electric and magnetic fields. Two ideas were advanced. The Curies favoured a wave nature, but Rutherford and others were at first inclined towards a particle nature, the particles having too large a velocity to show a deflection in a magnetic or an electric field.

We now know that γ-radiation consists of electromagnetic waves of wavelength around 10^{-11} m, i.e. equal to the wavelength of hard X-rays. The penetrative power of γ-rays is very great. From some sources the rays will pass through great thicknesses of lead. The rays also produce ionization.

Radioactive series

The idea that the atom had a definite structure became well founded early in this century, and the study of radioactivity provided answers to many of the problems which arose. The emission of α- and β-particles indicated that these were constituents of atoms. Rutherford and Soddy found that one radioactive element, in emitting a radiation, was transformed into another. For example, an atom of radium (atomic mass 226), in emitting an α-particle, becomes a new element, radon (atomic mass 222). The radon so formed emits another α-particle and becomes radium A (atomic mass 218). In this way radium eventually becomes lead, which is a stable element. Such a radioactive series is produced by an atom, which

Radiation	α		α		α		β		β,α		β		β		α	
Element	Ra	→	Radon	→	RaA	→	RaB	→	RaC	→	RaD	→	RaE	→	Po	→	Pb
Mass number	226		222		218		214		214		210		210		210		206

Radioactive change of radium into lead showing the radiations and the changes in atomic mass number.

TABLE SHOWING PROPERTIES OF THE RADIATIONS

	α-radiation	β-radiation	γ-radiation
Nature	Consists of stream of charged particles—helium nuclei.	Consists of stream of charged particles—electrons.	Consists of electromagnetic waves.
Mass	$4 \times$ mass of hydrogen atom.	$\frac{1}{1850}$ of mass of hydrogen atom.	Mass nil. Wavelength 10^{-11} m.
Velocity	Velocity up to about 2×10^7 m/s.	Velocity up to 99% of velocity of light (i.e. up to 2.97×10^8 m/s).	Velocity: 3×10^8 m/s.
Charge	Positively charged.	Negatively charged.	Uncharged.
Effect of field	Slightly deflected by magnetic and electric fields.	Greatly deflected by magnetic and electric fields.	Unaffected by magnetic and electric fields.
Ionizing properties	Strongly ionizing.	Weakly ionizing.	Weakly ionizing.
Penetrative powers	Small (30–80 mm of air).	Large (1 mm of lead).	Very great.

is too large to be stable, emitting parts of itself until it becomes small enough to be stable. The emission of an α-particle lowers the atomic mass by 4 and at the same time reduces the positive charge on the nucleus by 2. The β-particle is so tiny that no significant change in atomic mass occurs when this is emitted. At the same time, such an emission removes a negative charge, so that the net effect is to increase the positive charge on the remaining atom by unity.

This emission from heavy atoms takes place spontaneously and at a rate peculiar to each element. This rate cannot be influenced except by abnormal extremes of temperature and pressure.* Emission continues indefinitely, the rate of disintegration being proportional to the number of atoms present.

Detection and measurement of the radiations

Scintillations—the spinthariscope

α-radiation was observed to make a zinc sulphide screen luminescent. Careful observation showed the glow to be due to scintillations produced by individual particles striking the screen.

---

* The extremes of temperature and pressure existing within stars upset completely the binding forces which control the emission of particles from nuclei. We can now produce these conditions in some degree in atomic explosions and in apparatus being developed in laboratories. (*See* Fig. 43.19, ZETA.)

Observations of these individual splashes of light can be observed using a simple device known as a spinthariscope .(Fig. 42.3). As the intensity

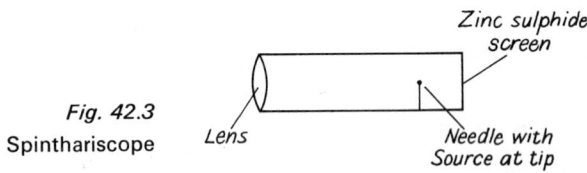

Fig. 42.3
Spinthariscope

of the scintillations is very low, the eye has to become accustomed to darkness before the effect can be seen. A small source of α-particles is mounted in a short tube at one end of which is a zinc sulphide screen. A magnifying glass at the other end provides the means of viewing the scintillations produced on the screen.

This principle was employed by Rutherford and his co-workers in many of the early experiments on α-particle emission. The same principle is used in modern scintillation counters, which enhance the effect of a single scintillation by using a photo-multiplier tube, in which the single effect is magnified to become a measurable electrical pulse. The pulses are counted electronically.

Ionization chamber—the Geiger-Müller counter

In early experiments a modified electroscope was used to measure the ionizing effect of radioactive radiations, but the need for an instrument to detect and to count individual particles was soon felt. Rutherford and Geiger devised the ionization chamber, which is shown diagram-

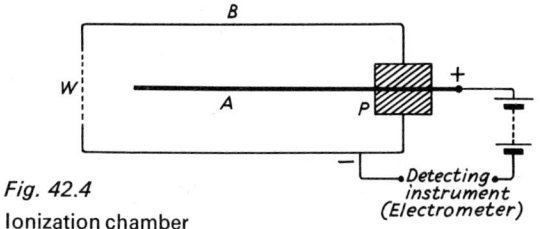

Fig. 42.4
Ionization chamber

matically in Fig. 42.4. *B* is a metal cylinder provided with a thin aluminium window *W* to allow the radiations to enter. A wire *A* is supported through an insulated plug *P* along the centre of the cylinder. Across *A* and *B* a high potential difference is applied, with a sensitive current detector in series. The radiation entering the chamber renders the enclosed gas (usually at low pressure) conducting, and this is indicated by the meter or galvanometer. The modern form of this instrument is known as a Geiger–Müller counter. It employs a much improved type of ionization chamber connected to an amplifier. The electric pulse resulting from the entry of a single particle into the chamber, when amplified, can be made audible by passing it through a loudspeaker, or it can be recorded by a counting device.

There are two kinds of counting device in common use. The *ratemeter* registers the number of particles received in the G.M. tube per minute, while the *scaler* sums up the total number of particles received. There is an obvious limitation to the rate at which the G.M. tube can function. The particles must not enter at too high a rate, as time is required for the quenching of the charge received from one particle before the tube is ready to register the next.

The pulse electrometer

Various modifications of the gold-leaf electroscope have, in the past, been devised for use as detectors of *α*- and *β*-radiations. A modern modification is now very popular for demonstrations and simple measurements. It is known as the *pulse electrometer* (Fig. 42.5). The leaf system is of special design. As charge is received, the divergence of the leaf increases until it touches

a side electrode and is discharged. The action is repeated continuously when charge is being supplied and a pulsing of the leaf results. The pulses are counted and the pulse-rate (number of pulses per second) found.

The inside system consists of a thin aluminium strip *A* fastened at the top end to a plate *P*, the lower end of the strip being coupled to the plate by means of a quartz fibre *F*. The effect of charging is indicated in Fig. 42.5. The position and also

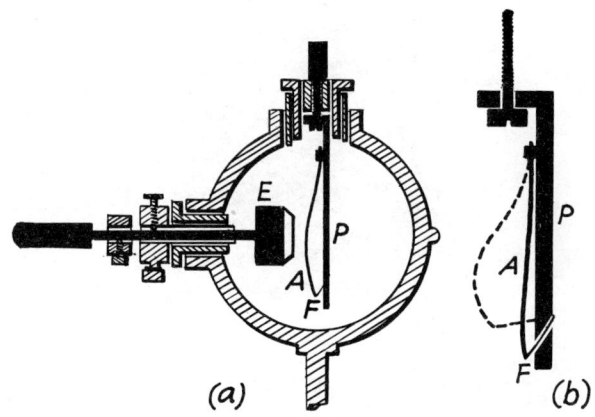

Fig. 42.5 The pulse electrometer

the potential of a side electrode *E* can be adjusted to alter the sensitivity of the instrument. When earthed and brought near to the leaf, this electrode can discharge the deflected leaf. Thus the leaf may be discharged in pulses. An ionization current is measured by counting the number of pulses in a given time and so determining the pulse-rate.

The dosemeter

This detector is another form of electroscope which was designed as a pocket instrument for the quick detection of dangerous radiation in an emergency. For laboratory use it is attached to an ionization chamber (Figs. 42.6 and 42.7).

The gold leaf of the ordinary electroscope is replaced by a loop of very fine metallized quartz fibre. This is attached to one end of a metal support, the other end of which protrudes downwards into the ionization chamber which forms the base of the instrument. The fibre system is viewed through an eyepiece. Charging is carried out by momentary connection to an H.T. battery

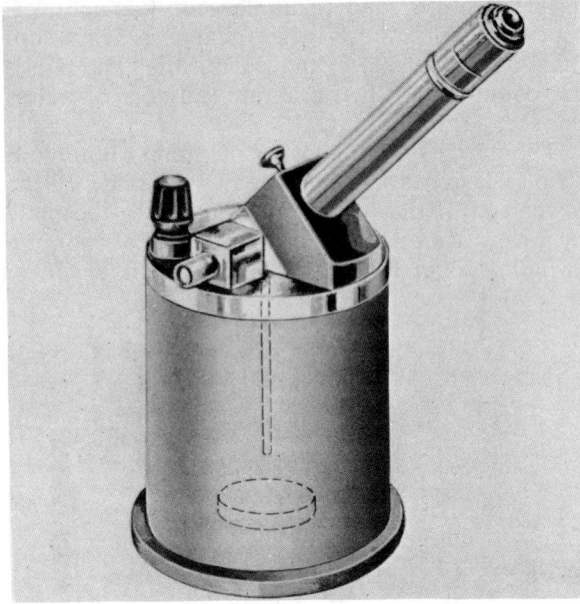

Fig. 42.6 Dosemeter

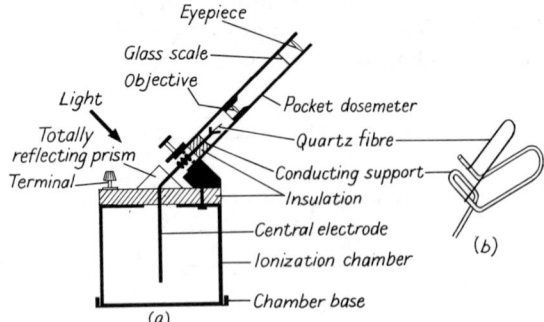

Fig. 42.7 Diagram of dosemeter showing essential parts

which is adjusted to bring the image of the quartz loop to the zero of a scale in the eyepiece.

The activity of a source is measured by placing it on the base of the ionization chamber where its effect causes a leaking of the charge from the inner system. The scale reading is taken at regular time intervals.

The solid state detector

Yet another application of semiconductors has been developed in the last few years. A silicon diode has been modified so that an ionizing radiation falling on it causes electrical pulses of short duration. These can be amplified and counted by a ratemeter or by a scaler.

The d.c. amplifier

The very small currents produced by the pulses of charge in the detectors can, in some cases, be measured conveniently using a d.c. amplifier. The output in such cases is registered on a micro-ammeter.

The cloud chamber

When a volume of air which is saturated with water vapour is suddenly expanded, the temperature drops, so that condensation takes place and a cloud of water droplets is formed. These droplets form easily if dust is present in the air, for condensation takes place readily upon small material particles. It is the condensation of water vapour upon particles of soot in the atmosphere that produces the 'pea-soup' fogs familiar to city-dwellers. If the air is free of dust, however, it may be supersaturated without condensation occurring. C. T. R. Wilson found in 1911, that in such cases droplets would readily form on any ions which were present. It thus became possible to show up the tracks of charged particles or ionizing rays in much the same manner as vapour trails reveal the path of an aeroplane through the atmosphere. A charged particle, such as an *a*-particle, ionizes the air molecules by collision, to form suitable nuclei for condensation. A strong track is thus

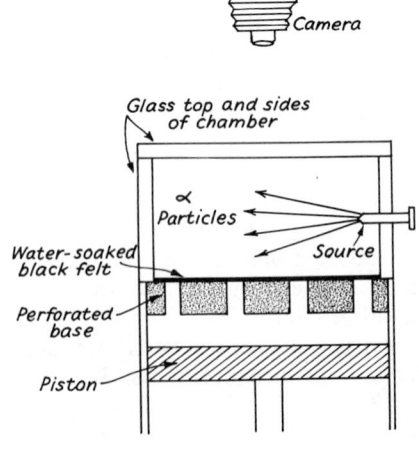

Fig. 42.8 Wilson cloud chamber

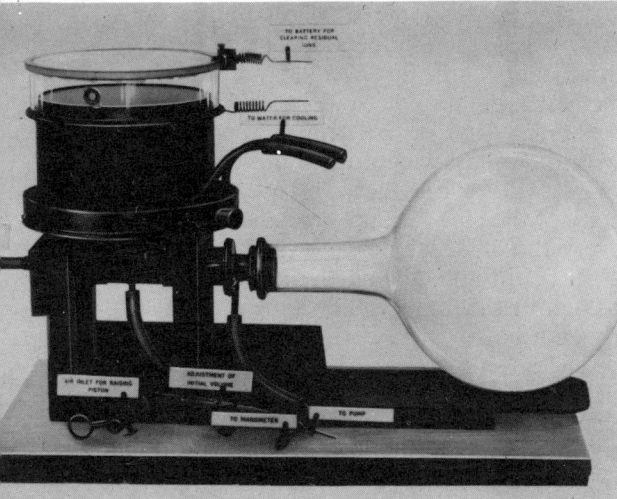

Fig. 42.9 Wilson's 1911 cloud chamber

Fig. 42.10 α-particle tracks showing deflection in a magnetic field (taken about 1920)

Fig. 42.11 (a) α-tracks and (b) β-tracks (in hydrogen) showing the difference in ionizing power of the particles

(a)

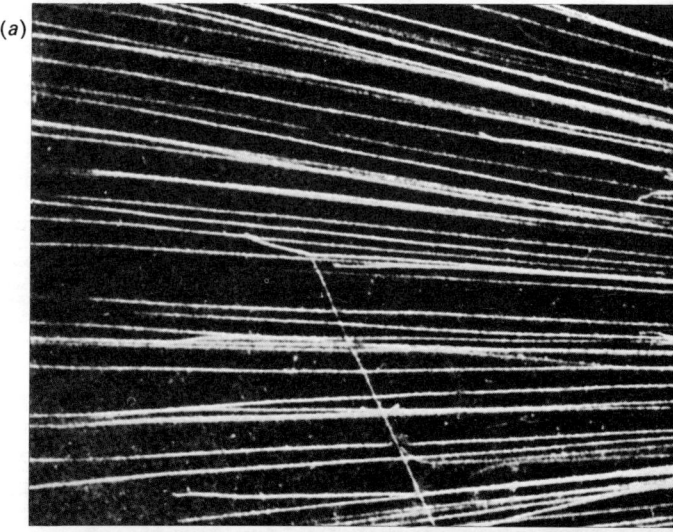

(b)

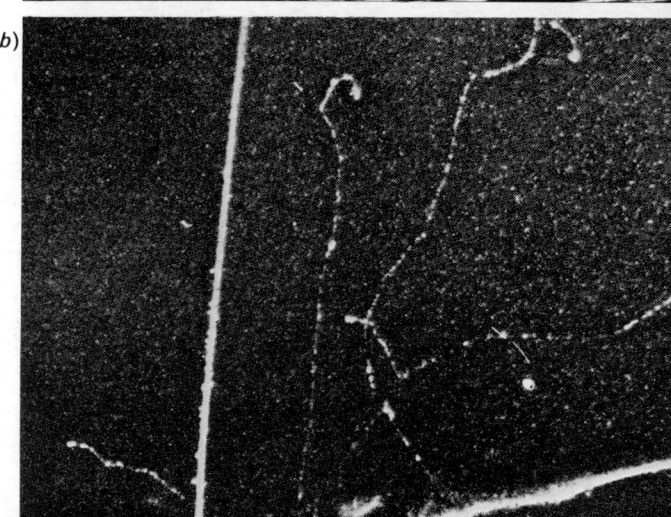

produced by the numerous droplets. The apparatus, called a '*cloud chamber*', used in the production and demonstration of these tracks of ionizing particles, is shown in Fig. 42.8. Fig. 42.9 is a photograph of Wilson's original cloud chamber made in 1911. The pressure in the chamber was suddenly reduced by the rapid lowering of its base. This was accomplished by allowing the air below the base to expand into an evacuated flask. Fig. 42.10 is a photograph of α-particle tracks taken by Dr Kapitza about 1920.* It shows the deflection of the particles in a powerful magnetic field. α-particles, being massive, produce a large number of ions along their paths, and so their tracks are thick and straight. β-particles, being much smaller, produce fewer ions and are deviated by the encounters. Their tracks are fine, irregular and short.

More recently other similar methods have been devised. The *condensation* chamber reveals ion tracks within a cooled and supersaturated enclosure. The *bubble* chamber consists of a large tank filled with liquefied nitrogen in which ion tracks leave a trail of minute bubbles in the superheated liquid. These newer chambers have advantages over the old form of cloud chamber, for the dense liquid helium reveals more complete tracks.

---

* Dr Kapitza worked for many years under Rutherford at Cambridge. He is now one of the leading physicists in the U.S.S.R.

Fig. 42.12 Bubble chamber at University of California, Berkeley

The chamber contains 675 litres of liquid hydrogen and is surrounded by a large magnet. It is used in conjunction with the Berkeley bevatron

Emulsion methods

Special photographic plates with a thick layer of emulsion, and solid blocks of emulsion, have been used for recording tracks of ions. The emulsion is exposed to the ions and then developed, thereby revealing tracks which can be seen under the microscope. Radiations outside the earth's absorbing layer of air have been studied by sending up emulsion blocks in balloons to high altitudes.

Radioactive decay

Radioactive decay consists of certain atoms undergoing a change by emitting a particle and changing into another substance. But not all the atoms change at once for, if they did, there would be a spontaneous emission of a large amount of energy. Thus in any radioactive specimen only a certain fraction of the atoms present will suffer a change in any second. Which particular atoms will change we cannot tell, for the choice is governed by *chance*. Some simple and interesting experiments can be performed to illustrate the laws of chance. Because we always deal with extremely large numbers of atoms, their behaviour obeys the laws of chance accurately. When we toss a coin, we know that the chances are even that it will turn up 'heads' or 'tails'. We also know that if the tossing is continued, heads and tails will not occur alternately. Although the chances are even, after

10 tosses, there may be 6 heads and 4 tails: after 1000 tosses, the numbers may be 552 and 448. The important point, however, is that the more tosses we make, the nearer we get to a 50-50 distribution. A similar kind of experiment can be carried out with dice. Throwing a large number of dice many times will reveal that, on average, about one-sixth of the dice turn up 1 (or any other number) at every throw. If we note down the number of dice at each throw showing the same number from 1 to 6, then the totals will become more nearly equal the more throws are performed.

These simple experiments show that when dealing with a chance happening a large number of counts should be taken. We know that by taking 4 counts instead of 1, we halve the deviation from the true value: the deviation is one-third by averaging 9 counts, and so on.

Consider again the statement that, in decay, a certain fraction of the number of atoms present suffers change each second. We must study a sufficient number of these changes, otherwise our counts will not be accurate. We shall see later how these counts are taken experimentally, but first let us see what happens in decay. If the number of atoms changing in one second is always the same fraction of the number of atoms present, the type of decay curve will be that shown in Fig. 42.13. The curve shows that half the atoms are disintegrating in one unit of time. Thus after the

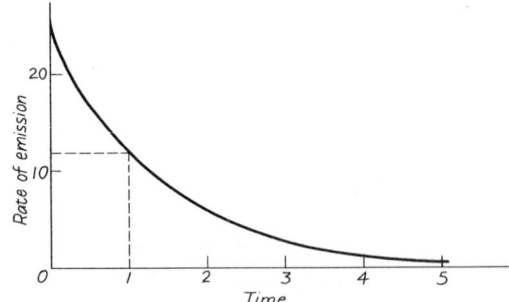

Fig. 42.13 Radioactive decay curve
Curve shows decay of a substance with half-life of unit time

lapse of one unit of time, only half the atoms will remain: after two units, only a quarter: after three units, only an eighth, etc. This graph is known as a *logarithmic* or *exponential curve* and it is a standard decay curve. It represents, for example, the decrease in a person's wealth if he spends a certain fraction of his capital each day.

The time taken for half the mass of a radio-active element to suffer disintegration is known as its *half-life*. For example, the half-life of radium is about 1600 years: hence 1 gramme of radium decays to 0.5 gramme in 1600 years and to 0.25 gramme in 3200 years. Of course, the radium does not disappear. It changes into its daughter products, as shown in the table below: the total mass does not change appreciably, for the loss in mass in the form of α- and β-particles is negligible.

The half-lives of the radioactive substances vary over a wide range; some are only a fraction of a second while others are thousands of years.

Logarithmic decay can be illustrated by a simple experiment. Water is allowed to run out of a long vertical glass tube through a small hole* in the bung at the bottom end. By plotting the height of the water in the tube against the time, a logarithmic curve is obtained. In this experiment the rate of effusion of the water varies as the head of water, which is, in turn, proportional to the quantity of water remaining.

A similar experiment can be performed with dice. A large number of dice are thrown and those showing 6 are counted and removed. The experiment is repeated with those dice remaining. A graph of *number remaining* after each throw against the *number of the throw* gives a log curve. It is

---

* A short length of capillary tube answers well. *See S.S.R.*, June 1960, page 439.

advisable to take the mean of a number of throws for each value. If the log of the number remaining is plotted instead of the actual number, a straight line graph is obtained. In these graphs the number of the throw corresponds to a time scale.

In radioactive disintegration, one substance changes into another which may also be radio-active, in which case it, too, disintegrates. In this way changes take place until a stable substance results. The table shows a radioactive series; active uranium eventually becomes stable lead. All the products remain in equilibrium with each other. Those with a long half-life are more plentiful than those with a short half-life.

A water experiment can be set up using two or more tubes, as described above, arranged so that the water runs from the first tube into the second and from the second into the third, etc. The rates of flow are made different by using draining holes of different bores. The water levels in the tubes eventually become steady (the level of water in the first tube has to be kept constant by supplying water at the appropriate rate). The head of water in each tube depends upon the rate of emptying, being smaller when this is greater. This state of dynamic equilibrium resembles that in a radio-active series when all the members of the family are in equilibrium. The mass of product present is large when its half-life is large. The half-life of a short-lived product can be estimated from the

THE URANIUM SERIES				
Element	Radiation	Atomic mass number	Atomic number*	Half-life
Uranium I	α	238	92	4.5×10^9 years
Uranium X$_1$ (Thorium)	β	234	90	24 days
Uranium X$_2$ (Protactinium)	β	234	91	1.14 min
Uranium II	α	234	92	2.5×10^5 years
Ionium (Thorium)	α	230	90	8.0×10^4 years
Radium	α	226	88	1622 years
Emanation (Radon)	α	222	86	3.82 days
Radium A	α	218	84	3.0 min
Radium B	β	214	82	26.8 min
Radium C	β	214	83	19.7 min
Radium C'	α	214	84	1.6×10^{-4} s
Radium D	β	210	82	22 years
Radium E	β	210	83	5 days
Radium F (Polonium)	α	210	84	140 days
Lead	(stable)	206	82	

* Atomic number is the number of protons in the nucleus and also the number of electrons circling the nucleus in the atom.

amount of it which is present when in equilibrium with its daughter products.

The idea of a radioactive series was first suggested by Rutherford and Soddy. As the table of the uranium series shows, some products must have the same atomic number but different atomic masses. Such were given the name *isotopes*. We have already seen that soon after this classification of radioactive products, J. J. Thomson showed the existence of isotopes of stable elements.

Carbon dating

The above principles are applied in the estimation of the age of historically important treasures and the technique employed is known as 'carbon dating'. Owing to the presence of neutrons in the atmosphere, a little nitrogen is changed into an isotope of carbon ($^{14}_{6}C$), *see* Chapter 43. This is a radioactive isotope with a half-life of 5600 years. An active plant absorbs some of this carbon-14 in the process of photosynthesis. Although the living plant may die, the carbon-14 continues to disintegrate at its constant rate in the decayed plant. A specimen of decayed wood or other organic matter can be tested and from its activity, measured with a G.M. tube and counter, its age can be estimated. This estimate assumes the rate of intake of carbon-14 to have remained constant over the years. The activity (counts per gramme) of the ancient specimen is compared with the activity of the new specimen. If counts were in the ratio of 1:2, then the specimen would be 5600 years old—the half-life of carbon-14. The rate of fall in activity is exponential.

Likewise the presence of radioactive material in rocks can lead to an estimation of their age. For example, the final product in the uranium series is lead. By measuring the relative mass of lead present to that of a suitable uranium product, the age of the rock can be estimated. For example, the half-life of uranium-238 is 4500 million years. By estimating the mass of uranium-238 to that of lead present in the same quantity of rock, the age can be calculated.

Measurement of radioactive decay—half-life

In order to demonstrate and measure half-life in the laboratory, it is convenient to choose an active substance having a short half-life. Thoron gas is suitable. An ionization chamber, as shown in Fig. 42.14 is attached to the pulse electrometer and the H.T. is applied across the central electrode and the outside of the case. The clips C_1 and C_2 are opened and, by means of the squeeze bottle, some thoron gas is introduced into the chamber. The amount used should be sufficient to produce

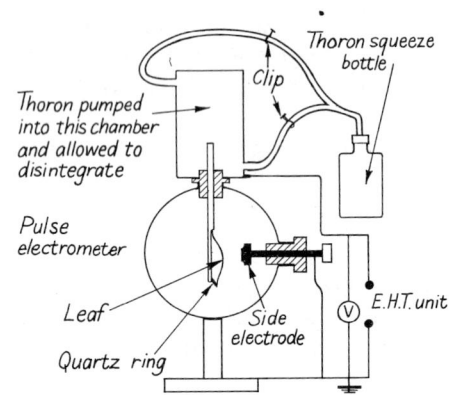

Fig. 42.14 Ionization and pulse electrometer for measuring half-life of thoron

a suitable pulse-rate. The times of successive pulses are noted and a graph of *number of pulse* against *time of pulse* is plotted. Readings are taken until the time between pulses, which progressively increases, is a few minutes. The graph

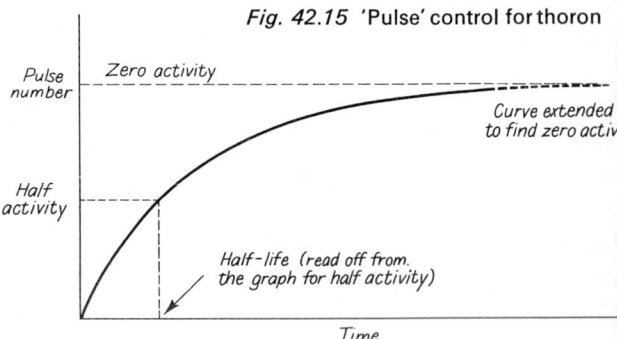

Fig. 42.15 'Pulse' control for thoron

has the shape shown in Fig. 42.15. It is an exponential curve and it approaches a maximum which is the pulse number at which activity ceases. The zero limit is obtained by extrapolation. The half-life is then read off from the graph as shown.

Range of a-particles in air

The range of a-particles in air can be measured using the pulse electrometer. For this experiment an ionization chamber is fitted on the top of the electrometer (Fig. 42.16).

The source is placed at S as shown and is directly connected to the leaf system of the electrometer. A p.d. is applied from an E.H.T. unit appropriate for the ionization chamber (about 3 kV). The telescopic lid is brought down to about

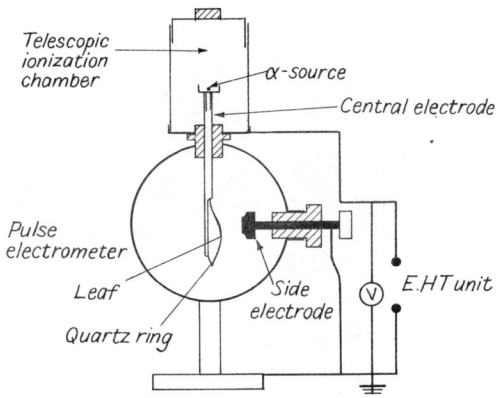

Fig. 42.16 Pulse electrometer with ionization chamber for measuring range of α-particles

1 cm from the source. This distance must be measured as accurately as possible, estimating the exact position of the source within its holder. When the E.H.T. unit is switched on, the electrometer should pulse at a slow rate. This rate is varied until it reaches a suitable value by adjusting the position of the side electrode. The pulse rate is then recorded and the count repeated for increasing separations of the lid from the source.

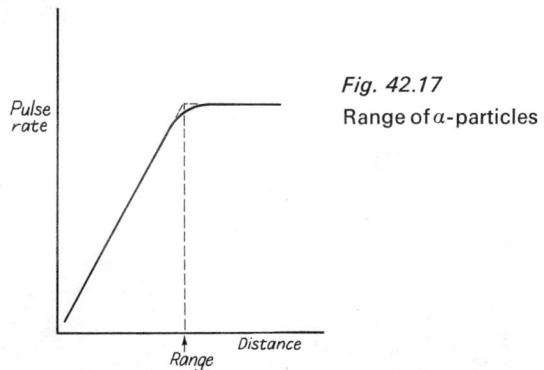

Fig. 42.17
Range of α-particles

A plot of the results is of the form shown in Fig. 42.17, which gives the range of the α-particles.

In this experiment increasing the distance of the lid from the source increases the working volume of the chamber. This increases the number ions participating and so the ionization current. When this becomes constant, it means that the α-particles cannot spread farther to ionize more air molecules and so their limit is denoted.

Magnetic deflection of β-rays

The deflection of β-rays in a magnetic field is easily demonstrated using a G.M. tube and ratemeter positioned as shown in Fig. 42.18. The stream of β-rays passing through the slit is arranged to pass through the second slit and to enter the

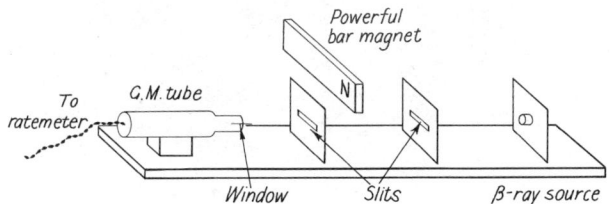

Fig. 42.18 Experimental arrangement for demonstrating deflection of β-rays

G.M. tube. This alignment is then slightly re-arranged so that the stream falls below the second slit. A bar magnet is brought up in a horizontal direction so that it creates a magnetic field perpendicular to the β-rays. When the magnet is at a particular distance, the ratemeter gives a high reading. Moving the magnet either nearer or farther away causes a reduction in this reading. This shows that the rays are deflected by the field.

A similar experiment can be done with α-rays, but a more elaborate arrangement is necessary. The rays have to be contained in a vacuum so that their range can be increased to allow for the production of a measurable deflection. The very strong magnetic field of a large electromagnet is also essential.

Measurement of absorption

The absorption of β-rays or of γ-rays is investigated by using a G.M. tube with ratemeter or with scaler. A source is mounted with the tube at a suitable distance so that the meter gives a nearly full-scale reading. Sheets of the absorbing material of known thickness are interposed between the source and the tube, and the new reading noted. Fig. 42.19 shows the apparatus arranged

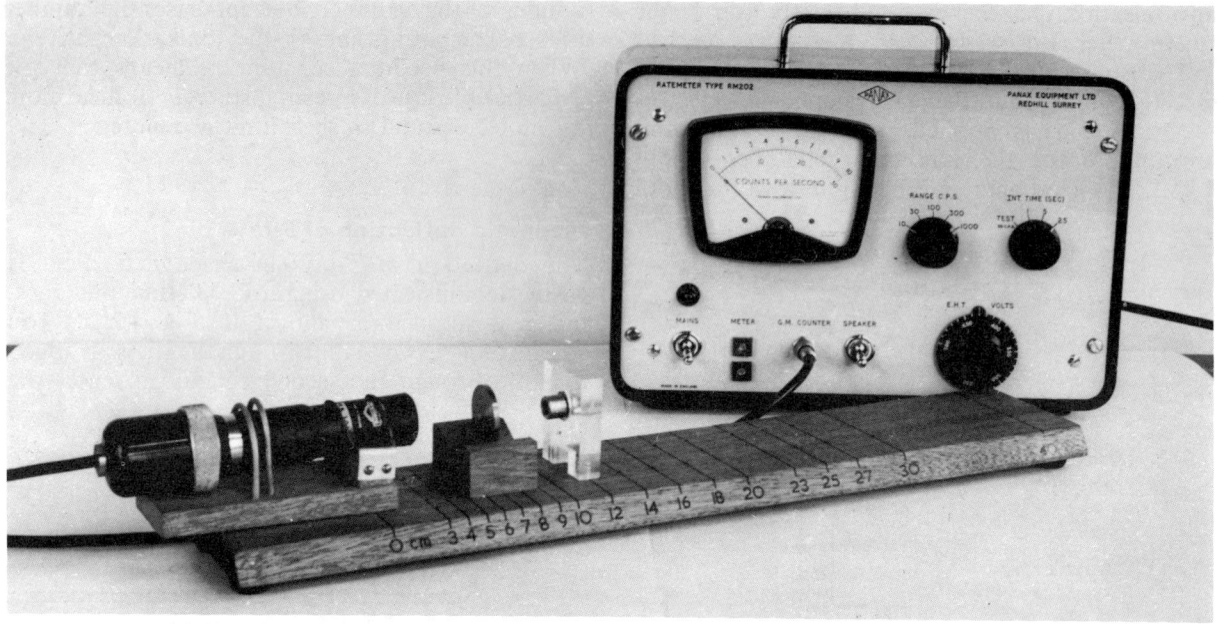

Fig. 42.19 Apparatus for laboratory experiments on radioactivity

This consists of a ratemeter connected to a Geiger–Müller tube. The arrangement permits of simple absorption experiments, and the investigation of the decrease in radiation with distance from γ-ray source

Fig. 42.20 (*below left*) Inside one of the 'hot cells'

A mechanic, wearing protective clothing, is adjusting equipment under the direction of the scientist in the background

Fig. 42.21 (*below right*) Mechanical handling of radioactive material

Radioactive materials in the two 'hot caves' shown here are handled by remote control. Each cave is a steel box surrounded by 0.9 m of concrete. The windows consist of glass tanks filled with zinc bromide solution

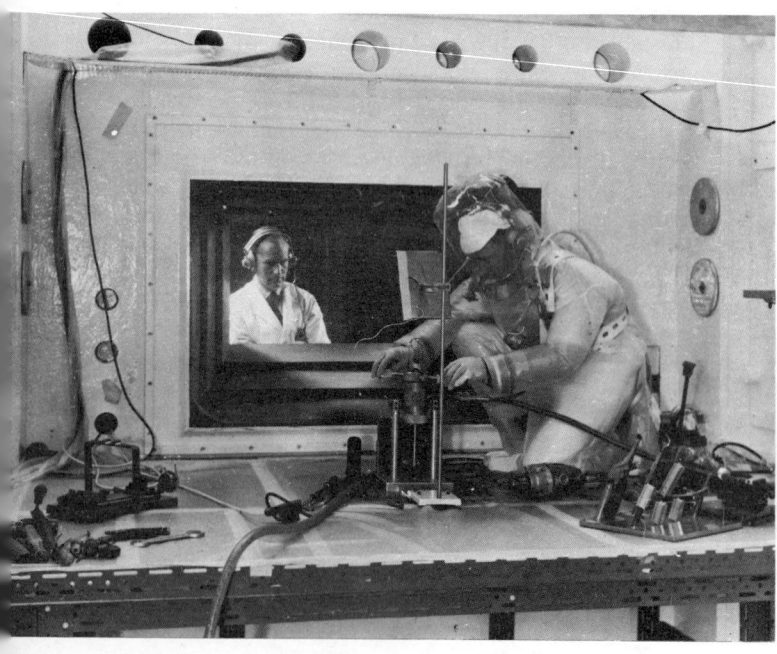

for such measurements. A plot of *meter reading* against *thickness of material* gives an exponential curve.

Variation of intensity of γ-rays with distance from source

As the distance from a source of γ-rays is increased, the intensity of the radiation decreases. This variation can be studied using the apparatus in the above experiment. In this case the distance between the source and the G.M. tube is measured and the corresponding ratemeter reading taken. A series of readings can thus be obtained. By plotting $1/\sqrt{(count\ rate)}$ against *distance*, a straight line is obtained. This shows that an inverse square law holds for the emission, i.e. the intensity of the radiation is inversely proportional to the square of the distance from the source. The graph should go through the origin if the *distance* plotted is precisely that between the source and the window of the G.M. tube. As the latter is usually protected by a gauze, the exact distance cannot be ascertained and, hence, the graph will show a small negative value for the distance when $1/\sqrt{(count\ rate)}$ is zero.

Safety precautions

Radioactive radiations and other radiations such as X-rays, are dangerous to the human body. They destroy cells and affect the blood.

There is no need to be afraid of performing experiments on these radiations provided simple precautions are exercised.* Such precautions include personal cleanliness and avoidance of the introduction of any radioactive matter into the body. α- and β-ray sources are easily shielded by a thin piece of metal and 6 mm thick slab of Perspex respectively. Powerful γ-rays can, however, penetrate many centimetres of lead. Such sources must be kept in suitable lead containers. It is important to remember that the inverse square law is applicable to a γ-ray source and so keeping some distance away from such a source is the first precaution. On no account must any source be handled: forceps must always be used.

---

* See *Safeguards in the School Laboratory* (John Murray for A.S.E.).

QUESTIONS

1. Describe how to investigate the properties of α-particles.

2. How can the deflection of β-particles in a magnetic field be shown experimentally?

3. What is a radioactive series? What information can be deduced from knowledge of the mass of each product present, when the products of a series are in equilibrium? Explain your answer.

4. Describe the method used by Wilson to make visible the tracks of ionizing particles. What other methods can be used to do this?

5. What precautions must be taken in dealing with radioactive substances? Why are some radioactive materials more dangerous than others?

6. Describe an experiment, giving full details of how to deal with the readings, to show that an inverse square law holds for the variation of the intensity of γ-radiation with distance from the source.

7. The activity of a specimen of the phosphorus radioisotope ^{32}P was measured every 10 days with a G-M counter and the following readings obtained.

Time (days)	0	10	20	30	40	50
Counts/s	200	124	77.0	47.8	29.6	18.4

From these readings plot the most suitable graph and from it deduce the half-life of the ^{32}P. Explain why you plot the readings in the way you do.

8. Explain how the half-life of a radioactive substance can be estimated from the mass present, when in equilibrium with other members of the series.

9. Successive products in a radioactive series emit α-rays and β-rays. What is the special connexion between the initial substance and the final product?

10. A radioactive isotope gives a corrected count of 200 counts/s on a ratemeter at 1 p.m. If the ratemeter reading is down to 50 counts/s at 5 p.m., what is the half-life of the isotope?

11. (a) Explain briefly how it has been established that the α-particle is (i) positively charged, (ii) very much heavier than the electron.

(b) A nucleus of radium (Ra) disintegrates by emission of an α-particle to form radon (Rn). Copy out and complete, by inserting subscripts and supercripts, the reaction involved:

$$^{266}Ra \rightarrow \,_{86}Rn + \,_2 He.$$

Where does the energy evolved in this radioactive disintegration come from?

(c) Explain how you would demonstrate that α-particles of a given velocity have a definite range in air at a given temperature and pressure. (O & C)

12. (a) What information about (i) the constitution, (ii) the stability of an atom of cobalt is contained in the statement '$^{61}_{27}Co$ is an electron emitter with a half-life of 1.65 hours'?

(b) What is the charge on the nucleus of the atom into which $^{61}_{27}$Co decays?

(c) Give the charge and mass number of a neighbouring isotope of cobalt.

(d) If a radioactive source contained 8×10^6 nuclei of $^{61}_{27}$Co at a given instant, how many of such nuclei would remain after 4.95 hours?   (O & C)

13. What is meant by (a) *radioactive decay*, (b) the *half-life* of a radioactive element?

A radioactive source, emitting alpha, beta and gamma radiations, is at the centre of a glass bulb. Describe the action of the radiations after they have left the source when the bulb contains (i) a vacuum, (ii) air. Describe how you would determine the range, in air, of the alpha particles from a given pure-alpha source.   (C)

14. Describe an apparatus by means of which gamma-rays from a radioactive material may be detected by a counting method.

Why do we refer to an 'average count' as something different from the actual count in a particular interval of time?

The average count from a radioactive preparation at a certain time is 256 counts per second. If the half-life of the material is 3 minutes, what will be the average count twelve minutes later?   (JMB)

15. (a) The following figures were obtained from the readings of a ratemeter for the α-particle emission from Thoron 220. Plot a suitable graph from the readings and deduce the half-life of the Thoron.

Time in seconds	0	20	40	60	80	100	120	140
Corrected counts per second	96	72	55	45	36	26	20	15

(b) Describe how radioactive radiations can be displayed by means of a cloud chamber.   (JMB)

16. Give full experimental details of how you would measure the variation in the intensity of γ-radiation with distance from a source. How would you plot your readings to show the relation between intensity of radiation and distance from source?

Could this method of investigation be used for β-radiation? Give reasons for your answer.   (JMB)

17. An α-particle of energy 8.5×10^{-13} J is emitted when a nucleus of polonium, Po, disintegrates to form a lead nucleus Pb, according to the relation

$$^{211}_{84}\text{Po} \rightarrow \,^4_2\text{He} + \text{Pb}.$$

(a) Deduce the atomic number of lead and the mass number of the isotope concerned.

(b) What happens to the energy which an α-particle originally possessed, as it passes through matter?

(c) If there are 6×10^{23} atoms per gramme atom, estimate velocity of emission of the α-particle.   (O & C)

18. Describe any *one* experimental arrangement which makes visible *either* the track *or* the impact of an alpha-particle. How would you use the arrangement to show that the emission of alpha-particles from a radioactive source occurs in a random manner?

What information concerning the atomic structure of radium is given by the symbol $^{226}_{88}$Ra? Radium (Ra^{226}) dis-integrates into radon (Rn^{222}) and an alpha-particle; express this in the form of an equation, using the symbols of the type given in the previous sentence.

What is the source of energy emitted in the disintegration? (Atomic mass: Ra = 226.05; Rn = 222.04; He = 4.003.)   (C)

19. Give an account of an experiment which shows the deflexion, in a magnetic field, of beta-particles from a radioactive source. Explain how the results of such an experiment indicate (a) the charge of the charged particles, (b) the fact that the individual particles have different speeds.

An element of atomic number 11 and mass number 24 disintegrates by beta-particle emission. State the atomic number and mass number of the new element formed; state, giving your reason, whether or not the new element is an isotope of the original element.   (C)

20. Describe briefly a simple experiment or observation which shows the *random* nature of emission from a radio-active substance.   (C)

21. What effect does a transverse magnetic field have on narrow beams of (a) γ-rays, (b) β-rays, (c) α-rays from a radioactive source, as they pass through an evacuated box? What conclusions can be drawn from observations of these effects?   (O & C)

22. An atomic nucleus A is composed of Z protons and N neutrons. What will be the composition of a nucleus B left when A emits an α-particle and of a nucleus C left when B emits a β-particle? What can we say about the masses of A, B and C?   (O & C)

23. Mention three characteristics of the alpha-particles emitted by a particular radioactive source. How do they differ from those of beta-particles from a particular source?

$$S. \qquad \mid D$$

S represents a small, very thin alpha-particle emitter and D is a section through the sensitive area of a detector which counts every alpha-particle reaching it from the source. Explain qualitatively how the number of counts per second changes when S is moved from a position close to D to one which is about 10 cm away (a) when S and D are in a vacuum, (b) when S and D are in air.

The range of the α-rays from S in air is 5 cm.   (O & C)

24. What is an *alpha-particle*, and what is meant by the *range* of an alpha-particle in a gas? Would you expect the range to depend on the pressure of the gas? Give your reasoning.

Describe any *one* experimental arrangement which makes visible *either* the track *or* the impact of an alpha-particle. How would you use the arrangement to determine the range of the particle?

Give a brief account of the effects of bombarding a thin metal foil, in a vacuum, with alpha-particles. What deduc-tions have been made from a study of these effects?   (C)

25. What is meant by a *radioactive element*?

State briefly the chief properties of alpha, beta and gamma radiations.

Explain the changes in the structure of an atom which result from the emission of (a) an alpha-particle, (b) a beta-particle. The element formed after the emission of an alpha-particle from radon (atomic number 86, mass number 222) is an *isotope* of polonium. Explain the meaning of the term isotope in this connection.   (C)

26. Describe the experimental arrangements you would use in order to investigate the effects produced by passing a fine beam of beta-particles from a radioactive source through:

(*a*) a magnetic field perpendicular to the direction of the beam,

(*b*) various thicknesses of thin metal foil.

Include a reference to the method of obtaining a fine beam of particles, and a description of the detecting device employed.

State the observations you would expect to obtain and discuss the conclusions you would reach from these results.

Comment on the fact that beta-particles, which usually possess higher velocity but lower energy than alpha-particles, also possess a longer range of penetration in air than beta-particles. (C)

27. The reading of a detector of radioactive radiation falls from an average rate of 1600 per minute to an average of 400 per minute in a period of 30 minutes. What is the half-life period of the radioactive source? In what *further* period of time would the average rate fall to 100 per minute? (L)

28. Carbon-14 is an *isotope* of carbon with a *half-life* of 6000 years and it decays by *β-emission*. What meaning is attached to the words italicized and to the number 14?

What is the number of ^{14}C atoms out of a sample now containing 10^6 atoms which will have disintegrated in 12 000 years from now? (O & C)

29. Outline *one* way of detecting alpha-particles and describe, so far as you can, how it works.

The counting rate recorded by a detector fixed in front of an alpha-particle emitter is 256 per s. This figure is an average rate worked out from a count lasting for several minutes. What is the average counting rate 20 days later, for the same arrangement, if the half-life of the emitter is 5 days?

If the number of alpha-particles were recorded at this time

for 1 s precisely, would you expect to find the number you have just calculated? Give a reason for your answer. (O & C)

30. A Geiger–Müller tube is connected to a counter which records the entry of ionizing particles in the tube. In successive minutes the counter records the arrival of the following numbers

4   3   4   5   3   5   (Average=4)

When a radioactive source is placed at a fixed distance from the Geiger–Müller tube the counts in successive minutes are

96   95   92   92   90   88   (Average=92)

Suggest the causes for each of these two sets of numbers, and state whether you consider them to show random emission, faults in the counter, radioactive decay, or some other cause.

At later times the average count for six successive minutes is as given in the following table:

Later time/min	30	60	90	120	150
Average count	71.7	55.5	43.8	34.8	27.5

Use this information, with the aid of a graph, to find the half-life of the source.

Give any *one* reason why it is advisable to continue observations of the counter for a while, with the source removed. (C)

31. What is meant by (*a*) *radioactive decay*, (*b*) the *half-life* of a radioactive element?

Thoron is a radioactive gas with a half-life of about 1 minute. When some of the gas is put in a cloud chamber, the tracks of alpha-particles can be detected. How would you use this fact to determine the half-life of the gas? Illustrate your answer by reference to the types of observation you would expect to make if, on first observing tracks in the chamber, there were 14 tracks visible.

What changes take place in the atomic structure of thoron (atomic number 86, mass number 220) if it emits an alpha-particle and then, later, another alpha-particle? (C)

43 Structure of atoms

Atomic models

We have seen in Chapter 41 how the idea of atoms being indivisible spheres was dispelled by Thomson's discovery of the electron. A new conception of an atom was then necessary and Thomson's first atomic 'model' (i.e. mental picture) consisted of a sphere of positively charged material with fixed electrons uniformly distributed within it. This model was known as Thomson's 'plum-pudding' atom, for the electrons were like currants in a spherical pudding.

Atomic structure is usually studied along with radioactivity because the topics are complementary. It was from experiments on radioactivity that Rutherford and his co-workers were able to discover much about the structure of atoms.

The scattering of α-particles

In 1911, Rutherford, Geiger and Marsden investigated the scattering of α-particles by thin metallic foils. In their experiment, illustrated in Fig. 43.1, they passed α-particles through gold foil.

The particles, on emerging from the gold foil, were caught on a zinc sulphide screen mounted on a microscope. The apparatus was contained in an evacuated enclosure so that the α-particles had a long range. It was found, however, that some particles did not pass straight on but were deflected on each side of the main beam. The numbers striking the screen in various directions

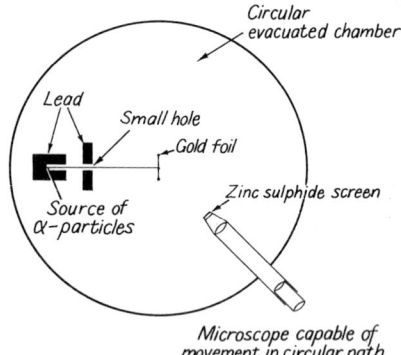

Fig. 43.1 Scattering of α-particles
α-particles scattered by thin gold foil are collected on a zinc sulphide screen

were counted by observing the scintillations. Complete investigation showed that some particles were even sent backwards, i.e. deflected through nearly 180° (Fig. 43.2). On studying this result critically, Rutherford came to the very important conclusion that the atoms in the foil must consist largely of empty space and have a small but heavy

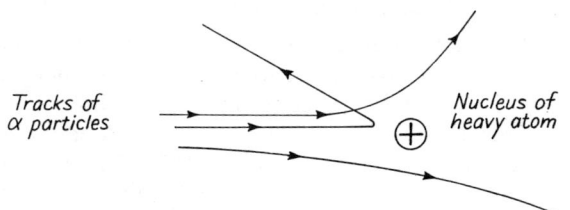

Fig. 43.2 Scattering of α-particles

The nearer the line of approach is to the centre of the nucleus the greater the deflection

central core or nucleus (Fig. 43.3). This must carry a positive charge to neutralize the negative charge carried by the electrons which surround

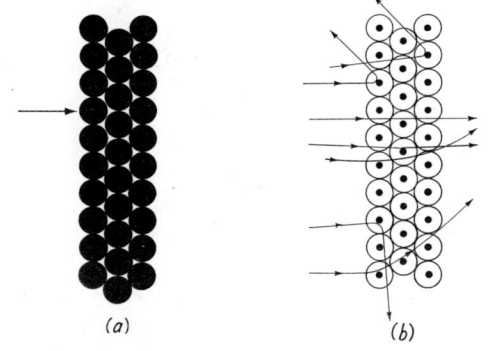

(a)   *(b)*

Fig. 43.3 Solid and nuclear atom models

(*a*) 'solid' atoms would not permit passage of α-particles
(*b*) Nuclear atom permits passage and illustrates scattering of α-particles

the nucleus. The positively charged α-particles were repelled by the positively charged nuclei of the gold atoms in the foil.

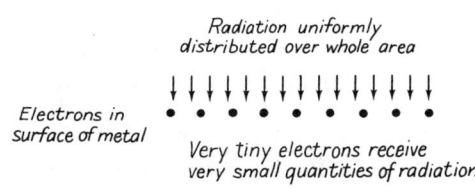

Radiation uniformly distributed over whole area

Electrons in surface of metal

Very tiny electrons receive very small quantities of radiation

The nuclear atom

These experiments led Rutherford in 1911 to suggest that the atom resembles a miniature solar system, and this is very much the picture of the atom which we have today. At the centre is the heavy positively-charged nucleus and round it circulate electrons. The total charge carried by the electrons is equal and opposite to that on the nucleus, so that the atom as a whole is neutral. If an orbital electron is detached, the remainder of the atom has a net positive charge and is then known as a *positive ion*.

The hydrogen atom, the simplest of all, consists of a nucleus bearing a positive charge and one orbiting electron bearing an equal negative charge. This special nucleus is presumed to contain a single particle which has been given the name of *proton*.

The quantum theory

A further and decisive development of Rutherford's atom came in 1913 when Niels Bohr, then a research student working under Rutherford, applied the quantum theory to the atom. This theory, proposed by Planck in 1901, states that energy exists in discrete units or *quanta*. It had been advanced to explain radiation from hot bodies, but had attracted little attention until Einstein in 1905 applied it to explain the photoelectric effect. This effect has been described in Chapter 41. The liberation of electrons from metals by radiation would take a very long time assuming the radiant energy to be uniformly distributed over the surface of the metal: the tiny electron would receive an inconceivably small share. By assuming the radiation to be in quanta, an impact of a quantum on an electron would result in the absorption of the quantum. In consequence, the electron would be ejected from the metal. The quantum of energy would be used to liberate the electron and what was left over would appear in the form of kinetic

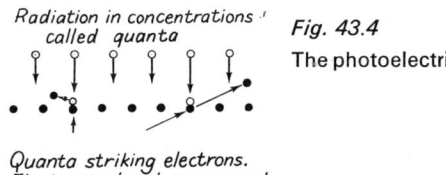

Radiation in concentrations called quanta

Fig. 43.4
The photoelectric effect

Quanta striking electrons. Electrons absorb energy and are ejected from metal.

energy of the electron (Fig. 43.4). In applying the theory to Rutherford's atom, Bohr gave the electrons precise orbits. These were confined to concentric spheres or shells about the nucleus and were called *stationary states*. Each shell could contain up to a certain number of electrons. The movement from one shell to another was possible, but no orbits could exist in addition to those in the shells. If one electron by some means or other (such as by being struck by another atomic particle or by heating) was knocked from an inner to an outer shell, another electron fell from a higher shell to take its place. The transition from a higher orbit to a lower one causes the loss of potential energy. This constitutes a quantum of energy which is emitted in the form of radiation. By considering the radiation emitted in this way by a hydrogen atom, Bohr derived values for the wavelengths of the bright lines in the hydrogen spectrum. The agreement with the measured wavelengths was fantastically close and so Bohr firmly established the quantum theory and the dynamic atom of Rutherford. Since 1913, the quantum theory has been extended and applied to numerous problems in physics.

The complete hydrogen spectrum is made up of three series—the Lyman, Balmer and Paschen series. Bohr's theory explained these series according to the transition shown in Fig. 43.5. The Balmer lines are in the visible region, the Lyman in the ultraviolet and the Paschen in the infrared.

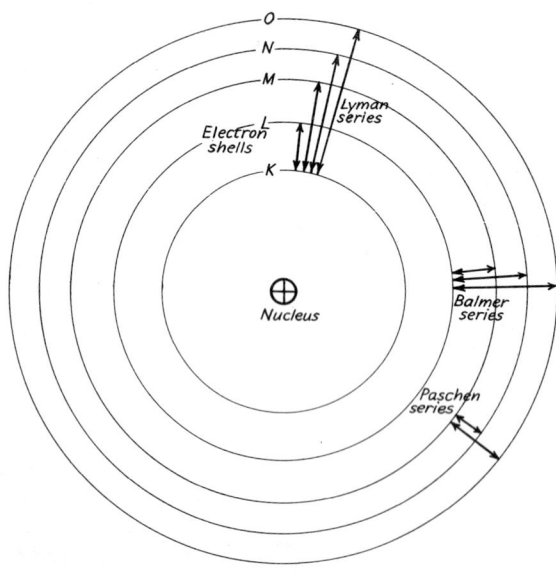

Fig. 43.5 Diagram showing possible transitions of electrons from higher to lower shells. The energy released in each transition is *hf*, where *h* is Planck's constant and *f* the frequency of the radiation emitted. The Lyman, Balmer, and Paschen series are groups of lines in the spectrum of hydrogen

Deflection—excitation—ionization

The electrons in an atom in a normal state are in orbitals which possess the lowest potential energy. When an atom is subjected to bombardment by electrons, there are three possible effects depending on the energy and direction of the bombarding electron. A near 'miss' causes the electron to be *deflected* due to repulsion by the orbiting electrons

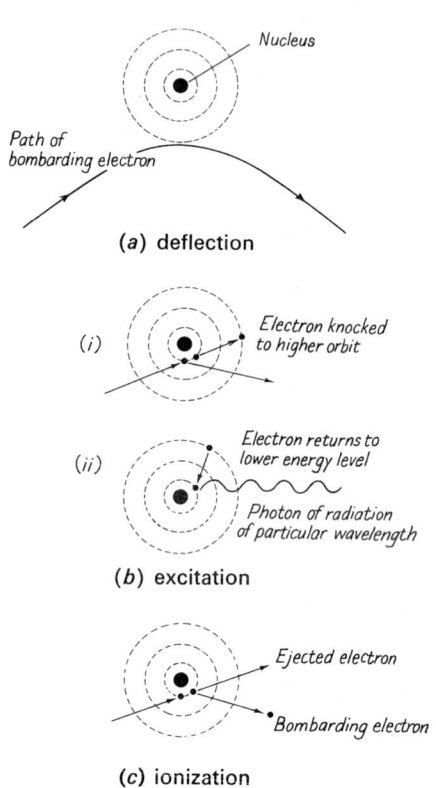

Fig. 43.6 Electron bombardment

(Fig. 43.6*a*). A direct 'hit' may knock an electron within the atom into a higher orbit or it may, if severe enough knock it clear of the atom. The first alternative is called *excitation* (Fig. 43.6*b*). The potential energy of the electron is greater in the higher orbit—energy is absorbed from the bombarding electron. The atom, however, immediately adjusts itself to regain its normal state. The electrons regain their original states by one or more moves. In doing so they lose potential energy which they give out in the form of radiation. Each fall from a higher to a lower orbit produces a pulse of radiation of a size dependent upon the particular transition. These 'pulses' of radiation are called *photons*, and are quanta of energy in the form of radiation.

When the electron within the atom is knocked out by the bombarding electron, the atom is left positively charged. This effect is *ionization* (Fig. 43.6c).

Orbitals and the periodic table

The limitations of Bohr's great achievement were quickly appreciated. The same principles could not readily be applied to more complicated atoms and molecules. However, by assigning to the elements a charge on their nuclei equal to their atomic number (which also gives the number of orbital electrons), a definite step forward resulted, in that the orbital electrons could be arranged in a manner similar to Mendeleeff's grouping.

Each electron shell or stationary state cannot contain more than a certain number of electrons. The maximum numbers for successive shells, going outwards from the nucleus, are 2, 8, 18, 32. Hydrogen has 1 electron in the first shell and helium has 2: this completes the shell. Lithium, with 3 electrons, has 2 in the first shell and 1 in the second. Successive atoms, in the order in which they appear in the periodic table, have 1 more each in the second shell until neon with 10 electrons—2 in the first shell and 8 in the second—completely fills it. The third shell is then built up in successive stages. The chemical affinity of a substance is determined by the outer electrons of an atom. When a shell is completed, an atom does not so readily lose or acquire an electron and so atoms with completed outer shells of electrons do not easily enter into chemical change. Such atoms are the inert gases, having atomic numbers 2, 10, 18, 36, 54 and 86. Atoms which require one electron to complete an outer shell display strong chemical affinity for those atoms which have one electron only in their outer shell. In this way the Rutherford-Bohr atom explains the classification of the elements and their positions in the periodic table.

X-ray spectra

Support for this classification came from another research worker in Rutherford's team, namely H. G. Moseley. As has been mentioned in Chapter 41, he investigated the characteristic X-ray spectra of metals and found that they followed a particular sequence. He showed this by plotting the graph of the energy of the X-rays against the square of the atomic number, which gave a straight line. These experiments defined the order of the elements and, moreover, they established the importance of the atomic number in the atom model. It represented the number of protons in the nucleus.

Theory of relativity

In 1905, Einstein put forward the theory of relativity. This theory is mainly concerned with bodies travelling at very high velocities such as the electrons orbiting about the nucleus. The full appreciation of the theory is much beyond the scope of this book, but one important result which we have to use is quite simple to comprehend. Einstein conceived the idea that matter, when moving, increases in mass. The increase is very small unless the velocity is approaching the velocity of light. A moving body possesses kinetic energy and so Einstein concluded that energy and mass are different forms of the same thing: a certain amount of energy is equivalent to a certain mass. This is expressed by the relation $E=mc^2$, where E=energy, m=mass, and c=velocity of light. For example, if 1 kg of matter were annihilated, energy equal to $1 \times (3 \times 10^8)^2$ J would be generated.

Wave mechanics

Although the application of quantum ideas to the atom model outlined above was enthusiastically received, certain aspects worried many scientists. An electron moving in a circular orbit is subjected to a constant acceleration and, according to classical mathematics, should be emitting energy continuously. Hence, the idea of stationary states could not be satisfactorily explained. However, the Rutherford-Bohr concept of the atom did enable scientists of the 1920s, as it does the beginner of today, to get a simplified picture which can be modified in the light of new discoveries.

The equivalence of mass and energy, stated in the preceding section, leads to the equivalence of mass and radiation, which is a form of energy. Thus we have a new idea, that particles are equivalent to waves and waves to particles. This dual state has led to a new mathematical treatment called *wave mechanics*.

With these ideas the stationary states within the atom can be more satisfactorily explained, but no longer can we think of electrons as definite particles. We now think more of a 'cloud' of charge in a stationary state.

A stream of electrons can, under certain conditions, display wave properties and the stationary states bear some resemblance to the nodes produced by stationary waves. (*See* Chapter 20.)

Transmutation

Many scientists of the past had an idea that there might be a common unit or brick which is a constituent of all atoms. The idea received a setback when it was realized that many atomic weights were not whole numbers. However, Thomson's discovery of isotopes of non-radioactive elements overcame this objection.* The unit brick was taken to be the nucleus of the simplest atom, viz. hydrogen. The hydrogen atom was visualized as having a simple nucleus, the proton, with one orbiting electron.

Radioactivity provides many examples of one element spontaneously changing into another by emitting *a*-particles. In 1919, Rutherford succeeded in accomplishing such a change or 'transmutation' from one element to another artifically. He used the apparatus shown diagrammatically in Fig. 43.7. The enclosure contained nitrogen at low

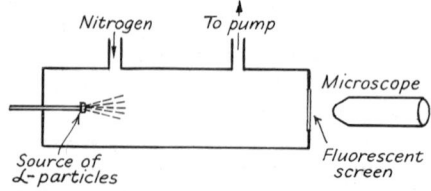

Fig. 43.7 Rutherford's transmutation experiment

pressure. The pressure was reduced to increase the range of the *a*-particles emitted from the source. Scintillations on the screen at the end of the enclosure were viewed with a microscope. The

---

* E.g. chlorine, with atomic mass 35.5, was found to consist mainly of isotopes 35 and 37 in the ratio 3:1.

range of the *a*-particles was about half the length of the enclosure, but scintillations were observed

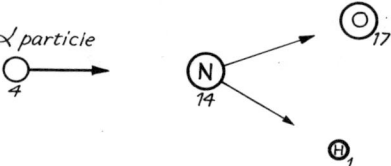

Fig. 43.8 Transmutation of nitrogen

on the screen. Rutherford proved these to be due to protons (Fig. 43.8). Thus some *a*-particles had struck nitrogen nuclei so hard that a change to

Fig. 43.9 Cockcroft and Walton's apparatus for the artificial disintegration of the elements

oxygen and hydrogen (proton) had resulted. (The equation is given on page 421.) This was not a transmutation of a measurable mass, but of comparatively few atoms.

The next step forward came when Cockcroft and Walton caused transmutation by producing their own missiles instead of using those naturally provided by radioactive sources. Such sources only emit particles with speeds up to definite limits. Cockcroft and Walton generated a supply of hydrogen ions (protons) in a discharge tube and then accelerated them by using a powerful electric field. The p.d. applied across the electrodes was 400 000 V. The fast-moving protons struck a lithium target (see Fig. 43.10), and the reflected

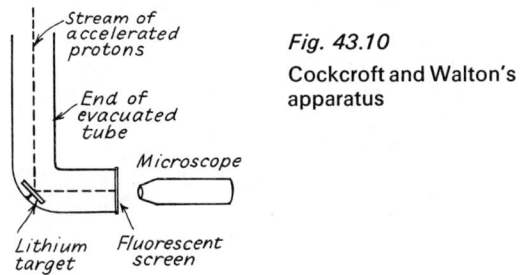

Fig. 43.10

Cockcroft and Walton's apparatus

rays were examined using a fluorescent screen. a-particles were detected which showed that, in some cases, a proton had united with a lithium atom and yielded two a-particles. (See equation on page 421.)

The neutron

Another decisive discovery came in 1932. Chadwick allowed a-particles to pass first through a thin layer of beryllium and then through paraffin wax (Fig. 43.11). The emergent radiation was received by a detector and found to consist of protons. The radiation between the beryllium and the paraffin

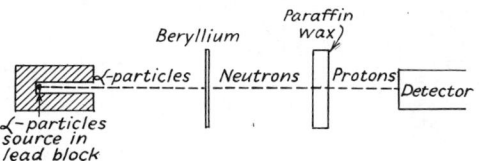

Fig. 43.11 Chadwick's neutron apparatus

wax was, however, an unknown one. Chadwick subsequently showed it to be a stream of particles each having a mass almost equal to that of a proton but carrying no charge. The existence of such particles had been predicted but, until then, not detected. They were called *neutrons*.

Constitution of the nucleus

Early attempts to build up a picture of the nucleus assumed that it contained both protons and electrons, but there were certain objections to this. Now, with the advent of another fundamental brick, a new design was possible and this we still retain today. Atomic nuclei contain protons and neutrons.* The forces which bind them together are still not completely understood. The number of electrons surrounding the nucleus is equal to the number of protons within the nucleus and gives the atomic number Z of the element. The atomic mass number A, which is always a whole number, is equal to the number of protons plus the number of neutrons. Thus the lithium nucleus, mass number 7, atomic number 3, contains 3 protons and 4 neutrons: surrounding the nucleus are 3 electrons. Uranium, mass number 238, has 92 protons and 146 neutrons (Fig. 43.12).

$$A = Z + N$$

where N = number of neutrons in nucleus.

---

* Protons and neutrons are collectively termed 'nucleons'.

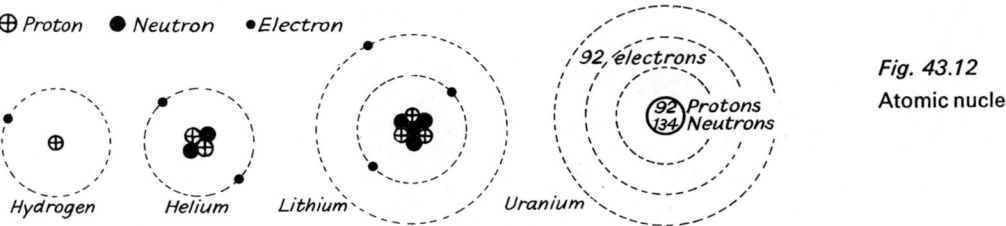

Fig. 43.12

Atomic nuclei

Fig. 43.13 The 390-cm cyclotron at Liverpool University

The helium nucleus (α-particle) comprising 2 protons and 2 neutrons, is an exceptionally stable and tightly bound group.

The nucleus of a radioactive atom is upset by the emission of α-particles and a rearrangement of its constituents may cause a neutron to be transformed into a proton and an electron. This change results in the emission of the electron and may also cause the emission of energy in the form of a quantum of radiation—a γ-ray.

Since 1932 when Cockcroft and Walton performed their experiment, tremendous efforts have been made to produce more and more powerful missiles to cause the fracture of nuclei. These atom-smashing machines employ potentials of a few million volts, such as are generated with large Van de Graaff generators. (*See* page 300 and Fig. 32.19.) The American scientist Lawrence devised the ingenious idea of generating high-velocity particles by sending them on a spiral path of increasing radius and giving them an urge forward twice every revolution. The apparatus for this purpose, called a *cyclotron*, incorporates a powerful electromagnet (*see* Fig. 43.13). From the cyclotron have been developed other accelerators such as the synchrotron and the bevatron (Fig. 43.14). The energy of the particles accelerated by these machines is measured in *electron-volts*.

The electron-volt is the energy acquired by an electron accelerated by a potential difference of one volt.

It is equal to 1.6×10^{-19} J.

The Berkeley bevatron (Fig. 43.14) produces particles with energies up to 6000 MeV (million electron-volts). The synchrotron developed by the Centre Européen pour la Recherche Nucléaire (CERN) produces 30 000 MeV.*

Fast-moving charged particles are often used to produce rapidly moving neutrons. These are the best missiles for, carrying no charge, they are not repelled by nuclei they are about to strike.

---

* A 200 000 MeV machine is planned for 1974 near Chicago and one producing 300 000 MeV is being considered by CERN for the future.

Fig. 43.14

General view of bevatron at University of California, Berkeley

Representation of radioactive changes

We can represent the changes we have been discussing by equations and we employ standard signs. The number attached to the upper left of the symbol, gives the mass number; the number to the lower left, the nuclear charge or atomic number: e.g. hydrogen ^1_1H (mass number 1, nuclear charge 1), lithium ^7_3Li (mass number 7, nuclear charge 3), the neutron ^1_0n (mass number 1, nuclear charge zero), the electron $^0_{-1}e$ (mass number zero, nuclear charge -1).

The transmutations of Rutherford, of Cockcroft and Walton (page 419), and of Chadwick (page 419) are represented respectively by,

$$^{14}_7N + {}^4_2He \rightarrow {}^1_1H + {}^{17}_8O$$

$$^7_3Li + {}^1_1H \rightarrow {}^4_2He + {}^4_2He$$

and

$$^9_4Be + {}^4_2He \rightarrow {}^{12}_6C + {}^1_0n$$

Mass changes

Up to the present we have used whole numbers for mass numbers, but it is now necessary to be more precise. Consider the transmutation represented in the second equation above,

	*atomic mass units**
Mass of lithium nucleus	$=7.0165$
Mass of proton	$=1.0076$
Total mass before transmutation	$=8.0241$
Mass of helium nucleus	$=4.0028$
Total mass after transmutation	$=8.0056$
Loss in mass	$=0.0185$

Thus the transmutation of one lithium nucleus causes the annihilation of 0.0185 atomic mass units, which is equal to $0.0185 \times 1.66 \times 10^{-27}$ kg or 3.07×10^{-29} kg. By applying Einstein's equation, $E = mc^2$, the energy liberated per single transmutation can be calculated.

It should be noted that in the transmutations described above, the actual amount of energy released during the experiments was *very* small, since only a few nuclei were affected. For the release of an appreciable amount of energy, some other method which affects vastly more nuclei is required.

---

* The carbon isotope $^{12}_6C$ is taken as the standard and equal to 12 atomic mass units. Hence 1 amu $=1.66 \times 10^{-27}$ kg.

Fission

Neutrons are effective missiles. They collide with many more nuclei than do positively charged missiles, such as a-particles, because they are not deflected away from the charged nuclei.

Transmutation experiments before 1939 produced new substances by knocking fragments, protons or a-particles, out of heavier nuclei. In 1939, Meitner and Frisch explained an experiment performed by Hahn and Strassmann as a splitting up of uranium-235 into two nearly equal pieces (barium and krypton) through the impact of a neutron. This effect is called *fission*. It was found that this fission was accompanied by the release of three neutrons in addition to γ-radiation. Further, there was a loss in mass resulting in a release of energy.

Here was something quite new—a particle of low energy releasing a large amount of energy, together with three neutrons each capable of repeating the action.

It is then possible to calculate the mass of a piece of uranium-235 (an extremely rare isotope of uranium) in which at least one of the three neutrons emitted by fission would be certain to hit, and to trigger off, another uranium nucleus. If such a mass of uranium were collected, the action of a single neutron would start a process in which the fission of the nuclei would take place at an ever-increasing rate (*see* Fig. 43.15). This is

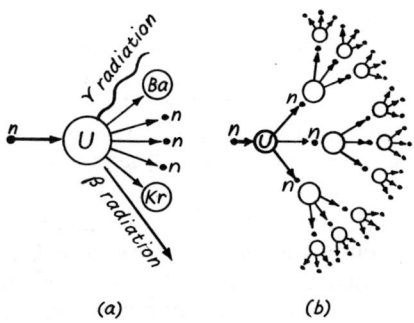

Fig. 43.15 Fission of uranium-235

(*a*) products of fission
(*b*) chain reaction resulting from production of three neutrons at each nuclear fission

called a *chain reaction*. In such a process a tremendous quantity of energy is released in a very short space of time.

Neutron + uranium-235 →

$$barium + krypton + 3\ neutrons + \gamma\text{-radiation} + energy$$

This was the principle underlying the first atomic bomb in which two pieces of 235U, each too small to produce a chain reaction, were brought together to form a mass large enough to ensure a chain reaction.

The nuclear reactor

Control of a chain reaction is necessary for it to be of practical use. Neutrons can be slowed down by allowing them to collide with *light* atoms, for such collisions simply cause the neutrons to lose some kinetic energy without producing any nuclear changes. A suitable substance for this purpose is called a *moderator*. The second isotope of hydrogen, deuterium ^2_1H, is suitable. Hydrogen ^1_1H will not do for it absorbs neutrons to become deuterium. Heavy water, that is water formed with deuterium ($^2_1H_2O$), is also a suitable moderator.

Very few 238U nuclei suffer fission when bombarded by neutrons, for most take up the neutron to become plutonium or 239U. Ordinary uranium metal contains 0.07 per cent of 235U. When it is bombarded by neutrons, some of the neutrons cause fission of 238U, some produce plutonium, some produce fission of 235U, which gives energy and releases more neutrons, and the remainder (the majority) do nothing. When ordinary uranium is enriched with 235U, a state can be reached when neutrons are being produced by fission of 235U at the same rate as they are being used. This critical state can be achieved by enclosing the uranium in a moderator. The first controlled moderator unit was built in 1942 in the U.S.A. (Fig. 43.16). This 'pile', as it was called, consisted of a huge sphere of graphite 5 m in diameter in which were embedded lumps of uranium. Cadmium rods were used as moderators: when some were removed, chain reactions commenced, and the temperature within the pile increased until it was

checked by further adjustment of the rods. In the large nuclear generating stations which are now dotted about the country, the heat generated by a controlled reactor supplies the steam necessary to rotate the turbines and, in turn, the generators.

Fusion

Nuclear fission is produced by the bombardment of heavy nuclei with suitable particles, but nuclear reactions can also be produced at extremely high temperatures, such as exist inside the sun and other stars. (The sun's temperature at its centre is estimated to be 15×10^6 K.) These are called *thermonuclear reactions*, and they result in the building up of heavier nuclei from lighter ones. This type of transmutation, which is termed *fusion,* occurs more easily between light than between heavy nuclei.

Fusion can be accompanied by a loss in mass and so a liberation of energy. The most important example of fusion is the synthesis of helium nuclei from hydrogen nuclei. This process takes place in stars and is the source of their energy. For equal masses about 500 times more energy is generated in the production of helium than in the fission of uranium.

The mechanism of the synthesis of helium by fusion is not definitely established, for it may be by the combination of four protons or two deuterons (deuterium nuclei) or even by some more complex method. As an example, we can take the fusion of two deuterons

$$^2_1H + ^2_1H \rightarrow ^4_2He + energy$$

Initial mass	$= 2 \times 2.0142 = 4.0284$ atomic mass units
Final mass	$= 4.0028$ amu
Loss in mass	$= 0.0256$ amu
Energy liberated	$= 23.8$ MeV $= 38 \times 10^{-13}$ J

Experiments are taking place to try to produce fusion directly in the laboratory. At Harwell, some

Fig. 43.16 The first nuclear reactor

This drawing shows the world's first nuclear reactor (pile) which came into operation in Chicago in December 1942. It was the result of a combined effort by scientists from Canada, the USA and Great Britain. In 1943 it was removed to the Argonne National Laboratory

Fig. 43.17 Calder Hall
The world's first nuclear power station, opened in 1965 Fuel elements for Calder Hall reactor

Fig. 43.18 Fuel elements for Calder Hall reactor

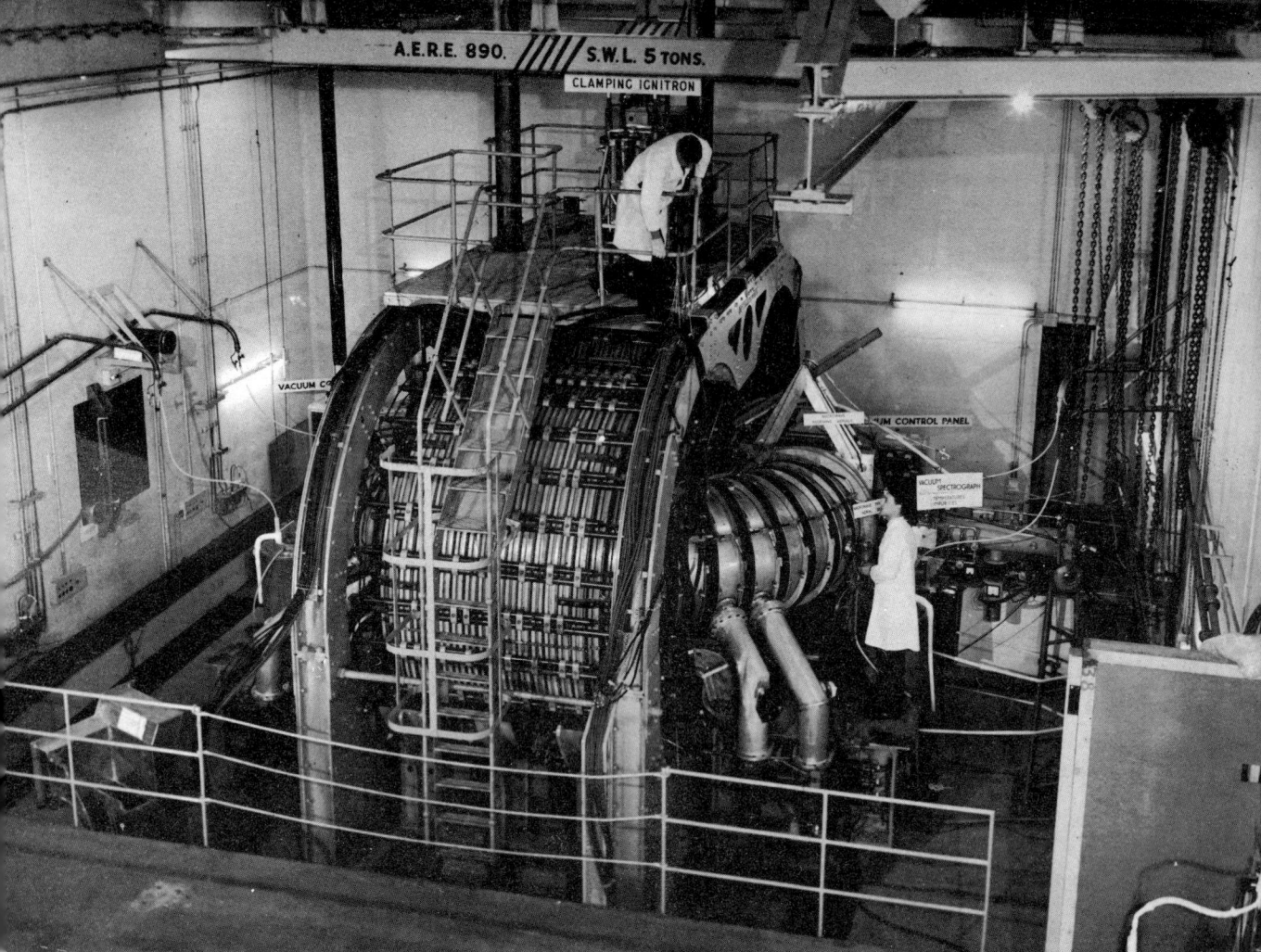

Fig. 43.19 ZETA (Zero Energy Thermonuclear Assembly)

This was the apparatus used for experiments leading to the control of fusion. Temperatures of 5 000 000 °C were achieved for millisecond periods

progress has been made using an installation, ZETA (Zero Energy Thermonuclear Assembly). This apparatus, shown in Fig. 43.19, produces a discharge through deuterium at low pressure contained in a circular tube. Electric pulses of high energy are induced in the gas at regular intervals. These cause the discharge to become constricted and exceptionally high local heating results. Thermonuclear conditions, however, have not been produced.

It is hoped that, eventually, not only will it be possible to employ fusion to produce energy, but that the energy will be produced directly in the form of an electric current and not in the form of heat which has to be converted by means of the conventional turbine and generator.

Artificial radioactivity

Experiments have shown that the bombardment of elements by fast-moving atomic particles produces isotopes which, in many cases, have proved to be unstable and radioactive. The bombardment by neutrons is easily accomplished in an atomic pile and all elements have responded to the treatment. The half-lives of these artificially radioactive substances may vary over a large range, and powerful sources of α-, β- and γ-radiations may be manufactured. These sources are now being put to many uses in research work, in industry and in medicine. For example, the functioning of a gland can be investigated by following the path of a radioactive substance

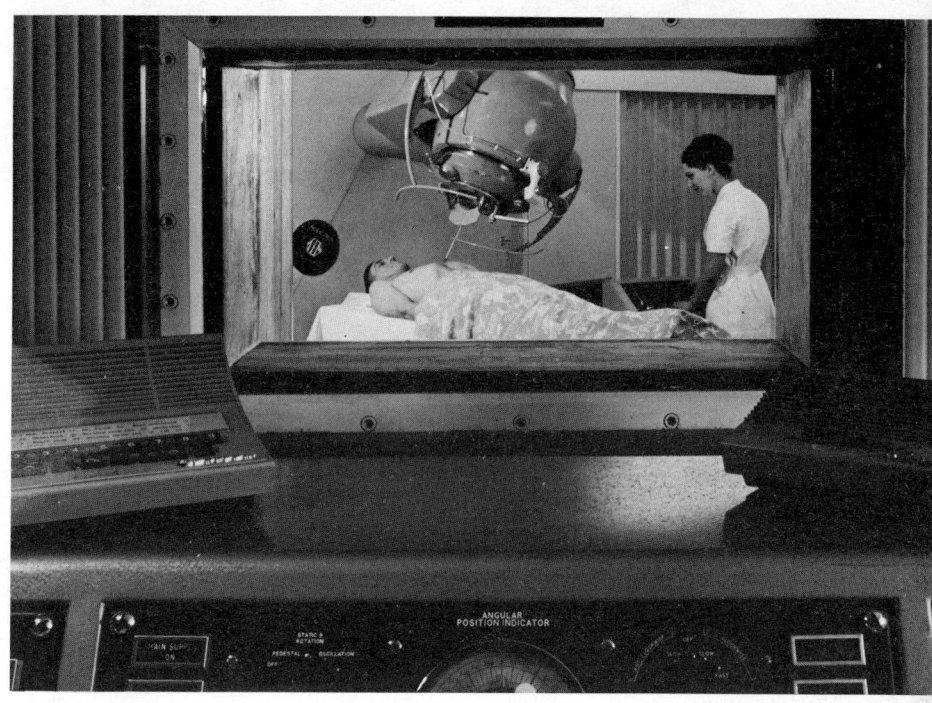

Fig. 43.20
View of the 5000 curie radiocobalt unit (Cobaltron) seen through a lead glass window 122 cm thick. The patient can be observed continuously during the course of the treatment

Fig. 43.21 Taking a radiograph at Stonehenge
In order to ascertain the depth of a crack in one of the Stonehenge stones, before lifting it, a radiograph was taken using a radioactive isotope (sodium-24) as source. The exposure lasted 36 hours

Fig. 43.22 Leak detection in water main using radioactive isotope as tracer

Fig. 43.23 β-ray thickness gauge
The gauge is checking the thickness of cloth. The source is in the container in the centre of the picture beneath the right-hand support of the machine

through it. Similarly, the spread of a substance through a plant may be explored. In such cases a radioactive 'tracer' of short half-life is used so that the subject is not affected by excessive radiation. A small but powerful source of γ-radiation, such as Co-60, can be used to investigate welding faults, in-stead of the heavy and cumbersome X-ray unit. Such sources are easily adapted for deep-ray therapy.

Experiments in this field have led to the production of new elements so that now the periodic table is extended to include over 100 elements (*see* Appendix 8).

QUESTIONS

1. Describe the Rutherford atom model and the contribution added to it by Bohr. In what ways have these ideas had to be modified?

2. How did the discovery of the neutron alter the ideas on the constitution of atomic nuclei?

3. Explain the liberation of energy produced by (i) fission, (ii) fusion.

4. Describe one application of artificial radioactivity in (i) industry, (ii) medicine.

5. What is meant by a radioactive 'tracer'? Give examples of the use of such tracers in (i) biology, (ii) medicine.

6. When fission occurs in a nucleus of U235, 200 MeV of energy is released. How many such nuclei must undergo fission in 1 s to maintain an output of 10 000 kW from a nuclear power generating station?

7. In a transmutation 0.020 a.m.u. of mass disappear. How much energy will this generate (i) in joules, (ii) in MeV?

8. In the experiment of Rutherford and Marsden, a beam of α-particles fell in a vacuum on a gold foil.
   (*a*) How did Rutherford and Marsden make observations upon these α-particles?
   (*b*) What observations led them to the conclusion that only a very small part of the volume of the foil was occupied by impenetrable matter?
   (*c*) How did they argue from the observation that a certain number of α-particles emerged from the foil on the same side as they entered? (JMB)

9. Describe the structure of an atom of *deuterium*, i.e. heavy hydrogen, atomic number 1, mass number 2. What difference is there in its structure when it exists in ionic form? (C)

10. When the nucleus $^{11}_{5}B$ is bombarded with a certain particle the nucleus $^{8}_{4}B$ is formed together with an alpha-particle.
    (*a*) The bombarding particle is.....................
    (*b*) The reaction is represented by the following nuclear equations:
    $$^{11}_{5}B + \ldots\ldots\ldots\ldots \rightarrow {}^{8}_{4}B + \ldots\ldots\ldots\ldots \quad (L)$$

11. Give a brief account of the structure of atoms, making use in your account of the terms, *neutron, proton, electron*, and *nucleus*. What information can you deduce concerning the structure of an atom of lithium, given that its mass number is 7 and that its atomic number is 3?

Give a brief account of the experimental evidence which suggests that atomic nuclei are very small and are positively charged. (C)

12. Describe an experiment which shows that alpha-particles (*a*) produce ionization, (*b*) are occasionally deflected through large angles, as they travel through matter.

An alpha-particle travelling with speed 10^6 m/s is deflected through 60° without change of speed. What is the magnitude of its change of momentum if its mass is 6.6×10^{-27} kg?

Imagine that the change of momentum occurs while the alpha-particle travels a distance of about 10^{-10} m. Estimate the order of magnitude of the average force acting.

How does this compare with the force of the earth's attraction for the alpha-particle? (O & C)

Appendix 1. Basic and supplementary SI units

Length—meter, m

The metre is the length equal to 1 650 763.73 wavelengths in vacuum of the radiation corresponding to the transition between the levels $2p_{10}$ and $5d_5$ of the krypton-86 atom. (1960)

Mass—kilogramme, kg

The kilogramme is equal to the mass of the international prototype of the kilogramme. (1901)

(The international prototype of the kilogramme is a certain piece of platinum-iridium kept at the Bureau International des Poids et Mesures at Sèvres.)

Time—second, s

The second is the duration of 9 192 631 770 periods of the radiation corresponding to the transition between the two hyperfine levels of the ground state of the caesium-133 atom. (1968)

Electric current—ampere, A

The ampere is that constant current which, if maintained in two straight parallel conductors of infinite length, of negligible circular cross-section, and placed one metre apart in a vacuum, would produce between these conductors a force equal to 2×10^{-7} newton per metre of length. (1948)

Thermodynamic temperature—kelvin, K

The kelvin is the fraction 1/273.16 of the thermodynamic temperature of the triple point of water. (1954)

Luminous intensity—candela, cd

The candela is the luminous intensity, in the perpendicular direction, of a surface of 1/600 000 square metre of a black body at the temperature of freezing platinum under a pressure of 101 325 newtons per square metre. (1948)

Amount of substance—mole, mol

The mole is the amount of substance which contains as many elementary units as there are atoms in 0.012 kilogramme of carbon-12. The elementary unit must be specified and may be an atom, a molecule, an ion, a radical, an electron, a photon, etc., or a specified group of such entities. (1969)

The following two units are supplementary basic SI units.

Angular measure—radian, rad

The radian is the angle subtended at the centre of a circle by an arc the length of which is equal to the radius.

(There are 2π radians in one revolution: one radian is equal to 57.296°.)

Solid angular measure—steradian, sr

The steradian is equal to the angle subtended at the centre of a sphere by an area of surface equal to the square of the radius.

(The surface of a sphere subtends an angle of 4π steradians at its centre.)

Appendix 2. Special names and symbols for certain derived SI units

Physical quantity	Name of SI unit	Symbol for SI unit	Definition of SI unit
frequency	hertz	Hz	cycles per second (/s)
energy	joule	J	$kg\ m^2/s^2$
force	newton	N	$kg\ m/s^2 = J/m$
pressure	pascal	Pa	N/m^2
power	watt	W	J/s
electric charge	coulomb	C	As
electric potential difference	volt	V	J/As
electric resistance	ohm	Ω	V/A
electric capacitance	farad	F	As/V
magnetic flux	weber	Wb	Vs
inductance	henry	H	Vs/A
magnetic flux density	tesla	T	Vs/m^2
luminous flux	lumen	lm	cd sr
illumination	lux	lx	$cd\ sr/m^2 = lm/m^2$

(The above list includes one or two units which are not mentioned in this text.)

Appendix 3. Relation between other common units (including British or fps units) and SI units

Physical quantity	Unit	Symbol	Equivalent SI unit
length	angström	Å	10^{-10} m
	inch	in	0.025 4 m
	foot	ft	0.304 8 m
	mile	mile	1.609 km
	international nautical mile	Int naut mile	1852 m
volume	litre	l	1.000 dm^3
	cubic foot	ft^3	0.028 32 m^3
	gallon	gal	0.004 546 m^3
mass	pound	lb	0.453 592 kg
	hundredweight	cwt	50.8 kg
	ton	ton	1016 kg
temperature	°Fahrenheit	°F	9/5 °C+32
	Rankine unit		5/9 K
	Fahrenheit unit		5/9 K=5/9 °C unit
angle	degree	°	1.745×10^{-2} rad
	radian	rad	57.30°
velocity	feet per second	ft/s	0.304 8 m/s
	miles per hour	mile/h	0.447 m/s; 1.609 km/s
density	pound per cubic inch	lb/in^3	27.68×10^3 kg/m^3
	pound per cubic foot	lb/ft^3	16.0 kg/m^3
force	dyne	dyn	10^{-5} N
	kilogramme-force	kgf	9.806 65 N
	poundal	pdl	0.138 N
	pound-force	lbf	4.448 N
pressure	atmosphere	atm	101 325 Pa (N/m^2)
	mm of mercury (torr)	mmHg	133.3 Pa
	bar	bar	10^5 Pa
	kilogramme-force per square cm	kgf/cm^2	9.8×10^4 Pa
	pound-force per square inch	lbf/in^2	6895 Pa
energy	erg	erg	10^{-7} J
	kilowatt hour	kWh	3.6×10^6 J
	calorie	cal	4.186 8 J
	foot poundal	ft pdl	0.042 J
	foot pound-force	ft lbf	1.356 J
	British thermal unit	Btu	1055 J
	electron volt	eV	1.6×10^{-19} J
power	British horsepower	hp	745.7 W
	foot pound per second	ft lbf/s	1.356 W

Appendix 4. Prefixes for SI units

Fraction	Prefix	Symbol		Multiple	Prefix	Symbol
10^{-1}	deci	d		10	deca	da
10^{-2}	centi	c		10^2	hecto	h
10^{-3}	milli	m		10^3	kilo	k
10^{-6}	micro	μ		10^6	mega	M
10^{-9}	nano	n		10^9	giga	G
10^{-12}	pico	p		10^{12}	tera	T
10^{-15}	femto	f				
10^{-18}	atto	a				

The above prefixes have been approved internationally. They represent decimal fractions or multiples of the SI units and also the specially named derived SI units.

The prefix is placed immediately before the SI unit.

When used with a unit bearing an index, the index applies to the prefix as well as to the unit.

$$\text{e.g.} \quad cm^2 = (0.01 \text{ m})^2 \qquad M\Omega^{-1} = \frac{1}{10^6 \ \Omega} \qquad \mu s^{-1} = \frac{1}{10^{-6} \text{ s}}$$

Appendix 5. Abbreviations for (a) common words, (b) units

Note: these abbreviations are printed in upright (roman) type

(a) Common words

alternating current	a.c.	experiment	expt.	specific gravity	sp. gr.
atmospheric	atm.	freezing point	f.p.	specific heat capacity	sp. ht. cap.
atomic weight	at. wt.	infrared	i.r.	specific latent heat	sp. lat. ht.
atomic mass unit	a.m.u.	maximum	max.	standard temperature	
boiling point	b.p.	melting point	m.p.	and pressure	s.t.p.
centre of gravity	c.g.	minimum	min.	temperature	temp.
coefficient	coeff.	potential difference	p.d.	ultraviolet	u.v.
direct current	d.c.	relative humidity	r.h.	vacuum	vac.
electromotive force	e.m.f.	root mean square	r.m.s.	vapour pressure	v.p.
equation	eqn.	soluble	sol.	volume	vol.
equivalent	equiv.	solution	soln.	weight	wt.

(b) Names of units

Note: (i) no full stop follows the abbreviation for a unit, (ii) some obsolescent units are quoted for reference

metre	m	newton	N	ohm	Ω
angström	Å	kilogramme-force	kgf	electron volt	eV
cubic centimetre	cm^3	joule	J	inch	in
litre	l	watt	W	foot	ft
millilitre	ml	watt hour	Wh	yard	yd
second	s	calorie	cal	pound	lb
minute	min	kilocalorie	kcal	hundredweight	cwt
hour	h	candela	cd	poundal	pdl
hertz	Hz	lumen	lm	pound-force	lbf
revolution per minute	rev/min	lux	lx	foot pound-force	ft lbf
gramme	g	coulomb	C	horsepower	hp
kilogramme	kg	ampere	A	British thermal unit	Btu
dyne	dyn	volt	V		

Appendix 6. List of symbols (printed in italic type)

length	l	density	ρ	coeff. cubic	γ
height	h	relative density	d	electromotive force	E, V
breadth	b	force	F	potential: p.d.	V
radius	r	weight	W	electric current	I
diameter	d	pressure	p	electric charge	Q
area	A, a	work	W, w	resistance	R
volume	V, v	energy	E, W	capacitance	C
time	t	power	P	Faraday constant	F
period	T	luminous intensity	I	charge on electron	$-e$
frequency	f	illumination	E	mass of electron	m
wavelength	λ	speed of light	c	atomic number	Z
velocity	v, u	temperature	θ, t	atomic weight	$A+$
angular velocity	ω	absolute temp.	T	(relative atomic mass)	
acceleration	a, f	quantity of heat	Q, q	molecular weight	$M+$
gravitational acceleration	g	specific heat capacity	c	Avogadro's constant	L
gravitational constant	G	specific latent heat	l		
mass	m	coeff. linear	a		

Appendix 7. The Greek alphabet

Letter	English	Greek small	Greek capital
alpha	a	α	A
beta	b	β	B
gamma	g	γ	Γ
delta	d	δ	Δ
epsilon	e	ε	E
zeta	z	ζ	Z
eta	e	η	H
theta	th	θ	Θ
iota	i	ι	I
kappa	k	κ	K
lambda	l	λ	Λ
mu	m	μ	M
nu	n	ν	N
xi	x	ξ	Ξ
omicron	o	o	O
pi	p	π	Π
rho	rh	ρ	P
sigma	s	σ	Σ
tau	t	τ	T
upsilon	u	υ	Y
phi	ph	φ	Φ
chi	ch	χ	X
psi	ps	ψ	Ψ
omega	ō	ω	Ω

Appendix 8. The periodic table

History

Following the work of such scientists as Döbereiner and Newlands, both of whom found certain relationships between the atomic weights of elements with similar properties, Mendeleeff in 1869 discovered that if the elements were placed in order of their atomic weights there was a recurrence of similar properties at regular intervals. Consequently, he constructed a table in which the elements were arranged horizontally in order of their atomic weights and vertically according to their similarity in chemical properties.

In order that the elements would fall into the correct groups as indicated by their chemical properties, he found it necessary to introduce blank spaces. He predicted that these spaces would be filled by the discovery of new elements and, from a knowledge of those elements above and below these gaps, also predicted their chemical properties. On the discovery of these elements it was found that his predictions were remarkably accurate.

General structure

Beginning with hydrogen, the element of lowest atomic weight, the elements are placed in order of their atomic weight. There are exceptions, however, e.g. argon (39.948) is placed before rather than after potassium (39.102). This arrangement, since justified by the discovery of isotopes, is to make the table correspond adequately with chemical properties. It is found that passing from left to right a new 'period' of elements begins after 2, 8, 8, 18, 18, and 32 elements. This gives vertical groups of like chemical properties.

Two series of elements, the lanthanide and actinide series, because of their unusual electronic structure each come under one vertical group.

Atomic weights

The atomic weights shown, except those italicized, are those listed by the International Commission on Atomic Weights (1963). Values given in italics are those of the isotope with the longest half life. All are based on $^{12}C \equiv 12$.

key

ANTIMONY	atomic number — 51
Sb	symbol
	name
	atomic weight — 121.75

Appendix 9. Glossary of terms used in atomic physics

Alpha particle: a helium nucleus emitted from certain radio-active substances.

Anode: the electrode from which the (conventional) current leaves.

Atom: the smallest unit of an element.

Atomic mass: mass of an individual atom. Usually expressed in atomic mass units (a.m.u.). *See also* atomic weight.

Atomic mass unit (a.m.u.): one twelfth of the mass of the carbon-12 atom.

Atomic number: a number assigned to all elements; it is the number of protons in the nucleus and also the number of orbital electrons, and is the same for all isotopes of the same element.

Atomic weight: should now be termed 'relative atomic mass'. All elements exist in more than one form with different atomic masses (isotopes), and the average mass (in a.m.u.) of the mixture of isotopes that usually occur is called the atomic weight.

Avogadro's number: number of atoms of an element contained in a mass of A grammes of a substance where A is the atomic mass of the element; it is 6.023×10^{23}/mol.

Beta particle: electron emitted from certain radioactive substances during decay.

Cathode: the electrode at which the (conventional) current enters.

Cathode rays: beams of fast moving electrons of various origins.

Chain reaction: a reaction where one event triggers off more than one similar event.

Cosmic rays: very high energy charged particles with their origin in outer space.

Cyclotron: a machine which accelerates charged particles to high energies.

Decay constant: fraction of radioactive atoms disintegrating in one second.

Deuterium: atom of heavy hydrogen, $_1^2H$.

Electron: basic unit of electricity; it is a negatively charged particle of mass approximately 1/1850 that of a proton.

Excitation: the state of an atom having an energy above its normal minimum, usually due to a change of electrons in their orbits.

Fission: disintegration of a heavy nucleus into roughly two equal parts, usually accompanied by the liberation of energy.

Fusion: the combination of light nuclei to form a single heavier nucleus; it is accompanied by the liberation of energy.

Gamma rays: electromagnetic waves of very short wavelength (about 10^{-11} m), having their origin in radioactive decay.

Ground state: the normal and stable state of an atom, when all the constituent electrons have their lowest energies.

Half-life: the time taken for half the atoms of a radioactive isotope to disintegrate; the activity is also halved.

Heavy hydrogen: an isotope of hydrogen having a mass of 2 (deuterium).

Heavy water: water molecules in which at least one of the hydrogen atoms is heavy hydrogen.

Ion: a charged atom or group of atoms.

Ionization: the process by which charged atoms or groups of atoms are formed.

Isobars: atoms with nuclei containing the same number of nucleons but a different number of protons, i.e. the same mass numbers, but different atomic numbers.

Isotopes: atoms with nuclei containing the same number of protons but a different number of neutrons, i.e. different mass numbers, but the same atomic numbers.

Mass number: the number of protons plus the number of neutrons in a nucleus.

Neutron: a fundamental particle with mass approximately equal to that of a proton.

Nuclear energy: energy produced by some change in a nucleus, usually during a transmutation.

Nucleons: collective name for protons and neutrons.

Nucleus: the central core of an atom; it contains protons and neutrons, and practically all the mass of an atom is in its nucleus. It is very small compared with the size of the atom and is positively charged.

Nuclide: an atom of an isotope characterized by its atomic number, mass number, and its energy state.

Photon: the fundamental unit of radiation.

Planck's constant (h): energy of photon $= h \times$ frequency of radiation.

Plasma: a gas in a highly ionized state; it is usually at a very high temperature.

Proton: a hydrogen nucleus; it has a mass of one a.m.u. (approximately) and carries a positive charge equal and opposite to that of an electron.

Radioactive series: descendant elements formed by successive disintegrations starting from a single parent element.

Radioactivity: the emission of α-, β-, and γ-rays from unstable atomic nuclei.

Radioisotope: radioactive isotope usually artificially stimulated.

Specific charge: ratio of charge to mass of a particle.

Tritium: heavy isotope of hydrogen, $_1^3H$.

Wave mechanics: a mathematical system particularly applicable to nuclear physics which incorporates particle-wave duality, i.e. the equivalence of matter and radiation.

X-rays: electromagnetic waves of short wavelength arising within the atom.

Answers

Chapter 2
10. 3.0×10^3 kg/m^3
12. 80%, 20%
13. 2.33 cm^3
14. 3.9 carats

Chapter 3
1. 13.2 N at 19° with 10 N
3. 17.3 N
4. 3640 N
5. 400 N; 80° with ground
7. 75 N
8. 57.7 N
9. 179 N; 1.46 m
10. 9.43 N; 1.20 N m
11. 104 N; 60 N
12. 16° 10′ to vertical; 20.8 N; 5.8 N
13. 88 N; 62.2 N; 750 N
14. 9 N; 37° with larger force
15. 30 N; (a) 0.19; (b) yes
16. 532 N
17. (a) 1730 N; (b) 1000 N
20. 104 N at 16° 7′ to vertical at A; 28.9 N normal to wall at B; 100 N downward at C
21. 80 N; 60 N
22. 57.7 N; 116 N
23. At A, 115 N normal to wall; At B, 415 N at 16° 7′ to vertical
24. 500 N; 870 N at 60° to ground
25. 577 N; 500 N up plane; 3000 J
26. 150 N normal to wall; 427 N at 20° 33′ to vertical

Chapter 4
1. 1.25 m
2. 111 N
3. 450 N; 1050 N
4. 114 kg
5. 100 g; 900 g
6. 555 N; 817 N
7. 480 N
8. 30 N
9. 750 N; 500 N; 0.6 m towards A
10. 9.21 g
11. 55 g
12. 100 cm; 50 g
13. 21.25 cm from B
14. 100 kg; 10 cm
15. 200 N
16. 4 kg
17. 0.287 cm
18. 115 N
19. 35 cm

Chapter 5
1. 1000 J
2. 4.7×10^4 J
3. 420 W
4. 500 s
5. (a) 24 kJ; (b) 600 kJ
6. 40 kJ/min; 133 kJ/min; 1550 W
7. 250 kJ; 2083 W
8. (a) 62%; (b) 20 kW
9. 30 kN; 450 kW
10. 4608
11. 50 MJ; 0.5 MW
12. 20 kW
13. 24 W
14. (i) 1.25 kW; (ii) 80%
25. 4.8 kW
16. 12.5 kW
17. 4200 J; 28 W
18. 1.5 kW

Chapter 6
1. 12.0 N
2. 24; 595 N
3. 2 N; 87.5%
4. 83 N; 60 m; 3000 J
5. 21.2 N
6. (a) 625 N; (b) 7500 J; (c) 50 s
7. 81.9 kg
8. 20; 56%; 11.2
9. (a) 2.25; (b) 54 kg; (c) 360 J
10. 0.18; 16%
11. 11.2 kJ; 71.4%
12. 0.25; 3.2 m
13. 1000 N; 18 kJ
15. 333 N
16. 125 N; 1000 J
17. 2.5 kJ

Chapter 7
1. 560 mmHg
2. 4.9 cm
4. 3.6 cm
5. 26.5 cm
6. 0.94 N; 2.83 N; 1.89 N +thrust due to atm. pressure
7. (a) 2.45 N; (b) 4.08 N; (c) 0.164 kg

8. 12 500 N
9. 0.8; 18.75 cm
10. 60
11. 490; 86%
12. (a) 10.34 m; 8.202 m
13. 16.25×10^3 kg/m^3
14. 1.778 N
15. (a) 18.52 cm; (b) 23.53 cm
16. 40 000 kg
17. (a) 0.516 N; (b) 0.28 N
18. (a) 0.42 N; (b) 0.32 N
19. 1.16 N
20. 25.4 g
21. (a) 15.7 g; (b) 5 cm
22. 14.8 cm from A; 3.08 N
23. 0.25 N; 0.8
24. (a) 10.1 cm^3; (b) 7.19 N
25. 120 N
26. (i) 5 cm^2; (ii) 10 cm
27. 600 kg; 450 N
28. 1 cm^2
29. (i) 8 cm; (ii) 4.1 cm
30. 15.7 cm
31. 1.2 cm^2
32. 90 g less
33. 6.5 cm^3; 0.065 N
34. 1.013×10^5 N/m^2
35. 6 m
36. 1.22×10^4 N
37. 55.6 cm^2; 27.8 cm^2
38. 9.5 mmHg
40. 15 000 J; 0.25 kW
41. 2 kW
42. 8.0 g; 0.625
43. 830 mmHg

Chapter 8
1. 185 m/s; 72.4 km/h
2. 4 s
3. (a) 5 s; (b) 4 s
4. 1.20 m/s^2
5. 2.3 cm/s^2; 14.1 cm; 1.17×10^{-3} J; 24.2×10^{-3} N s
6. 59.4 m/s; 134.2 m
7. 5 m/s^2; 8.9 m
8. 45 m; 6 s
9. 100 Hz
10. 20.6 min; 77° 10′
11. 11 m/s
12. 70 s; 34 m/s
13. 20 m
14. 0.375 m/s^2; 300 m
15. (i) 4.2 s; (ii) 8.5 m/s; 180 J
16. 10 m/s; from 53° 8′ W of S

Chapter 9
1. 2.25 m; 1.5 s
2. 1200 N
3. (a) 0.05 m/s^2; (b) 0.15 m/s; (c) 0.225 m; 0.0675 J
4. (a) backwards 6° 51′ to vertical; (b) forwards 6° 51′
5. 5 m/s^2; 0.45 s
6. 3.2 m/s
7. (i) 150 m/s^2; (ii) 15 m/s; 149.5 m/s^2
8. (i) 50 N; (ii) 40 N
9. 312.5 m; 1.56×10^4 J
10. 0.10 m/s^2; 0.63 m/s
11. (a) 20 m; (b) 2 s
12. 45 m
13. 5.8 m/s^2; 10.7 m
14. 20 m/s; 40 000 N
15. (a) 4.0 s; (b) 80 m; 10 m/s^2
16. 200 N; 1.25 m/s^2
18. (a) 2 m/s^2; (b) 1 N
20. (i) 3.2 m/s^2; (ii) 93.5 kg
21. 15 m
23. 880 N
24. 1.41 N s; 1.0 J
25. (a) 10 N s; (b) 4 N s; (d) 25 J; (e) 4 J; 6 J; (f) 15 J
26. 1.25 s; 7.81 m
27. 13.3 cm/s
28. 1.0 N s; 0; 1.0 J; 0

Chapter 12
10. 420 m/s

Chapter 13
20. (a) 8.6; (b) 0.83
21. 2.2
22. 5 N

Chapter 14
4. 0.22 K
5. 0.48 K
6. 233 K
7. 1.4 K
8. 10.6 g
10. (a) 1.5×10^3 J; (b) 3.46×10^2 m/s
11. 0.34

Chapter 15
10. 2.5 cm

Chapter 16
3. 0.445 mm
4. 0.70 m
5. 126 °C
7. $1.68 \times 10^{-5}/°C$
8. 18.8 cm^3
9. 86 °C
10. 9.75 cm^3; 171 °C
11. (a) 4.8×10^{-3} cm^3;
    (b) 19.2 cm
12. 765 °C
14. 2.2×10^5 N/m^2
15. 836 mmHg
16. $3.7 \times 10^{-3}/°C$
17. 1.86 atm
18. 294 mmHg
19. 62.5 dm^3
20. 1.06 dm^3
21. 45
22. 1092 °C; 15 dm^3; −91 °C
23. 58 °C
24. 1.27×10^5 N/m^2
25. 11.7 km
26. 2.9 cmHg
27. 5 atm; 1200 dm^3
28. 40 °C; −40 °C
29. 60 N; 627 °C

Chapter 17
1. (a) 3120 J; (b) 11 000 J
2. 0.875×10^3 J/kg K
3. 24.5 °C
4. 455 °C
5. 665 °C
6. −86 °C
7. (a) 60 480 J; (b) 252 J/s;
   (c) 6.6 kg
9. 30.2 kJ
10. 244×10^3 J/kg
11. 30 g of ice melt,
    mixture at 0 °C
12. 6791 J; 4158 J;
    2213×10^3 J/kg
13. 46.7 g
14. 5.1 g
15. 113 g
16. 120 °C
17. 8.5 min; 40.4 min
18. 0.92×10^3 J/kg K
19. 40.4 g
20. 17.5 g
21. 16.0 g
22. 1.19×10^3 W;
    0.527 g/s
23. 11.2 g/min
24. 180×10^3 J/kg
25. 1.2 cm
27. (a) 2300×10^3 J/kg;
    (b) 100 W
28. 1178 J/s; 1336 J/s;
    4.9 A

29. 0.179 kg
30. 20 K; 38 °C
31. 31×10^3 J

Chapter 18
11. 223 cm^3
12. 144 mmHg
17. 747 mmHg
18. (a) 118 mmHg;
    (b) 25 °C

Chapter 20
11. 15 Hz
12. (a) 8 Hz; (b) 1.0 cm;
    (c) 20° 42′
15. 1.3 cm; 66° 47′
16. 6.3 Hz

Chapter 21
4. 0.90 cm
8. 20 m
11. 0.22
12. 26.7 cd; 1.94 m
13. (a) 10 s; (b) 2.5 s
14. 45 cd
15. 4.0 m

Chapter 22
7. 4° or 7 cm

Chapter 23
1. (a) 36 cm in front;
   (b) 12 cm behind
2. 6 cm from mirror
3. 8 cm; 16 cm
4. 9 cm
6. 7.5 cm; 5 cm from pole
7. 2.67 cm; 5.33 cm
8. 8 cm; 24 cm in front
9. 1.8 cm behind; virtual;
   0.6 cm high
10. (a) 20 cm; (b) 1.2 cm;
    3.0 cm
11. 2 cm

Chapter 24
3. 54 cm
4. 1.1 : 1
5. 12 cm below surface
7. 30° 10′
8. 18° 35′ towards base
10. 1.33 cm below top
    surface
11. 18° 35′
13. 2.28 m

Chapter 25
4. 12 cm
5. 7.2 cm from lens; 0.4 cm
   high
6. 16 cm; 8 cm
7. 17.6 cm to 25 cm from
   lens
8. 3 mm
9. 24 cm
10. 12 cm; 6 mm
11. 9 cm from lens
12. 30 cm; 3.0 cm; 0.4 cm
13. 0.5 cm
14. 90 cm on same side;
    10×; 11 cm; 110 cm

Chapter 26
3. Convex; 66.7 cm
4. Convex; 37.5 cm
6. 7.14 cm; 3.5×
7. (i) 0.46 cm; (ii) 10×
8. 3.0 cm; 5.3 cm
9. (a) 25 cm; (b) 24.17 cm
10. 16 cm
11. 0.83 cm; 24×
12. 9.6 cm

Chapter 27
7. 24° 16′
8. 7
10. 36.0 cm
11. 9.8×10^{-5} cm
12. 10.0 cm

Chapter 29
2. 6640 m
5. 15.8 km/h
6. 133 m
8. (a) 333 m/s; (b) 100 m
10. 750 Hz
11. 333 m/s; 4.4 m/s
12. 324 m/s
13. 1.73 km
15. (a) 0.25; (b) 1.5

Chapter 30
1. 199 Hz
2. 576 Hz
3. (a) 420 Hz; (b) 280 Hz;
   (c) 35
4. (a) 384 Hz; (b) 296 Hz
5. 20 cm away from A
6. 56.25 N
7. 384 Hz
8. 42 cm
9. 0.67 of original length;
   2.25 of original tension
10. 40 cm
12. 322 Hz
14. 200 Hz; 42.5 cm
16. 256 Hz

18. 576 m
19. (i) 66.7 cm; (ii) 270 NzH
21. 330 Hz; 396 Hz; 528 Hz

Chapter 32
16. 4×10^{-3} C; 0.67 μA

Chapter 34
1. 2.5×10^{-3} Ω
2. 39.3 cm
3. 1.25 A; 5.33 A
4. 5.33 V; 6.67 V
5. 48 Ω
6. 5
7. 2.23 m
8. 16.3 Ω
9. 1 Ω; 4 Ω; 2:1
10. 12 V; 0.02 Ω/m
11. 2 A; 20 V
12. 0.5 A; 0.56 A
13. 8.8×10^{-3} mm
14. 0.50
15. 2.4 A; 4.8 V
16. 0.3 A; 0.27 A
17. Ammeter will read
    approx. 0; voltmeter
    will read approx. 2 V

Chapter 35
1. 4:1
2, 7.2 MJ
3. 96 Ω
4. 14 min
5. (a) 4 A; (b) 4.375 min;
   (c) 0.0875p
6. 35p
7. 143 min; 10½p
8. 21 A; 9.5 Ω
9. 1.52 A; 1.26 MJ; 2.1p
10. (b) 0.8 A; P, 2 A;
    Q, 1.33 A
11. (c) 2:3, Q greater; 3:2,
    P greater
12. (a) 48 W; (b) 22 W;
    (c) 8 W
13. (a) 9.7 V; (b) 37.2 W;
    (c) 8.14 W
14. 5.25 hours
15. 0.187p
16. 70.5 Ω; 1.35 MJ
17. 15 A; 13.3 Ω; 4.32 kW
18. (a) 529 Ω; (b) 50 W
19. (a) 12.6 kW; (b) 4.2 Ω
20. (a) 400 Ω; (b) 100 W;
    (c) 360 kJ
21. (a) 50 A; 4 A; (b) 0.004 Ω
    0.625 Ω
22. (b) 0.2 A; 0.1A; 0.2 A;
    (c) 80 W; (d) 6.25 hours;
    (e) (iii)
23. (a) 15 kJ; (b) 5.04 kJ;
    (e) 9.96 kJ; 332×10^3 J/kg

Chapter 36
1. 0.74 cm^3
2. (*a*) 0.46 A; (*b*) 385 cm^3
3. 2.2×10^{-3} cm
4. 0.11 A
5. (*a*) 0.5 A; (*b*) 400 Ω;
   (*c*) 0.59 g
6. 62.5 minutes
7. (*a*) 0.5 A;
   (*b*) 3.3×10^{-4} g/C
8. 42.3 cm^3 H$_2$; 21.2 cm^3 O$_2$
9. (*a*) 1.14 A;
   (*b*) 0.51×10^{-3} m^3
10. (*a*) 2 Ω; (*b*) 2×10^{-3} kWh
11. (*a*) 0.84 A;
    (*b*) 0.26×10^{-3} m^3
12. 5.36 Ω
13. 7.13 g
14. 0.84 A; 1.75×10^{-4} m^3
15. 2.4×10^{-29} m^3
18. 38 Ω
19. (*a*) 6.9 Ω, 0.8 Ω;
    (*b*) 0.87 Ω
20. 0.1 A; 0.8 V
21. 120 V; 116.8 V

22. 60; 0.25 Ω
23. 3 Ω; 0.66 V; (*a*) 0.3 A;
    (*b*) 0.46 A
24. 12 Ω
25. 0.6p
26. (*a*) 6 V; (*b*) 10 Ω
27. 1.6 Ω; 1.2 V
28. (*a*) 0.025 A; (*b*) 0.15 A
29. (*a*) 0.05 A; (*b*) 1.25 V;
    (*c*) 0.0625 W

Chapter 37
7. 3.36×10^{-2} Ω in parallel
11. (*a*) 992 Ω in series;
    (*b*) 0.008008 Ω in
    parallel
12. 999 Ω in series
14. 30 W

Chapter 38
1. 4.4×10^{-7} Ω m; 2.2 Ω
2. 37.5 V
3. 2.19 V; 0.38 Ω
4. 1.15×10^{-6} Ω m

7. 4×10^{-7} Ω m
8. 4.98×10^{-7} Ω m
9. 2.5×10^{-6} Ω m
10. 0.67 V; 1.33 V; 0.4 V;
    0.8 V
11. (*a*) 10.0263 Ω in parallel;
    (*b*) 999.5 Ω in series;
    108 V
12. 110 V
13. 2 V; 0.5 Ω
14. 1.95 Ω
15. 3 mV
16. $12\ 000$ Ω; 0.125 Ω
17. 267 Ω; 15 Ω

Chapter 39
9. 153 V
10. 83.3%; 60p; 2.5 A

Chapter 40
1. (*a*) 18.9 kw; (*b*) 10 W

Chapter 41
10. 2 W

12. 1.6×10^{-19} C
16. 7.5×10^{14}; 4.8×10^{-16} J

Chapter 42
7. 14.4 days
10. 2 hours
11. 88; 222; 4; 2
12. (*b*) 28; (*c*) 27; 62;
    (*d*) 10^6
14. 16 counts/s
17. (*c*) 1.6×10^7 m/s
27. 15 minutes; 30 minutes
28. 0.25×10^6
29. 16 counts/s
30. 80 minutes

Chapter 43
6. 3.12×10^{17}
7. (i) 2.99×10^{-12} J;
   (ii) 18.7 MeV
12. 3.3×10^{-21} N s;
    3.3×10^{-5} N; 5×10^{20}
    times greater

Index